Information Technology for Energy Managers

Information Technology for Energy Managers

Compiled and Edited by
Barney L. Capehart, Ph.D., CEM

Associate Editors
Paul J. Allen
David C. Green
Lynne C. Capehart

LONDON AND NEW YORK

Published 2020 by River Publishers
River Publishers
Alsbjergvej 10, 9260 Gistrup, Denmark
www.riverpublishers.com

Distributed exclusively by Routledge
4 Park Square, Milton Park, Abingdon, Oxon OX14 4RN
605 Third Avenue, New York, NY 10158

First published in paperback 2024

Library of Congress Cataloging-in-Publication Data

Capehart, B.L. (Barney L.)
Information technology for energy managers/Barney L. Capehart.
p. cm.
Includes bibliographical references and index.
ISBN 978-0-8247-4617-9 (print) ISBN 978-8-7702-2240-2 (electronic)
1. Intelligent buildings. 2. Buildings--Mechanical equipment--Automatic control.
3. Buildings--Energy conservation--Automation. 4. Information technology.
5. Web servers. I. Title

TH6012.C35 2004
658.2'6'dc22

2003062363

Information technology for energy managers/Barney L. Capehart
First published by Fairmont Press in 2004.

Routledge is an imprint of the Taylor & Francis Group, an informa business

Publisher's Note
The publisher has gone to great lengths to ensure the quality of this reprint but points out that some imperfections in the original copies may be apparent.

978-0-88173-449-2 (The Fairmont Press, Inc.)
978_8_7702_2240_2 (online)

While every effort is made to provide dependable information, the publisher, authors, and editors cannot be held responsible for any errors or omissions.

ISBN: 978-0-8247-4617-9 (hbk)
ISBN: 978-87-7004-594-0 (pbk)
ISBN: 978-0-203-91319-2 (ebk)
ISBN: 978-1-003-15111-1 (eBook+)

Table of Contents

Foreword

Michael G. Ivanovich
Editor-in-Chief, HPAC Engineering Magazine
Penton Corporation

Congratulations on picking up this book. You have taken the first step in refining your knowledge and skills in meeting the demands of the fastest-growing and most promising field in the buildings industry.

Information technology is a meeting of the minds involving computer scientists and the engineering community. The melding of these two professions will spark a renaissance in the buildings industry that has the potential to make life—not to mention the economy and environment—better for everyone.

Information technology is your tool for knowing if a building is working, how to make a building work better, and how to take the lessons learned from those buildings and apply them to the design, construction, operation, and maintenance of new buildings, as well as to the retrofit of existing buildings.

As a computer scientist, building scientist, and seven-year veteran of the trade press, I am well aware of the state of our industry from its foundation to its skylights. Believe me: You need this book. Let me list a few reasons why:

- Equipment and systems are becoming highly integrated with computer and information technology. Ask any chiller, and it will tell you how much information it uses to function and how much of that information it can make accessible to you at any time from anywhere. How does it do that? It uses networked, web-accessible, equipment-level controls tied into a networked, web-accessible building automation system. Your chiller might even decide to tell you itself that it is about to shut down if you have it set up to alarm you via pager, e-mail, telephone, or fax when some threshold has been crossed.

- Engineers and operators are challenged to keep up with emerging and evolving technologies, and none is emerging and evolving faster than information technology. The reasons are many. Among them: tighter staffs are answering more alarms; legacy equipment and controls are running longer because of smaller capital budgets; and staff turnover and shrinking training budgets are making it difficult to keep personnel up to speed. This is especially true in campus systems.

- Many different types of information technology are coming to market simultaneously, and they are highly complex at first blush. The ones I know of include:
  1. Automated diagnostics for continuous building commissioning, and early warning of impending or just-occurring failures;
  2. Integrated and interoperable building automation systems that are being combined with fire/life-safety, security, and lighting networks,
  3. Enterprisewide integration of building-generated data from the aforementioned systems and the application of the data to all aspects of business operation (not just building operation);
  4. Self-learning and self-adapting chillers that optimize their own energy-consumption patterns while also monitoring and protecting their equipment health;
  5. Highly sophisticated graphical user interfaces for operator workstations, chillers, and compressors;
  6. Meters, submeters, dataloggers, and hand-held instruments that have the sole purpose of generating and pre-processing data that can be used for commissioning systems, monitoring energy use, and optimizing systems by adjusting schedules, set points, and other settings for highest energy efficiency.

- The line between manufacturer and service provider has blurred to the point where the person selling you a control system might also offer to buy your whole facility and run it for you while financing the deal in such a way you that pay little, if anything, up front. Or they will offer to manage the construction process from design through turnover. Information technology arms you with the tools needed to analyze such deals, to negotiate a new utility contract and to benchmark the performance of a building or system in advance of a capital project. And after it is completed, you can use the same tools for contractor-performance validation and new-status-quo documentation.

- Information technology can help you sleep better at night. Your building will call if it has a problem. Your engineering staff responding to an alarm will be better advised as to the nature of the problem and potential fixes. Service contractors will be advised in advance of the problem's characteristics and should arrive with an array of replacement parts and a more efficient work regimen that cuts their response time (having been guided to your door by GPS transponders and web-accessible maps).

Reading this book will not make you an IT expert any more than reading a book on thermodynamics will make you an engineer. However, increasing your awareness is the first step in developing a strategy to learn more of what you need to know to apply relevant, practical, and affordable information technologies.

As a break from the rigors of graduate school, I took a one-week course in log-cabin building in the mountains of Colorado. We peeled logs and learned how to safely start, operate, and maintain a chain saw and shape and notch logs so they stack up to form building elements. The mixture of classroom and experiential education was the best way to indoctrinate me on a topic that was totally foreign. The learning environment didn't hurt either. That was 15 years ago, and although I cannot say that I will ever build a log cabin, I can say there have been times when I was glad for the instruction on how to use a chain saw and I know that someday I might buy a log cabin or contract a builder. I also know more about how trees grow and how to judge wood quality. I know more about forestry as a field of science and industry. I became more interested in woodworking, which I since have taken up as a hobby. Not all of this was learned during the course. What I learned during the course led me to learn other things either through osmosis, extrapolation, or other educational processes.

My point is that, in picking up this book, you not only have embarked on a path of learning of immediate value, but initiated a process of learning that will compound like interest over time. You will notice things you never did before and learn from them. And you will be more open-minded and accepting about new technologies and be able to see where they can help or hurt your efforts to improve your performance or that of your staff and buildings.

But this book is only one piece of the learning formula. You have to try things out; put your hands on them; kick the tires, so to speak; experiment. After reading the chapter on web-accessible control systems, call up a sales rep and have him or her show you a few installations. You might want to make a list of the information technologies discussed in this book, prioritize them by relevance to your work, and develop a strategy for deepening your learning that eventually leads to a business plan for incorporating some of them into your profession and/or building.

The editors and associate editors have done a tremendous service to the buildings industry by creating this book. The many authors have done a tremendous service by contributing. Pioneering is tough work, and that's what all of these people are doing. They've set out into the wilderness to establish a new field of knowledge and products that are meant to make your job more efficient and effective. From research scientists to consulting engineers to product-application engineers to website owner/editors, this book is the melding of a lot of diverse and expert minds from the fields of computer science, energy and building-systems engineering, and product manufacturing, sales, and service. Driving their activities and yours are the economics and decisions that govern all professions. Somehow, all of that has come together so you have this book in your hands. Now it is your responsibility to put it to use.

So congratulations on picking up this book. See you on the other side.

Cleveland, Ohio
June 2003

Foreword

Ken Sinclair
Editor/Owner www.AutomatedBuildings.com

As the self-designated collector, connector and catalyst for large building automation industry information, I am extremely pleased to share my thoughts in a foreword to this book. The information presented here on web-based energy information and control systems is very timely as the building and facility automation industry is in a stage of rapid evolution. The valuable information in this book will start you on your journey to understanding web-based ways. Linking to web sites like ours at AutomatedBuildings.com, and the collected and connected web-based information found there will provide an ongoing sequel to this book.

A lot of industry groundbreaking concepts have been painted on our AutomatedBuildings.com website by over 160 authors over the last four years, attesting to the rapid transformation of tired text-based, proprietary DDC systems into leading edge web-based energy and facility information and control systems with graphic presentations. The transformation from old modem-based DDC technology to TCP/IP communication is a time-warping sensation. Standardized browser-based graphical presentation replaces the awkward proprietary-based interfaces known to a select few, and makes critical facility energy use and operating information available to a much larger audience. This includes all levels of management—and even the world, if that is desired.

Over the last four years, I have seen what I have been writing about turning into reality. This is extremely gratifying. As I was assembling articles and interviews about the building and facility automation industry evolution to the web in our on-line magazine, I often wondered if I was just collecting the thoughts of early adopters. "Will this ever happen or am I just documenting small pockets of evolution?" It was clear to me that making this come about would take a metamorphosis of industry thought. I can tell you that this is now happening! The April, 2003 BuilConn event in Dallas, TX, was the first ever meeting that laid serious groundwork to allow us to hasten our rate of evolution of web-based facility management automation systems.

The ability of a web-based energy and facility automation system to provide mammoth changes to buildings with only grams of substance, and the ability to be easily reconfigured to grow and adapt with the building makes the web-based system the essential backbone of leading green building design. Building Automation has evolved from simple control of equipment to a higher level of occupant and even architectural interaction. The meshing of web-based Building and Facility Automation with Broadband and Telephony has created a powerful new virtual architectural fabric. I call this fabric "Enviromation," a combination of the words Environment, Communication and Automation.

This new web-based fabric provides the greatest strength of all building materials, a powerful connection to the building occupants. There are additional strengths in the fact that this new virtual fabric can insure that building energy is purchased at the lowest cost from the environmentally correct source. Once the energy arrives at the building, expert operators can insure that this energy is used to create the greatest comfort for the least environmental impact. To insure that the original system design intent is achieved, this same web-based virtual fabric can provide real-time feedback and interaction to the original designers. The ongoing virtual interaction with the building by both occupants and designers will lead to improved sustainability while increasing the overall real and perceived values. It is vitally important that all of those purchasing, operating and possibly modifying these web-based energy and facility management systems have some basic understanding of the underlying Information Technology base that is critical to designing and operating these systems.

As the economic pressure continues to mount to do more with less manpower and to reduce the operating cost per square foot of our buildings and facilities, these powerful, anywhere, web-based facility management systems with graphical presentation interface upgrades provide cost effective vehicles to achieve this goal. These interfaces are important to management, which receives input from several sources and can then insure the lowest cost operation in today's complex buildings. Utilizing intellectual input from valuable resources that provide information to systems can only be done with this technology. New upgraded servers can be expanded to include many forms of building information, all web-based. These cost-effective upgrades often determine if the existing DDC system is capable of growing into the new web-based operational model that is evolving for most of our buildings and facilities.

In my opinion, the appearance of new facility management system hardware products will follow the personal computer end device market by creating low-cost, high-functionality, multiple-use devices that will evolve as commodities. This is all possible because of hardware, software , network and communication standards. Once market share is captured (likely by new manufacturers from the computer industry), the production of specialized control devices in small numbers and at high cost will no longer be viable. This statement may seem bold, but I think if you view it in a 5- to 10-year future window you will agree. This is both bad and good news. The bad and good news is that the building and facility automation industry will continue to undergo radical change for the foreseeable future. For those not willing to change, it will be a scary time, but for those ready to move on, opportunities will abound.

Are we, as professionals in this building and facility management industry, doing everything we can to educate users, suppliers and consultants to the tremendous advantages of these cost effective upgrades to web-based energy and facility information and control system adoption and operation? Our success depends on moving quickly to provide this education about Information Technology and about the powerful web-based energy and facility information and control systems that greatly expand the functionality of the existing building and facility DDC systems.

As you read this book and each of the author's approaches I hope this foreword will help set the scene for their presentation of a technology in transition, and for their vision and their method of how best to increase the adoption of this innovative, web-based information and control technology.

Vancouver Island
June 2003

Preface

Barney L. Capehart, Ph.D., CEM
Editor

Advances in new equipment, new processes and new technology are the driving forces in improvements in energy management, energy efficiency and energy cost control. Of all recent developments affecting energy management, the most powerful new technology to come into use in the last several years has been Information Technology—or IT. The combination of cheap, high-performance microcomputers together with the emergence of high-capacity communication lines, networks and the internet has produced explosive growth in IT and its application throughout our economy. Energy information and control systems have been no exception. IT and internet based systems are the wave of the future. Almost every piece of equipment and almost every activity will be connected and integrated into the overall facility operation in the next several years.

The internet, together with the World Wide Web—or web—has allowed the development of many new opportunities for energy and facility managers to quickly and effectively control and manage their operations. The capability and use of IT and the internet in the form of web-based energy information and control systems continues to grow at a very rapid rate. New equipment and new suppliers appear almost daily and existing suppliers of older equipment are beginning to offer new web-based systems. Energy managers, maintenance managers and facility managers are all having to deal with this rapid deployment of web-based equipment and systems, and need to be prepared for current and future applications of internet based technologies in their facilities. In some cases, facilities are developing their own information and control systems, or at least subsystems, and are trying to understand how to connect and interface new IT equipment to their older energy management or facility management systems.

The purpose of this book is to help prepare energy managers, maintenance managers and facility managers to understand some of the basic concepts and principles of IT. We hope that they can successfully apply IT to their facility, and at least have the knowledge to supervise the IT work of a consultant or a vendor. Knowing what is going on and what is involved is important information for the energy manager if they are going to successfully purchase, install, operate, and possibly modify these complex, web-based energy information and control systems. Information Technology is a very comprehensive area, and this book's purpose is to address the most significant concepts and principles that the typical energy or facility manager might need. The emphasis of this book is on energy and operational data input, computer networking, use of facility operation databases, and sharing data using the web and the TCP/IP communications protocol. Early chapters introduce basic principles, structures and definitions needed for most facility IT applications; and later chapters cover general application principles and will discuss specific software and hardware requirements for typical energy information and control systems.

It has been my pleasure to work with my Associate Editors on this important contribution to the IT education and training of working energy managers and facility managers. Mr. Paul J Allen, Mr. David C Green and Ms. Lynne C Capehart have all played a major role in getting this book prepared and completed. My most sincere thanks go to each of these three people who have made my job much easier than it could have been. I also want to thank each of the 37 individual authors who have written material that appears in this book. Without their kind and generous help in writing these detailed chapters, this book would not have been possible. Each of these authors is identified in the alphabetic List of Authors following the Table of Contents.

Finally, I hope that all four of us on the Editorial Team have helped contribute to the successful application of new Web Based Energy Information and Control Systems in many of your facilities.

Gainesville, Fl
June, 2003

List of Authors

Paul J Allen
Walt Disney World

George G. Barksdale
Science Applications International Corporation

Michael Bobker
Association for Energy Affordability

Barney L Capehart
University of Florida

Martel Chen
Siemens Building Technologies

Alex Chervet
Echelon Corp

Dale Fong
Itron Inc.

Lynn Fryer-Stein
E Source Energy Information and Communication Services

David C Green
Green Management Services Inc.

Karen Herter
Lawrence Berkeley National Laboratory

Steve Herzog
Obvius LLC

Michael Ivanovich
HPAC Magazine

Robert E. Johnson
University of Florida

Leland Keller
E Source Energy Information and Communication Services

Brendan Kiernan
E Source Energy Information and Communication Services

Satkartar Kinney
Lawrence Berkeley National Laboratory

Hans Kranz
Siemens Building Technologies, Ltd.

Jim Lewis
Obvius LLC

Fangxing Li
ABB Inc.

John J McGowan
Energy Controls, Inc.

Jim McNally
Siemens Building Technologies

Gerry Mimno
Advanced AMR Technologies, LLC

Naoya Motegi
Lawrence Berkeley National Laboratory

Klaus Pawlik
Accenture

Curt Phillips
erg/Technicon

Mary Ann Piette
Lawrence Berkeley National Laboratory

Partha Raghunathan
i2 Technologies, Inc.

Rachel Reiss
E Source Energy Information and Communication Service

Rich Remke
Carrier Corporation

Allan Schurr,
Itron Inc.

Jim Sinclair
AutomatedBuildings.com

Miles D. Smith
Science Applications International Corporation

Michael R. Tennefoss
Echelon Corporation

Steve Tom
Automated Logic Corporation

Deepak Wanner
Precidia Technologies

Joel Weber
Weber and Associates

Tom Webster
Lawrence Berkeley National Laboratory

Valerij Zvaritch
Ukranian National Academy of Sciences

A short biography for each author is provided in the section About the Authors that appears on page 415 at the end of the book.

Section One

Introduction to Information Technology for Energy Managers

Chapter 1

Understanding Web-based Energy Information and Control Systems

Barney L Capehart, Editor
University of Florida

THE NEW INFORMATION TECHNOLOGY WAVE

One of the most powerful technological waves of our time is sweeping through the development and application of energy information and control systems. This powerful wave is information technology, or IT as it is commonly known. IT has brought with it the internet, the World Wide Web, TCP/IP, web browsers, relational databases, and a host of highly capable, time saving software tools. In addition, IT has promoted the development of inexpensive, high speed microcomputers as well as high capacity communication links, networks and intranets. IT is revolutionizing our entire economy, and it has finally come to our doorstep in the form of web-based energy information and control systems.

We have two basic options in my opinion. One is to stand by and let the IT and web-based system wave roll over us, drown us, and thereby relieve us of any future concern. The other option is to climb on the IT and web-based system wave and ride it to previously unknown heights of energy efficiency, energy cost reduction, improved occupant comfort, improved service and manufacturing performance, and finally to a state of optimized performance of our entire facility. This second option is clearly the most desirable one, and the one that the authors in this book believe is available today for those energy and facility managers who can make the transition to the full use of IT capabilities in their buildings and facilities.

WHY SHOULD YOU IMPLEMENT A WEB-BASED SYSTEM?

A web-based energy information and control system can significantly enhance your facility operations. Steve Tom, from Automated Logic, makes an excellent argument in Chapter 2 that using web-based energy information and control systems results in "Doing more with more." He says that "web-based systems can provide more access, more flexibility, and more interoperability; and they can provide these benefits over a wider area than conventional control systems can. Thus the common phrase 'doing more with less' does not apply here. Web-based systems give you more to work with; they just don't have to cost more." This is a tantalizing view of the possibilities in the world of web-based energy information and control systems. The purpose of this book is to try to aid this transition to web-based systems by providing energy and facility managers a collection of resources that help explain basic IT and web terms, as well as describe the basic features and system structures of web-based information and control systems.

The use of information technology, mainly in the form of TCP/IP and the web, has revolutionized the facility and building energy control systems industry. The control systems industries approach to providing "open systems" was to develop BACnet and LonWorks. The ASHRAE BACnet standard 135 was first adopted in 1995, but by then the information technology standard of TCP/IP was well on its way to becoming the "open" system standard for many of the new players in the energy control systems area. The World Wide Web is the largest open system in history, and the technology that makes it possible to access sites around a university campus, or around the world, can make your building's or facility's data available to authorized users anywhere in the world. There is still value to using BACnet and LonWorks, as well as many of the proprietary communications protocols in existing and new building and facility automation systems. This issue of open systems is addressed in many of the chapters of this book, and addressed from different viewpoints of the users and the information and control system suppliers.

WHY IS IT IMPORTANT TO LEARN TO SPEAK INFORMATION TECHNOLOGY?

There are a number of reasons why the energy managers, facility managers, and facility decision makers at all levels need to learn to speak IT. Without question, the major reason is that the wave of IT for information and control systems from energy management alone to overall facility and enterprise management is here. There is no real choice: either embrace it and ride the wave to its heights; or let the wave roll over you, and most likely your job will then be taken over by someone who has learned to speak IT. It is really a matter of survival at this level. Not only will the IT wave roll over people who do not embrace it, but it will roll over buildings, facilities, campuses, manufacturing plants, schools, hospitals and other institutions that do not fully adopt it and reap the benefits of more efficient, more productive and more cost-effective operation. Competitive economic forces will make these decisions for you and for your facility.

It is also important for the operating personnel at facilities to learn to speak IT in order to do an effective job of installing, operating and maintaining these web-based energy information and control systems. I was talking with Steve Tom from Automated Logic as I started preparing this book, and Steve related the following account of why it is important for all energy and facility managers and operators to be able to speak IT.

"I will say that most energy managers I've met are desperately in need of IT knowledge. A few years ago we switched from a traditional energy management system using proprietary PC software in the front end and a BACnet over Ethernet communication backbone to our current web-based system using a BACnet/IP communications backbone. We knew we would need to give our dealers a lot of training on our new software and hardware, but we greatly underestimated the amount of IT training we'd have to do. We naively assumed that if our system was going to be using an existing IP network, the installers could simply ask the IT department for the necessary IP addresses and subnet masks. After all, we'd been hanging our equipment off existing networks for years. It turned out that our move to a web-based system changed everything, at least in the minds of the IT professionals. Before, they regarded our equipment as being about as much of a threat as a new printer—hang it on the network anywhere and it won't hurt anything. Once we added a web server and IP routers to the list of equipment we were going to hang on their network, everything changed. Suddenly the IT folks were grilling our dealers about bandwidth, ports, and a host of other details that were way beyond their knowledge level. Typically, the on-site representative for the new web-based system was an HVAC technician who just wanted to hook up the wires and get out of there. The fact that our technicians couldn't immediately answer the IT department's questions made the IT folks even more suspicious. In many cases key pieces of hardware were locked up in IT vaults where the HVAC technicians had no access. All programming, start-up, and commissioning activities had to be scheduled through the IT department, and sometimes required an IT employee to escort the technician throughout the work. In extreme cases, the Facilities Maintenance staff wound up running their own, dedicated IP network throughout the building or the campus because it was easier than trying to get permission to use the existing network.

"At first, I thought maybe we just had the bad luck to run into a few oddball IT staffs, but gradually I realized this was a major problem. At a 2002 CABA conference (Continental Automated Buildings Association) I heard similar horror stories from several other control system manufacturers, consulting engineers, and end users. More than one was firmly convinced that the existing IT staffs were the biggest obstacle to building automation today. I think that's unfairly placing all the blame on the IT staffs, but I do think the gulf between the IT staff and the facility maintenance staff is a major problem. Web-based systems typically take up a minuscule portion of the available bandwidth, and the security concerns can be addressed with a wide array of networking tools, but the IT folks won't know this if the facility staff doesn't 'speak IT.' Many times the energy manager is one of the driving forces behind a new building automation system, so if you can bring him up to speed on IT issues the entire project goes smoother. I'm very hopeful that your book will help bridge this gap."

WHO IS INVOLVED IN LEARNING ABOUT WEB-BASED ENERGY INFORMATION AND CONTROL SYSTEMS?

Originally, this book was envisioned as a resource solely for working energy and facility managers, to help them prepare for, and succeed in their specification, purchase, operation, and maintenance of these complex web-based energy and facility information and control systems. As the development of this book proceeded, and various authors came forward to describe their interest in this new technological area, it became clear that there were many other people who would benefit from this book. Besides the actual users at the buildings and facilities, there has been a keen interest from the developers, suppliers, integrators and consultants working with these web-based information and control systems. Software vendors, hardware vendors, computer service and support organizations, and consulting firms have all shown a tremendous interest in this book, and its potential to help them educate themselves and their customers and clients. There are so many new terms, concepts and operational aspects of these web-based systems that there is a great need for an educational and training resource to assist in this technology transfer process from suppliers to end users. I have been amazed and impressed with the widespread recognition of the need for a comprehensive, yet basic and readable book that introduces this topic in a way that is understandable to the average person working in the energy and facility management area that is not an IT trained professional.

WHO WERE SOME OF THE EARLIEST DEVELOPERS OF WEB-BASED INFORMATION AND CONTROL SYSTEMS THAT INFLUENCED MY THINKING?

One of the earliest applications of web-based energy information systems was created in 1992 by Dr. Dale Kirmse, an associate professor in the Department of Chemical Engineering at the University of Florida, Gainesville, FL. Gator Power—a joint venture of the University of Florida and Florida Power Corporation—was installing a new 43 MW combined cycle cogeneration power plant on the UF campus, and Dr. Kirmse was a member of the Faculty Oversight Committee for that project. His most significant contribution to this project was suggesting that the cogeneration system be fully instrumented and used as a teaching laboratory for the College of Engineering at the University. He pursued this goal, and was successful in getting this teaching component into the funding and installation of the cogeneration power plant.

Dr. Kirmse then had the additional foresight to recognize that the newly emerging World Wide Web, using the TCP/IP Internet Protocol and the Mosaic Web Browser (at the time), could be used to make the data from the cogeneration plant visible and accessible to students anywhere on the campus, as long as they were using a computer connected to the internet and were using a web browser. He then followed this recognition with the programming work to get the sensor data from the SCADA system at the cogeneration facility delivered over the internet as web pages to any user with a web browser. Not only were these operational data on the cogeneration power plant now accessible to UF engineering students, but they were also accessible to anyone with internet access and with a web browser on their computer.

I was impressed by this system, and certainly with the work that Dr. Kirmse did himself, without the benefit of many of the time-saving software system tools that we have today. However, I have to honestly admit that I did not fully appreciate the significance of this system or of Dr. Kirmse's contribution, and certainly not the extent to which this would eventually become the dominant technology for most information systems—not just energy and facility information systems. This information technology solution to the design and operation of energy information systems has progressed as a major technological wave. In management organization terms it is a paradigm shift. It is at least as powerful and encompassing as the shift from pneumatic and electrical analog control systems to DDC—direct digital control—systems in the 1980's. This new IT wave, or paradigm shift, has brought in a whole new group of companies working in energy information and control systems, and has forced major changes in the business strategies of the established control companies.

Around the same time as this project at the University of Florida began, Paul Allen at Walt Disney World in Orlando developed his DOS-based METER program to collect monthly utility billing data from the utility billing system at the Reedy Creek Improvement District (RCID)—Walt Disney World's utility provider. Paul formatted a monthly utility data report that was sent out to all of the various cost centers at WDW to inform them of their utility usage. The report was sent via a e-mail in what Paul termed a "Virtual Utility Bill." Continuous improvements in the METER program included added hourly electrical interval data that were available from RCID's SCADA system, and additional utility reports to provide better insight into energy usage patterns at WDW. In 1996, Paul envisioned an upgrade for the METER program to a web-based client-server program he called the "Utility Reporting System." Most of the utility analysis programs on the market at that time were Windows-based. Paul's vision was to create a program to make WDW's utility information available on every PC desktop within the Walt Disney Company—worldwide. Using the Disney intranet and using inexpensive tools such as a relational database, a web-server and HTML programming was the simplest way to make this vision a reality.

In early 1997, Paul started creating the Utility Reporting System with the help of Dave Green. Together Paul and Dave created a Foxpro and Foxweb application that took the Foxpro tables found in the original METER program and put them on the web. At that time, only a few people at WDW had web browsers on their computers, and all of those were Netscape. It was in this time frame that Paul said "I hope someday all WDW computer users will have a browser so they will be able to use the new program I am developing." With Paul's knowledge of the WDW organization and equipment, and their need for information on utility costs, he came up with the format for the utility reports; and Dave Green came up with the overall layout of the Utility Reporting System software design for the web.

All of this work in designing and creating the WDW Utility Reporting Systems was still accomplished well before the commercial appearance of most of the current web-based systems available in today's market.

Other facilities and other people have also made significant progress in developing and implementing their own web-based energy and facility information and control systems. I am not advocating that every facility should develop their own web-based system, but it is clearly possible for some facilities to accomplish these same results at reasonable costs if they so choose. Many of the chapters in this book have how-to information that can greatly assist in this process. For other facilities, the idea of developing their own web-based energy or facility management system is simply not an economically or technically viable plan. For those facilities, the goal of this book is to try to provide sufficient background in the IT area so that successful discussions can be held with the commercial developers and providers of these comprehensive information and control systems.

ACKNOWLEDGMENTS

First and foremost, I would like to acknowledge the technical and organizational help provided by my associate editors Paul J. Allen and David C Green. Paul is the one who first turned me on to web-based systems as practical and powerful tools that could be used by the average building or facility. Much of my initial education in the area of web-based systems came from discussions with Paul. His willingness to help me understand the value and operation of web-based energy information and control systems, as well as supporting my interest in editing and producing this book, has made the task much easier and much more enjoyable. He is the author or co-author of six chapters in this book. Paul was a student in the first energy management class I taught at the University of Florida, and now he is my teacher for many of these new energy applications of information technology. This is one of the true rewards for being a professor at a university.

I met Dave Green through his work with Paul Allen at Walt Disney World in Orlando, FL, and Dave has become a tremendous resource for me, and for this book. Dave is the author or co-author of six chapters in this book. Dave's real specialty is the design of web pages that offer quick and intuitive user interfaces. This is a skill that is sorely missing in most of the web applications I am familiar with. I hope some of the people designing web pages for my financial institutions, stock accounts and most other web services will eventually read this book and use some of Dave's ideas and principles in designing much more user friendly human interfaces.

Also serving as an associate editor is my wife Lynne Capehart, who has contributed much of her spare time in reading, editing, formatting and sometimes even writing parts of chapters for this book. She has handled the myriad details involved with keeping track of the chapters for the book, and their state of completion as the process went on. She has worked tirelessly to help insure the quality of the material throughout the book. Because of her there is a much more consistent format and style of the various chapters written by the different authors. Without her, I would probably not have been able to complete this book in any reasonable amount of time.

Last, but not least, I would like to thank each of the 37 other authors who have contributed chapters to this book. The field of information technology itself is far broader than any one person can hope to fully understand. When the application of energy information and control systems is added to this, there is simply no one person who can blend these two areas together and present detailed information about each of the relevant pieces and topics. Thus it is only possible to produce a book on these diverse subjects with the help of many knowledgeable and skilled authors writing chapters in their individual areas of expertise. All of the authors have full time jobs that require the majority of their attention. Writing a chapter for this book means taking time each night and each weekend to create something that is related to their day-time work, but which still requires a separate effort to put together their ideas and explanations for the benefit of others. This time consuming task has been generously undertaken with the only reward being the knowledge that they have helped other people find their way in this complex field of IT and web-based energy information and control systems. My most sincere compliments and thanks to each of you.

Finally, it should be noted that most of the book chapters are individual, stand-alone contributions to the book. As such, there is subject matter that is repeated in several of the chapters to make them stand-alone pieces. I have not attempted to remove this duplication, since I believe that it is useful in this context, and that also it is beneficial to have different authors explain some of the same concepts in different words. I find it valuable myself to read things on the same topic that are written by different authors so that I can get a good feeling for whether I understand the overall message being presented by each author. Also, the stand-alone style allows readers to go into the book at almost any point and read a chapter without having to read all of the preceding chapters.

Chapter 2

Introduction to Web-based Information and Control Systems*

Steve Tom, Director of Technical Information
Automated Logic Corp.

THIS CHAPTER IS ABOUT ways in which a web-based information and control system can enhance a facility's operations. It introduces the concepts involved in developing a web-based control system, and presents the view that using a modern, web based system results in doing more with more, rather than doing more with less. Significant advances in using these web based information and control systems are associated with more access, more flexibility and more interoperability. Benefits of wireless access and of open systems are also discussed.

INTRODUCTION

Doing More with More

Let me begin by explaining the subtitle "Doing More with More." Web-based systems can provide more access, more flexibility, and more interoperability; and, they can provide these benefits over a wider area than conventional control systems. Thus, the common phrase "doing more with less" doesn't apply here. Web-based systems give you more to work with; they just don't have to cost more.

Since everyone seems to have their own definition of what a web-based system is, let me state my definition at the outset. A web-based system is one in which the primary user interface is provided through web pages which are accessed on a standard web browser. Monitoring conditions, running reports, changing setpoints, changing schedules, receiving and responding to alarms, downloading updated control programs and graphics—all the typical activities an operator may perform on a day-to-day basis are handled through a browser. There may be other tools for off-line preparation of major engineering changes, creating new graphics, etc., but there is no "operator workstation" other than the browser interface.

A "web-enhanced" system, on the other hand, still uses a conventional operator workstation as the primary operator tool but provides some features through a web browser. As you might have guessed, a web-enhanced system will provide some, but not all, of the benefits of a web-based system. Both will provide more access, more flexibility, and more interoperability over a wider area than a conventional control system. How do they do this? Let's look at these benefits one at a time:

More Access

Conventional control systems require special software to be installed on each operator workstation. The primary workstation or workstations are typically located in a maintenance office, which can range from a high-tech 24/7 operations center to a salvaged desk in the corner of a mechanical room. Additional copies of the workstation software may be installed in other locations—the maintenance superintendent's office, the facility engineer's office, a laptop computer that the on-call technician takes home with him, etc. The number of these workstations is limited by the fact that workstation software is not free, and few customers can afford to install this software on every computer that might be used to access the system.

A web-based control system, on the other hand, only requires special software to be installed on a single computer, which becomes a web server. (Multiple computers can be used for system redundancy, but conceptually it's still a single computer.) This server generates web pages that serve as the user interface to the system. Any computer with a standard web browser can be used

*This chapter was taken from articles originally published on AutomatedBuildings.com and used with their permission.

as an operator workstation. This means all of the computers listed above, and many more that weren't listed, can be used as an operator workstation if you allow it. (For security reasons you may want to restrict the computers that can access your system, and you will certainly want to restrict the people who can log in, but that's a topic covered by other authors in other chapters.)

How does this increased access allow you to do more with your system? By allowing everyone who has a legitimate need to work on the system to do so, from whatever location they wish, at any time of the day or night. The facility engineer, the maintenance superintendent, and the maintenance technician can all access the system from their office, from their home, or from any other computer they have available to them. The office administrator can access a restricted portion of the building, say, to adjust the schedules and temperatures in the engineering offices. The shipping foreman can have similar privileges in the shipping department.

Letting people control their own departments gets you out of the business of running a call center for minor temperature adjustments, but of course it doesn't mean you abdicate your role as the energy manager. You can see every change they make, set limits on their adjustments, and monitor departmental energy use. If your facility is too small to justify a full-time energy engineer, you can hire a consultant who can use your web-based system to analyze your energy use and improve your operations from the comfort of his office—even if that office is on the other side of the world.

More Flexibility

Because standard web pages are used as the operator interface, any "gadget" that surfs the internet can be used as an operator workstation. Certainly any Windows PC or laptop can be used, but the choices are wider than that. Mac users surf the internet. So do Linux users. Many cell phones and personal digital assistants (PDAs) can surf the internet as well. Your building automation system may need to generate special web pages for the small screens on these non-PC browsers, but that's not a problem for a well-designed web-based system. Conventional building automation systems have long been able to buzz a pager when something goes wrong in your building. With a web-based system you can now use a cell phone to respond to that page, browse the system, and make whatever changes you want, from wherever you are. If you happen to be at a movie when the page comes in, you can make these changes without leaving the theater. (If the plot gets a little slow you can even call in on your own, just to see how things are going.)

The advantages of wireless communication are not limited to cell phones. Wireless Ethernet transmitters and receivers are a low-cost way to give roaming access to a technician's laptop anywhere in your building. Ever listen to a radio conversation between a technician on a ladder, looking at a VAV box, and his buddy sitting at an operator workstation? "OK, open the damper. Is it open? Try closing it. You sure you're on the right VAV box? No, I said I was across the hall from the conference room. Now try it. Is it open?" etc., etc. With a web-based system and wireless Ethernet, the technician on the ladder can test the system himself, without all the confusion and delays of the radio calls. More importantly, his buddy is free to do something a little more productive than sitting in front of a computer screen, pushing buttons on command.

More Interoperability

The use of internet standards makes web-based systems interoperable at the hardware level. To those of us who don't have our heads in the Ethernet these standards may be a confusing "alphabet soup" of acronyms—TCP/IP, HTML, XML, WML, HTTP, IEEE 802.11b—all of which ensure web devices can connect to one another. The WAP cell phones, standard web browsers, wireless Ethernet networks, and similar technologies already described would not work without these standards. Simply connecting devices does not guarantee interoperability, however, any more than the phone on my desk guarantees I can communicate with people in China. That phone can connect me to a telephone in China, but if I want to communicate I'd better learn to speak Chinese. The same concept holds true for building automation systems.

The internet standards inherent in a web-based system ensure connectivity, but for true interoperability the system also needs to use a standard "language" like BACnet. BACnet allows equipment made by different vendors to talk to one another over the communication channels provided by the internet or a local intranet. Before BACnet it was common for HVAC control vendors to create custom translators or "gateways" to connect their system to a chiller, a boiler, or some other isolated piece of equipment. BACnet eliminated the need for the custom translators. Now the building automation system can talk to the equipment directly, and what's more, it can talk to a building automation system made by a competing vendor. Configuration tools like "autodiscover" make it possible for a programmer or a knowledgeable end-user to see what information is available in the other system, and to establish new links between systems. These links are not limited to just exchanging point data, either. BACnet includes high-level system functions, like scheduling, alarming, and trending, which make it possible to manage the entire system from one vendor's workstation. Combining a powerful language standard, like BACnet, with the connectivity standards of the internet, makes web-based systems very interoperable.

The use of internet standards is also bringing a new level of interoperability to building automation—one that will extend well beyond the physical plant. A new class of applications called web services will make it possible to link a building automation system with other computer systems on the internet. This would make it possible, for example, to use weather forecasts and real-time utility prices as inputs in the control algorithms for your building. A tenant billing computer could connect to a building automation system and run a report on energy used by one specific tenant. A tenant billing system could query the building automation system about hours of operation, setpoint adjustments, or metered energy use. Similarly, an ice storage system could connect to a weather forecasting computer to retrieve a weather forecast. These types of interactions are possible today, but only if you're willing to invest the time and money required to develop a custom application to link the two computers. Web services will expedite this process, providing tools to discover which computers on the web have the information you need, the tools to extract the data, and a standard format for the electronic data exchange. Since web-based control systems are already on the web and are based on internet standards, web services are a logical way to enhance their interoperability.

Web Services

Web services will greatly simplify computer-to-computer transactions. It will allow "generic" applications to connect, locate, and exchange data automatically. Some experts are saying it will prove to be much more powerful than anyone yet suspects, just as the PC turned out to be more than just a smaller version of a mainframe computer, or the internet proved to be more than just a better way to run an electronic bulletin board.

One problem that may slow the growth of web services is the lack of standardization of the data that is being exchanged. Web services provide a standard way to locate and gather data, but if every computer application has a different data model, the data may be of limited use. This proved to be a problem in some of the early internet based business-to-business transactions that predate web services. XML provided a universal way to format the data, but if the data structures used by the two systems were fundamentally different there was still a lot of hand coding required. Web services require the marriage of existing internet standards such as XML (extensible modeling language) for data exchange and standard data models like BACnet. The task of creating these standards is formidable, but the potential benefits are mind-boggling.

One suggestion has been to develop "vertical standardization," that is, standard data models within each industry. A data model that works in the Electric Utilities industry would probably not fit the needs of the Real Estate industry, but standardization within the utilities industry would make it much easier to develop a web service that would work with multiple utility companies. Great in concept, but where would such a standard come from? ASHRAE's BACnet committee might be a good starting place, and they have discussed an XML extension of BACnet that could be used with web services. Perhaps a question that needs to be answered first is, "If the dream of web services were to become a reality, and you could exchange data with any other computer system in the world, how would you use this tool to improve your building automation?"

Wider Area

An airport manager uses his building automation system to check the temperatures and schedules in multiple concourses. A physical plant employee at a university enters a "winter storm—no classes" schedule into buildings throughout the campus. An energy manager uses trends and reports to compare the performance of government facilities throughout the southwestern US and Hawaii, activating emergency measures when required. Nothing you haven't heard before? Maybe, but in each of these examples the data are coming from multiple building automation systems made by competing vendors, and is being integrated into a single user interface by a web-based front end. The interoperability promises of standard communication protocols have come true, and the new web-based control systems extend this interoperability across the globe.

The benefits that web-based control systems bring to a wide area operation are staggering. The World Wide Web is by definition the widest area network on the planet, and web-based systems are designed to operate over the World Wide Web. National retail stores, hotel chains, large corporations, and government agencies can exercise better control over their far-flung facilities by utilizing the web. Detailed energy analysis, fault detection, 24-hr alarm centers, and similar functions which could not economically be implemented in each individual building become very attractive when managed over the web. School districts, university campuses, medical centers, and others who manage facilities within a smaller area have long used conventional control systems to provide this centralized control. For them the benefits of a web-based control system are that it uses the networks they already have in place, without requiring a special facility network or dedicated phone lines for dial-up access. Even a small family business with a single facility can benefit. For them, "wide area access" means they can check on their building from home and receive alarms through their home computer if anything goes wrong after hours.

Of course, the idea of putting your facilities on the web does raise security concerns. There are a few people surfing the web whom you most certainly do not want accessing your building. In this, you are not alone. Banks, hospitals, government agencies, military units—there are many organizations who depend on the web for day to day operations but who don't want to share their data with anyone who stumbles across their URL. Fortunately, there are many tools to safeguard data on the internet: firewalls, secure sockets layers, virtual private networks. The details of these security provisions are beyond the scope of this chapter, but suffice it to say that they are sufficient to protect the sensitive data of these users. Chances are "hackers" are much more interested in these data than in your setpoints and schedules, but it is comforting to know that you can have the same level of protection as these high-threat targets.

More access, more flexibility, more interoperability, wider area integration, but not more cost. There are many more ways in which a web-based control system enables you to do more with more, but this is a start. The exciting part is that this is not a "pie in the sky" prediction of things that might be. There are hundreds of web-based control systems in use today, and they are living up to their promises.

Wide Area Interoperability

The concept of interoperability is not new, but getting proprietary control systems to talk to one another used to be so difficult that it was usually confined to retrieving a few nuggets of information from major pieces of equipment like chillers or boilers. The emergence of standard point-level protocols like LonWorks and Modbus made this job much easier, but the exchange was still primarily limited to individual data points. BACnet raised the ante by including higher-level building automation functions like scheduling, alarming, and trending. This greatly simplified the process of making the entire system interoperable, that is, using a front end made by one manufacturer on a system made by another.

To generate a trend graph using point-by-point interoperability, for example, the front-end computer must be connected to the system for the entire time period being trended so it can read each value as it occurs. Alternatively, a special piece of hardware must be installed on the system to gather and store the trend values in a format that can be read by the front-end computer. With BACnet, the original vendor's hardware stores the trend data in an interoperable format to begin with. Now any other vendor's BACnet front end can connect to the system, retrieve the BACnet trend file, and display it. Similarly, turning a piece of equipment "on" at a predetermined time can be handled by point-by-point interoperability if the front end computer is connected at the time the change needs to be made, but if you want to schedule it to turn "on" at 6:00 AM next Tuesday you need to use an interoperable scheduler, like BACnet's.

BACnet is a logical choice for wide-area interoperability because it is impractical to keep the front-end computer connected to every building in the system 24 hrs a day. The choice then becomes one of using an interoperable protocol for the high-level functions or installing a proprietary piece of hardware in every building. In a web-based system, a web server replaces the front-end computer but the choice is the same. Either the server uses BACnet to communicate with each building as required, or a proprietary piece of equipment needs to be installed in every building. Both approaches provide interoperability, but BACnet provides a cleaner solution.

Open Systems

The idea of open systems has become so popular that nearly every manufacturer claims to offer one, and the corporate "spin doctors" are twirling to new heights in their attempts to create definitions of open systems which include their products. Much attention has been focused on communication protocols, with BACnet supporters battling the LonWorks camp for the title of "most open." In reality, communication protocols are only one part of a building automation system, and a truly open system must be open at all levels. This means it needs to use an open communications protocol, be accessible from any computer on the network, be written in a platform-independent language, use an industry standard database, and provide the user with all necessary programming tools. That's a pretty tall order, but fortunately technology is available to make this possible.

The World Wide Web is the largest open system in history, and the technology that makes it possible to access sites from Arkansas to Zanzibar can make your building's data available to authorized users anywhere in the world. The word "authorized" is important, and the security encryption that protects your credit card transactions on the web can protect your facility data. Web technology includes more than just transmitting data across the web, it also involves providing data in the form of web pages. This allows any computer with a web browser to view and edit the data, not just a computer loaded with the control vendor's software. This is especially important to customers who have many technicians supporting a wide network of controls, such as a typical college campus. Purchasing building automation software for multiple computers can be expensive, and it ties the technicians to those computers. With web technology, technicians can monitor, troubleshoot, and repair the system from any computer on the network. If that network happens to be the internet, technicians can work from virtually any computer in the world. For maximum flexibility, the web pages themselves should be based on open standards like HTML and JavaScript. This allows the data to be accessed by computers running Windows, Unix, Linux, Apple OS, or other operating systems commonly found on large systems.

In addition to "platform independent" web pages, the software that generates these pages should be written in Java so the web server itself can run on any operating system. The server should also use an industry standard database that supports CORBA (Common Object Request Broker Architecture) to store vital system data so the data will be available to many different software applications. Configuring the system and programming the controllers requires special programming tools, so the control system vendor should provide the customer with all tools and documentation required to allow the customer to modify or expand their building automation system. With these, the customer is not "locked in" to a single vendor when changes are needed.

Finally, an open system must use an open communications protocol. Since the corporate spin doctors can't agree on what constitutes an open protocol, let's use a neutral definition, provided by a recognized authority on the telecommunications revolution. MIT Professor Nicholas Negroponte says "A truly open system is in the public domain and thoroughly available as a foundation on which everybody can build." BACnet, which was developed by ASHRAE, has been adopted by ANSI, and adopted by ISO as an international standard, fits this definition to a "T." It's important that this protocol be the "native" language used throughout the control system. Gateways, or custom-programmed "translators" which provide an open protocol at one or two points are better than nothing, but the end result is the vendor who controls the gateway controls the system.

Building Controls and BAS

Monitor and Control

The BAS products currently on the market have the ability to properly monitor and control buildings. The current generation of products is not perfect by any means, but they provide a pretty impressive list of building automation functions. Take a look at any critical facility operation today, from high-speed data centers to biohazard research labs, and you'll probably find a BAS. Some customers have extended their BAS to include manufacturing process control because they liked the power, speed, and flexibility of the BAS. In the future, these systems will get even better. As high-speed internet connections become commonplace and new technologies such as web services mature, BAS will be able to improve control algorithms by incorporating data from computers outside the BAS; data like weather forecasts and real-time utility rates. It's possible to do this now with custom programming, but these new technologies will make it practical to implement these strategies as a matter of course.

Recent Innovations

Standard communication protocols like BACnet and LonWorks are becoming the norm, rather than the exception, and customers are expecting more and more integration of their building systems. Web-based control systems have really started to catch on, and almost all BAS vendors now offer at least some level of internet access. These are really complementary developments, but at first glance it's easy to mistake them for competing technologies. The fact that internet standards like HTML make it possible to use a single browser to look at two independent systems makes it appear these systems are interoperable, but that illusion disappears the first time you try to apply a new schedule to both systems. In the near future, these developments may merge; perhaps the BACnet committee will develop an XML schema to promote interoperability over the internet.

Internet Platforms and Displays for Building Control

The growth and acceptance of the internet is probably the most exciting development since direct digital control (DDC). Use of the internet has opened up entirely new vistas for BAS. For one thing, it has greatly expanded the interoperability of BAS. Multiple buildings with control systems made by multiple vendors can be converted into a single, integrated system that extends over vast geographic areas. Such projects would not have been practical without the internet. This trend will accelerate even faster as existing BAS protocols like BACnet become integrated with internet standards like XML. There are also a lot of "spinoff" benefits to the BAS, as internet related products such as internet browsing cell phones, wireless Ethernet laptops, and high-speed web servers become part of the building manager's toolkit. The net result is that BAS capabilities are progressing much faster than when the development was limited to a few small R&D staffs working for control system manufacturers.

Using Existing IT Infrastructure to Accommodate BAS Functions

The use of the IT infrastructure has increased significantly, but there's a lot of learning that needs to take place on both sides. IT professionals are understandably suspicious of "outsiders" who want to tie into the systems they've worked so hard to build. They want to know more about the technical aspects of these systems, how much bandwidth they'll consume, and whether or not it will increase the vulnerability to "hackers." BAS professionals need to learn a lot more about IT systems so they can answer these questions, and so they can make the best use of a shared network. Experience has shown that a BAS can be integrated into an existing network with no adverse affects. The bandwidth consumed by a BAS is negligible compared to the other systems on the network, and a properly designed BAS does not open new security threats.

BAS and Wireless Technologies

Wireless technologies add a powerful new tool to the BAS toolkit, especially when used to augment the user interface. WAP enabled cell phones, personal digital assistants, and wireless Ethernet links for technician's laptops can add significant new capabilities to a BAS. On the other hand, wired systems are not going away any time soon. Wired networks of controllers and sensors have significant cost, performance, and reliability advantages over wireless networks in many applications. In some situations, wireless networks make sense, but they need to be evaluated on a case-by-case basis. Some facilities have a mixture of wired and wireless networks for the entire IT structure, including the BAS

Problems with BAS

Some older BAS couldn't meet the owners' needs or were improperly designed and installed, but the biggest problems with BAS are usually the "classics": lack of mechanical system maintenance, lack of operator training, and a lack of support for the facility management operation. The best BAS in the world is going to fail if it's connected to a set of dampers that are rusted solid, through years of neglect. A BAS can be an extraordinarily useful tool for finding these mechanical problems if it's used correctly, but many of the older BAS had complex or confusing user interfaces. Operators are always "under the gun" to fix problems, and if the system is confusing they will be encouraged (forced?) to bypass the BAS and manually override the system. There's an old saying that a simple system that works will outperform a complex system the operators don't understand, because the operators will "simplify" the complex system into something they do understand. This axiom is also true of BAS.

There comes a time in the life of any BAS when the features and capabilities of a new system justify a complete replacement, but a BAS should last much longer than five years. Companies must make their new products backwards compatible because customers have a right to expect support for older systems

Need for Simplified Control Methods

The control method should be as simple as possible, but still meet the needs of the user. Some functions, like adjusting setpoints or entering schedules, need to be very simple because they will be used by all facility managers, regardless of expertise. More complex applications, like optimizing control of an ice storage system, will of necessity require a more highly trained operator. These operators need to be able to configure trends, run reports, set up alarms, and perform other high-level control functions, but the user interface still needs to be as simple as possible. A well-designed BAS will provide a very simple interface for day-to-day tasks while still offering "power user" tools for advanced control functions.

Specifying Building-control Solutions

Specifying a BAS has got to be one of the toughest jobs in our industry. There is no such thing as a generic BAS because there is no such thing as a generic building or a generic customer. The specifying engineer has to carefully evaluate the building owner's needs in terms of cost, features, interoperability, expandability, and a host of other considerations. It is almost impossible to do this without involving BAS vendors because they are the best sources of information on what's available. If you don't take a close look at systems that are currently on the market before writing a spec you run the risk of either writing a spec that's impossible to meet or specifying something that's so basic it will lack technology and features the customer really needs. BAS vendors can be very helpful in providing sample specifications, but of course those specifications will favor their system. If you're going to sole-source the system you want, the vendor's specs might be exactly what you want. If you want competitive bids, you'll need to gather several specs and find common ground.

Retrofitting an Existing Facility

The challenges of retrofitting an existing facility with a BAS are really no different than the challenges of retrofitting an existing facility with a conventional control system. You need to do a complete survey of the existing mechanical system to see what's actually there (vs. what's on the drawings) and to make certain it's still in usable condition. The existing sensors and actuators should be reviewed at this time to determine which, if any, can be re-used with the new BAS. Finding the original design data for VAV boxes, control valves, etc. can sometimes be a challenge, but again this is not unique to a BAS retrofit. Finally, it's important to take a look at the overall design to make certain the mechanical systems are still appropriate for the current operation. Sometimes there's been a change in occupancy or a change in function that makes the entire mechanical system ill suited for the new use. If that's the case, a new control system is not going to solve the problem, even if it is a BAS.

CONCLUSION

The use of web based energy and facility information and control systems truly offers the potential to "do more with more." Utilizing the power of the internet, high speed computers and communications networks, and graphical displays through a web browser provides a powerful and innovative approach to new energy and facility management systems. This chapter has hopefully provided the reader a useful educational introduction to many of the important concepts and operational features of these web based systems.

Section Two

Overview of Web-Based Energy Information Systems and Energy Control Systems

Chapter 3

How a Web-based Energy Information System Works*

Barney Capehart, University of Florida
Paul Allen, Walt Disney World
Klaus Pawlik, Accenture
David Green, Green Management Services Inc.

ADVANCES IN NEW EQUIPMENT, new processes and new technology are the driving forces in improvements in energy management, energy efficiency and energy cost control. Of all recent developments affecting energy management, the most powerful new technology to come into use in the last several years has been information technology—or IT. The combination of cheap, high-performance microcomputers together with the emergence of high-capacity communication lines, networks and the internet has produced explosive growth in IT and its application throughout our economy. Energy information and control systems have been no exception. IT and internet based systems are the wave of the future. Almost every piece of equipment and almost every activity will be connected and integrated into the overall facility operation in the next several years.

The internet, with the World Wide Web—or web—has become quickly and easily accessible to all facility employees. It has allowed the development of many new opportunities for energy and facility managers to quickly and effectively control and manage their operations. The capability and use of IT and the internet in the form of web based energy information and control systems continues to grow at a very rapid rate. New equipment and new suppliers appear almost daily and existing suppliers of older equipment are beginning to offer new web based systems. Facility managers, maintenance managers and energy managers are all having to deal with this rapid deployment of web based equipment and systems, and need to be prepared for current and future applications of internet based technologies in their facilities. In some cases, facilities are developing their own information and control systems or at least subsystems and are trying to understand how to connect and interface new IT equipment to their older energy management or facility management systems.

The purpose of this chapter is to help prepare energy managers to understand some of the basic concepts and principles of IT. We hope that they can successfully apply IT to their facility, and have the knowledge to supervise the IT work of a consultant or a vendor. Knowing what is going on and what is involved is important information for the energy manager if they are going to successfully purchase, install and operate these complex, web-based energy information and control systems.

ENERGY INFORMATION SYSTEM (EIS)

The philosophy, "If you can measure it, you can manage it," is critical to a sustainable energy management program. Continuous feedback on utility performance is the backbone of an energy information system. A basic definition of an energy information system (EIS) is: equipment and computer programs that let users measure, monitor and quantify energy usage of their facilities and help identify energy conservation opportunities.

There are two main parts to an EIS: (1) data collection and (2) web publishing. Figure 3-1 shows these two processes in a flow chart format.

Everyone has witnessed the growth and development of the internet—the largest computer communications network in the world. The internet and the World

*A preliminary version of this chapter appeared in *Strategic Planning for Energy and the Environment,* Vol. 22, No. 3, 2003 and is used with permission of Fairmont Press.

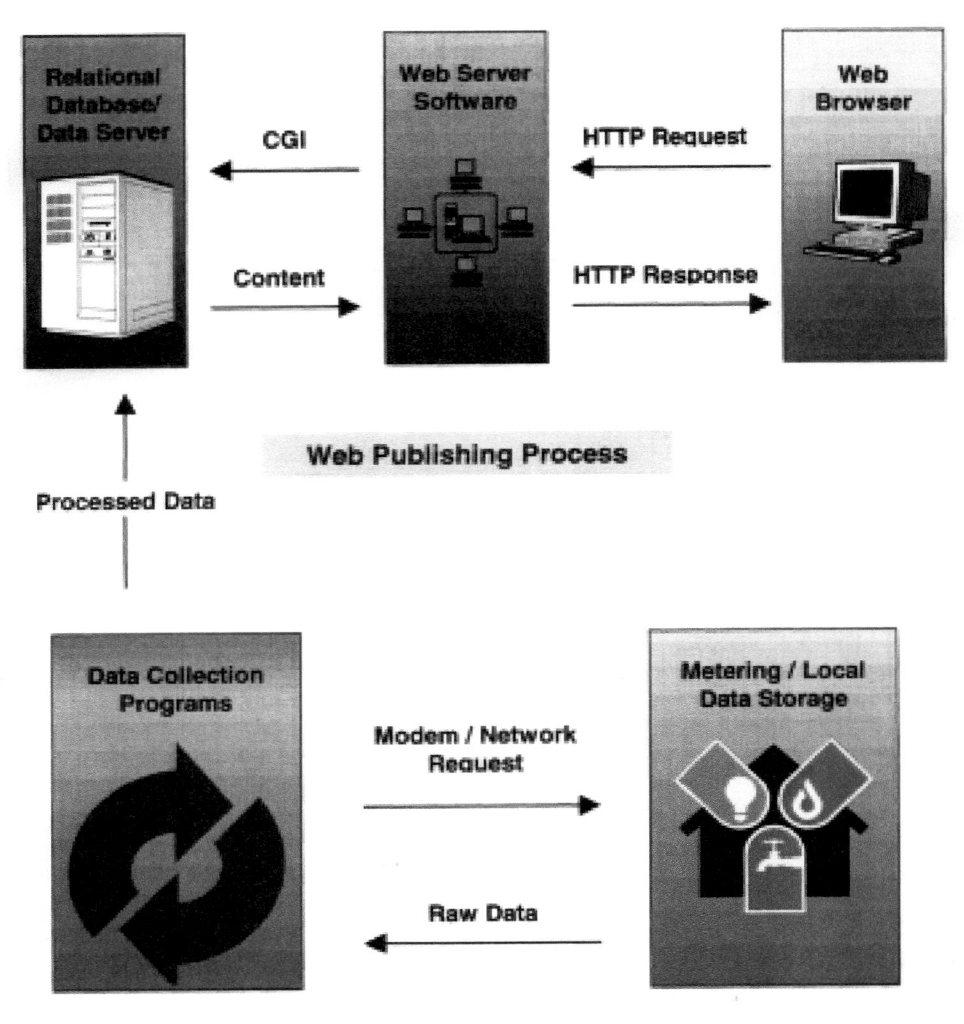

Figure 3-1. Energy Information System Functional Layout

Wide Web (Web), using the TCP/IP communications protocol, has made it much easier to access and distribute data. Using a web browser, one can access data around the world with a click of a mouse. An EIS should take full advantage of these new tools. Here are definitions for the terms intranet, TCIP/IP, and web browser.

Intranet

An intranet is a small, self-contained, private version of the internet, using internet software and internet communications standards. Companies are increasingly using intranets to give their employees and other approved people easy access to facility and corporate data.

TCP/IP

Transmission Control Protocol/Internet Protocol (TCP/IP) is a family of industry standard communications protocols that allow different networks to communicate. It is the most complete and accepted enterprise networking protocol available today, and it is the communications protocol of the internet. An important feature of TCP/IP is that it allows dissimilar computer hardware and operating systems to communicate directly.

Web Browser

A web browser is a program used to display data transferred over an intranet or the internet. It lets users select, retrieve and interact with resources on the web. The most commonly used web browsers are Internet Explorer and Netscape.

METERING EQUIPMENT

The first task in establishing an EIS is to determine the sources of the energy data. Utility meters monitored by an energy management system or a dedicated power-monitoring system are a good source. These systems provide a good means to collect energy data. Metering equipment collects the raw utility data for electric, chilled water, hot water, natural gas and compressed air. Information from these meters is collected by either direct analog connections, pre-processed pulse outputs or by digital, network-based protocols.

Direct Analog Connection

The utility meter produces a 4-20mA or 0-10V signal for the instantaneous usage continuously monitored by an energy management or process control system. The control system calculates the utility usage integrating the instantaneous values over time.

Pulse-Output

The utility meter outputs a pulse for a pre-defined amount of energy usage. A local data storage device accumulates the pulses.

Digital, Network-based Protocol

The utility meter calculates and totals the energy used within the meter. A dedicated local area network (LAN) links the energy meters together. A local data storage device polls these meters at predefined intervals to store the energy data until retrieved by the server.

DATA COLLECTION PROGRAM

Data gathered from all of the local data storage devices at a predefined interval (usually on a daily basis) are stored on a server in a relational database.

Relational Database

A relational database is a collection of tables, rows and columns used to store data, and is organized and accessed according to relationships between data. Examples of relational databases are FoxPro, SQL and Oracle.

There is a variety of methods used to retrieve these data:

Modem Connection

A modem connection uploads the data from the local data storage device to the energy data server. Typically, the upload takes place on a daily basis, but could be transferred more frequently, if needed.

LAN or WAN Network Connection

A local area network (LAN) or a wide area network (WAN) connection, established between computers, transfers energy data files to the energy data server.

FTP Network Connection

File transfer protocol (FTP) is an internet protocol used for transferring files from one computer to another. It is used to move the energy data files from the local data storage devices to the energy data server.

Once the energy data have been transferred to the energy data server, an update program reads all of the various data files and reformats them into a format that is used by the web publishing program

An essential feature of this process is to design the programs and tables so they will update quickly. Typically, utility meters can generate an enormous amount of data, so it is important to keep the table records manageable. Breaking the data up into yearly, monthly, weekly and daily files keeps the data manageable.

WEB PUBLISHING PROGRAM

To publish the energy data on an intranet or the internet, client/server programming is used. The energy data are stored on a central computer, the server, and waits passively until a user makes a request for information using a web browser, the client. A web publishing program retrieves the information from a relational database, sends it to the web server, which then sends it to the client that requested the information.

Web Server

A web server is a program that runs on a network server (computer) to respond to HTTP requests. The most commonly used web servers are Internet Information Server and Apache.

HTTP

Hypertext transfer protocol (HTTP) is an application layer protocol used to transfer data across an intranet or the internet. It is the standard protocol for moving data across the internet.

HTTP Request

An HTTP request is data sent from a web browser to a web server.

HTTP Response

An HTTP response is web browser content sent from a web server to a web browser in response to a specific HTTP request.

A CGI interface program coordinates the activity between the web-server and the web publishing program and allows for simultaneous multi-user access.

Common Gateway Interface (CGI)

CGI is a method used to run conventional programs through a web browser.

The client/server process for an EIS uses the steps below (See Figure 3-1):

1. A user requests energy information by using their web browser (client) to send an HTTP request to the web server.
2. The web server on the network server activates the CGI interface program. The CGI program then starts up the web publishing program.
3. The web publishing program retrieves the information from the relational database, formats the data in HTML and returns it to the CGI interface program.
4. The CGI interface program sends the data as HTML browser content to the web server, which sends the content to the web browser requesting the information.

This entire process takes only seconds depending on the connection speed of the client's computer to the web.

HTML

Hypertext markup language (HTML) is the format of the file containing web browser content. Developers use HTML to create web pages, and build graphical documents that contain images, formatted text, and links to other documents on the web.

Browser Content

Browser content is data—text, numbers, images and links—displayed in a web browser. Content is the new name for information carried over a network. It may be words, numbers, pictures, sound, or any form of what we used to call information.

PROGRAMMING CHOICES FOR EIS WEB PUBLISHING

The following sections provide additional detail on programming choices used in the development of the web publishing programs. These are only a few of the many programs available. Both server-side and client-side programming environments are applicable.

Using Server-side CGI Programs to Create Content

Server-side programs are programs that run on the network server. CGI programming is the most fundamental way to access relational database information over the internet or intranet and display dynamic content. This is important to any EIS since the amount of data involved will undoubtedly require processing with a relational database of some kind.

Dynamic Content

CGI programs create dynamic content as a web page that is displayed in a web browser at the same time it is created.

Static Content

Developers create static content as a web page that is stored in a file and displayed later. The web browser retrieves the data and displays the static content.

Although CGI is a necessary piece of the puzzle in creating dynamic content it may just be a stepping-stone to some other program. Small CGI programs often times just launch secondary programs that run on the web server and return content to the browser. A secondary program could be nearly any custom program that has the ability to send data back to the web server.

FoxWeb is a small CGI program that connects to a Visual FoxPro database application. The application then calls any number of custom designed queries and procedures to return results to the browser as HTML. More information about FoxWeb is available at http://www.foxweb.com.

PERL (Practical Extraction and Report Language) is an application used for CGI programming. One reason is that it is free. Another is that it is portable across operating systems. In other words, applications written in PERL will work on any operating system. It also has the ability to connect to many types of databases. A good source for information on PERL is at http://www.perl.com.

ColdFusion is a server-side application that uses a ColdFusion server. The server executes templates containing a mixture of HTML and ColdFusion instructions, and then returns the results to the browser as pure HTML.

Active server pages (ASP) are also a popular choice recently. The ASP program on the Windows web server will automatically interpret web pages ending with the extension ."asp." The web pages are a mixture of ASP instructions, Visual Basic code and HTML. ASP is not portable across operating systems. It only works with Microsoft web servers.

Java servlets are Java programs that run on a web server and build web pages. Java is the latest of a long line of "higher level" programming languages such as FORTRAN, Pascal and C++. It is also portable across operating systems. Java servlets written for one web server on one operating system will run on virtually any web server and operating system.

Java server pages (JSP) are similar to active server pages except that the pages consist of Java code in place of Visual Basic code. This makes the code portable across operating systems. The web pages typically end with the extension ."jsp." This tells the web server to interpret the embedded Java code like any other Java servlet. Information about Java servlets and Java server pages is available at this Johns Hopkins University web site: http://www.apl.jhu.edu/~hall/java/.

Any organization wishing to develop an EIS should carefully consider which server-side applications to use. The decision should be a practical one rather than a popular one. All of the criteria below should be a part of the evaluation process:

- What operating system is predominantly available to the facility?
- What programming languages are the support personnel willing to work with?
- What applications are compatible with the existing database?
- How much of the budget is available to spend?

Using Client-side Applications to Enhance Content

Client-side applications can create a deeper level of interactivity within web pages. Scripting languages such as JavaScript and VBScript are less complex versions of other languages like Java and Visual Basic, respectively. They reside within the HTML of a web page and provide a great deal of functionality that HTML itself cannot. Scripts such as these can validate input, control the cursor programmatically and much more.

Dynamic HTML (DHTML) is the result of scripting languages taking advantage of the extensions common to the latest browsers to make the pages change after they are loaded. A good example of this is a link that changes color when the user places the mouse cursor over it. Much more dramatic effects are possible using DHTML. However, the two most popular browsers Internet Explorer and Netscape interpret DHTML differently. Good information about DHTML is available at http://www.dynamicdrive.com/.

Cascading style sheets (CSS) are special HTML features that allow much more flexibility to format elements of a web page. The ability of CSS to describe the style of an element only once rather then every time you display the element provides a separation of content and presentation. This makes your web pages less complex, much smaller and therefore faster to load. Beware that CSS is only fully supported in the latest versions of browsers (4.0 and above).

Java applets are small Java programs that are stored on a web server and are called for from the HTML in a web page. Statements in the HTML pass parameters to the applet affecting its functionality. Unlike Java Servlets, the browser downloads the applet and runs it using the browser operating system rather then the operating system of the web server. Free Java applets are widely available on the internet. KavaChart applets are very useful for charting trends in data. KavaChart applets are available online at www.ve.com.

Extensible markup language (XML) is a meta-language and has a number of uses. A meta-language is a language used to explain another language. XML organizes data into a predefined format for the main purpose of sharing between or within computer systems. Furthermore, its uses include data organization and transfer, data presentation, data caching, and probably some that we have not invented yet. More information about XML is available at http://xml.com/.

Developers can use any or all of these to enhance the content of an EIS. Three important points to remember about using client-side applications to enhance your browser content are:

1. Many client-side applications require later versions of browsers to work correctly. Be sure all of your users are using the required browser versions.
2. Many client-side applications are available free. Search the internet before spending resources developing your own custom client-side applications.
3. Client-side applications will make your web pages more complex. This adds to development and maintenance costs. Be sure to weigh the benefits of these enhancements against their costs.

Support for EIS Web Publishing

Support for an EIS is quite involved. It includes Help pages on the application itself to guide the user. There must be an email address and telephone number of someone for users to contact. Adequate system support personnel must be available to help with system or network problems. Developers should track and document updates to the application. Undoubtedly, at some point, the tools you use to develop your EIS will require support as well.

Help pages should be simple and to the point. Avoid long discussions about how the EIS is developed and why. Save that for the "About This System" section or something similar. Each menu item or function should have its own help page. That information should be accessible from the main menu or at anytime within the function itself. Contact information about where to get further help should be available on every page.

Email addresses and telephone numbers published for help should be those of the local data expert. Many problems arise due to missing data or misunderstanding of the data.

These issues should be resolved or discounted before sending the problem on to the programming or system support personnel.

IT system support may not be as responsive as we would like. These personnel have a tremendous job to do. It is important to maintain a good relationship with them since their expertise is critical to the success of the EIS. However, at times when they are not able to respond it helps to have as much control over the EIS system as possible. Perhaps locating the server or the data collection machine in a location that is more accessible can help. Many times simply re-running a data collection routine or making a minor change to the code is all that is needed to solve a problem.

The programming support personnel should track and document all changes to the code. A good version tracking system is necessary to preserve versions that may be useful later. Keeping a maintenance log will help to recall changes made to the EIS system as well as contact information for sources of outside help.

Outside support from vendors comes in various forms. There is "free," "free only by email," "paid by hour," "paid by the call" and many other variations. Due to the nature of the EIS data, these applications require a constant level of support. Not necessarily a great deal of support, but they seem to be in a constant state of evolution. There may be a new report for someone, new data available, anomalies in the data, or new versions of tools to install. The value of outside support that is available for the tools used may be a deciding factor in choosing components of the EIS.

Choosing EIS Web Publishing Tools

We might define "tools," in this case, as utility applications that require more configuration effort than programming effort. There are some exceptions of course. In any case, tools fall into three categories:

- open-source or free,
- purchased,
- developed.

Tools are needed to perform such functions as batch emailing, charting, scheduling application run times, and database to web connectivity. The relational database and web server are also tools.

Batch email applications can be of any category. There are good free ones and purchased ones. Some email servers have batch processing capability and some do not. Purchased batch email applications are relatively inexpensive. They have some variations in features such as whether or not they will send HTML or attachments. They are also easy to develop.

Charting tools are really too complex to warrant developing. There are some very good free and open-source charting tools that have all the features of purchased ones. If not, one could probably purchase extensions to the standard features of a free charting tool.

The EIS needs scheduling programs to launch data collection applications at predefined times. Some operating systems have scheduling programs built in. They may be difficult to configure though. A purchased version is probably the best choice since their cost is quite low.

There are some applications that do much of the database to web connectivity for you. These either require purchasing or, in the case of the free ones, a great deal of programming. The purchased ones are a good choice since much of the error reporting is part of the application. The purchased versions can be very expensive or relatively inexpensive. Each database has its own connectivity options so much of this decision rests on which relational database is used.

The database used is likely a purchased one or may actually be open-source. There are open-source databases like MySQL competing with the best of the others. Commercial relational database systems range in cost from very inexpensive like MS Access to very expensive like Oracle.

Open-source web servers like Apache are available for some operating systems and are very widely used and reliable. Others are free with the operating system like MS IIS. Some are also commercially available for a few hundred dollars. The web server choice depends mostly on the operating system of the server itself.

The EIS tools will likely be a mix of open-source, purchased and developed applications. It is important to consider budget constraints, operating systems and support when deciding which and what type of tools to use. Always plan ahead of time for the compatibility of all the tools the EIS will use.

ENERGY INFORMATION SYSTEM EXAMPLE

This section will "get under the hood" and describe the specific programming tools used to create an actual energy information system. The goal here is to relate some of the basic concepts presented earlier in the chapter to a specific example of an energy information system.

URS overview

In 1997, an innovative intranet-based computer program called the Utility Reporting System (URS) was developed by Reedy Creek Improvement District (RCID) to provide a means to "publish" utility metering information and track the results of energy saving efforts at the Walt Disney World Resort using the Disney intranet. The URS provides continuous feedback on utility performance and pinpoints energy waste for further investigation.

The URS is an EIS that is based on Microsoft's Visual Foxpro database management system and uses custom programs to (1) gather the data from all data sources and (2) publish the data on the Disney intranet. The advantage of this approach is that the programs can be customized to collect all utility data—no matter its source—from a variety of existing utility data sources.

The URS was developed to make sub-metering more effective. By continuously "shining a light" on utility usage at each facility, utility costs are minimized by providing timely and informative reports. Continuous feedback on utility performance pinpoints problems in the energy management system that need attention.

The URS and nine other spin-off programs continue to be used everyday by users throughout the Walt Disney Company. The URS is now used at the Disneyland Resort in California and the Disneyland Paris Resort, which is a testament to its simplicity and low cost of operation.

URS Program Overview

The URS was initially created to publish the RCID monthly utility billing data. Each month the RCID utility billing data are output from the RCID utility billing system as a comma delimited file for each meter showing the account number, utility, consumption and cost. By making the RCID billing data available electronically each month, the time required to update the data is reduced significantly and data input errors are eliminated.

A Visual Foxpro program was developed to read the RCID monthly data into the database tables used by the URS. Aside from updating the monthly billing data, the program also determines if any new RCID utility accounts were added and automatically updates the URS meter account definition table.

To speed up web-browser access to the URS, the meter-level monthly billing data are summed to different hierarchical levels. Besides the meter level, which provides the finest level of detail, the accounts for each building or group of buildings are totaled to a subarea level data tables. Likewise, the accounts for each business unit are totaled to produce an area level data table.

Besides the monthly billing data, the URS also provides access to a wide range of hourly submetering data. These data can be very useful for determining how energy is used on a near-real time basis. They provide a finer level of detail and help energy managers quantify their energy saving efforts on a hourly/daily basis. Problems can be pinpointed quickly and controls adjusted to keep energy consumption minimized.

The URS updates the submetering data tables on a daily basis. The data are recorded hourly by the respective data collection system and is transmitted to the URS web server on a nightly basis. A Visual Foxpro program reads the various raw submetering data files into common submetering database tables. Once the tables are updated with the prior days submetering data, they are copied to the appropriate subdirectory on the web server and are then available for viewing using the URS.

Relational Database Table Structure

The URS uses several data tables to organize the massive amount of utility data. Shown below are the basic tables:

Account Definition Table

Includes meter account level definitions and grouping variables used by the program.

Monthly Data Tables

Includes the monthly billing data from RCID Utility Billing System.

Daily Data Tables

Includes data recorded on an hourly basis from power monitoring systems and energy management systems.

Email Data Table

Includes names, email addresses and the utility reports sent.

Data Collection Programs

Data collection programs are used to pull the utility data into the data tables. Visual FoxPro reads the various data sources and organizes the data into common data tables. Shown below are the various data collection tasks in the URS:

Monthly Utility Data

An FTP transfer from the RCID billing system downloads this data in an ASCII comma delimited file on a monthly basis. The database administrator creates a new record in the account definition table each time a new billing account is available.

Power Monitoring Hourly Data

RCID's Supervisory Control and Data Acquisition (SCADA) power monitoring system records max, min and average hourly data for electric meters and outputs this data to an ASCII file each day. A Visual Foxpro program reads this data and reformats it into an hourly data table.

Energy Management System (EMS) Hourly Data

The EMS produces files that include trends of analog/digital points and consumable data from utility meters connected to the EMS. Data collection programs copy these files from the EMS servers to the URS server after the EMS creates them each night. A Visual FoxPro program reads the data from these reports, reformats it, and then adds it to the hourly data table.

Veris Industries Power Monitoring Data

Hourly data are recorded in local data collection devices called Eservers. On a nightly basis, a program supplied by Veris Industries automatically collects this data from each Eserver into an ASCII comma delimited file. A Visual FoxPro program pulls the data from these reports, reformats it, and then adds it to the hourly data table.

It is important to point out that an enormous amount of data can be generated from hourly utility data. To keep the data manageable, the hourly data are broken up into separate monthly data files.

Once the data collection program finishes updating all of the data from the various hourly data sources, a Visual FoxPro program creates additional files that total the hourly data to sub-area and area levels. This step helps speed up web-browser access to the URS. Finally, the data are copied to the URS server data directory where it is ready for viewing on the Disney intranet.

URS Data Reports

To make the URS easy to use, it sends HTML-based reports via email on a daily basis (to report on hourly data collected) and a monthly basis (to report on monthly billing data). A daily utility report is created for each business unit and emailed to the business unit distribution list. Using HTML-based email reports allows the tabular report to link to graphs showing daily and monthly utility profiles. Users view the reports using their email program (Microsoft Outlook) and are able to produce graphs by simply clicking on a hot-links in the email. Sending email on utility usage helps to increase employee participation in reducing their facilities energy consumption. (See Figure 3-2 for a sample daily email report.)

Email also increases the likelihood that the user views the utility data. Instead of waiting for the user to visit the URS web site and figuring out how to generate the same report, the URS delivers the report via email.

Web Publishing Program

Visual Foxpro is used as the program language used to generate the web pages for the URS. FoxWeb is the CGI program that interfaces Visual FoxPro with the web server. The URS uses several reports to view and graph both the monthly utilities billing information and the hourly utility data. Kavacharts Java applets, called from the Visual FoxPro programs, generate the graphs used in the URS.

The challenge of producing an effective EIS is to create reports that are both informative and easy to use. The program should be designed so that a user can easily produce reports and graphs with a few clicks of the mouse. The URS makes extensive use of embedded links to sub-reports and graphs. This programming interface makes the URS intuitively easy for the user to navigate.

URS Results

The URS, like all EIS programs, does not result directly in energy reductions. Instead, the knowledge, operational insight, and experience gained from utility data can result in operational changes and corresponding energy savings.

The most significant result of the URS has been increased awareness of utility usage. Wasteful practices are corrected and energy-efficient systems are showcased as best practices. Individual have the URS to track

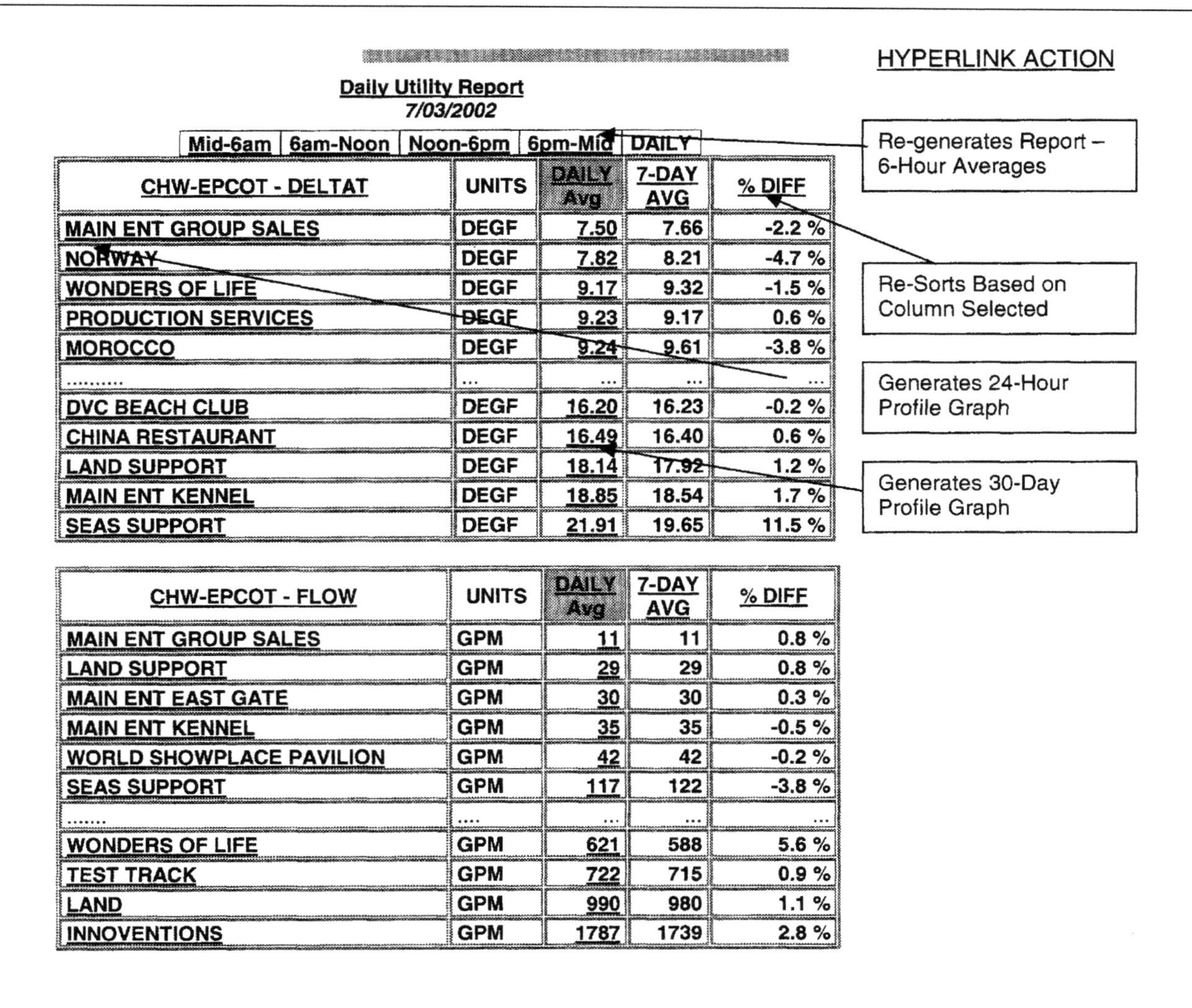

Daily Utility Report
7/03/2002

Mid-6am	6am-Noon	Noon-6pm	6pm-Mid	DAILY

CHW-EPCOT - DELTAT	UNITS	DAILY Avg	7-DAY AVG	% DIFF
MAIN ENT GROUP SALES	DEGF	7.50	7.66	-2.2 %
NORWAY	DEGF	7.82	8.21	-4.7 %
WONDERS OF LIFE	DEGF	9.17	9.32	-1.5 %
PRODUCTION SERVICES	DEGF	9.23	9.17	0.6 %
MOROCCO	DEGF	9.24	9.61	-3.8 %
....	...	...	...	...
DVC BEACH CLUB	DEGF	16.20	16.23	-0.2 %
CHINA RESTAURANT	DEGF	16.49	16.40	0.6 %
LAND SUPPORT	DEGF	18.14	17.92	1.2 %
MAIN ENT KENNEL	DEGF	18.85	18.54	1.7 %
SEAS SUPPORT	DEGF	21.91	19.65	11.5 %

CHW-EPCOT - FLOW	UNITS	DAILY Avg	7-DAY AVG	% DIFF
MAIN ENT GROUP SALES	GPM	11	11	0.8 %
LAND SUPPORT	GPM	29	29	0.8 %
MAIN ENT EAST GATE	GPM	30	30	0.3 %
MAIN ENT KENNEL	GPM	35	35	-0.5 %
WORLD SHOWPLACE PAVILION	GPM	42	42	-0.2 %
SEAS SUPPORT	GPM	117	122	-3.8 %
....	....	...	...	...
WONDERS OF LIFE	GPM	621	588	5.6 %
TEST TRACK	GPM	722	715	0.9 %
LAND	GPM	990	980	1.1 %
INNOVENTIONS	GPM	1787	1739	2.8 %

Figure 3-2. Utility Reporting System HTML-Based Report Example Daily Chilled Water Report

and monitor their energy use and this has translated into lower utility bills.

Shown below are other significant results from the URS:

- *Energy project savings* are quantified using monthly utility billing data.
- *A Utility Awards Program* was developed based on the monthly billing data. Awards are presented to those areas that show the largest percent change from prior year consumption levels. Competition to "win" the award helps motivate individuals to reduce consumption.
- *Utility billing errors* have been identified using monthly billing data. A comparison report is emailed each month to responsible individuals in each business area. The report is sorted based on the change from prior year data. Problem meters sort to the top or bottom of the report and many times are the result of billing errors.
- *Utility cost reimbursement* is used to quantify operating participant utility usage. Utility submeters are automatically read each day and a daily report emailed to the operating participant.
- *Redundant metering* is used to provide a daily check on meter accuracy. Individual facility utility usage is totaled and compared to plant production each day. Defective metering is quickly identified and fixed.
- *Chiller and boiler plant efficiency* is measured by comparing the utility input and output. This helps identify poor performing plant operation.
- *Inefficient chilled water usage* is shown in a daily report that pinpoints facilities with low chiller water differential temperatures.
- *New facility utility estimates* are generated using utility data from comparable facilities.

CONCLUSION

Today's energy manager needs to be knowledgeable about the basic principles and concepts of IT since this is a fast growing area of new systems and services. This chapter described the basics of how an energy information system is set up and how it works.

Additional Information

More information about energy information systems as well as links to many of the tools discussed in this chapter can be found at http://www.utilityreporting.com/.

Bibliography

Allen, Paul J and David C. Green, Managing Energy Data Using an Intranet—Walt Disney World's Approach, Proceedings of the 2001 World Energy Engineering Congress Conference & Expo, Atlanta, Ga., October, 2001.

Bos, Bert, Cascading Style Sheets[article on-line] (2002, accessed 8 July 2002); available from www.w3.org/Style/CSS.

Hall, Marty, Java Programming Resources[article on-line] (1999, accessed 8 July 2002); available from/ www.apl.jhu.edu/~hall.

INT Media Group, Incorporated, TCP/IP[article on-line] (2002, accessed 6 June 2002); available from webopedia.internet.com/TERM/T/TCP_IP.html.

Ireland, Blair, Introduction to Perl—Become a Guru[article on-line] (accessed 6 June 2002); available from www.thescripts.com/serversidescripting/perl/tutorials/introductiontoperl—becomeaguru/page0.html

Kington, Max, ColdFusion—An Introduction[article on-line] (accessed 6 June 2002); available from www.thescripts.com/serversidescripting/coldfusion/tutorials/coldfusion-anintroduction

Marshall, James, HTTP Made Really Easy[article on-line] (August 15, 1997, accessed 8 July 2002); available from jmarshall.com/easy/http

Murdock, Robert, ASP Basics—What is ASP?[article on-line] (accessed 6 June 2002); available from thescripts.com/serversidescripting/asp/tutorials/aspbasics/page0.html

NCSA Software Development Group, CGI: Common Gateway Interface[article on-line] (1999, accessed 6 June 2002); available from www.w3.org/CGI

Richmond, Alan, Dynamic HTML[article on-line] (2002, accessed 6 June 2002); available from wdvl.internet.com/Authoring/DHTML

Software References

Apache HTTP Server Project, Apache Software Foundation, www.apache.org

ColdFusion MX, Macromedia, Inc., www.macromedia.com/software/coldfusion

Eserver Energy Information Server, Veris Industries, Inc., Portland, Oregon, www.veris.com

FoxWeb, Eon Technologies, Alameda, California, www.foxweb.com

Java, Sun Microsystems, Inc., java.sun.com

Kavacharts, Java Applet for graphing data, www.ve.com

MySQL, MySQL AB, www.mysql.com

Perl, Perl Mongers—The Perl Advocacy People, www.perl.org

Visual FoxPro, Microsoft Corporation, msdn.microsoft.com/vfoxpro

Richmond, Alan, Dynamic HTML[article on-line] (2002, accessed 6 June 2002); available from wdvl.internet.com/Authoring/DHTML

Chapter 4

The Evolution of Building Automation Systems Toward the Web*

Paul Allen, Walt Disney World
Rich Remke, Carrier Corporation
David Green, Green Management Services Inc.
Barney Capehart, University of Florida
Klaus Pawlik, Accenture

THE PURPOSE OF THIS CHAPTER is to help prepare energy managers to understand some of the fundamental concepts of web-based building automation systems (BAS). We thoroughly examine each component of a BAS in today's BAS technology and what a BAS might look like in the future. The capability and use of information technology (IT) and the internet in the form of web-based energy control systems continues to grow at a rapid rate, and it is imperative that facility managers, maintenance managers and energy managers become ready to work with current and future applications of internet based control technologies in their facilities. The emphasis here is on the use of computer networking, use of facility operation databases, and sending and receiving control system data over the web using the TCP/IP communications protocol, and other features that make up a successful BAS.

INTRODUCTION

The combination of low cost, high performance microcomputers together with the emergence of high-capacity communication lines, networks and the internet has produced explosive growth in the use of web-based technology for building automation systems. Many of these current BAS systems use a proprietary information structure and communications protocol that prevents the easy modification of the BAS operation, and greatly limits the plug and play application and addition of interchangeable components in the system. Control solutions such as BACnet and LonWorks have helped this situation somewhat, but they have also introduced their own levels of difficulties. The BAS of the future will integrate state-of-the-art information technology (IT) using TCP/IP and the web that is available to almost all users and facilities. Open and standard approaches from IT using the internet and the web, and in particular, using the standard internet communications protocol TCP/IP, have become popular in the BAS area, and new players have developed BAS systems that are not as reliant on proprietary information structures and communications protocols. These new IT based systems are rapidly overtaking the older BAS systems. All of the established BAS companies are quickly developing ways to interface their systems with the web and with TCP/IP allowing the use of standard web browsers such as Internet Explorer and Netscape.

The internet standard communication protocol is TCP/IP, and all IT people understand it well. Network administrators commonly use TCP/IP as the communications protocol for the local area network (LAN) in facilities. This high-speed backbone is the heart of the network interconnection of computers, systems, and components of the facility's control and information system. TCP/IP continues to grow in use as the local communications protocol as well as the LAN protocol. Every day, more devices and controllers using TCP/IP directly are becoming available for our use, as is evident from advertisements and catalogs we see. As more use is made of TCP/IP directly, there is less need for other protocols, and especially less need for proprietary protocols. There will most likely always be a continued use of some proprietary protocols at the farthest points of our local control units, boards and components. These proprietary protocols used at this level of our energy control system may represent the least cost and fastest communication capabilities.

*A preliminary version of this chapter appeared in *Strategic Planning for Energy and the Environment*, Vol. 22, No. 4, 2003 and is used with permission of Fairmont Press.

The goal of this chapter is to provide a description of the principles and application of IT using TCP/IP to the design and operation of web-based BAS. However, since most of the BAS systems on the market today use exclusive or more extensive use of non-TCP/IP proprietary communications protocols, a significant part of this chapter describes the current operation of many of these BAS systems, and identifies the areas where IT and web-based BAS can be applied. As time goes by, more BAS systems will utilize web-based control and communications, and eventually—in the opinion of the authors—most BAS systems will utilize web browsers and TCP/IP communications for the majority of the operations and data exchanges in the overall system.

THE BASICS OF TODAY'S BAS

At a minimum, a BAS is used to control functions of a heating, ventilating, and air conditioning (HVAC) system, including temperature and ventilation, as well as equipment scheduling. Substantial additions to these basic functions are usually required to comprise a "true" BAS. They include monitoring utility demand and energy use, building conditions, climatic data, and equipment status. Often, the BAS reports results provided in the form of utility load profiles, trends and operation logs of equipment, and generation of maintenance schedules. Even basic BAS are generally expected to perform control functions that include demand-limiting and duty cycling of equipment.

More elaborate BAS can integrate additional building systems, such as, video surveillance, access control, lighting control and interfacing with the fire and security systems. However, in large organizations and campuses today, it is still more common to see dedicated systems for these additional building systems due to divisions in management functional responsibility, code issues, and features/performance of dedicated systems.

Today's BAS are expected to receive and process more sophisticated data on equipment operation and status such as from vibration sensors on motors, ultrasonic sensors on steam traps, infrared sensors in equipment rooms, and differential pressure sensors for filters. Top of the line BAS today also have additional capabilities, such as Chiller/Boiler Plant Optimization, Time Schedule/Setpoint Management, Alarm Management and Tenant Billing to name a few. Most BAS Manufacturers today have started to offer some form of web-based access to their existing control systems and are actively developing web-based capability for their future products.

The following sections introduce the hardware and software that make up some of the functions of a basic BAS commonly used today.

Control Unit Hardware

The control units used for a BAS provide the inputs, outputs and global functions required to control the mechanical and electrical equipment. Most BAS manufacturers provide a variety of control units tailored to suit the specific need. Shown below is a list of the most common control units:

Universal Processor Unit

Works with the Universal I/O unit and contains the control logic and programs for the application. Module usually includes some on-board I/O.

Universal Input/Output Unit

Provides expansion I/O for the Universal Processor Unit. Inputs include temperatures, relative humidity, pressures, and fan & pump status. Outputs include on/off, and valve/damper control.

Primary Controller Unit(s)

Provides global functions for the BAS control network that can include communication interface between PC front-end software and lower-tier controllers, real-time clock, trend data storage, alarms, data transfer between lower-tier controllers, and higher-level programming support. Some BAS manufacturer's combines all these functions into one primary controller while other BAS manufacturers have separate controllers that are dedicated to each global function.

VAV Box/Fan Coil Controller

Self contained controller with integral processor and I/O designed to control a VAV box or a Fan Coil Unit.

DX Controller

Controller designed to control a multistage cooling and heating direct expansion (DX) air conditioning system.

For further reference, the Iowa Energy Center has an excellent web site (www.ddc-online.org) that shows a complete overview of the designs, installations, operation and maintenance of most BAS on the market today.

Control Unit Programming

Control units typically contain software that can control output devices to maintain temperature, relative humidity, pressure, and flow to a desired set point. The software programming can also adjust equipment on-off times based on a time-of-day and day-of-week schedule to operate only when needed.

The software used to program the control units vary by BAS manufacturer basically fall into three categories:

1. Fill-in-the-blank programming standard algorithms
2. Line-by-line custom programming
3. Graphical custom programming.

Fill-in-the-blank

Uses pre-coded software algorithms that operate in a consistent, standard way. The user fills in the algorithm configuration parameters by entering the appropriate numbers in a table. Typically, smaller control devices use this type of programming, like those that control a fan coil or VAV box controller. These devices all work the same way and have the same inputs and outputs.

A few manufacturers have used the fill-in-the-blank programming for devices that are more complex where a variety of configurations can exist, such as air handlers. Standard algorithms are consistent for each individual component. As an example, the chilled water valve for an air handling unit is programmed using the same standard algorithm with only the configuration parameters adjusted to customize it for the particular type of valve output and sensor inputs. Programming all of the air-handler devices using the appropriate standard algorithm makes the air-handling unit work as a system.

The advantage of fill-in-the-blank standard algorithms is that they are easy to program and are standard. The downside is that if the standard algorithm does not function the way you want or there is not a standard algorithm available, then the system requires development of a custom program.

Line-by-line Custom Programming

Programmers use this to create control programs using the BAS Vendors controls programming language. They start the code from scratch and customize it to control the system. In most cases, programs can be re-used for similar systems with modifications as needed to fit the particular application.

The advantage of the line-by-line custom programs is that technicians can customize them to fit any controls application. The disadvantage is that each program is unique and trouble-shooting control problems can be tedious since technicians must interrogate the code line-by-line.

Graphical Custom Programming

Vendors developed this to show the control unit programs in a flow chart style, thus making the programming tasks more consistent and easier to follow and troubleshoot.

Below are some additional issues to consider regarding control unit programming:

1. Can technicians program the control units remotely (either network or modem dial-in) or must they connect directly to the control unit network at the site?
2. Does the BAS Manufacturer provide the programming tools needed to program the control units?
3. Is training available to learn how to program the control units? How difficult is it to learn?
4. How difficult is it to troubleshoot control programs for proper operation?

Control Unit Communications Network

The control unit network used by control units varies depending on the manufacturer. Several of the most common control unit networks used today include RS-485, ethernet, ARCnet and LonWorks.

RS-485

Developed in 1983 by the Electronic Industries Association (EIA) and the Telecommunications Industry Association (TIA). The EIA once labeled all its standards with the prefix "RS" (Recommended Standard). An RS-485 network is a half-duplex multi-drop network, which means that multiple transmitters and receivers can exist on the network.

Ethernet

The Xerox Palo Alto Research Center (PARC) developed the first experimental ethernet system in the early 1970s. Today, ethernet is the most widely used local area network (LAN) technology. The original and most popular version of ethernet supports a data transmission rate of 10 Mb/s. Newer versions of ethernet called "Fast Ethernet" and "Gigabit Ethernet" support data rates of 100 Mb/s and 1 Gb/s (1000 Mb/s).

ARCNET

A company called Datapoint originally developed this as an office automation network in the late 1970's.

The industry referred to this system as ARC (attached resource computer) and the network that connected these resources as ARCNET. Datapoint envisioned a network with distributed computing power operating as one larger computer.

LonWorks

Developed by the Echelon Corporation in the 1990's. A typical node in a LonWorks control network performs a simple task. Devices such as proximity sensors, switches, motion detectors, relays, motor drives, and instruments, may all be nodes on the network. Complex control algorithms are performing through the LonWorks network, such as running a manufacturing line or automating a building.

Control Unit Communications Protocol

A communications protocol is a set of rules or standards governing the exchange of data between control units over a digital communications network. This section describes the most common protocols used in a BAS.

BACnet

Building Automation Control Network is a standard communication protocol developed by ASHRAE specifically for the building controls industry. It defines how applications package information for transportation between building automation system (BAS) manufacturers. The American National Standards Institute has adopted it as a standard (ASHRAE/ANSI 135-1995)

LonTalk

An interoperable protocol developed by the Echelon Corporation and named as a standard by the Electronics Industries Alliance (ANSI/EIA-709.1-A-1999). Echelon packages LonTalk on their "Neuron chip" which is embedded in control devices used in a LonWorks network.

Proprietary RS-485

The protocol implemented on the RS-485 network is usually proprietary and varies from vendor to vendor. The Carrier Comfort Network (CCN) is example of a proprietary RS-485 communications protocol.

Modbus

In 1978, Modicon develop the Modbus protocol for industrial control systems. Modbus variations include Modbus ASCII, Modbus RTU, Intel® Modbus RTU, Modbus Plus, and Modbus/IP. Modbus protocol is the single, most supported protocol in the industrial controls environment.

TCP/IP

Transmission Control Protocol/Internet Protocol (TCP/IP) is a family of industry standard communications protocols that allow different networks to communicate. It is the most complete and accepted enterprise networking protocol available today, and it is the communications protocol of the internet. An important feature of TCP/IP is that it allows dissimilar computer hardware and operating systems to communicate directly.

Proprietary vs. Open Protocol

With the introduction of LonTalk and BACNet protocols, the virtues of "open" and "interoperable" networks have been the buzz words of late. The term "open" has taken on a positive connotation while the term "proprietary" has taken on a negative connotation. This section will discuss this perception in more detail.

In recent years, LonTalk has emerged as an "open" protocol and been embraced by numerous BAS manufacturers. However, to obtain a truly open BAS, Echelon's vision is that all BAS devices (sensors & actuators) be intelligent Lon compatible devices interconnected in a LonWorks network. However, most BAS manufacturers that have developed LonWorks compatible products continue to use a control unit with I/O connected to "dumb" sensors using proprietary control unit software. The only thing that has changed is that the system uses LonTalk as the communications protocol between the control units and other LonWorks compatible devices in the BAS. Therefore, the degree of "openness" of these systems is debatable.

The negative connotation of using a BAS with a proprietary protocol may be offset by other benefits that may be more important to an end user, such as:

- Reliability
- Serviceability
- Standardization
- Ease of programming
- Expandable
- Low Cost
- Speed

Providing the most flexible BAS at the lowest cost is important to end users. To reach this goal, it is the opinion of the authors that BAS systems in the future will utilize the open and standard approaches found in web-based control and communications for the majority of the BAS operations and data exchanges.

Client Hardware/Software

Normally, a PC workstation provides operator access into the BAS. The PC workstation may or may not connect to a LAN. If a server were part of the BAS, the PC workstation would need LAN access to the server data files and graphics. Some smaller BAS use standalone PC's that have all the BAS software and configuration data loaded on each PC. Keeping the configuration data and graphics in-sync becomes problematic with this approach

A graphical user interface (GUI) is one of the client-side software applications that provide a window into the BAS. The GUI usually includes facility floor plans that link to detailed schematic representations and real-time control points of the building systems monitored by the BAS. The GUI allows technicians to change control parameters such as set points, time schedules, or temporarily override equipment operation. Other client-side software applications include:

- Alarm monitoring
- Password administration
- System setup configuration
- Report generation
- Control Unit programming and configuration

Server Hardware/Software

Servers provide scalability, centralized global functions, data warehousing, multi-point access and protocol translations for a mid to large size BAS. Servers have become more prominent in the BAS architecture as the need has grown to integrate multi-vendor systems, publish and analyze data over an intranet or extranet and provide multi-user access to the BAS. While having a central server on a distributed BAS may seem contradictory, in reality, a server does not take away from the stand-alone nature of a distributed control system. Servers enhance a distributed control system by providing functions that applications cannot perform at the controller level. In fact, a BAS may have several servers distributing tasks such as web publishing, database storage, and control system communication.

Operating System and Server Hardware

The key to any server operating system (OS) is robustness. Whether the server operates on Unix, Linux, Mac, or Windows, it is imperative that the server OS has the ability to handle all of the tasks required of it. This usually requires a true multi-tasking, multi-threaded OS with very flexible networking capabilities. Windows 2000 is a very popular choice for this task, not only because it fits the criteria, but also because many IT departments support it. IT support is critical to the success of any mid to large scale BAS installation. When designing a BAS, it is in all parties' best interest to involve the IT department and designate a BAS champion from that department early on. Even though the manufacturer of BAS server software may be able to operate on a wide variety of hardware platforms, many times the IT department will dictate the server hardware and OS. Since backing up the server database usually falls under the IT department's responsibility, it is imperative that the server hardware meets the necessary criteria. This may include hot-swappable hard drives, dual Network Interface Cards, rack-mounted PCs and server versions of a particular OS.

Control System Programs

Servers provide the ability to globally control a BAS. Facility-wide time scheduling, load-shedding, or set point resets are examples of global functions a BAS server can perform. Since these types of functions are overrides to the standard BAS control programs, having them reside in the server requires that steps be taken to insure continued control system operation should the server go down for any length of time. The distributed BAS should have the ability to "time out" of a server override if communications with the server is lost. When the server comes back on line, the BAS should have rules that govern whether the override should still be in effect, start over, or cancel. Servers also can perform computational tasks, offloading this work from the BAS control units.

Custom Programs

Most server based control programs are a combination of custom and standard control applications. Since there are individual needs at every site, server-based control programs must be flexible enough to meet those needs without a complicated set of procedures for the end user. A common approach is to provide some basic level of control with a standard, vendor supplied software product, then having the vendor provide the "hooks" into the software for custom application development. Since many software products work with the Microsoft Windows OS, standards such as Dynamic Data Exchange (DDE), Object Linking and Embedding (OLE) or Dynamic Link Libraries (DLL) are common interfaces to standard applications.

Server to BAS Network Communications

Servers communicate with the BAS through a variety of electrical and network protocols. The most common communication method is a direct RS-485 multidrop connection to the lower tier controllers. Depending on the size of the BAS, the server may have several communication ports connected to multiple RS-485 communication busses, or it may have one communication port connection. Some servers may use a combination of TCP/IP network connections and RS-485 communications busses to lower tier controllers. When TCP/IP is used, usually the server encapsulates the same communication messages that would normally travel over the RS-485 wire into a standard internet protocol (IP) packet. Once the packets reach their destination IP address (either another computer or a dedicated ethernet to RS-485 converter), the network transfers the packets to the RS-485 bus for communication to the lower tier controllers. See Figure 4-1 for a schematic of today's BAS network.

Servers may also support modem communication to both the BAS lower tier network of controllers and to off-network monitoring sites. Through an OS standard such as remote access communication (RAS), modems can be set up to dial directly into the server and access many or all functions that the server provides, such as dynamic graphics, historical data, and controller interrogation or programming. Modems can also be used as dedicated communication interfaces to the lower tier BAS network, providing users access directly to the BAS controllers for system status information, schedule changes, or even programming changes. Whether the modem accesses the server or the BAS network directly, the connection must be secure to insure that unauthorized users cannot dial into the system and cause harm. Many BAS manufacturers utilize a combination of password levels and communication encryption to accomplish an acceptable level of security. RAS connections can employ the same security measures installed for the IT network security since it is actually a dialup network connection. Modems serve one other purpose for BAS systems, they can be used to dial out alarms when they occur to a workstation or pager, and they also can be used to dial out and transfer historical data to a remote site based upon rules set up for transferring the data.

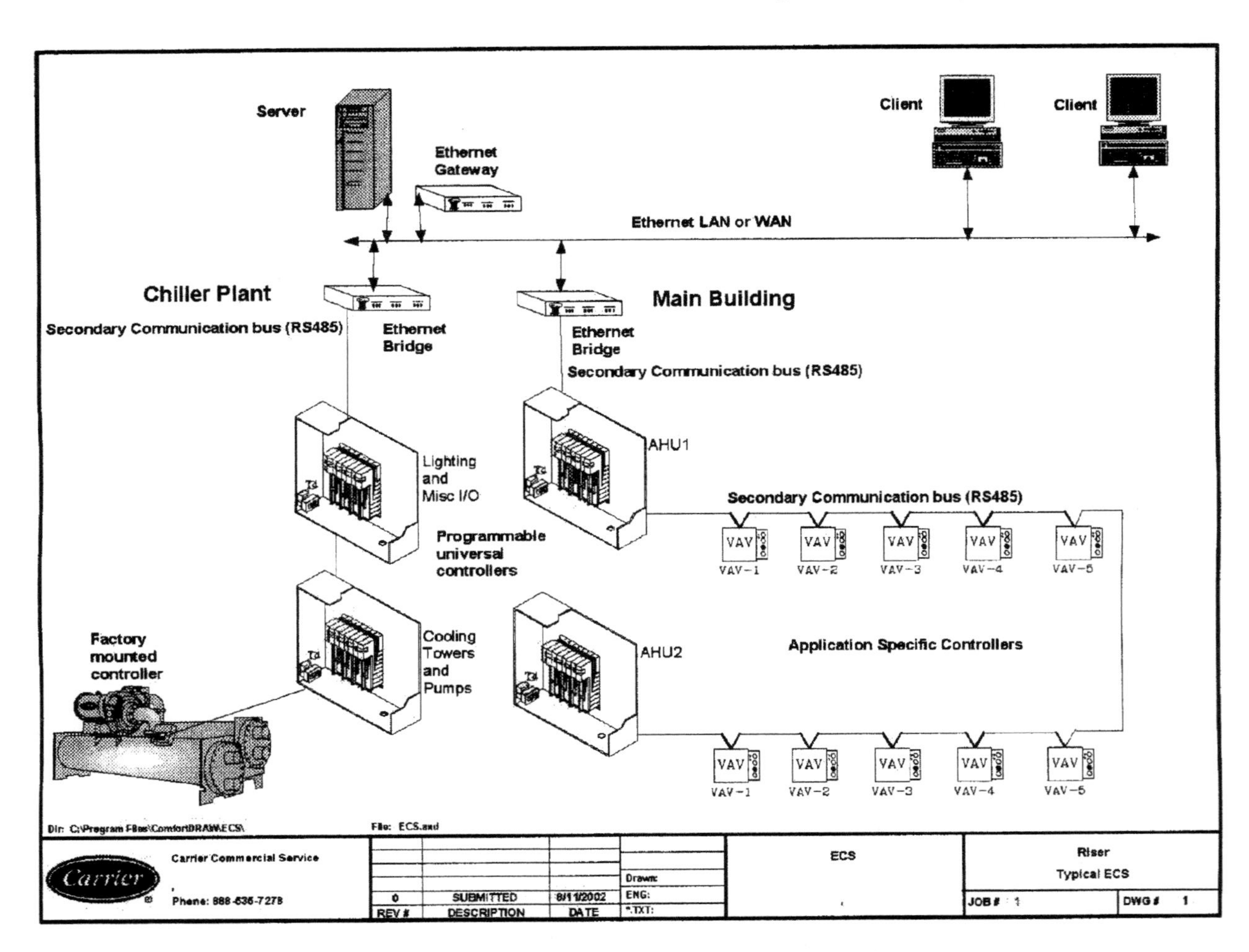

Figure 4-1. Today's BAS Network Schematic

INTEGRATING THE WEB INTO BAS

Servers and the Web

Web-based server applications are becoming the standard for most mid to large BAS installations. Applying web technology and standards to a BAS provides several advantages. Since the web and the protocols that access the web have become mature, defined standards, the BAS can inherit these standards, making connectivity and interoperability between enterprise level applications and the BAS easier to achieve. Current XML standards allow a BAS manufacturer to publish their control network data over an intranet or internet in a standard format that another application or control system vendor that supports XML can read.

Extensible Markup Language (XML)

XML organizes data into a predefined format for the main purpose of sharing between or within computer systems. Furthermore, its uses include data organization and transfer, data presentation, data caching, and probably some that we have not invented yet.

Web-based servers can share real-time data, historical trends, control signals, as well as provide multi-vendor connectivity by following internet standards. Some BAS manufacturers have implemented an open JAVA standard for Building Automation and control. Whether the BAS uses HTML, XML, JAVA, or any other internet standard, a key advantage of using the internet is that all of the standards are hardware and operating system independent. Web-based servers also have the advantage of allowing multiple user access to the BAS with a standard web browser, making the overall installed cost of the BAS very attractive.

Many BAS manufacturers today provide some sort of web connectivity. Some achieve connectivity through conversions of existing software products and platforms that export information to the web, while others have taken the approach of providing a stand-alone web server device that takes their proprietary protocol or multiple protocols and ports them to TCP/IP. The dedicated stand-alone web server can have several advantages over the software conversion approach:

- The server is typically written in a portable language such as JAVA, making it hardware independent
- The server becomes a "Black Box" with an IP address on the LAN/WAN, making it easier to administer and troubleshoot
- The server is upgradeable over the LAN/WAN without regard to the currently running operating system.
- The server becomes a two-way portal between the BAS and enterprise level applications, reducing the number of layers between the BAS and the enterprise.

JAVA-based servers provide the closest thing to open protocol. JAVA provides a hardware and OS independent platform for development of communications and data transfer between IP based objects. Most operating systems support a JAVA virtual machine, allowing JAVA code to run unmodified from one type of system to another. This greatly reduces development costs, as the programmers need only write the JAVA code once. In addition, since the JAVA virtual machine does not touch the client or server's hardware directly, it provides the necessary security required in today's networking environment. JAVA applets can perform all kinds of BAS functions. Here are just a few examples:

- Displaying dynamic data on a graphical HTML page
- Converting data
- Trending data and storing it on a remote server
- Performing calculations on data from various sources and sending the result to the BAS controllers
- Accessing XML or SQL databases to display data graphically or to perform data analysis

Since JAVA is part of the toolset used with TCP/IP based systems, it communicates directly with HTTP, XML, and any other web-based standard in existence today. A good example of a JAVA-based web server that incorporates all of the above-mentioned technologies is the CarrierOne Comfort Integrator (www.carrier.com), powered by the Niagara Framework™. The Niagara Framework™ is a JAVA-based object model that provides web access between multi-vendor BAS's and enterprise level applications.

Future Trends in BAS Web Integration

The BAS of the future will likely see web technology implemented further down the wire. As the cost of 32-bit microprocessors continues to fall, many manufacturers will find it cost effective to put the TCP/IP protocol directly into their controllers in the field.

In the future, we may see BAS controllers that use TCP/IP protocol directly. A BAS manufacturer could either create plug-ins or generate an XML schema that would define the data communications. The idea is to share the plug-ins among systems requiring access to the BAS. They would typically be JAVA applets. Each controller would be, in essence, a mini web server. HTML graphic pages for that particular device could be stored directly in the controller's memory, serving up dynamic controller data, allowing edits to controller configuration, or having hyper-links to other controllers or systems. Tasks such as data collection, global communication, scheduling, etc. could be handled by either a controller designated as a master server, or another TCP/IP device on the network. See Figure 4-2 for a schematic of the BAS network we might see in the future.

XML Schemas

This is a language, written in XML, which describes the data structures and constraints found in a XML data file.

There are some technical and network management challenges that come with this future BAS vision however. TCP/IP over ethernet is the current standard for transporting data between systems. Ethernet has several drawbacks when trying to push this technology down to the BAS controller level. First, when compared with 3-wire twisted shielded pair cable, the wiring is costly to purchase and install. It must be home run to a hub or router, which would cause raceways and conduits to grow exponentially on a project. In addition, it becomes a management nightmare for the IT department, since every device would require greater bandwidth and security supervision. The TCP/IP protocol will need to evolve to support a multi-drop network to economically compete with RS-485 proprietary communication. Several manufacturers are already either experimenting with or providing TCP/IP over an RS-485 connection, but some hardware, such as RS-485 to ethernet converters and RS-485 routers have yet to make it to the mainstream electronics marketplace. Once all of these products mature, along with the management tools IT will require to maintain network security and uptime, web servers at the controller level will become mainstream technology.

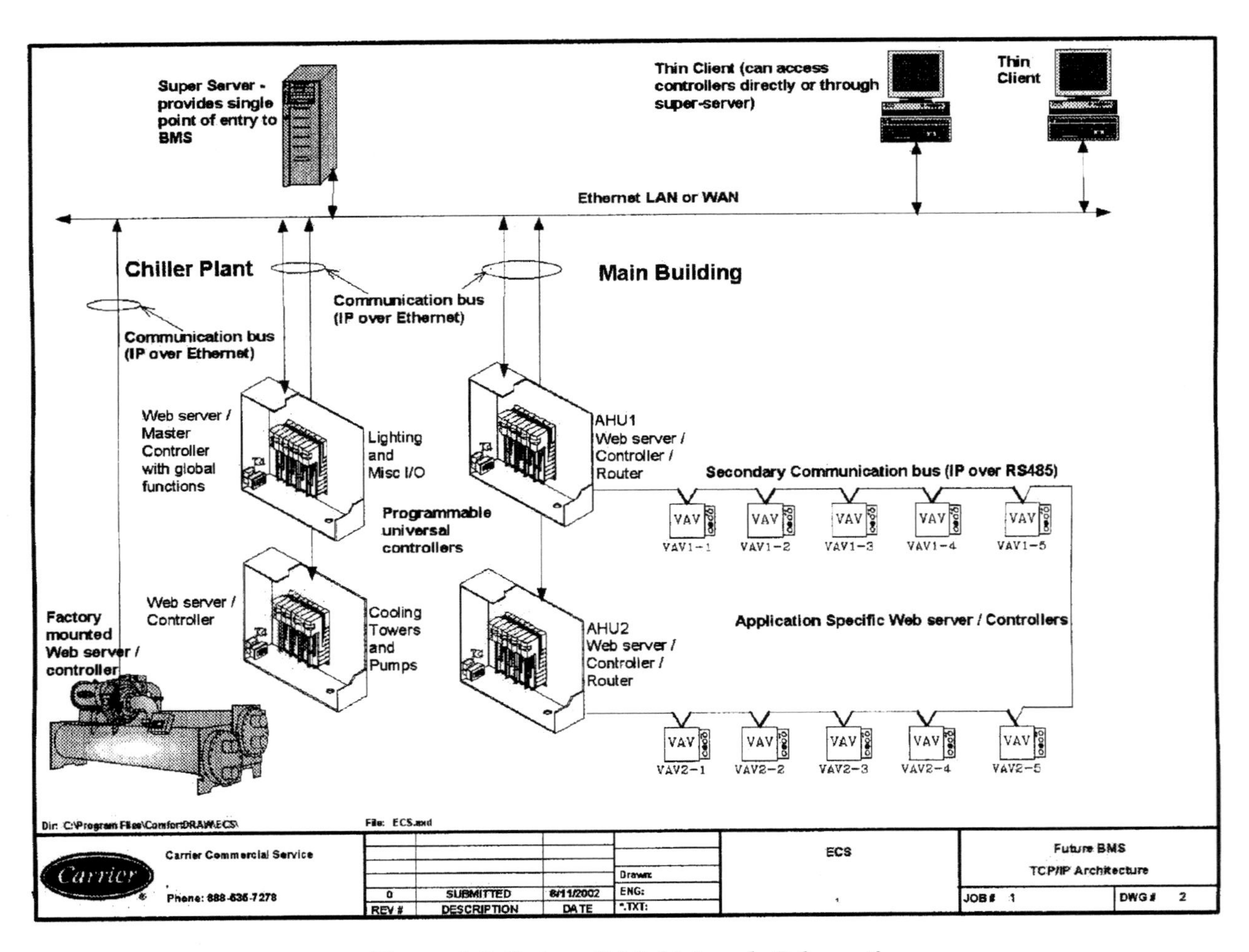

Figure 4-2. Future BAS Network Schematic

CONCLUSION

The BAS of old relied heavily on a collection of separate systems that operated independently, and often with proprietary communication protocols that made expansion, modification, updating and integration with other building or plant information and control systems very cumbersome, if not impossible. Today the BAS is not only expected to handle all of the energy and equipment related tasks, but also to provide operating information and control interfaces to other facility systems, including the total facility or enterprise management system. A BAS can only accommodate these expanded functions using the speed, capacity and interoperability of the internet, the web and the standard TCP/IP communications protocol. These requirements, together with the need for remote entry of data requests and operational changes through standard web browsers, among other features, are driving the BAS of today to the web, and into the domain of information technology. It is imperative that all energy managers and facility managers become prepared to operate, modify and improve their web-based energy information and control systems; as well as become comfortable with specifying and purchasing new web-based BAS and other facility enterprise management systems.

Measuring, monitoring and maximizing energy savings is a fundamental task of all BAS, and is the primary justification for many BAS installations. Improving facility operations in all areas, through enterprise information and control functions is fast becoming an equally important function of the overall BAS or facility management system. The web provides the means to share information easier, quicker, and cheaper than ever before. There is no doubt that the web is having a huge impact on the BAS industry. The BAS of tomorrow will rely heavily on the web, TCP/IP, and high-speed data networks. If you have not done so already, it is a good time for energy managers to get to know their IT counterparts at their facility. The future BAS will be here sooner than you think. Get ready—and fasten your seat belts!

Additional Information

More information about energy information systems as well as links to many of the tools discussed in this chapter is available at www.utilityreporting.com.

Bibliography

Proprietary Perspectives, Joanna R. Turpin, *Engineered Systems Magazine*, August, 2002.

Software References

RS-485, www.engineerbob.com/articles/rs485.pdf
Ethernet, www.techfest.com/networking/lan/ethernet.htm
ARCNET, www.arcnet.com
LonWorks & LonTalk, www.echelon.com/products/Core/default.htm
BACnet, www.bacnet.org/
Modbus, www.modbus.org/default.htm
XML, www.xml.com
Iowa Energy Office, www.ddc-online.org

Section Three

The Technological Benefits of Modern Web-based Energy Information and Energy Control Systems

Chapter 5

What Network Building Control Can Do for End Users

Lynn Fryer-Stein
Brendan Kiernan
Leland Keller
Rachel Reiss
E Source Energy Information and Communication Service

FACILITY MANAGERS and energy service providers (ESPs) are beginning to reap the benefits of networked building control—the practice of integrating building management systems (BMSs) with corporate intranets or the internet. Traditionally, BMSs have been capable of saving about 10 percent of overall building energy consumption by making sure that equipment runs only when necessary, that it operates at the minimum required capacity, and that peak electric demand is minimized. BMSs may also help save energy by recording equipment operation data that can be used for diagnostics and troubleshooting. By leveraging the benefits of networking BMSs, energy managers gain remote access for monitoring and controlling their buildings, extensive operating benefits at individual locations, as well as the strategic benefits of company-wide energy management and aggregated procurement.

INTRODUCTION

The proliferation of the internet and information technology (IT) hasn't stopped at the outside of buildings—it's actually changing the way that buildings are operated. Facility managers and energy service providers (ESPs) are beginning to reap the benefits of networked building control—the practice of integrating building management systems (BMSs) with corporate intranets or the internet.

There are two types of products available for network-based building management: BMSs made accessible over the internet or through a local area network (LAN) and add-on web-based systems.

Building Management Systems

Systems designed to monitor, manage, and control building equipment go by many names: building automation systems (BASs), energy management systems (EMSs), energy management and control systems (EMCSs), central control and monitoring systems (CCMSs), facilities management systems (FMSs), and so on. Throughout this chapter, we'll use building management systems (BMSs) as an umbrella term.

A BMS is an integral part of facility management. Although they are most prevalent in large buildings, BMSs are available for structures of most sizes. They control the operation of HVAC equipment (when that equipment starts and stops) as well as its running capacity, adjusting fan speed and supply air temperature to maintain comfortable conditions while optimizing energy use. Many BMSs also control fire response systems, security for access and control, video surveillance systems and lighting systems.

Traditionally, BMSs have been capable of saving about 10 percent of overall building energy consumption by making sure that equipment runs only when necessary, that it operates at the minimum required capacity, and that peak electric demand is minimized.[1] BMSs may also help save energy by recording equipment operation data that can be used for diagnostics and troubleshooting.

Add-on Enterprise Management Systems

The long reach of the internet has also given rise to a new class of products—add-on enterprise energy management systems. These systems are typically used to manage multiple sites at the same time, to access building information from the web, or for load curtailment. Vendors of such products include WebGen Systems, CMS Viron, Triduim, Enflex, Envenergy, Itron, Cimetrics, and others. Add-on systems are different from BMSs because they are designed to be integrated with an existing system that is unable to perform the specific network-based functions we've been describing. A full-scale BMS has those capabilities built in.

Benefits of Network Building Control

Some benefits of network building control are operational in nature, aimed at improving performance at individual locations. Others are more strategic in nature, making it possible to implement company-wide initiatives for cost-effective energy management and procurement. Network control makes it possible to:

- Improve ease of use
- Provide remote control
- Participate in demand response programs.
- Detect equipment problems early.
- Have expert, 24/7 building control and management.
- Save money by eliminating unnecessary service calls.
- Access the full capabilities of the EMS.
- Adjust building environments to meet changing business needs.
- Respond quickly to system problems.
- Diagnose problems before calling for service.
- Commission HVAC systems in new construction or for repairs.
- Plan and carry out preventive and predictive maintenance.
- Store and analyze data.
- Use benchmarking to identify opportunities for improving energy management.
- Create a single interface for all building systems.
- Aggregate load profiles for energy procurement.
- Manage portfolios of buildings to minimize energy procurement cost risks.

Improve Ease of Use

Older, hardware-based BMSs were clunky and often difficult to master. Each vendor's system had its own look and feel and unique graphical front end. Staff turnover was also a problem, because if a departing operator was trained when the BMS was first installed, that person's successors would not necessarily have received similar training. By contrast, most facility managers are familiar with web browsers and would probably consider them relatively easy to navigate.

Provide Remote Control

Using a web browser or a virtual personal network (VPN), a facility manager can monitor and control several buildings from computers in different locations, rather than being tied to a stationary workstation dedicated to BMS applications. Access could be possible from several points within the same building, across a campus, or across the world. All it takes is a computer with web browser or a corporate network connection, such as a VPN.

For example, at 40 of its stores and at corporate headquarters, PETCO Animal Supplies added a gateway with internet capabilities to existing HVAC and lighting equipment control systems to take advantage of remote control capabilities. The company also installed complete BMSs at 87 other stores that had no prior energy management controls. These technologies now allow Tim Speller, PETCO's national energy manager, to remotely monitor, manage, and change thermostat setpoints and to control individual lighting circuits in every facility. Speller said, "The new controls allow me to better manage the day-to-day use of energy in the monitored stores from wherever I happen to be at the time. In the past, in-store energy usage varied dramatically from one store to the next because employees would manually change the temperature setpoints. In-store temperatures are now automated and controlled from company headquarters in San Diego."[2]

Participate in Demand Response Programs

Using a BMS, building managers can reschedule lighting, HVAC, water heating and other major loads in a building in response to price signals or curtailment requests from their energy service provider. Network building control allows one energy manager to reschedule multiple buildings and aggregate the demand response to negotiate a better deal with the energy service provider sponsoring the load curtailment. Using the same PETCO example, Speller used a California Energy Commission grant to install the network building control system that helped PETCO save $65,000 monthly and shed 5.5 megawatts of curtailable load in the 2001-2002 Small Commercial and Industrial HVAC and Lighting Demand Response Program.

Without the network building control, the energy manager must convince each individual store manager to participate in a demand response program, and then trust them remember to curtail loads properly. Some companies would not be large enough to be able to participate in such programs if they couldn't aggregate the response from multiple company locations.

Detect Equipment Problems Early

If it's easier to monitor more often and you can do it from anywhere, building managers will probably check on how their systems are functioning more often. And as they become more familiar with a facility's loads and peaks, they'll be able to spot a piece of equipment that is operating poorly before it fails completely. As Paul Allen told us, "Networked building control makes it easier to detect equipment problems. It's very visible when things aren't working. It's like shining a light on something."[3] Having advance warning that equipment will need to be replaced makes it possible to demonstrate to the finance department that capital is needed for specific repairs and upgrades.

Furthermore, equipment repair or replacement can cost significantly less when planned in advance to avoid a business interruption. With this advance notice, energy managers have a better opportunity to purchase equipment based on life cycle equipment and operation costs, rather than being forced to choose equipment that is immediately available.

Have Expert, 24/7 Building Control and Management

Remote monitoring and control gives energy customers access all day, every day to dedicated experts whose sole job is managing building control systems—a "luxury" that is usually available only to larger and more complex energy-intensive facilities. Remote monitoring can bring expertise to sites where no one has been specifically responsible for managing energy use. In addition to handling such straightforward tasks as programming thermostats, the experts—who usually include specialists in lighting or HVAC—may be able to identify system improvements that could lower operating costs or extend the life of building equipment.

All too often, the retail manager of a local chain store, restaurant, or school (whose job it is to sell sweaters, serve meals, or educate students, and who may or may not have an understanding of how building equipment works) is also responsible for keeping the facility clean and comfortable. Whenever the building gets too hot or too cold, that manager winds up calling the HVAC service contractor. The contractor must also be summoned when adjustments are needed to accommodate seasonal variations (like the beginning and end of daylight savings time) and changes in business needs (for example, extending store hours for holiday shoppers). When a business signs on for remote monitoring services, managers at all locations are supplied with a phone number they can call for help anytime. Such services are often referred to as "24/7" because they are available 24 hours a day, 7 days a week.

Western Building Services, and Honeywell, and the Murphy Company are examples of companies that run 24/7 call centers staffed with qualified building technicians. Although the majority of calls are for schedule-related thermostat adjustments ("Please keep the school cooled until 10:00 p.m. tonight, as we're having the annual school play"), many calls are comfort-related ("It's too hot in here!"). Staff at the center respond to each call as necessary, obtaining data directly from the building's EMS and then correcting the problem remotely (by changing a schedule or a setpoint, for example) or dispatching service contractors to the site.

Save Money by Eliminating Unnecessary Service Calls

In many chains, the branch manager calls service companies even for simple tasks such as rescheduling lights and thermostats or changing temperature setpoints. Most of these changes could be programmed remotely, avoiding the $100 to $150 expense of a service call. That may not be a great deal of money for a single facility, but it adds up quickly for chains with thousands of stores.

Access the Full Capabilities of the EMS

In theory, EMSs save about 10 percent of overall annual building energy use.[4] In practice, savings are often much smaller. A large part of the savings attributable to EMSs comes from scheduling equipment to turn off when it is not needed. But occupancy, schedules, and internal loads change over time. Unless the EMS is programmed to account for these changes, conditions will become uncomfortable. And once the occupants of a stuffy meeting room figure out how to override the controls, the system is likely to be left operating 24 hours a day in a mode that isn't necessary once the meeting is over.[5]

One study of 11 buildings in New England with EMSs found that five of the buildings were producing less than 55 percent of expected savings, and one site was producing no savings at all. The systems were failing to produce savings in large part because the EMSs were being used only for tasks that far simpler controls (such as time clocks and programmable thermostats) could carry out, and the more sophisticated control strategies, such as outdoor air reset and optimal start/stop, were not programmed and implemented when the systems were installed.[6]

To achieve the savings an EMS promises, someone needs to be watching it, upgrading programs on an ongoing basis as building conditions change, and making sure that it is keeping the building comfortable so occupants don't try to override or disable it. With network building control, an energy manager can grant limited system access to facility managers or others, reducing their workload of generating and distributing reports.

Most EMSs generate alarms when preset limits are exceeded, but because facilities staff often lack the time or training to respond, alarms may be ignored or the function completely disabled. When an EMS sends an alarm to a remote monitoring center, an expert can log on to the system, determine the nature of the alarm, and fix the problem. If possible, the expert will make the necessary change directly by issuing a programming command; when that isn't possible, the call center may page someone at the facility (or at home, if a critical alarm comes in during nonbusiness hours) or dispatch repair contractors to the site.

Western Building Services (Western), acquired by Comfort Systems in 1996, is a controls and HVAC service company that operates a remote monitoring center at its headquarters in Denver, Colorado. Western installs EMSs in school buildings, hospitals, banks, pharmaceutical research and biotech companies, large commercial and office spaces, and hotels. The controls systems come with a one-year warranty that includes access to Western's 24-hour Systems Support Center for remote monitoring. Western also monitors access systems, such as key card entry systems. During that year, all alarms are sent to Western's call center, as well as to a terminal in the monitored building.

Typical calls to Western include chiller, fan, and pump alarms, and requests to adjust lighting controls and reload EMS software. Historically, only about 5 percent of the calls received at Western's Systems Support Center require the dispatch of service people. About 60 percent are for standard EMS control, such as resetting temperatures and schedules, and can be handled directly from the keyboard. About one-third of the calls are automatic alarms for minor situations that require no action, such as a door being open for more than 10 seconds during the day (which happens commonly when people pass each other on their way in or out and stop to chat) or a space temperature being high when Western knows the equipment is being worked on or cannot supply more cool air.[7] About 80 percent of Western's warranty customers choose to continue the company's monitoring services after the initial year.

Adjust Building Environments to Meet Changing Business Needs

In some cases, reprogramming building systems to different temperature settings or to different schedules can actually improve a business's performance. For example, a clothing store might sell sweaters over the winter holidays and bathing suits a few weeks later. Networked building control allows building EMSs to be easily and quickly reprogrammed to keep the store cooler during the sweater sale and cruise-warm for the swimwear sale. Similarly, schedules can easily be changed to accommodate extended shopping hours between Thanksgiving and Christmas or a one-time midnight sale. Via remote control, experts can adjust building systems to correlate with activities on specific days in a way that most store managers would be unlikely to do.

Respond Quickly to System Problems

In the sales-driven retail environment, perhaps the greatest benefit of networked control is that problems are responded to very quickly. A remote monitoring and control center can program a change in temperature or schedule to take effect immediately, whereas it can take hours for a service technician to arrive and correct a problem. Hot, stuffy conditions in a store can have an immediate effect on shoppers, who are likely to pass on by or leave a store before making a purchase. The proper functioning of HVAC systems is most critical during the busiest and most profitable shopping times—when the system is operating near full capacity.

With remote monitoring and control, thermostat setpoints and equipment schedules can be reprogrammed and corrected in a matter of minutes. Equipment malfunctions can also be quickly detected and corrected. In many cases, problems can be corrected even before building occupants are aware that a problem exists.

Based in Overland Park, Kansas, CMS Viron is an energy service subsidiary of HVAC manufacturer York International. Viron maintains direct information links via phone lines with a chain of some 250 department stores. In the event of an HVAC fault—for example, a temperature that goes out of range—the in-store EMS automatically calls Viron's headquarters and notifies computers there of the problem. Screening software verifies the problem and alerts Viron's technical staff, who may be able to diagnose and readjust the building systems before store temperatures reach noticeably high or low levels.

On one occasion, Viron was notified by the EMS in

a department store that a chiller that was supposed to be on was not working. It would have taken a few hours for temperatures in the store to rise to uncomfortable levels with the chiller out of service. Viron was able to dispatch York service to repair the chiller before the temperature problem became apparent to in-store staff or customers.[8]

Diagnose Problems Before Calling for Service

Remote monitoring can often identify what is wrong with a system before a repair truck is dispatched, and that helps make the repair process more efficient. For example, Field Diagnostic Services, Inc., (FDSI) develops software and hardware specifically designed for HVAC monitoring and control, with special emphasis on packaged air-conditioning equipment. Its ACRx Controller system measures and records suction air temperature and pressure, liquid line temperature and pressure, supply and return air temperatures, ambient air temperatures, and condenser air temperature. This information is used by FDSI and the customer to create proactive maintenance programs, diagnose problems, identify service needs, dispatch the right service personnel, and verify that service has been properly performed.[9]

How does a service technician know when a condenser coil is truly dirty enough to warrant pressure washing? Visual inspection is subjective and is not easily quantified, but the ACRx system examines eight measurements and pays special attention to the difference between the condensing and ambient temperatures and the air temperature rise across the condenser. This tells a customer when the heat exchanger needs to be cleaned and that it was cleaned well enough to have the needed effect on the refrigeration cycle. Todd Rossi, vice president of science and technology at FDSI, notes that, "in the short term, the customer sees a higher-efficiency unit with lower utility bills. Over the longer term, the equipment works better, lasts longer, and service costs are expected to go down."[10]

Commission HVAC Systems in New Construction or for Repairs

Networked building control can be used to verify that HVAC systems are operating according to control system instructions. HVAC systems can be configured remotely for example, and remote monitoring and control features can confirm that the system was installed correctly or that service work has actually been done correctly.

Similarly, remote monitoring can help resolve repair problems. One hears stories about HVAC contractors who have made service calls without ever getting up on the roof to do the work. FDSI tells of a case in which an HVAC mechanic returned three times to clean a condenser coil, each time claiming to have cleaned the coil. FDSI's ACRx system indicated that equipment performance had not returned to normal as a result of those visits. In the end, the original mechanic was fired and a new one discovered that dandelion spores had become stuck between the two condenser coils. He cleaned them properly, and the ACRx confirmed that the equipment had returned to normal operation.

Plan and Carry out Preventive and Predictive Maintenance

Remote monitoring can help determine when maintenance will be needed. Maintenance—if done preventively at all, rather than waiting for failure—often is based on an estimate of scheduled run hours. Data collected through monitoring services can better indicate optimal times for:

- making filter changes (when pressure difference across the filter or fan energy reaches a certain level);
- performing routine relamping or motor lubrication (based on the actual number of run hours);
- tensioning fan belts (a very low power factor indicates that a motor is lightly loaded, which may indicate that a belt has slipped off or has broken); or
- rebuilding motors or replacing bearings on the basis of vibration analysis (excessive vibration can indicate worn bearings or a bearing failure) or actual number of starts (which may be more relevant to motor condition than the number of run hours).

In-house facility managers can make the same preventive maintenance plans if they have the appropriate sensors and data loggers on site.

Store and Analyze Data

Many building managers believe that they already have control of their building through an EMS. However, as discussed previously, EMSs are designed to control the building—that is, to turn components such as chillers, pumps, and fans on and off at certain temperatures, or to open and shut dampers—not to generate energy management reports.

Use Benchmarking to Identify Opportunities for Improving Energy Management

Data from a large group of sites paint a different

picture from data analyzed site by site. Analyzing data from many buildings can yield insights into energy management practices, making it possible to take practices that work well at one location and implement them at others. Benchmarking is not limited to working with interval or near-real-time data, and it often makes sense to begin benchmarking using billing data, for a relatively coarse comparison of energy use and costs among sites. Refining the benchmarking with hourly data provides a deeper understanding of how and when inefficient usage occurs. It might show, for example that energy use at one location increases to a certain level and stays there until morning, while at other locations energy use increases briefly from 10 to 11 p.m. but then drops back to low nighttime levels. Analysis of the data in this example might suggest a way to avoid this situation: ensure that the night cleaning crews turn off the lights before they leave.

Several service companies create databases of monthly billing information that can be sliced and diced in a variety of ways. They include Alliant Energy Integrated Services (EnergyTRAX), Ameresco (EnfoTrak), Avista Advantage (FacilityIQ), Cadence (Cadence Enterprise), Good Steward Software (Energy CAP and FASER), and Save More Resources (The Utility Manager Suite). Comparing an individual site's energy consumption with group averages makes it easy to pinpoint the buildings with the best and worst performances and to identify the sites that most need operational improvement.

These programs can provide a variety of reports designed to meet the needs of different people. For example, an energy manager will receive data and reports related to energy and operating costs and perhaps tenant billing expenses based on hours of operation; a plant engineer would see reports identifying equipment that is in need of preventive maintenance; and the CFO would receive a big-picture summary of energy costs. These reports are all generated from a single database, but the information is filtered and presented in a way that is appropriate to each individual's needs and level of authority.

Aggregate Load Profiles for Energy Procurement

When businesses are able to purchase their power competitively, those with better-than-class-average load profiles (that is, high load factors) that can provide potential suppliers with actual load profiles will likely get better prices than those that cannot. Actual load profiles let suppliers know what requirements they will need to meet, and therefore they do not have to assume as much risk as they might in the absence of load data. Of course, the bigger savings come from being able to respond to time-differentiated prices, which requires not only a meter that records consumption in hourly (or smaller) increments, but also a communications link so the customer can view the data frequently in near-real time. Real-time data access is rarely cost-effective—those who may need it are typically sophisticated energy managers managing very energy intensive operations (like blast furnaces) on time-varying tariffs. With competitive energy procurement a reality in only a handful of U.S. states, a national load profile may be of less value than a state or regional profile.

Of course, it is not necessary to invest in full-blown network building control in order to generate load profiles. Many utilities and ESPs also make interval data available on their web sites to those customers with interval meters. Many of the sites with meter data information have rate libraries and tools for calculating the cost of supplying that load under various tariffs.

Manage Portfolios of Buildings to Minimize Energy Procurement Cost Risks

Managing a group of facilities as a portfolio can have far-reaching strategic benefits. Consider, for example, an energy manager who is responsible for a group of buildings in California. The manager has contracted with a supplier for a fixed price, up to a set demand level. Above that level, the manager will buy power at the spot market price. Unfortunately for the manager, the times when the portfolio of buildings is likely to exceed the preset demand cap are precisely the times—very hot days, for example—when the spot market price of power can be highest.

A manager who can remotely monitor and control a group of properties could see demand rising, anticipate exceeding the demand cap, and take action—perhaps raising the temperature setpoint in all of the buildings on a rolling 15-minute basis—to cut overall demand without having a major negative effect on occupant comfort in any one building. This kind of coordinated action would be impossible without networked control systems.

CONCLUSION

Building energy management systems, once networked via the internet and corporate intranets, provide a broad variety of time and energy-saving advantages. These include energy monitoring and control to the energy manager and other outsourced service providers, operating benefits at individual locations, as well as the strategic benefits of company-wide energy management and aggregated procurement.

References

[1] David Wortman, Evan A. Evans, Fred Porter, and Ann M. Hatcher, "An Innovative Approach to Impact Evaluation of Energy Management System Incentive Programs," Proceedings, American Council for an Energy-Efficient Economy Summer Study (August 1996), pp. 6.163—171.

[2] Florence Lu, "Attention Shoppers: Demand Response in Retail," EIC Currents, no. 5 (March 2003), p. 2.

[3] Paul Allen, personal communication (March 4, 2003), Chief Engineer Energy Management, Walt Disney World Company, Lake Buena Vista, FL, tel 407-824-7577, e-mail paul.allen@disney.com.

[4] E source, "Energy Management Systems," prepared for Southern California Edison's Energy Design Resources Program (December 1998).

[5] J. Yago, "Survey Shows Users Often Fail to Make the Most of Automation," *Energy User News*, v. 4 (April 1992), p. 25.

[6] D.N. Wortman, E.A. Evans, F. Porter, and A.H. Hatcher, "An Innovative Approach to Impact Evaluation of Energy Management System Incentive Programs," Proceedings of the ACEEE 1996 Summer Study on Energy Efficiency in Buildings (Washington, DC: American Council for an Energy-Efficient Economy, 1996), pp. 6.163—6.171.

[7] Kerry Kirby, personal communication (November 20, 1998), Operations Supervisor, Western Building Services, Inc., 6820 N. Broadway, Unit G, Denver, CO 80221, tel 303-429-9219, fax 303-429-8728.

[8] R. LaCombe, Vice President for Marketing, CMS Viron Corp., 12980 Foster Drive, Overland Park, KS 66213, tel 913.563.3500.

[9] Todd Rossi, personal communication (November 25, 1998), Vice President for Science and Technology, Field Diagnostic Services, Inc., 825 Towne Center Drive, Suite 110, Langhorne, PA 19047, tel 215-741-4959, ext 15, fax 215-741-4995, e-mail rossi@fielddiagnostics.com.

[10] Todd Rossi[9].

Chapter 6

Life-Cycle Considerations for "Smart Equipment"*

Michael G. Ivanovich, Editor-in-Chief,
HPAC Engineering

THIS CHAPTER DISCUSSES the information technology components of smart HVACR equipment and real-world life-cycle parameters for considering smart equipment for applications in building systems.

INTRODUCTION

Market forces, such as staff reduction, and technology innovations, such as miniaturization and the development of open standard protocols for controls networking, have worked together to increase the intelligence of HVACR equipment. The results are greater reliability, speed, efficiency, responsiveness, and precision for not just the equipment, but operators as well. "Smart equipment" and the capabilities it empowers constitute some of the most important and unheralded innovations in the HVACR industry today.

Smart equipment is mechanical equipment integrated with diagnostics and communications capabilities manifested as electronic circuits, sensors, and, potentially, embedded processors. Smart technology offers tremendous potential to optimize the operation of stand-alone equipment, as well as entire HVACR systems. The resulting improvements in reliability, energy efficiency, diagnostics, and reporting, along with the fact that smart equipment may be directly connected to, or networked with, local and remote workstations, are cause for celebration in the buildings industry. These features can produce both direct and indirect economic benefits for owners, attract "Net-generation" workers into engineering and technical professions, and help resolve chronic issues such as maintenance and service lapses, poor indoor-air quality, and comfort.

*Reprinted with permission from Networked Controls 2001, a supplement to *Heating/Piping/Air Conditioning Engineering*, July 2001, www.hpac.com.

Smart equipment may include some or all of the following:

- *Sensors*. Types include on/off, position or open/closed indicator, temperature, moisture, vibration, voltage, flow, and more.
- *Microprocessors*. These are programmable computer chips that process, store, and forward sensor inputs and send alarms and other information to control panels and/or network destinations such as web servers and operator workstations.
- *Control mechanisms*. Including motors, springs, valves, relays, and actuators, these provide on/off or self-tuning capabilities.
- *Control panels*. Installed on or near equipment or at a distant location, these provide the human-machine interface. Although they are not "new," giant strides have been made in refining them.
- *External data-communication capability*. This is provided by a serial port for connecting a portable computer used to download data, a wireless transmitter and/or receiver, or a cable connection to a local area network (LAN).

Some equipment, such as chillers and boilers, became "smart" long ago. This led to advancements in their energy-efficiency, operability, and serviceability. The relatively high cost of these system components, environmental considerations, and safety factors drove the requirements for inherent computer control and self-diagnostics. However, recently, there has been a proliferation of smart equipment in the form of smaller components that can be installed in greater quantities in buildings (e.g., a steam trap as shown in Figures 6-1), or as smart-OEM devices in custom and off-the-shelf packaged units (e.g., motor/drive assemblies). Consequently, designers and owner-engineers need new, systematic procedures and resources for evaluating the IT-related features of traditionally "unenlightened" equipment so they can decide whether or not to spend the extra money to accommodate smart features. Building practitioners need training on how to maintain the equipment, how to avoid impairing performance, and how to address issues that have not applied historically. Smart equipment requires new learning, especially for those who began working with buildings before computers became commonplace.

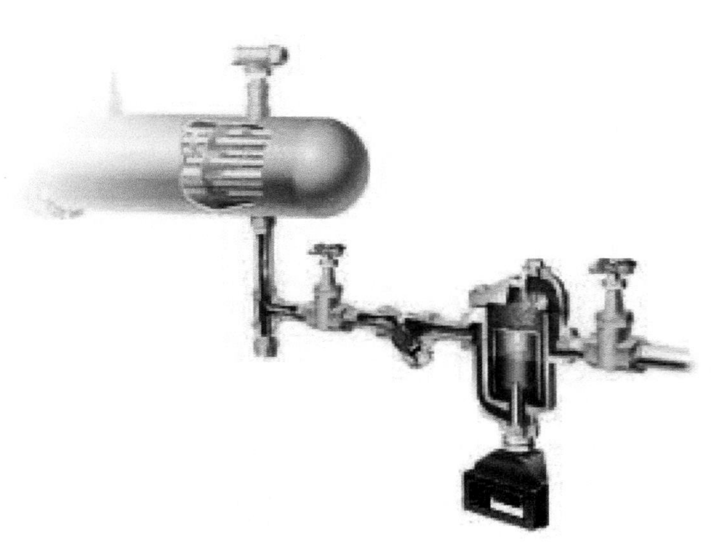

Figure 6-1. Smart steam traps check for temperature reductions caused by a backing up of condensate, as well as steam loss through the trap. (Image, taken from Bulletin No. 180, courtesy of Armstrong International Inc.)

LOOK BEFORE YOU LEAP

New features such as equipment-mounted "information systems" potentially can provide tremendous benefits; however, costs and risks accompany them. Before describing the exciting benefits of smart equipment, it is important to reveal the real-world considerations that come with them.

The drive for higher productivity from reduced work forces throughout the buildings industry has many looking for panaceas. If design and O&M requirements are not carefully considered, owners could get stuck with devices that either do not work or that have fragile components that break often.

The new sensors, microprocessors, mechanisms, and communications hardware (and software) have to be reliable, accurate, and long-lived. They must be installed, tested, and programmed with care. They have to be checked periodically and recalibrated or replaced as needed. Cabling, power, and environmental issues, such as temperature, moisture, airborne chemicals and particulate matter, and electronic noise also may play a role in the smart-equipment decision. Because the "smart stuff" is likely to be made and possibly installed by a third-party OEM supplier, engineers may need to investigate that supplier's reputation for quality, as well as the supplier's quality-assurance practices. Engineers also need to investigate how well the smart features are integrated with the mechanical equipment. How are sensors and mechanisms attached? Is the complete (integrated) unit tested prior to shipment, and, if so, by whom? What is the warranty for the integrated equipment and its smart components?

As an example of what could go wrong, if 500 smart widgets were installed in a new system, and the diagnostic sensor was too sensitive, 500 false alarms would flood the operator log periodically. If the control logic were set to automatically turn off a widget upon an alarm condition, 500 widgets would have to be reprogrammed or retuned.

As with any equipment decision, designers and owner-engineers have to define requirements and perform life-cycle cost-benefit analyses to ensure that owners get the greatest value. This includes considering a "smart" steam trap versus a conventional steam trap and figuring how to get the associated data and alarms back to the operator station and integrate that information into the ebb and flow of O&M procedures. This extra design time costs money, and the additional thinking time may be difficult to work into a contract. Furthermore, all of the typical specifications have to be considered for each component. Then, for each component, the decision to upgrade to smart equipment has to be made. That decision will impact the design of the building automation system (BAS), operation-and-maintenance routines, documentation, and training requirements for the building staff and service contractors.

These additional considerations may seem formidable at first, but with standardized evaluation procedures and experience, they can be worked through efficiently for applications with requirements that merit the extra expense.

For existing facilities, it might be best to swap a single piece of equipment with its IT-empowered counterpart before committing to an all-out replacement program. If more than one manufacturer offers smart-equipment options, it might be possible to install a variety so that quality and features can be compared. Additionally, sensors that can be upgraded from off-the-shelf OEM models to sensors with longer life, greater precision, and the ability to hold calibration longer also may be worth a careful look. This is especially true for equipment being installed in hard-to-reach places or where functions such as billing, life-safety, and product/process quality are involved.

BENEFITS OF SMART EQUIPMENT

The primary and secondary benefits of having smart equipment will depend on the application. In a district energy system serving a large university campus, having self-testing/alarming steam traps in the distant steam lines could reduce trap-testing visits and increase trap-failure responses. The man-hours saved could be applied elsewhere, and the energy savings from having continuously tested traps could be substantial. In a hospital, having a strainer send an alarm when it needs to be cleaned could prevent an untimely backup. Pumping energy and costs could be reduced by regularly servicing the strainers.

Primary Benefits of Smart Equipment Include:

- Optimized (self-tuning) performance, which leads to higher equipment and system energy efficiency, longer equipment life, and, ultimately, reduced service and replacement costs.
- Early warning of abnormal conditions requiring operator intervention, which leads to greater system reliability, reduced downtime due to failures, and scheduled service calls (rather than higher-cost emergency calls).
- Storage of operating run-hours, run times, and performance/condition trend logs. This data and information could be used for faster troubleshooting, predictive maintenance, documentation-backed warranty claims, rescheduling, and the fine-tuning of alarm thresholds.

Secondary benefits result from primary benefits. For example, the cost savings (a measurable, primary benefit) resulting from better and longer equipment performance can be invested in other facility improvements, compounding the direct benefits further. Some secondary benefits may be recognized in the short term, while others may be recognized in the long term and be vaguely discernible—they tend to blend in with the myriad other things happening in a building. Secondary benefits of smart equipment may include:

- Greater staff morale and efficiency, leading to the easier recruitment and retention of talented workers. For example, smart equipment used throughout a building may imply a modern and "intelligent" facility—especially if a sophisticated BAS also is in place. This could be used to attract net-generation workers who might otherwise be turned off by a building considered old-fashioned.
- Increased tenant productivity resulting from better indoor-air quality and comfort.
- Increased process productivity and better product quality resulting from an improved production environment and increased staff morale and staff retention/recruitment.

Quantifying or basing decisions on anticipated secondary benefits of smart equipment is tricky and hard to defend because benefits may result from many other types of actions, such as:

- The replacement of a tyrannical or incompetent manager with a benevolent and competent one.
- The implementation of employee-friendly management policies and practices.
- The increasing of pay or the offering of performance incentives.
- Investment in non-HVACR equipment such as modern plant machinery and ergonomic office furniture.

Moreover, direct and ancillary benefits may result from investing in higher-quality HVACR equipment that does not have smart features. The bottom line is that analyses have to be broad, deep, and fair.

CONTROL PANELS

In the vernacular of information technology, control panels are human-machine interfaces. They range from simple on/off switches and LED indicators of operating status (on, off, warning, critical alarm, etc.) to full-blown computers featuring graphical user interfaces and complex control programming. Some include digital readouts of current conditions, while others store trends ranging from "previous reading" to weeks or months of readings. Some provide a programming interface, security features (such as password protection), and data communications interfaces.

A feature that may be attractive to owner-engineers responsible for multiple pieces of equipment provided by a single manufacturer is the portable control panel. Such equipment has a plug-in port for a control panel that is inserted only when needed and that can be used for a variety of equipment models.

When Evaluating Control Panels, Consider:

- The suitability of the operating environment and the operators (are they computer savvy?).
- Environmental factors such as air quality (oil mist, dust, corrosive chemicals, and moisture), electronic and audible noise, lighting quality, temperature, and vibration.
- Physical factors such as line of sight; the ergonomics of the panel's buttons, dials, and switches; and the usability of the control panel's software user interface.

Control panels can either enhance operator access or limit it. Given the variety of these parameters, operators should help evaluate panels and determine where to put them. Additionally, having operators, owner-engineers, and design engineers consult with one another and with applicable manufacturers/suppliers could lead to innovations in control-panel design.

MICROPROCESSORS

Microprocessors essentially are computers, so consider the following when evaluating smart equipment with embedded processors:

- *Existence*. Does a microprocessor exist, or is the suite of smart components meant to sense and alarm or perform very simple control operations (e.g., turn on an alarm at the operator workstation)?
- *Functionality*. If a microprocessor exists, what is it meant to do? Can its performance be customized or refined upon installation or in the future?
- *Timing*. Does the microprocessor have its own clock, or does it get timing from an external device?
- *Memory*. Can the microprocessor store equipment-performance trends, the system's operating conditions (air or water temperature passing through, rate of flow, amount of flow, etc.), and the operating-environment conditions (ambient temperature and relative humidity)?
- *How much data can be stored*? What happens when the memory is full?
- *Sampling and data processing*. The parameters that can be measured (sampled) and calculated must be determined early on. Adjustable sampling rates enable operators to take readings as often as necessary to provide reliable monitoring and control without abusing sensors and wasting memory.

Having the equipment's microprocessor perform calculations can increase the diagnostic capability of the smart equipment and help reduce data to extend its logging capacity. Sample rates and "pre-processing" data often are considered together—for example, selecting 10-min sample times leading to hourly averages. Only the hourly average is recorded, meaning that one value for the hour is appended to the log rather than the six taken during the hour. For some or all readings, the interval's maximum, minimum, and standard deviation sometimes also can be stored. This information can be used for troubleshooting and predictive maintenance.

DATA COMMUNICATIONS

Smart equipment can be connected to networked building automation systems, or it can have direct connections to central operator workstations and local field panels. For communication with individuals or monitoring/dispatch centers outside of the building, dedicated phone lines, radio transmitters/receivers, and cellular technologies can provide access to wide-area networks (WANs). In some cases, as part of the purchase, the equipment supplier may provide connectivity. In other cases, third-party equipment suppliers may provide connectivity if they are contracted to handle messaging, data retrieval, storage, processing, and, in some instances, scheduled reporting.

The type of network connection appropriate for smart equipment is a function of:

- The equipment's role in the building system (critical, non-critical).
- The existence, condition, and adaptability of a pre-existing BAS and/or network.
- The operating conditions (electronic interference, humidity levels, temperature, etc.).
- The operator's or contractor's comfort level with computers and software.
- First cost and operating budgets.
- Response-time and level-of-detail requirements for emergency service.

INTEGRATION WITH A BAS

Integrating widely dispersed smart equipment into a building automation system is, without question, the most difficult challenge facing a designer. Minimizing the variety of communications or connectivity approaches makes integration easier. For example, if some equipment offers a voltage signal or a modem connection while other equipment offers a telephone modem or a radio network, selecting the modem option for both equipment types will simplify the integration task. Expert assistance may be needed to develop an integration strategy when many types of smart equipment—or many pieces of one type—are deployed.

CONCLUSIONS

Smart equipment can provide many benefits, including increased equipment and system energy efficiency and early warning of failures. However, innovations also represent new challenges and potential downsides.

The IT-related components of smart equipment add complexity and, therefore, additional risk of failure. Sensors and control mechanisms have to be properly selected, installed, and maintained; microprocessors have to be programmed correctly; real-world conditions for control panels and data-communications networks have to be thoroughly investigated; and operator education and training has to be built into procurement and O&M processes.

Chapter 7

Introduction to Web-based Energy Information Systems for Energy Management and Demand Response in Commercial Buildings*

Naoya Motegi
Mary Ann Piette
Satkartar Kinney
Karen Herter
Lawrence Berkeley National Laboratory

THIS CHAPTER PROVIDES an overview of web-based energy information systems and demand response systems. We begin with an overview of various types of systems, followed by a description of past literature on these technologies. We then provide a discussion of the market for EIS and DR systems in commercial buildings. We also introduce the concept of "web-based EMCS" to describe web-based building control systems. Each individual type of EIS is described in detail. The layers of technology are characterized as well.

INTRODUCTION

Energy information systems (EIS) are composed of software, data acquisition hardware, and communication systems. They are used to provide energy information to a variety of people including energy managers, facility managers, financial managers and electric utilities. Though EIS have a wide variety of uses and capabilities, the basic uses of EIS are:

- To meter and collect building energy consumption data,
- To provide time-series visualization of the data,
- To help users understand the energy trend of buildings and find energy saving opportunities.

A web-based EIS also has a user-friendly web browser interface that can be accessed from anywhere via the internet.

Data types processed by EIS include energy consumption data; building characteristics; building system data, such as heating, ventilation, and air-conditioning (HVAC) and lighting data; weather data; energy price signals; and energy demand-response event information. This chapter summarizes the key features available in today's EIS, and provides a categorization framework to understand the relationship between EIS, energy management and control systems** (EMCSs), and similar technologies.

There are two types of energy information systems: "web-based" and "non-web-based." Web-based EIS products access and manipulate building facility data via the internet, while non-web-based EIS manipulate data stored onsite. In this chapter, we focus only on web-based EIS products.

In a common EIS architecture, shown in Figure 7-2, building energy consumption data are collected by metering devices installed at building sites (A), and dispatched via a gateway or other communication device (B) through an internet connection or telephone line to a database server located at an EIS service provider's physical site (C). The offsite database server stores and

*This chapter is part of a report for the California Energy Commission, Public Interest Research Program, that was originally published by the Commercial Building Systems Group, Lawrence Berkeley National Laboratory, as Report No. LBNL—52510 (April 18, 2003)

**EMCSs, also known as EMS (energy management systems), BMS (Building Management Systems), or building automation systems (BAS) are systems that, through a series of sensors and controllers, facilitate operation and control of end-use equipment within a facility, usually HVAC equipment. Because the majority of currently installed EMCS are not web-based, we use the term EMCS to refer to conventional EMCS without internet connections unless otherwise indicated.

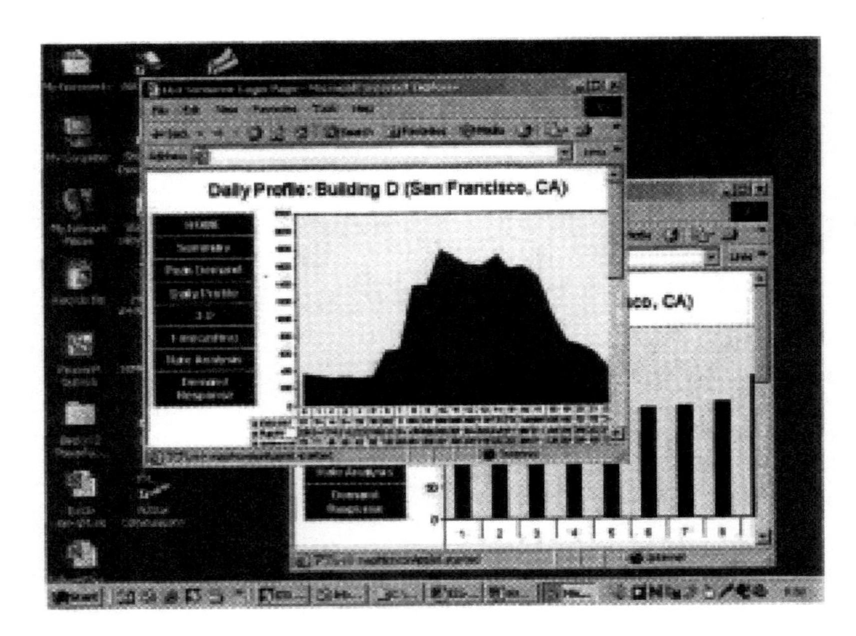

Figure 7-1. Example of EIS desktop image

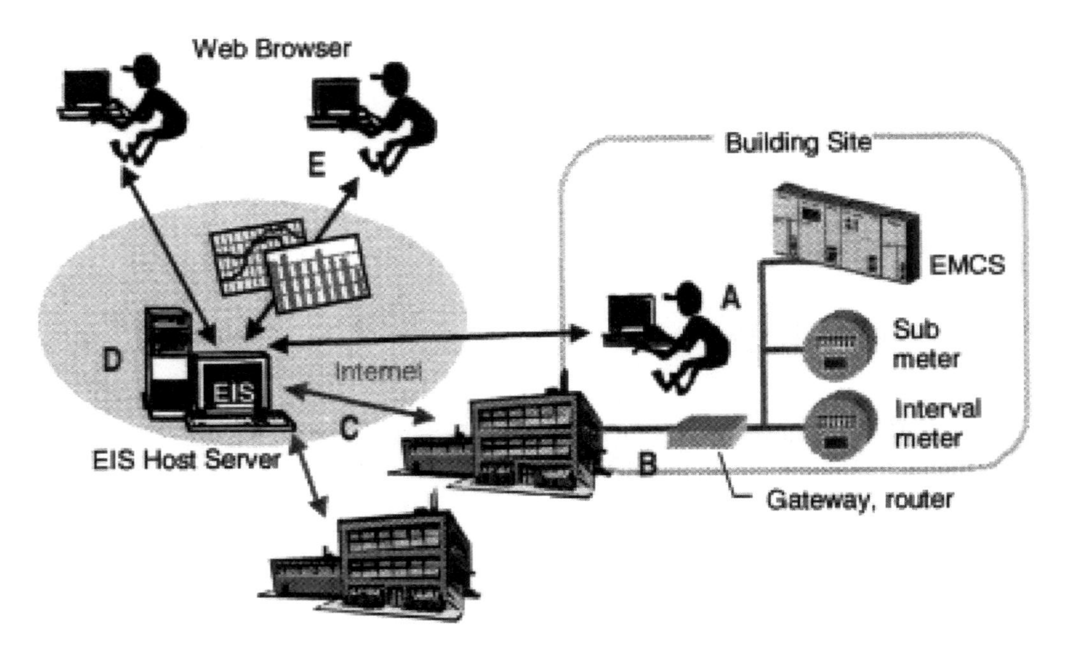

Figure 7-2. Typical Architecture of an EIS

archives this data (D) and EIS users access the database server remotely using a web browser (E). The application program installed on the database server provides a user-friendly interface to facilitate energy management and utility programs. The software commonly provides data visualization and may include additional features such as the ability to download raw data.

EIS have only been in development since the middle of the 1990's. During that time, their capabilities have been improved and expanded to include a wide variety of operability and connectivity[1]. One catalyst for these improvements is that major utility companies, to cope with electricity reliability problems in recent years, have created demand response (DR) programs, which offer facilities cash incentives for reducing peak loads. To make the DR programs feasible and efficient, utility companies have adopted and promoted EIS products as communication systems with their customers.

Today, EIS are used by utility companies, energy service companies (ESCOs), and facility owners and operators. Because these different users are likely to use an EIS for different purposes, it can be confusing for them to choose the most appropriate EIS from many similar products.

BENEFITS OF EIS FOR OPERATORS

The primary benefit of an EIS is to assist facility operators, owners, and other decision-makers in managing building energy use. An EIS helps operators and energy managers understand the energy use patterns of their building or buildings, including issues such as:

- Timing and magnitude of peak electric demand,
- Daily load shapes,
- Historic baseline energy use,
- Unexpected operation schedules,
- Cost variations by hour, day, week, month, and year.

Real-time or daily updating of hourly energy consumption data allows users to evaluate building performance issues that have been difficult to observe. It also enhances the retro-commissioning process[2]. Since most EIS products provide real-time or daily updates of hourly trend data, facility operators can check the impact of an operational strategy immediately following or within a day of the operation. In the absence of an EIS, an impact evaluation would have to be postponed until the monthly utility bill arrived. An EIS also allows facility operators to see the hourly detail of the impact, whereas a monthly utility bill would show only the monthly total. With internet capabilities, an EIS can help manage hundreds of geographically spread sites. Although energy managers may be able to assemble and organize monthly utility bill data for hundreds of sites, this work is painfully time-consuming. EIS products can facilitate such multi-facility energy management tasks.

EIS are also strongly coupled with demand-response programs and strategies. Some EIS products allow energy managers to participate in DR programs and execute shedding strategies. Most of these systems also provide immediate feedback on the DR event.

Literature Review and EIS History

A recent report by Levy Associates describes EIS development history from the view of metering technologies and applications[1]. This report surveyed the historical development of automated meter reading (AMR) with a focus on the customer interface (including EIS) and hardware systems. According to the report, utility metering and information practices have not changed much in the last 50 or 60 years. Standard metering systems were designed to measure only total monthly electricity usage and provided little or no value to the customer. There was some awareness, however, that metering information could support highly valued customer services. Utility field trials have repeatedly demonstrated valuable meter data applications for every type of customer.

In 1996, E-SOURCE released a series of reports outlining the basic capabilities of EIS products available at that time. Their first report categorized data sources and defined key features of EIS products, complete with practical examples and graphics[4]. Later E-SOURCE reports covered technical specifications, potential benefits, market strategies, and prospects of EIS for large commercial energy-consumers, ESCOs and utility companies[5]. The EIS industry was (and still is) immature, and the variation and use of EIS products have further developed since then.

In recent years, the capabilities of EIS have improved, increasing its functionality from visual reports to remote control of equipment. A recent article in the *ASHRAE Journal* noted an increasing market for energy-related information services[6]. In this report, the authors briefly introduced some of the advanced capabilities of EIS products, including dynamic utility rate analysis (for time-of-use* or real-time-pricing** rates), remote control, energy efficiency analysis, and DR program bidding capability.

To assist in the 2001 California energy crisis, numerous DR programs were promoted. To enhance DR program participation, the California Energy Commission released a guidebook for demand response technologies[7]. This guidebook familiarizes facility and operations managers with enhanced automation technologies including EIS and EMCS as well as HVAC and lighting controls technologies. The guidebook provides cost and savings estimates for common enhanced automation options. It presents strategies for system selection, project planning, and implementation.

---

*Time-of-use (TOU) rate charge preset rates for energy and demand that are lower during off-peak and higher during seasonal/daily peak demand periods. TOU rates are often mandatory for very large customers, and voluntary for smaller customers[7].

**Real-time pricing (RTP) is hourly pricing of electricity where the cost per kWh varies by hour and by day. For example, the utility gives customers a 24-hour price forecast each day for the following days, allowing them to adjust usage daily to minimize costs. Typically, RTP is tied to the wholesale market price[7].

Market Categorization

Energy information systems have evolved out of the electric utility industry in order to manage time-series electric consumption data. EIS products have been developed quickly with various features and complexities in order to satisfy the wide variety of client needs. In this section, the market for EIS is characterized into several categories by client and by EIS type.

EIS features and capabilities have been designed to serve a variety of markets and different types of customers. The key segments of customer groups are:

- Multi-Site Clients
- Individual Buildings
- Energy Providers
- Energy Service Companies (ESCOs)

Figure 7-3 shows a classification of the EIS clients.

Multi-site Client Market

Multi-site clients are defined to be organizations that manage multiple buildings. These clients include large corporations, retail chains, restaurants, office buildings, governmental organizations, and educational facilities. This is a key market segment for EIS products. One advantage of an EIS is the ability to easily look at, and optionally control, multiple sites spread over a large geographical area. In this chapter, we use the term "site" to mean either a standalone building or a retail space within a larger building.

EIS are reported to be cost-effective tools for multi-site clients, since a single EIS covering multiple sites is typically less costly than having an operator at each site. For example, chain grocery stores, which generally do not have a facility operator or energy management systems, often waste energy due to inefficient operation, system malfunctions, or unnecessary scheduling. Today most multi-site clients have only a few energy managers to oversee hundreds of sites. With an EIS, multi-site clients can save significant expenses on human resources by using off-site energy managers to manage their buildings remotely.

Individual Buildings Market

Most buildings have limited data acquisition, visualization and management tools, which can be powerful tools for facility managers to understand their building energy consumption trends. Some EIS products provide large volumes of energy and related data, including whole-building electricity usage information, but also end-use energy consumption, temperature, on/off status, or other system condition data, in hourly or shorter intervals. In most cases, however, energy managers and facility operators are too busy to look at such detailed data. Advanced EIS products allow processing and simplification of this data, allowing the operator to save time by reviewing simplified metrics and key criteria. For clients needing more detailed diagnostic analysis, some EIS vendors also offer remote manual diagnostic services.

Energy Providers and Electric Utility Markets

An EIS can provide powerful interactive communication between customers and energy providers. EIS vendors sell customized products to energy providers. The energy providers then redistribute the EIS to their customers as an optional service, or they may provide it free to participants of their load management programs.

An EIS can be used by electric utilities to manage DR programs that request customers to reduce load when the electric grid is taxed and energy supplies are limited[8]. In California, major energy suppliers, including Pacific Gas & Electric, Southern California Edison, San Diego Gas & Electric, and Sacramento Municipal Utility District (SMUD), have purchased EIS products to operate their customized peak demand programs. While

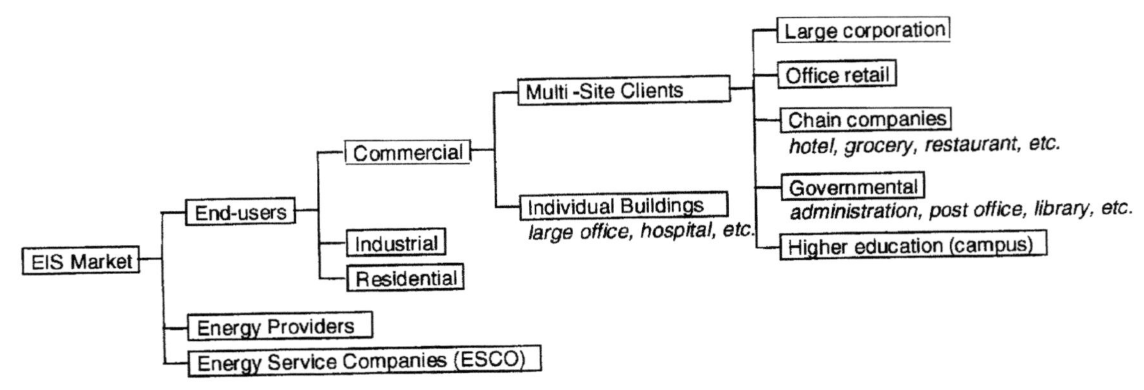

Figure 7-3. Classification of EIS Clients

the implementations were successful, these utilities are still exploring additional capabilities and benefits of EIS in the current California energy markets context.

Energy Service Companies Market

Energy service companies (ESCOs) typically provide two major services to their clients: building diagnostics and performance contracts. An EIS can help with both of these tasks. By using an EIS, ESCOs can diagnose their clients' buildings remotely and in real time, instead of repeatedly visiting sites, as is required in the absence of an EIS. This saves both time and human resources.

For performance contracts, ESCOs track building energy use to verify energy savings against the baseline. They also track any changes in environmental conditions and occupational settings such as occupancy schedules, temperature setpoints, and equipment runtimes; this will assist the building operator in planning for future electricity curtailments. An EIS is an essential tool for tracking such data.

Types of EIS

Web-based energy information systems have evolved out of the electric utility industry in order to manage time-series electric consumption data. However, other energy management technologies have also expanded their capabilities, and are beginning to merge with the EIS technology. Since EIS products are relatively new technologies, they are changing quickly as the market unfolds.

Figure 7-4 is a Venn diagram showing the relationships between EIS and related fields. The "demand response" field has developed systems that enable utility-operated demand response programs or other demand curtailment measures (e.g. responsive thermostat*, direct load control devices**). The "EMCS" field has developed energy management and control systems including non-web-based systems. The technologies in these fields are different from each other, yet there are a number of overlaps between the fields.

In this chapter, EIS are categorized into four types as shown in Figure 7-4. Demand Response System and Enterprise Energy Management are commonly used terms[7, 9, 10]. Basic-EIS and web-EMCS are terms developed by the authors to assist in comparing the key attributes of various EIS.

*A thermostat that can receive external signals and respond by adjusting temperature settings.

**Devices that allow the utility system operator to interrupt power to individual consumer appliances or equipment.

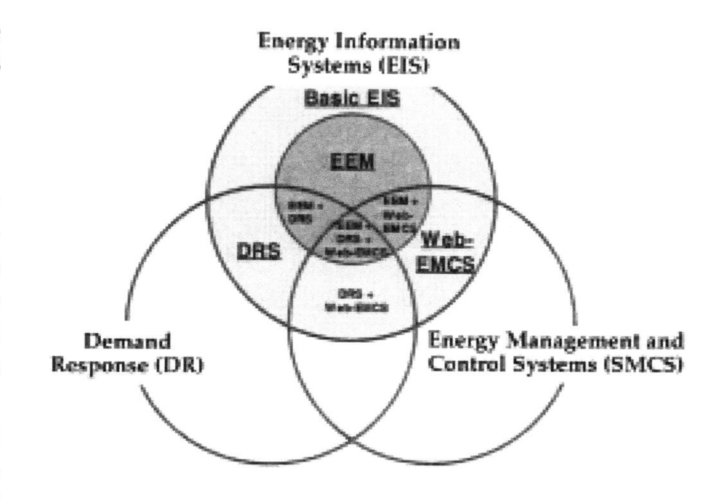

Basic Energy Information Systems (Basic-EIS)	Gather, archive, summarize and display whole-building electricity data.
Demand Response Systems (DRS)	Communicate between utilities and customers to facilitate demand response programs.
Enterprise Energy Management (EEM)	Manage overall energy costs by facilitating energy benchmarking and procurement optimization over a business enterprise.
Web-base Energy Management and Control Systems (Web-EMCS)	Integrate multiple building systems (e.g. HVAC, lighting, generation) and/or monitor and control building systems at the component level by communicating with the EMCS via the internet.

Figure 7-4. Types of EIS and Related Fields

Many EIS products are designed to perform various functions overlapping multiple fields. DRS, which provide both EIS and DR functionality, fit in the overlap between EIS and DR fields. Web-EMCS, which have both EIS and EMCS features, fit in the overlap between those fields. Most EEM, which are considered to have functions of Basic-EIS, fit inside the EIS field. Some EIS may fit into more than one category, but an EIS with many features and overlapping categories is not necessarily more advanced. Some EIS products that include only one feature may be more advanced and complex than those with multiple features.

Basic Energy Information Systems (Basic-EIS)

In this chapter, we have defined "energy information systems" as a broad range of web-based tools to monitor, archive and analyze building energy data, and control building systems; however, common industry terminology for "energy information systems" refers to a narrower range of systems that provide only fundamental data acquisition, management and visualization of utility electric metered data[7]. To distinguish the common industry terminology from wider range of EIS products, we will refer to these fundamental tools as "Basic Energy Information Systems (Basic-EIS)."

The capabilities for Basic EIS are:

- Whole building electricity data collection
- Hourly or 15-minute interval data collection
- Data acquisition
- Historical and real-time data visualization

Basic-EIS emerged early in EIS development. Some traditional Basic-EIS products were provided by utilities to their customers to offer easy access to historical energy consumption data as a value-added service. A Basic-EIS retrieves and plots hourly or sub-hourly trend data, but does not provide detailed data analysis or allow remote system control. If such functionality is desired, users must scan the data plots for inconsistencies and manually adjust building systems as needed. Because of the simplicity of Basic-EIS products, their main advantage is that they are less expensive than other EIS products.

Demand Response Systems (DRS)

Demand response systems (DRS) are notification and response* tools used to simplify the execution of DR programs offered by electricity providers[7].** DRS basically work as real-time communication gateways between energy providers and their customers. For customers with multiple facilities, DRS enable energy managers to organize their energy data simultaneously and remotely, and enable users and program managers to implement control strategies and then verify the participants' demand savings (kW reduction). The recent rapid development of EIS products was partly due to the sudden need for DRS products in 2001 during the California electricity crisis.

Here we provide an example of how a DRS might be used in a "demand bidding" type DR program. When a demand curtailment event occurs, the utility company program manager notifies DRS users by e-mail, pager, fax, phone or other method. The users can then respond to the event through a password-protected EIS website. Users decide whether to participate in the demand reduction event (in case of a voluntary program), and if they accept, they can bid how much they will shed and

*The notification-and-response communication is often called "two-way communication." Monitoring and control function is also called "two-way communication" (in this case, monitoring-only function is called "one-way function"). This report doesn't use the term "two-way communication" to avoid this confusion, though it is very common industrial terminology.

**For example, with a single click, an energy manager can respond to a DR event in hundreds of buildings simultaneously, according to a preset start time, duration time, and load response.

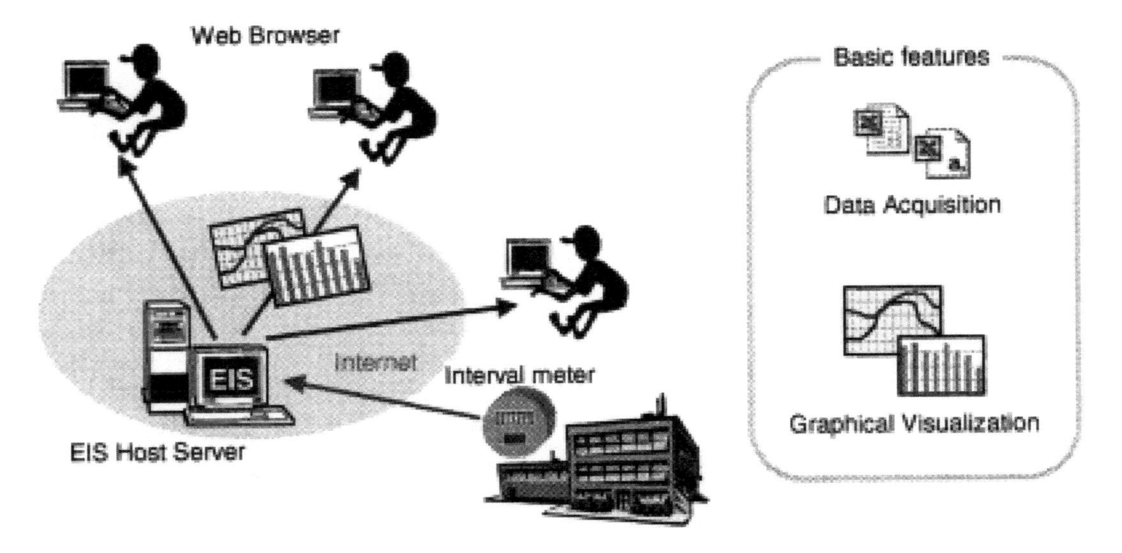

Figure 7-5. Basic Energy Information Systems

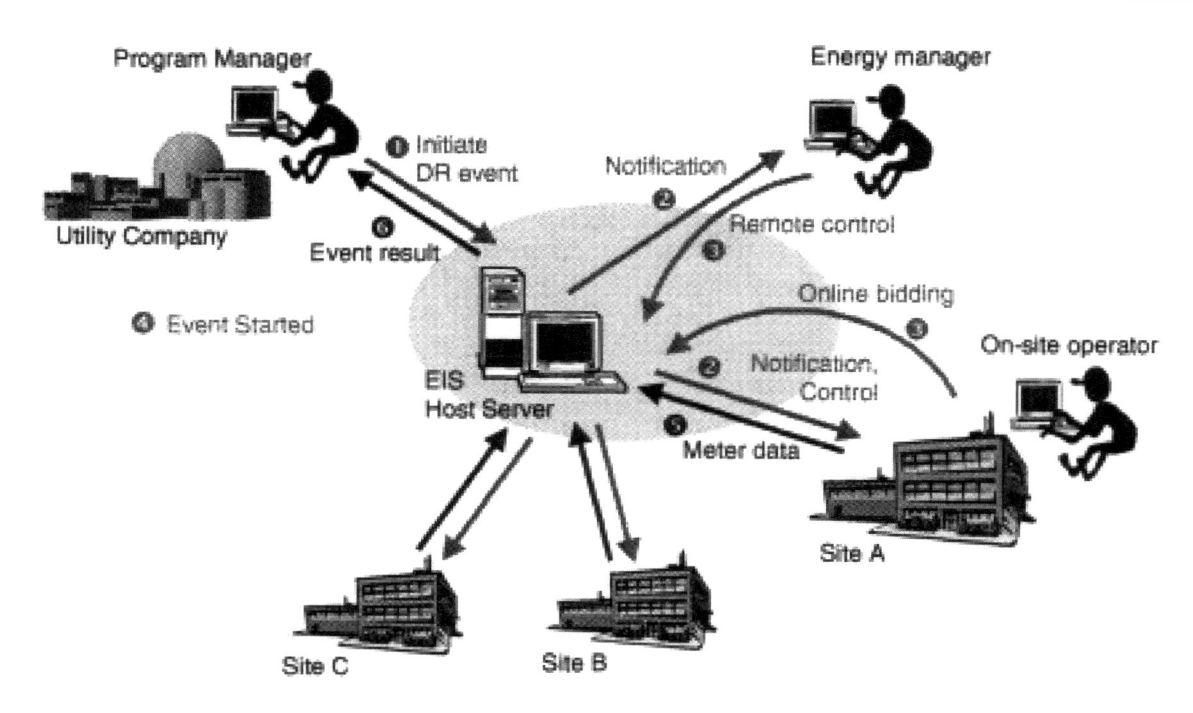

Figure 7-6. Demand Response Systems

at what price. Since most DRS products possess the data acquisition and visualization features of a Basic-EIS, users can monitor their demand reductions in real-time. These trend data are also used for curtailment verification after the event is over.

DRS products facilitate the implementation of DR programs, which can benefit both energy providers and customers. The utility, as an energy retailer, meets electric demand by producing or purchasing electricity at or near cost. During peak demand periods, the utility is often forced to purchase supplemental electricity at a more expensive rate. To avoid purchasing such expensive electricity, the utility offers rewards for customers to reduce their demand during the peak period. A large customer with a number of buildings in the affected area can pool demand reductions from these buildings to affect a large aggregate reduction and reward. Such a large reduction potential can even qualify a customer as a small power producer in the electricity wholesale market.

There are several types of DR programs. "Pay-for-performance" programs offer participants a utility-determined price for kW reductions below a given historical baseline. "Demand bidding" programs allow participants to bid the price at which they would be willing to forgo the electricity, and the utility can then compare between the available DR and supply bids. DR program participants benefit both from the DR program rewards and from energy cost savings at the standard rate. Such programs require complex communications between the utility and the DR program participants. Thus, the notification-and-response communications provided by DRS products play an important role in DR program implementation.

In addition to the notification-and-response communication features, many DRS products have remote control functionality. A large number of DRS clients are multi-site facilities that do not maintain an operator at each site. For such facilities, the remote control capability of DRS products is necessary for energy managers to participate in DR programs. Loads commonly targeted for monitoring and remote control include chillers, fans, thermostats and lighting. These devices can be controlled individually or via a gateway installed at each site. Popular control strategies include modifying thermostat settings, disabling or reducing chiller or fan operation, and dimming interior lighting[1].

Another important feature of DRS is verification and analysis of DR events using baseline or forecasting techniques. Baseline techniques require calculation of an energy consumption referent to estimate demand savings. Forecasting techniques predict future energy consumption to plan demand response or peak avoidance. The two calculation techniques can be similar in method, and can vary from simple averages to complicated algorithms.

Enterprise Energy Management (EEM)

Enterprise Energy Management (EEM) products integrate multiple business processes involved in energy decision-making into a consistent information infrastructure[9], and are capable of multi-site aggregation[10]. In this chapter EEM is defined as an EIS that facilitates energy benchmarking and energy procurement optimization manually or automatically over several sites. Multi-site clients who have numbers of sites spread over a wide geographical area are the major target users of EEM products.

Energy Benchmarking

EEM products are used to compare and benchmark energy use among a portfolio of sites by plotting energy-use data for multiple buildings and normalizing by area or weather. These normalized comparisons allow operators and energy managers to determine how buildings compare to each other. They also indicate if a month, day, or hour has high energy use or unusual operation relative to a baseline. Based on the results from the EEM, energy managers can identify energy saving opportunities, and make decision for retrofits or energy saving investments.

Energy Procurement

EEM has a set of available utility rates of number of utility companies. Energy managers can compare which utility procurement options are most beneficial for their specific energy use.

Energy Cost Evaluation

Energy usage is converted into energy cost, and the cost information is useful to justify retrofit costs. The cost evaluation also ensures that energy bills are correctly charged. EEM products consolidate the cost information and generate reports that are easy to understand. Since EEM users may include non-technical personnel as well as energy managers or operators, EEM features have to be user-friendly.

Web-based Energy Management and Control Systems (Web-EMCS)

Approximately 75 percent of commercial and industrial facilities over 50,000 square feet have energy management and control systems (EMCSs)[7]. EMCSs are control systems to optimize operations of end-use equipment, usually HVAC, through a series of sensors, communications, and controllers. Unfortunately, EMCS data are rarely utilized for optimizing operations, mainly because they do not have adequate data analysis infrastructure. The main limitation of many EMCS products is the lack of interoperability among building systems. Today's building control systems are becoming increasingly complex, integrating the needs of HVAC systems, lighting systems, on-site energy generation, and security systems. However, many EMCS are designed to control HVAC only, and have limited capabilities in integrating with other building systems.

In this chapter, we define a web-EMCS as an EIS that emphasizes system integration capabilities and has

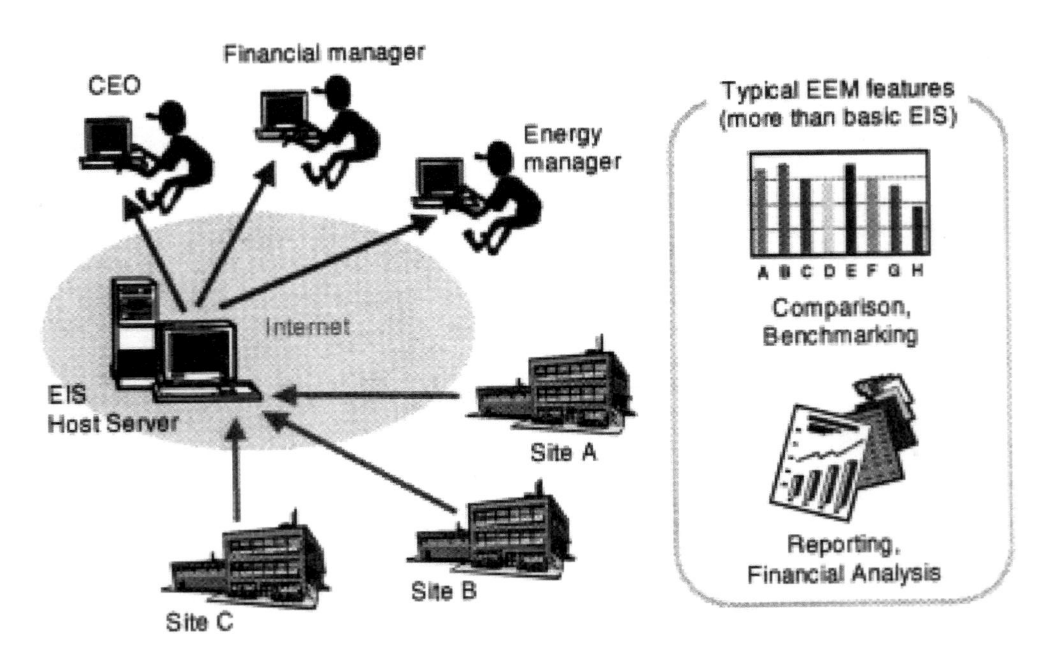

Figure 7-7. Enterprise Energy Management

the ability to monitor and retrieve data from an EMCS or similar system via the internet. Thus, web-EMCS products:

- Monitor and control building systems at the component-level by communicating with EMCS or similar technology via the internet.

- Integrate multiple building systems, such as HVAC, lighting, generation, and security, using a gateway* or similar technology.

The system architectures of web-EMCS products vary from simple to complex. A simple system might consist solely of a meter that sends and receives signals to monitor and control the equipment on which it is installed. A more complex web-EMCS might integrate multiple systems, including meters, lighting, security systems, on-site generation, and non-web EMCS. Systems with a gateway can translate different protocols to an open protocol such as BACnet. Where systems use the same protocol, a simple router can be used instead of a gateway. One important distinction is how deep into the building systems the web-based control systems reach. Some web-EIS provide a single contact point to an EMCS or other controller, while others provide deeper levels of control access to individual buildings systems such as air handlers or perhaps even zone controls.

Gateway systems can be divided into monitoring-only and monitoring-and-control types. Monitoring-only (one-way) gateways translate data from devices or systems to a form usable by the EIS database, but are not capable of translating the other way. Monitoring-and-control (two-way) gateways can translate in both directions. Web-EMCS products that focus on underlying system connectivity can work as a system platform for other EIS products that focus on visualization and/or data analysis applications. In such cases, the EIS application will be highly customized, allowing integration with the other EIS application products.

Like their EMCS cousins, many web-EMCS products control building systems automatically according to user-programmed algorithms. This feature can be useful for responding to DR program events. DR program par-

*Most existing systems use proprietary protocols. A conventional EMCS is unlikely to have an open protocol such as BACnet or LonWorks, so gateways are typically capable of translating between various proprietary and non-proprietary protocols.

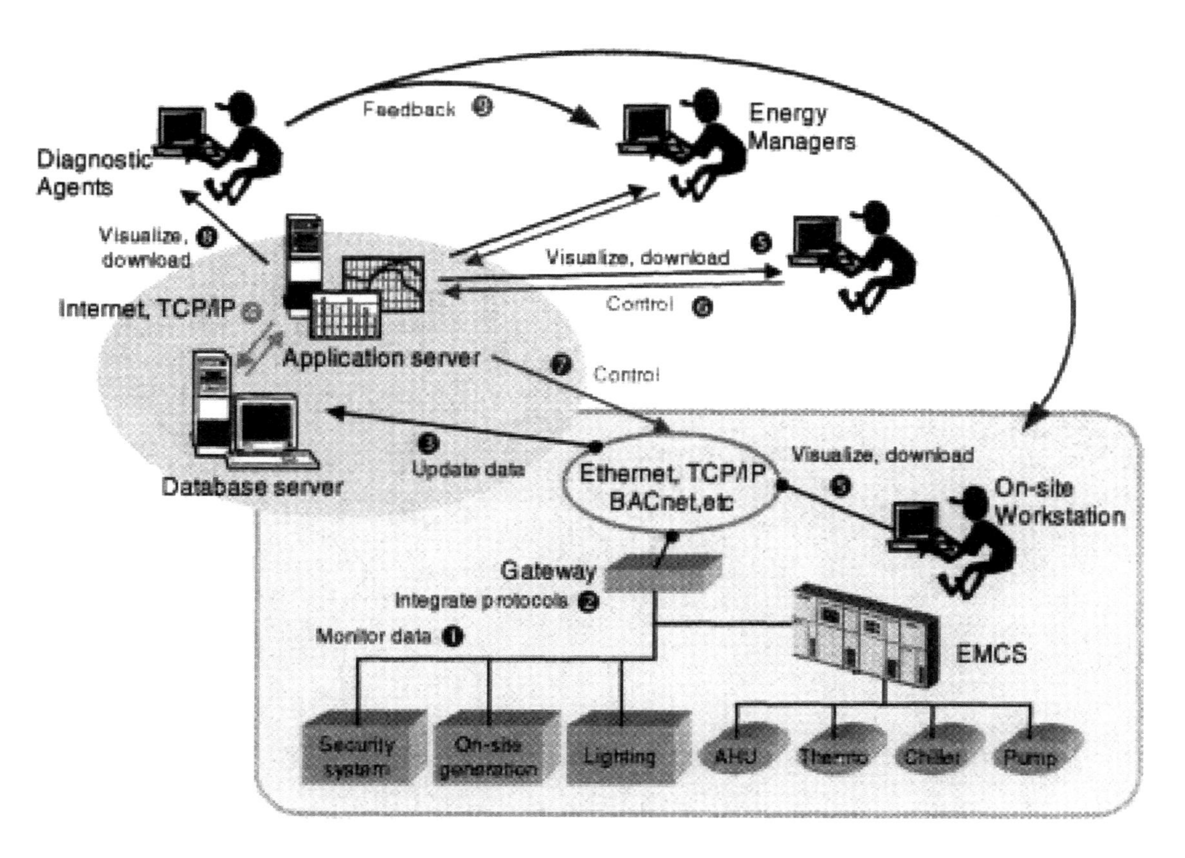

Figure 7-8. Web-based Energy Monitoring and Control Systems

ticipants with a web-EMCS can program the system to automatically modify thermostat settings, reduce chiller or fan loads, and dim lighting when a curtailment call is received. In this sense, web-EMCS product functionality can overlap that of DRS products.

Web-EMCS can enhance real-time and continuous building diagnostics and commissioning, though it is not yet common. Building systems data typically targeted for monitoring and diagnostics include chiller power, air temperature, airflow, and end-use electric loads. With the data visualization capabilities of web-EMCS, these detailed data can be utilized for component-level system diagnostics. Web-EMCS products can provide graphics to analyze time-series data to help facility operators analyze their building systems, but such activities are time-consuming. While most web-EMCS have analysis and reporting capabilities, these capabilities do not actually diagnose or make decisions. This is the common limitation of EIS in general.* To make up for this limitation, some EIS vendors and ESCOs provide manual diagnostic services for web-EMCS clients.

TECHNOLOGY LAYERS

EIS are a combination of technologies spanning the distance between metering and user-interface. Each EIS has its own emphasis and unique characteristics. For instance, an EEM tends to focus on user-friendly visualization software, while a web-EMCS tends to focus more heavily on system integration. These pieces of technology can be divided into stages of data transfer from metering data at each component (Sensors and measurement) to transferring trend data to a data server (Connectivity) to storing and archiving data (Server configuration) to analyzing and displaying data for users (Application). These EIS technologies are further described in Figure 7-9.

*A few web-EMCS products automatically select optimized control without human interaction, but these advanced capabilities are not common.

Sensors and Measurement Layer

"Sensors and measurement" refers to the equipment used, the method of collecting data, and the type of data collected. Common data sources include electric interval meters, EMCS, and sub-meters. Whole-building power data can sometimes be read from utility meters; however, end-use data require additional meter installation. Data can be collected and stored by the existing EMCS, by the EIS, or by some other secondary monitoring system. At a minimum, whole-building power data are required by EIS, while some also use end-use power data and weather data.

Connectivity Layer

The ability of an EIS server to communicate with existing building systems is a key factor in the adoption of the technology. Three networking scenarios apply to EIS:

- Single Meter: A meter sends/receives signals directly to/from the internet server.
- EMCS: The internet server communicates with the EMCS to monitor and control building systems.
- Gateway: A gateway (or router) integrates multiple building systems.

Without a common communication protocol, the EIS cannot communicate with the EMCS. In this case, a gateway is required to translate the protocols and inte-

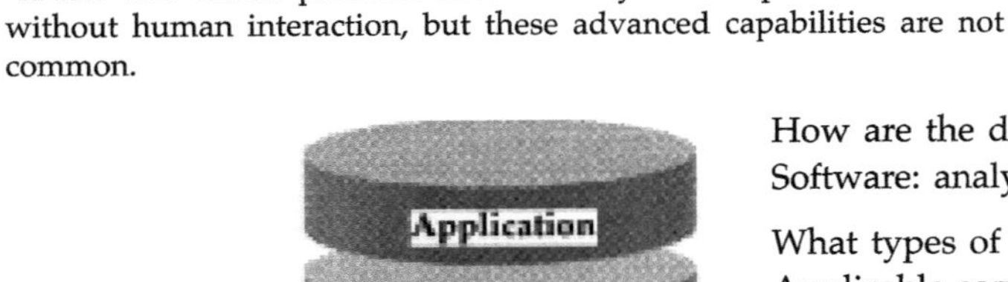

How are the data manipulated and displayed?
Software: analysis and display

What types of server or database systems are available?
Applicable server: Unix, Linux, WinNT, etc.
Database type: SQL, OBDC, etc.

How are the trend data transferred to the server?
Protocol: BACnet, LonTalk, hybrid
Hardware: Gateway system, system integration, etc.

What types of sensors, measurement devices/systems are used?
Measurement device: sensor, interval meter, EMCS, sub-meter, etc.

Figure 7-9. Technology Layers of EIS

grate multiple systems used by the EIS. If all the systems to be controlled are using same protocol, a router, which is less expensive than a gateway, can be used to integrate the systems. Communication between systems may be monitoring-only or monitoring-and-control. Communication between the EIS and the internet is accomplished through standard communications pathways including Ethernet, telephone line, or wireless communication.

Server Configuration Layer

The two main database server configuration options are[1] the database type (e.g. ODBC, Microsoft SQL, developers proprietary system, etc.), and[2] the PC operating system the database system can run on (e.g. Unix, Linux, WinNT, etc).

Application Layer

Application software accesses and manipulates historical data stored in the database server. The application software can be installed on the user's PC; however, most EIS application software is stored on the database server so users can access it from any PC. Methods of visualization and analysis vary widely depending on users' needs. Rate analysis, other financial analysis features, and DR event analysis are also included in this layer.

SUMMARY

This chapter presented a categorization framework for energy information systems and provided an overview of common EIS features.

Table 7-1 summarizes common features of EIS. Each feature can be briefly categorized in one of the layers of technologies that are defined in the section above.

Table 7-2 summarizes common features of each EIS type. These are not restrictive categories and there is significant variation among the various systems. This framework allows us to understand the general nature of these tools as they currently exist on the market and the key features driving the architecture of each tool.

Table 7-1. Common Features of EIS

Metering	
Internet download	Download data via Internet.
EMCS download	Download data from EMCS (e.g. electricity, temperature, tons, flow).
Sub-metering	Electricity metering of portions of sites (e.g. individual end-uses).
Real-time	Update at least every 15 minutes.
Connectivity	
DR Notification	Notification of DR signal for manual response.
Internet control	Control thermostat or other HVAC settings via Internet.
Integration	Integrate systems that have different protocols, manufacturers, and other configurations.
Application	
Load profile	Display hourly electricity usage.
Summary	Aggregate by week or month.
Benchmark	Compare among multiple buildings or historical data.
Rate analysis	Estimate and/or predict energy costs based on the existing tariff.
Forecasting	Forecast near-future load profile.
Diagnostic	Automatically diagnose building systems.
Automated Control	Automatically control building systems according to a preprogrammed strategy.

Table 7-2. Feature Summary of Each EIS Type

Type of EIS	Metering				Connectivity			Application						
	Internet download	EMCS download	Sub-meter	Real-time	DR Notification	Internet control	Integration	Load profile	Summary	Benchmark	Rate analysis	Forecasting	Diagnostics	Automated control
Basic-EIS	●			◐				●	●		●			
EEM	●							●	●	●	●			
DRS	●			●	●	◐		●			◐	◐		◐
Web-EMCS	●	●	●	●		●	●	●	◐				◐	◐

●: Usually covered ◐: Optional [12]

References

[1] Levy Associates, August 2001. "Advanced Metering Scoping Study," California Energy Commission.

[2] Price, Will and Reid Hart, 2002. "Bulls-Eye Commissioning: Using Interval Data as a Diagnostic Tool." Proceedings of the 2002 ACEEE Summer Study on Energy Efficiency in Buildings.

[3] Hyde, Frank, February 1995. "Software Helps Keep Customers Tied to the Line. (Enerlink)," *Electrical World* v209, n2: 43.

[4] Fryer, Linn, 1996. "Tapping the Value of Energy Use Data: New Tools and Techniques." E SOURCE Exclusive Reports on Energy and Efficiency Strategic Memo. SM-96-3.

[5] Komor, Paul, May 1996. "Online Energy Services For Commercial Energy Users." E SOURCE Exclusive Reports on Energy and Efficiency Strategic Memo. SM-96-4.

[6] Kinter-Meyer, Michael, M. Burns, August 2001. "Energy-Related Information Services." Building Connections Article, *ASHRAE Journal.*

[7] Xenergy and Nexant, 2002. "Money and Energy Saving Resources from the Enhanced Automation—Technical Options Guidebook." California Energy Commission publication #400-02-005F. http://www.consumerenergycenter.org/enhancedautomation.

[8] Goldman, Charles A., Michael Kintner-Meyer, and Grayson Heffner, 2002. "Do "Enabling Technologies" Affect Customer Performance in Price-Responsive Load Programs?." Proceedings of the 2002 ACEEE Summer Study on Energy Efficiency in Buildings.

[9] Friend, William R., 2002. "Energy Management—The Last Untamed Resources." Food CIO Forum. http://www.siliconenergy.com/newsroom/articles.htm.

[10] Thompson, Olin, 2002. "Enterprise Energy Management Software—The Key to Effective Energy Utilization." TechnologyEvaluation.com

Acknowledgements

The authors are grateful for assistance from Grayson Heffner (LBNL), and our sponsors Martha Brook, California Energy Commission (CEC), and David Hansen, US Department of Energy (DOE). We are also grateful to the numerous EIS and web-EMCS vendors who assisted in this effort. This work is part of LBNL's High Performance Commercial Building's Program, Element 5 (Integrated Commissioning and Diagnostics). This program is supported by the CEC and by the Assistant Secretary for Energy Efficiency and Renewable Energy, Office of Building Technology, State and Community Programs of the U.S. DOE under Contract No. DE-AC03-76SF00098.

Chapter 8

Trends Affecting Building Control System (BCS) Development*

Tom Webster, P.E.
Center for the Built Environment
University of California, Berkeley CA
Lawrence Berkeley National Laboratory, Berkeley, CA

INTRODUCTION

THIS CHAPTER and Chapter 27, "BCS Integration Technologies—Open Communications Networking," will focus on building control system (BCS)** technology trends, emphasizing the impact of emerging technologies on energy management systems and products. The purpose of these two chapters is to provide energy practitioners with basic informational tools to aid in decision making relative to energy management systems design, specification, procurement, and energy savings potential. This chapter provides an overview of the factors affecting the development of BCS technology.

In the "big picture" all types of commercial building control, management, and operations software and hardware, can be put under the umbrella of building control systems (BCS). However, the focus of these two chapters is primarily on those (sub) systems that have the largest impact on energy consumption, i.e., HVAC and lighting.

BACKGROUND

While adding yet another item to the acronym soup is not desirable, none of those currently used adequately reflect the nature of the emerging facility energy related systems. These systems are no longer simply EMCSs, or FAS/BASs, or EMS/BMSs. They perform a mixture of control, energy and facility management functions*** and their scope is being expanded in at least one important new way—information exchange.

Information exchange as it is emerging goes beyond reporting of monitoring and control information to a workstation front end; it includes the transformation of basic control system data into information suitable for a number of functions ranging from data visualization for support of operations to sophisticated facility energy management analysis and reporting. Thus the following comprises a more accurate and comprehensive terminology for the emerging products by BCS vendors:

- Energy—all energy related monitoring and control;
- Management—meaning building/facility wide management of energy;
- Control—real-time DDC and supervisory control of HVAC, lighting;
- Information—energy, operations, and performance information;
- System—network of control and monitoring devices.

EMCISs fit into a larger context of energy related offerings. Since the early 1990's, in response to deregulation and energy crises, a broad range of new energy information system (EIS) products have been introduced that augment or expand the basic functions covered by

*Although this chapter was based on an analysis of the impact of BCS developments on the federal sector, it is broadly applicable to the commercial sector as well. This chapter was originally published as a report for the Federal Energy Management Program, New Technology Demonstration Program, Report No. LBNL—47650.

**Please refer to the Glossary at the end of this book for a complete listing of all acronyms and their definitions.

***Within this context automation is implicit since it encompasses both control and management functions.

BCS's. These include utility EIS, demand response systems (DRS), and enterprise energy management systems (EEM).[2] There is some degree of overlap between these and the EMCIS products offered by traditional BCS vendors that are the focus of this series. The EMCIS definition includes the functions of traditional EMS and EMCS but does not include those functions of facility management systems that go beyond energy related capabilities, e.g., computer aided facility management and maintenance systems (CAFM/CMMS), access/security, and fire. A complete facility automation system usually builds on the core HVAC and lighting systems to support these other functions via systems integration.

In this chapter, BCS and EMCIS will be used somewhat interchangeably, with BCS generally referring to all types of building control and automation systems and EMCIS more specifically for emerging HVAC and lighting control systems; i.e., those that have the largest impact on energy and comfort.

It should also be noted that most federal buildings use technology that is available commercially off the shelf (COTS) so the focus of both chapters will be on technologies that are readily available and are not specialty or custom solutions.

FEDERAL SECTOR POTENTIAL

Analysis of EIA/DOE statistics[3,4] yields the penetrations of BCSs into the commercial and federal sector building stock shown in Table 8-1.

The data indicate two trends: (1) A greater fraction of federal buildings (both by inventory and floor area) have BCSs than does the overall U.S. commercial building stock; and (2) a high percentage of large buildings have BCSs.

Other EIA/DOE data indicate that small buildings in the federal sector represent a large fraction of the total number of buildings (~77%) but a smaller fraction of the total floor area (~33%). This is counter to the situation of the commercial sector where both the number and floor area of small buildings are the larger fractions.

These results suggest that there still is significant untapped potential in both large and small federal buildings; i.e., the percentage of buildings not yet fitted with BCSs amounts to 28% of overall floor space for small and 33% for large buildings.* However, this should be qualified by a study of relative cost and energy impact considerations.

TRENDS AFFECTING BCS DEVELOPMENT

The following sections highlight some of the main factors that are influencing the development of BCS products.

Enabling Technologies

Figure 8-1 illustrates the primary technical developments that are driving change in energy management technology as well as a host of consumer and commercial product and service offerings.

Computing Power

Because of advancements in processors, storage and software, computing power continues to increase at a rapid pace enabling applications that previously were

*This results from the fact that although small building floor area is only 33% of total, 84% of the small building floor area has not been fitted with BCSs.

Table 8-1. Penetration of BCSs in Building Stocks

	U.S. Commercial Stock	*Federal Stock*
By Total building inventory	5%	21%
Small buildings*	4%	16%
Large buildings	34%	38%
By Total building floor area	24%	39%
Small buildings	16%	
Large buildings	51%	

*Small buildings are defined as those under 50,000 sf.

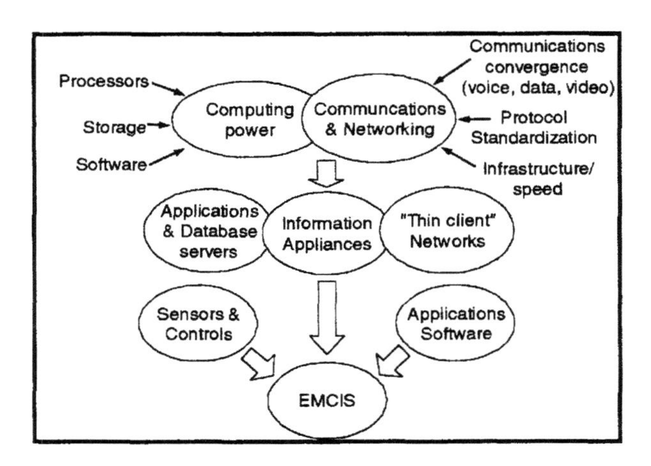

Figure 8-1. Enabling Technologies

not practical or even possible. Smaller, faster, cheaper is the continuing trend in processing power. Two general classes of processors are important to building applications; micro-controllers typically used for embedded applications such as HVAC unit controllers and increasingly used in specialized network capable information appliances, and microprocessors used in PCs, servers, and routers. The power to cost ratio for both will continue to increase for the foreseeable future. Advances such as the copper computer being developed by IBM exemplify this trend. The use of copper allows the feature size to be reduced to 0.2 microns (from about 0.35 for aluminum) and increases speed by as much as 15% due to lower resistance.[5]

Some experts argue, however, that the PC revolution is over in the US; the market is becoming saturated and so competitive that there is little incentive to invest in this area. Attention instead has shifted to embedded processing and network computers that rely heavily on internet connectivity and applications service providers (ASP). It is predicted that by 2005 the number of worldwide web connected embedded processors (~10 billion) will exceed PCs by 10:1.[6] This will result from a proliferation of special purpose devices, the so-called information or network appliances that will be used to accomplish a variety of tasks. If these devices achieve new levels of portability, ruggedness, reliability and ease of use, they will have the potential to transform the way buildings are operated and managed.

Advances in software are also continuing to transform computing. While it is estimated that software represents 30% of the value in a device today, in the future it will be 70%[6]. Advancements in operating systems, especially real-time operating systems (OS) for embedded processors, but also for PCs with the advent of such products as Linux and Transmeta's new approach to emulation computing. Object based applications programming languages such as Java, Visual Basic, C++ as well as scripting languages such as XML for web based applications, finite state programming techniques, neural nets, genetic algorithms, and the open source movement* are all aimed at the creation of better and more capable software at both the systems level and the applications level.

Developments by IBM also exemplify advances in another important area for embedded systems—small form factor storage devices. The IBM Microdrive is a 1.8 inch drive that stores 170-340 Mbytes of data. These drives can be used in portable devices and are now offered in a variety of devices such as the Matchbox PC currently offered by Tiqit Computers[7]. The use of flash memory (i.e., non-volatile re-writeable passive memory) also is increasing at a rapid rate.

Communications and Networking

"In the race between computing and communications, communications won," states Tannenbaum[8]. This statement refers to advances in fiber optics that are rapidly progressing to the point where bandwidth will be virtually unlimited—except for the ability of computers and the electronics on each end of the fiber to keep up. In 1995 a technique called wavelength division multiplexing (WDM) was invented that allows light waves of different frequencies to be multiplexed in a manner similar to FDM techniques used in electrical systems. WDM is made possible by another technology called MEMS where tiny mirrors/prisms are being machined into chips using semi-conductor technology. In 1998 as a result of these technologies, optical transmission capabilities increased from 8 channels per fiber at 2.5 Gbps to 96 channels at 10 Gbps each.[9]. These limits are currently being extended to over 1000 channels per fiber.** When these fibers are bundled into a cable with over 400 fibers, total bandwidth increases to several hundred terabits per second (Tbps). The bottlenecks in this technology are at the interfaces and switches and routers needed to make a network function. Even the capabilities of these devices are being advanced so that "terabit routers" were introduced in 2000.[10]

The use of the internet is driving this need for bandwidth. It is projected that the convergence of data, voice, and video will increase the bandwidth require-

*Open source refers to non-proprietary software offerings best exemplified by Linux.
**Ciena Corporation offers (8/2000) optical transport products using DWDM technology with a capacity of 2 Tbps over a single fiber.

ments of the internet by over 4600% during the next few years (after 1998).[10] Although the number of internet users in the US has leveled out slightly to about 165 million between late 2000 and early 2002 the size of the internet as measured by the number of hosts doubled during the period of 2000-2002 to about 150 million. The demand for bandwidth will continue due to increased usage, more sites, and the switch to broadband access by many users.

Wireless communications is likewise a technology that may result in significant changes in how many tasks are accomplished. Not only wireless cell phones for voice communications, but high bandwidth WAN and LAN data services are being deployed. Newer developments such as Bluetooth and the IEEE 802.11 "wireless ethernet LAN" or WiFi are encouraging growth in wireless services that will enable many types of mobile computing/mobile access applications that potentially could have a considerable impact on building operations. Likewise, new versions of "self-configuring" wireless networking are being developed to support large, flexible sensor arrays. (See below and[11].)

Supporting the development of network infrastructure are corresponding developments in network architecture and standardization. Building network architectures are being "flattened," i.e., becoming less hierarchical, due to protocol convergence as well as more "open" due to standardization. Flatter and more open networks in building systems are being fostered by proliferation of the internet TCP/IP protocol and Ethernet as well as protocols such as CORBA, OPC, XML, and BACnet, and LonTalk. "Thin client" networks where computing devices do less on-board computing relying instead on access to centralized application servers via networking are being used to perform complex analysis and data archiving. A whole industry of ASPs is being formed to support this concept.

The central point of this discussion is simply that these advances in communications and computing are providing virtually unlimited bandwidth, and unlimited access to computing and information resources.

Sensors and Controls

Besides the software developments described previously, advances in controls, including adaptive tuning* and statistical process control techniques promise to improve the accuracy and reliability of controls.[12] In addition, a whole new generation of devices is being developed from MEMS technology in very small form factors with integrated sensing, wireless communications, and power scavengers (i.e., battery-less). These devices have the potential to provide much more flexibility and redundancy, and greater amounts of information than is currently feasible. This information could be used for monitoring and diagnostics as well as for control purposes (see a preview of the future by Pister[13]).

Other developments such as occupant feedback techniques aim to use "people as sensors" to provide another level and type of information. Two-way information flow makes these systems a rough kind of control loop.[14]

New Demands

The list of influences on BCS development would not be complete without mentioning the demands from potential new building technologies and information access. New space conditioning technologies such as task ambient conditioning (TAC), underfloor air distribution (UFAD), and mixed-mode* systems may require new approaches for sensors, controls, and networking. Closely related is the emergence of occupant control and interaction as a means to support better comfort and improved operations. An emerging emphasis on distributed power and demand responsiveness to help manage utility grids and keep costs low may require different information capabilities. The continued development of commissioning, FDD, and life cycle information systems most likely will require new approaches to networking, new software tools and control algorithms. Legacy system support, i.e., interfacing of new analysis and data management tools to older systems will be essential to facilitating the use of new technologies. And finally, system interoperability, one area that continues to be interesting but elusive, may require additional communications and software technologies to facilitate its development.

Other Influences

For building control and operations, like many other aspects of modern business systems, reliance on technology is paramount. Although the pace of incorporation of new technology is uneven, it seems clear that there is constant pressure to improve productivity and lower operating costs to fuel growth and revenues and that these improvements increasingly depend on technology. In addition, three other factors at least indirectly influence BCS development:

---

*The pattern recognition adaptive controller (PRAC) technique developed by J. Seem of JCI is a good example of this technology.

**The combination of natural ventilation and mechanical HVAC systems.

- Utility deregulation
- Organizational change
- Industry trends

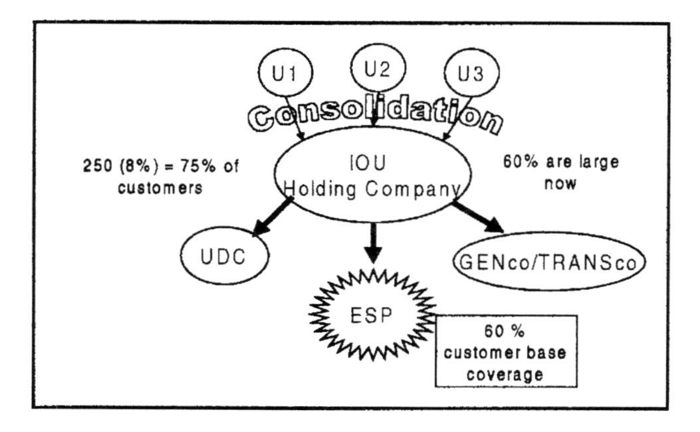

Figure 8-2. Utility Restructuring

Utility Deregulation

Figure 8-2 attempts to capture the essence of what has been occurring in the electric and gas utility industry in the past few years. Most of the restructuring in response to deregulation efforts has occurred in investor owned utilities (IOU); in the 1990s the IOUs represented 8% of all utility companies, but accounted for 75% of $250B in total US energy revenues. In 2002 this process was significantly altered due to the problems encountered in California and the collapse of Enron. Just prior to these events, the consolidation process (shown in Figure 8-2) characterized the utility industry.

Besides illustrating consolidation, Figure 8-2 emphasizes the importance of the formation of affiliates. In addition to utility distribution (UDC) and generation (GenCos) functions, energy service provider (ESP) affiliates are providing impetus to energy related (and even non-energy related) business ventures that impact buildings most directly. ESPs have had an impact on many industries including the buildings industry and the federal sector primarily through ESPC programs. They have key advantages due to their size, access to customers,* resources, and infrastructure expertise. Their primary motivation is profits and not just from energy markets. Since they are unregulated they have been exploring a broad range of products and services both within and outside of their traditional focus on energy supplies and services. Recent events in California and competition from BCS vendors and other players** suggest that continuation of this process as shown is not assured. Nonetheless, any changes in the goals of the utility industry or its regulators are likely to influence the development of BCS products, either directly or indirectly.

Organizational Change

Any change in the commercial sector that fosters development of building technology indirectly affects the federal sector. For example, in the commercial sector where high productivity, and employee retention are key, there is a significant focus on space flexibility, comfort and IEQ to support the needs of dynamic organizations. This tends to drive the development of new communications (e.g., wireless) and sensing and control technologies (as well as new approaches to space conditioning such as underfloor air distribution). Because the federal sector often leases facilities in private commercial buildings and purchases COTS BCS equipment for their own buildings, these changes in commercial buildings become changes that benefit the federal sector as well.

Industry Trends

Technological advances and proliferation of communications technology driven by the IT industry are breaking industry barriers. The technologies of industrial process control, building control and automation, and business information systems are converging, leading to lower costs, wider usage, increased reliance on standardization, and an impetus for integration of both IS and building systems networks.

Utilities (via ESPs) are seeking ways to supply services beyond energy procurement. Along with the strong desire among building owners and managers for flexibility and vendor independence, this movement fosters open systems solutions.

Finally, the building industry market is beginning to model the organization of the industrial process market; the old vertically integrated orientation of the major BCS suppliers is being restructured into:

- Commodity hardware suppliers
- Software specialists
- System integrators

*As shown in the Figure 8-2, current ESPs, although at arms length, "cover" about 60% of the customers serviced by their holding companies.

**Although some IOUs are very large they still are relatively small compared to the true global energy companies like Exxon with revenues of about $100B. Some believe these will ultimately be the primary consolidators of IOUs, large and small. This is evidenced by the entry of Shell Oil and, more recently, Chevron into the energy services arena. Estimated revenue for energy services is a number similar to energy procurement revenues.

IMPACT OF BCS DEVELOPMENT ON IMPLEMENTATION

Potential of Advanced Technology

The effect of the developmental trends described above could have significant impact on buildings in at least the following areas:

- Operations
- Performance

Operations

Analyzing and diagnosing systems is a key issue for operations. This requires great skill, high quality and reliable information, and a means for managing and quickly presenting it in a meaningful way. Since the EMCISs and analysis tools are increasingly complex, remote monitoring and analysis by central service bureaus is emerging as a powerful way to accomplish these functions.[15] These bureaus, whether agency owned or outsourced to an ESP, could significantly increase analysis and monitoring capabilities by centralizing sophisticated tools and highly skilled practitioners who can focus on high level issues.

Energy metering/monitoring and load profiling could provide facility managers the ability to better control energy consumption and negotiate energy supply contracts. Enterprise-wide access to energy data can facilitate information dissemination and management oversight.

Fault detection and diagnosis (FDD) tools have the potential to augment and extend operators' expertise and increase system performance. Load management, automatic curtailment/DRS applications could make systems more responsive to price and utility demand signals to control operating costs.

Increased involvement of the occupants in the operation of the facility via two-way web-based occupant interaction systems could improve operations responsiveness, occupant satisfaction, and operating costs.

Performance

Energy and comfort performance could be improved by advanced control techniques such as adaptive tuning, better mode switching (finite state programming) by eliminating simultaneous or sequential heating and cooling, and better sequencing in general. Increased use of graphical programming could make it easier to create control sequences, which is key to better performing systems. Optimized control fostered by open standards, communications networking, and more capable processing makes possible better integration and coordination between sub-system component unit controllers and the EMCIS.

Impact on Effectiveness

Despite the advances in technology described above, it is not apparent that these changes will improve the effectiveness of BCSs in actual practice. There is a perception (based on both anecdotal information and various studies[16]) that energy management systems are not as effective as they could or should be in performing their intended functions of reducing energy use and lowering the cost of operations while maintaining comfort. Understanding this better will be key to determining if increased reliance on advanced technology will make any difference in performance.

There are at least four ways "effectiveness" can be compromised:

Design and specification

If the system is improperly designed and/or specified, its capabilities and functionality will be compromised, not to mention the increased potential for problems.* This includes the issue of proper specification for the intended application. Most building control systems have traditionally been designed around monitoring and control functions; to use them for data acquisition and information purposes imposes new requirements on sensing, data quality, and data storage as well as network architecture and performance.

Integration of BCS and IS networks has not occurred to a significant degree although there is increasing impetus to do it to reduce cost and simplify maintenance of these systems. However, this integration could add to the complexity of these systems since the data processes are fundamentally different. Control networks require more small, deterministic, and time critical data transfers while IS systems tend to have large transfers that are less time critical. Also, proliferation of the types of devices of different functional domains could hamper performance as well as maintenance and operations.

Installation

Due to the highly competitive and cost constrained market in which the BCS industry operates, there is a tendency to minimize installation costs by providing

*One example of inadequate specification is the Dirksen Courthouse in Chicago that was originally specified with a LonTalk backbone that was subsequently changed to Ethernet due to performance problems.

only the minimal basic functions or simplified functions that are "canned" solutions that have been "proven" in other installations. This results in solutions that may not meet, or only barely meet, the specified sequence of operations. Since full commissioning of these systems is rare, identifying these deficiencies is difficult. This situation also drives the users to rely more on the system vendor to support changes and upgrades, which locks the users into expensive (but lucrative for the vendor) modifications.

Utilization

Many systems come with functionality that is never used (or conversely, beneficial capabilities that are available but never purchased) due to lack of operator awareness, training, or lack of initiative on the part of management. Many operators are directed to focus only on operational problems rather than using the BCS as a tool for improving energy and comfort performance. Even systems that are well designed may not be fully utilized unless operators are properly trained and there is a supportive management structure.

Performance

To detect whether a system actually saves energy and/or cost requires analysis. These analyses tend to be complex and their interpretations can be ambiguous and/or controversial (e.g.,[17]) which tends to inhibit motivation to conduct them. Although reduced energy use is an important goal, it should be remembered that energy is not the only or necessarily the major driver for improving systems. The goal is better expressed by the combination of improved overall operations, occupant satisfaction and productivity, and lower total cost of operations.

CONCLUSION

The advances in enabling technologies as well as other influences could foster a fundamental but evolutionary paradigm shift in how buildings are operated and managed. Advances that might have an impact include those in communications and computing, sensors and controls, new demands for information (e.g., EIS, and DR), and support of new conditioning and diagnostics technologies. As evidenced by current trends, integration with EIS and increased reliance on web-based technologies are likely to be key aspects of building control systems of the future.

The changes in infrastructure capabilities augmented by advances in applications software that support more efficient and sophisticated analyses, can potentially provide better system control and functionality, and better support for building practitioners and managers through remote centralized services and data and information management tools.

However, a number of challenges remain: without a better understanding of why current systems do not perform to their expected level, the impact of new technology is likely to result in minimal change in the performance (energy, comfort, and cost) of EMCISs. The issues of effectiveness must be addressed before or as a part of implementation of new systems. In particular there must be assurance that new products can accommodate legacy systems and can be effectively used by busy practitioners and technicians via better analysis and data presentation capabilities.

Since the federal building sector contains a sizable fraction of buildings that are not yet equipped with EMCISs and those that are probably have a significant number of existing EMCSs that could be improved, the potential for energy and costs savings from emerging systems based on advanced technologies appears to be large if the primary challenges are addressed.

Some of these challenges and other issues related to emerging EMCIS products and applications are addressed in Chapter 27, "BCS Integration Technologies—Open Communications Networking."

Trademark Notices

LON, LonTalk, LonWorks, LonMark, SNVT, Echelon, and Neuron are trademarks of Echelon Corp.

ARCNET is a trademark of ARCNET Trade Association

BACnet is a trademark of ASHRAE

All other product, trademark, company or service names used are the property of their respective owners

References

[1] Webster, T.L., "Trends in Energy Management Technology: BCS Integration Technologies—Open Communications Networking." DOE/FEMP, LBNL-47358, December 2000.

[2] Motegi, Naoya and Mary Ann Piette. "Web-based Energy Information Systems for Large Commercial Buildings." National Conference on Building Commissioning: May 8-10, 2002, Lawrence Berkeley National Laboratory.

[3] EIA, "CBECS: Commercial Buildings Characteristics 1995." DOE/EIA-E-0109, Energy Information Administration (EIA), USDOE, August 1997.

[4] "Federal Buildings Supplemental Survey 1993." SR/EMEU/95-02, Energy Information Administration, USDOE.

[5] "Back to the Future: Copper Comes of Age." IBM Research Magazine, April 12, 1999, www.research.ibm.com/resources/magazine/1997/issue_4/copper497.html.

[6] Bayne, J.S. "Integration Opens New Portals." Buildings Systems Innovation, Johnson Controls, January 1999, pp. 12.

[7] "The Matchbox PC." TIQIT Computers, website: www.tiqit.com.

[8] Tannenbaum, A.S., 1996. Computer Networks. Prentice Hall, New Jersey, 1996, pp. 87.

[9] "Through a Glass Quickly." Business Week, December 7, 1998.

[10] "A Trillion Bits Per Second is within Reach." Business Week, February 21, 2000, pp. 32D

[11] Rabaey, J., J. Ammer, T. Karalar, S. Li, B. Otis, M. Sheets, T. Tuan. "bwrc.eecs.berkeley.edu/Publications/2002/presentations/isscc2002/pico_for_wrlss_senr_ntwrks_isscc2002.htm" PicoRadios for Wireless Sensor Networks: The Next Challenge in Ultra-Low-Power Design," Proceedings of the International Solid-State Circuits Conference, San Francisco, CA, February 3-7, 2002.

[12] Seem, J.E. "Implementation of a New Pattern Recognition Adaptive Controller Developed Through Optimization." ASHRAE Transactions, V.103, Pt. 1, 1997.

[13] Pister, K. "My view of sensor networks in 2010," Berkeley Sensors and Actuators Center (BSAC), 2000. Available from website: www.eecs.berkeley.edu/~pister/SmartDust/in2010" http://www.eecs.berkeley.edu/~pister/SmartDust/in2010

[14] CBE. "Using Occupant Feedback to Improve Building Operations." Center for the Built Environment (CBE), 2002. Available from website: http://www.cbe.berkeley.edu/RESEARCH/briefs-feedback.htm

[16] "GSA Energy and Maintenance Network (GEMnet) Concept Design." General Services Administration Region 9, Final Report No. 30135.02, July 31, 2000.

[16] "Potential for Reducing Peak Demand with Energy Management Control Systems." WCDSR-137-1, Wisconsin Center for Demand Side Research, Madison, 1995.

[17] Diamond, R., T. Salsbury, G. Bell, J. Huang, and O. Sezgen, R. Mazzucchi, J. Romberger, "EMCS Retrofit Analysis Interim Report," LBNL-43256, Lawrence Berkeley National Laboratory, March 1999.

Acknowledgements

The author would like to acknowledge the dedicated support and guidance provided by Bill Carroll of Lawrence Berkeley National Laboratory (LBNL) and DOE/FEMP/NTDP for providing funding for this work.

Chapter 9

How to Get the Most Out of Your EIS: A New Organizational Management Paradigm

George G. Barksdale, Director of Federal Programs
Energy Solutions Group
Science Applications International Corporation, McLean, VA

Miles D. Smith, Program Manager
Energy Solutions Group
Science Applications International Corporation, McLean, VA

OVER THE LAST FORTY plus years, technological advancements have systematically replaced human beings with electronic and digital controls to improve the efficiency of equipment, systems, and processes. Unfortunately and over time, most of these electronic minders have been left unattended and unmanaged. Changes to management structure, necessary to ensure that the machines are controlling as designed, have generally not been instituted. To achieve and remain at design efficiency levels requires that the right information be generated, analyzed by experts, provided to the appropriate management personnel in an understandable format, and acted upon appropriately by management.

INTRODUCTION

In the microcomputer age, energy monitoring is feasible to previously unimaginable degrees. It is generally accepted that what can be monitored can be managed. But "good management" still depends on human beings. Specifically, it depends upon successful, continuous interactions among the managers and those who are managed, and between the managers and the machines that they rely upon to provide useful information.

Collecting and disseminating energy data does not constitute effective energy management in an organization, nor can it cause systemic change. To achieve beneficial change, energy data must first be converted to information, then analyzed, then provided to appropriate staff in a useful format, and finally acted upon in a positive manner. This chapter describes a new approach to energy management that involves a change in the organization's overall management philosophy. It also proposes involving higher-level management in energy decision-making through use of a web-based information dissemination system.

CURRENT ENERGY MANAGEMENT STRATEGY

Facilities, and the buildings within a facility, are usually in a continuous state of design and redesign, construction and retrofit. This results in the use of varying types of energy management and control equipment. Some buildings have no control systems; some have minimal and unsophisticated control systems, while others have modern energy management and control systems (EMCS) and direct digital control (DDC) systems.

Most sophisticated control systems are found in newer buildings and in buildings with relatively high energy usage. These systems are programmed, usually through a personal computer (PC) terminal, to operate a multitude of energy-consuming devices and building systems. However, despite their sophistication, EMCS and DDC systems operate autonomously. Many monitor and control a single building or a small group of buildings, and do not possess the ability to access data or promote energy efficiency at the campus or multi-campus level. Furthermore, a significant percentage of existing control systems do not perform at their designed levels.

Contemporary energy management strategies are usually founded upon three components: capital investment projects involving equipment replacement, improved energy awareness (manager/user/occupant), and changes or improvements in operations and maintenance (O&M). While most energy programs include these three areas, the project component receives the majority of attention because 70 percent or more of total energy savings comes from such capital investment projects.[1]

Because projects have historically generated the most savings, ongoing program development focuses on identifying new projects to implement.

Project Component

The project component is often viewed as a cure-all for shortcomings within the overall energy management program. Project development typically begins when an energy services company (ESCO) is invited to perform a preliminary facility analysis and/or an independent energy audit.

Projects are categorized as:

- Direct-funded projects (including new construction and repair/replacement)
- Projects funded by a performance contract

Engineering analysis, typically following an energy audit, is usually employed to select those projects with the greatest return on investment. This is generally a narrowly focused plan that has been developed by the organization's management or an ESCO. These audits are:

- Autonomous—i.e., not part of an overall strategy or plan
- Generally not undertaken with a direct tie to project financing
- Usually treated as drivers rather than tools for the accomplishment of a strategy

Attractive (i.e., short payback time or high payback rate) projects have become more elusive over time, particularly for aggressive energy management programs and newer facilities. In fact, the pool of attractive projects is almost depleted at many facilities. A reasonable estimate of the value of those projects that remain at a given facility is between 10 percent and 30 percent of that facility's annual utility funding over 10 years. This is based on experience and averages derived from all federal projects. Where few projects have been accomplished over the previous ten-year period, the value rarely exceeds 30 percent. Where projects have been implemented across a campus and applied consistently to all facilities over the previous 10-year period, the value of future projects is estimated at closer to 10 percent.[1]

Realistically, the overall savings effect of new and renovation work may be even lower than anticipated because capital investment projects exhibit a high degree of savings degradation over time. This degradation derives from reductions in the efficiencies of equipment and/or controls due to average or below average maintenance performance. Some facilities perform maintenance on a "replace when broken" basis; others outsource maintenance. Although outsourcing is a definite improvement over "replace when broken," it only provides maintenance at specific single points in time rather than on a continuous basis. It is not usually process related and may have no inherent feedback loop to assure continued efficient performance.

The only ongoing, viable solution to maintenance-related degradation is the establishment of an outsourced maintenance contract that is performance based. This requirement of continued acceptable performance provides the critical feedback loop. Although this approach is important for all equipment, it is most critical for control systems. Control systems must operate near top performance most of the time because inefficiencies at the control level can have far reaching effects on the performance of many other system components.

New construction and renovations are required to follow ASHRAE 90.1-1999, *Energy Standard for Buildings Except Low-Rise Residential Buildings*. Buildings constructed under this ASHRAE standard will yield approximately 15 to 20 percent greater initial energy savings than those constructed before 1975.[2] There are two savings-related components covered by the ASHRAE standard: Structural and Heating Ventilation & Air Conditioning (HVAC). Structural components will provide relatively consistent lifetime savings. The HVAC and control components, however, tend to demonstrate savings initially but are very likely to degrade over time. Where commissioning is not performed, or not properly performed, controls and equipment will never achieve their efficiency potential. The conclusion is that, while the ASHRAE standards have improved significantly, ***long-term sustainability of effort is dependent upon improved O&M practices***.

Awareness Component

Top energy managers generally agree that a visible energy awareness component can produce savings or cost avoidance of 5 to 10 percent overall.[3] The awareness component, though difficult to quantify, should never be ignored or overlooked.

As an example, during the summer of 2001, California Governor Davis directed the Department of Consumer Affairs to conduct a statewide media awareness campaign to inform the public about the importance of electrical usage reduction. Known collectively as the *Flex Your Power* campaign, these efforts are attributed with a reduction in overall energy consumption of 6.7 percent, a record 14 percent during peak hours in June 2001, and an overall demand reduction of 8.9 percent adjusted for growth and weather factors. According to the California Energy Commission, these results were "far greater than expected." Preliminary survey findings suggest that these savings were the result of voluntary actions taken by about 70% of households.[4]

The energy awareness component includes the following:

- Continuous determination of the existing level of attitudes and awareness
- Programs that are designed to improve attitudes and awareness
- Effective program publicity
- Management oversight and involvement
  - Direct—Rewards or penalties aimed at energy related activities
  - Indirect—Management involvement attributed to awareness.
- During
- O&M Component
- Recurring maintenance
- Operations
- Equipment replacement through maintenance
- Day to day programs that impact equipment

Inadequate maintenance has been an ongoing issue for most organizations. For example, at the Department of Defense, studies in 1985 and 1986 discovered that, on average, only half of the officially reported maintenance was actually being performed. This study was based on a survey of 20 bases across the Strategic Air Command. Among other problems, recurring maintenance or recurring work programs (RWPs) would be scheduled and a craftsman would be dispatched to perform the work. The craftsman would sign off that the maintenance was complete; however, the maintenance was often not performed or only partially completed. Overall, tasks were incomplete, incorrectly performed, or not performed for more than 50 percent of the scheduled entries.[5]

The only measured parameter in these studies was "scheduled maintenance." This indicates that the maintenance problem was actually worse, because necessary performance management protocols such as benchmarking or baselining were not measured in the study. This means that additional tasks that should have been included on the schedule were not included. In addition, more maintenance resources were available during the mid-1980s than are today.

PROBLEMS WITH CURRENT STRATEGY

Limited Project Selection Criteria

Facilities generally do not select their projects by using a comprehensive energy management plan. For example, the federal government primarily depends on projects to meet their energy savings goals.[6] This approach is straightforward and easily understood, but it does not require a well-defined strategy. Instead, this method encourages a piecemeal approach—projects are tossed into "the program" and then scheduled based on audits or ESCO evaluations.

Audits and evaluations, like most studies, tend to generate multiple approaches and/or projects each time they are performed. Those projects that are adopted tend to be fast payback, high return on investment projects, which focus only on the most obvious savings regardless of the overall number of projects available. Because ESCOs must limit their engineering costs (which are their burden unless the project is adopted), they generally limit engineering or detailed analysis. These projects are rarely comprehensive; they are usually the lowest hanging fruit.

No Upper-Management Level Support or Control

Ultimate responsibility for energy programs currently rests solely upon the energy managers, and few programs have significant higher management involvement. The net effect is that the energy manager must develop and market programs on his or her own. The technical and organizational limitations of the position, compounded by a lack of direct management support, often result in program stagnation.

The energy management role has been relegated to a junior position in many organizations. This position offers little opportunity for upward mobility and is often a part-time position or a "named" duty without defined responsibilities. Where a permanent or part time position exists, it is filled, in many cases, with a technician. The technician is buried too deeply within the reporting chain to garner the visibility that is critical to effecting change.

Ineffective Organizational Structure

The energy manager's attempts to effect change in an organization or facility are often perceived as complaints. Energy issues brought to maintenance are often resented and viewed as intrusions into their way of doing business. The energy manager's goal is to increase efficiency, while the maintenance manager's goal is to reduce complaints. This difference in viewpoint has created a certain animosity between energy managers and their associated maintenance shops. Similar problems exist between energy managers and their engineering sections.

The traditional management structure is composed of bureaucratic layers. These layers, by their nature, discourage change and are embedded in a status quo that results in an inability to reduce utility costs.

No Analysis of Historical Data and No Benchmarking

Traditional EMCS and DDCs, and the personnel that support them, face informational constraints that restrict the promotion of broad energy management concerns. Data are retrievable only at DDC or EMCS source locations and is in a multitude of formats. Analytical and visualization tools are not provided and data are rarely archived. Because no long-term data exists, seasonal and trend analyses are not possible, and even simple tasks such as usage comparisons of similar buildings can be arduous. These conditions severely limit opportunities for real-time observation, long-term trend and seasonal analyses, and the realization of campus-wide energy savings.

Installation of an EIS allows data from the legacy (existing) system to be transmitted via internet or local intranet to a central server for processing and archiving. Here, it is stored and manipulated to provide analysis, graphical presentation, and reporting. An archiving mechanism stores data from the entire network continuously and for periods measured in years. In contrast, typical EMCS and DDCs store data for 90 days or less, making historical comparison, trend, and seasonal analyses impossible.

With large operations, this analysis, presentation, reporting, and storage would involve vast amounts of data from multiple and diverse sources. Any installed system should provide analyses and presentations in a variety of forms that are intuitively understandable to the energy manager and other staff. Connection through the internet or a local intranet would offer facility-wide accessibility to the system, its data, and its functional capabilities. Modern energy information systems (EIS) provide a unique and unparalleled capability to resolve the unique challenges of geographic dispersion, high variability among legacy energy management and building control systems, security, growth, and organizational complexity.

Controls Applications Not Functioning as Designed

Mechanical time clocks were the original tool associated with the largest effect on energy savings. Properly installed and used, this is still true today. However, systems inspected years ago displayed a universal problem; the control pins were routinely removed and never replaced for a variety of "convenience" reasons. Removal of the pins, of course, negates any time-clock derived savings. Over time, the supposedly infallible (no pins to pull) electronic time clock was developed. This device was further enhanced, its capabilities were broadened, and today we reference it by a variety of names. The most common of these is "EMCS" for energy management and control system. Unfortunately, and despite the technological advances, the problem continues. On average, more than 75 percent of EMCS "timed" equipment has been switched into the hand (manual) mode and operates continuously.[7]

In many cases, the non-functioning DDC or EMCS application was supposed to directly control large sections of a particular facility or installation. In one instance, a 20-installation regional operation was found to have 90 percent of its control applications turned off.[8] This problem is obviously not due to a lack of technology; it is because people will act in ways that serve their immediate needs and they will often override that which they do not understand or cannot quickly make functional. To resolve this problem, the root cause must be determined and corrected. Without doing so, an EMCS is no more useful than a pinless time clock.

DoE studies have determined that, even when EMCSs are operating perfectly, they seldom meet their projected savings. A lack of adequately trained operators and maintenance personnel primarily causes this shortfall.[9]

SOLUTIONS

Shift the Paradigm

Consolidate the Right Information

Vast quantities of data are available from sources such as meters, host utilities, DDCs, and EMCSs. The true value of these data is almost never realized for lack of access, storage, or compilation. Without these three elements, data cannot be converted into useful information. Data itself is not a proxy for useful information, but manual conversion of available data into information has exceeded the skills and time constraints of all but the most talented of energy managers. For most, static manual comparisons are arduous and dynamic manual analyses border on the impossible. Fortunately, contemporary EISs monitor, compile, and analyze tens of thousands (or hundreds of thousands) of information points, and efficiently convert this data into usable information.

To make effective decisions, managers must be provided the appropriate level of information and it must be presented in an appropriate format. The data must be screened and sorted so that the information required for a specific application is provided in formats that are appropriate for different management levels. When data are properly presented and in the right hands, managers at all levels can be active participants in the energy analysis process and the best possible decisions can be made.

This approach is patently different from what exists in most organizations. In the traditional approach, EIS output (assuming any exists) flows directly to the energy manager. With web-based energy information systems using the internet or an organization's intranet, the opportunity is available for higher-level management to have easy access to energy use data. A click of the mouse could bring up display screens filled with relevant information.

The energy manager must attempt to influence those around him to improve their performance within their particular supportive role, and he/she must do so without any real authority. This leads to frustration, friction, animosity, and a lack of teamwork, which are diametrically opposed to the division or branch manager's vested interest in interdepartmental teamwork and harmony toward the goal of energy reduction. The web-based solution allows the energy manager to disseminate important energy information and relate it to other managers' specific areas of concern. Periodic emails to the managers with links to the information screens that they should see allows focus on shared goals and facilitate the energy manager's job.

Provide Expert Analysis

Energy program analysts must understand how the various energy systems are designed to work, then validate that they are operating as designed. This validation process must include many cross checks that are often overlooked. Periodic system reviews must ensure that the correct data are being retrieved and displayed (e.g., that a given temperature readout is accurate and is truly reading temperature). Non-validated systems often display incorrect information such as fans or motors that indicate "on" when they are not. Many crosschecks are simple; the fan indicates "on" but a separate screen indicates no airflow. Software that supports continuous commissioning exposes these anomalies as they occur and provides the energy manager with the confidence that derives from dependable information.

To be effective, a skilled analyst will understand the basic applications or packages of applications that are available from a typical EIS. There are six, natural (user-perspective) functional groups of applications or packages as shown in Table 9-1. Their commonality derives from their normal day-to-day use and in the approaches and/or the techniques that an analyst would use to understand them. Unfortunately, EIS software packages are rarely organized to address these natural groups. In fact, there is little or no industry consistency among software suite packages, which are sold by software application, control application, subgroups of applications, points, or other operational aggregations. The user has the difficult task of organizing and understanding what the various software suppliers are offering, and then determining the correct mix of features or modules that meet their requirements by functionality or by natural group. For example, one functional group may require the purchase of three or more software modules.

Skilled analysts understand control applications and how maintenance efforts (specifically lack of maintenance) will affect them. They understand how the client uses the applications and how they are executed by the control system. They also understand how to leverage a variety of analytical techniques, which themselves are founded upon the client's unique operations.

With these basic understandings, even partial information from the systems can be evaluated against basic assumptions that may be generic to that particular operation. This becomes a tool that can be used to make comparisons and to benchmark facilities. It also allows the analyst to calculate the value of benchmarking, to compare one facility to another, to gain a better understanding of core problems, and to justify projects that support that facility.

Change the Management Approach

Analyze the Information and Operations, then Re-commission the Applications

One of the first undertakings after installation of an EIS is the re-commissioning of all existing applications. "Re-commissioning" has become a buzzword, particularly in the arena of mechanical equipment and controls where its implementation is obviously important. But

Accounting

Utility Bill Verification
Client Sub-billing
Utility Accounting

Rate Opportunities

Demand Management
Real-Time Purchasing

Direct Use Reduction

Setbacks
Shutdowns
Awareness

Maintenance

Reduce Unscheduled Maintenance
Reduce Lag Time for Maintenance
Improved Report Generation
Lagging Reporting
Electronic Reporting
Avoided Down Time
Improved Occupant Satisfaction

System Maximization

Heating and Cooling Overlap
Min/Max Off Time
Duty Cycling
Optimum Start/Stop
Hot Water Reset
Chilled Water Reset
Economizer Cycle Cooling/Heating
Hot/cold Deck reset
Boiler Optimization
Chiller Optimization
Boiler Efficiency Evaluation
Chiller Efficiency Evaluation
System Efficiency Evaluation
Other Maintenance Evaluations

Comparisons

Baselining
Macro Benchmarking
Benchmarking by Category
Comparison of Benchmarked Facilities
M&V for Existing Projects
Identification and Justification of Energy Projects

Table 9-1. Natural Groups of EIS Applications

the term is misleading because, so often, the original commissioning was never performed or it was performed incorrectly. With or without competent initial commissioning, applied resources are usually inadequate to properly maintain systems throughout their lifetimes. Evidence of both of these problems abounds.

A survey of newly installed Air Force control systems in the mid-1980s found that systems did not work as designed in more than 80 percent of the facilities.[5] This is understandable—commissioning standards were woefully lacking in the 1980s. But, as cited, the same problems are evident in systems installed in government and commercial sites even today.

The 1990s promised a "return to basics" in RWP. That mandate, however, was short-lived. Most facilities have returned to repairing equipment on an "as it breaks" basis with little or no scheduled maintenance. Even companies that maintain their equipment in-house or through service agreements have little beyond routine maintenance. They are certainly not enjoying progressive or continuous commissioning maintenance processes and the inherent savings that would accrue. Sadly, these themes are consistent and, even when equipment has been initially commissioned correctly, the odds of preserving design settings through traditional maintenance are slim.

Change the Management Approach Paradigm

As discussed, the energy manager shoulders the burden of the entire energy cost control program. This is unfair and counterproductive because quality, successful energy programs depend on broad involvement. An individual with responsibility but no authority will never succeed in achieving energy reduction goals. The pyramid must be reversed so that program responsibility is placed with management. The energy manager then becomes a resource to provide critical information to other managers, enabling all to make better and more timely decisions. As mentioned above, using web-based methods for distribution of energy information make the energy manager's job easier.

Develop Team Mentality

Inverting the pyramid demonstrates that management relies upon information and input from all of the personnel that support the energy program. The energy manager still retains valuable information, which in turn supports the entire program and all team members. The EIS provides information to staff members at all levels allowing them to make informed decisions. The establishment of this team approach allows everyone to focus on goal accomplishment.

Continuously Commission the Applications

As stated previously, commissioning occurs at a single point in time. The moment commissioning of a device or system is completed is the moment degradation begins. Optimum efficiency can only occur with continuous commissioning supported by ongoing maintenance. Applications must be continuously commissioned or they will quickly depart from the baseline design settings and efficiencies. Studies demonstrate a continuous degradation of even the finest systems from poor maintenance practices. This is troubling because it shows that savings projections generated through capital investment energy projects are often short lived.

Develop and Prioritize Projects

Once most major systems are connected, baselined, and monitored, energy comparisons may begin. These comparisons provide insights toward potential savings and can help establish better/best solutions when "best and worst in category" outliers are identified. These conclusions, in turn, lead to the development and prioritization of energy service performance contracts (ESPC) and other projects.

PROJECTED BENEFITS AND BENEFIT COMPONENTS

General Savings

Direct dollar savings from the proper implementation of an EIS range from 5 percent to 20 percent or more depending on the breadth of modules applied and the quality and amount of information that is available to the system.[10] Key information for one installation may be useless to another, so applications must be individually assessed. Once key parameters are identified, returns on investments may be estimated.

Savings are generally broken into two categories:

1. ***Direct Cost Savings***

- Direct-funded savings or savings generated by avoiding utility costs (including energy and demand costs for electricity).
- Cost avoidance due to manpower savings. Manpower savings are only generated through reductions in force (RIF) or reductions in hours (minus RIF charges).
- Reductions in contract requirements for manpower or services.
- Higher productivity that results in a lower cost-to-production ratio for a manufactured product.

2. ***Indirect Cost Savings***

- Efficiencies realized through higher productivity without associated decreases in personnel costs or production costs
- Improved awareness that provides:
  - — Improved overall job performance,
  - — Personnel empowerment toward improved client responsiveness; and
  - — Indirect savings through equipment or energy projects.

Savings Dependencies

Savings are directly dependent upon the facility's initial and ongoing project goals. Generally, they look for a software module or application that matches some functional requirement. This project or software requirement is thus "need focused" and there are many software applications that could be employed. Typically, a user searches for a particular application designed to address that one need. Yet, if the user broadened their search to additional software applications, the number of options might increase tremendously and the economic analysis, though more complex, might generate a lower life-cycle cost. Cost efficiency requires a broader review of applications that could generate savings. The shift in view from basic requirements to strategic needs will provide the highest possible payback.

For example, these are just a few of the many specific software modules/applications that can provide both direct and indirect savings:

Utility Billing Validation

This software module or application is often one of the prime modules for companies to move into the EIS realm. Utility bill-validation modules may save costs associated with high bill review overhead costs, and may also reduce resource requirements by increasing the accuracy and timeliness of the bill evaluation itself. Beyond this, the module does not, by itself, reduce direct costs. This application is universal and does not depend on utility rate structures or state laws.

Utility Accounting

These systems are generally credited toward indirect savings and are probably the second most used application. Associated savings categories are related to man-hours and fiscal responsibility. Agencies are sometimes able to reduce personnel if the process was previously performed manually. Utility accounting can also save direct dollars for agencies where this service was previously performed by a sub-contractor. This program can allow the energy manager to allocate savings to the appropriate account centers so the money saved can be used as an incentive to fund specific quality-of-life programs. This application is usually teamed with bill-validation applications and is also universal and not dependent on utility rate structures or state laws.

Demand Management

These programs allow the energy manager to evaluate the demand levels and restructure loads. This is related to specific rates where a campus pays a premium for demand charges. The user can reduce their loads when they approach peak use for the monthly intervals and save a lot of direct costs. This requires a real time analysis of load and a breakdown of use so the energy manager can shift or defer the load to achieve a better demand charge for the month. Significant direct savings can be achieved if rates and loads allow the use.

Load Curtailment

This is similar to demand management except the utility provides a price concession for the overall rate if the user agrees to curtail or reduce load during peak or other demand critical periods. Like demand management, this is a direct cost savings and requires real-time (or near real-time) management of the loads.

Real-Time Purchasing

Probably more than any other, this module has driven more consumers to web-based systems. The direct savings can be tremendous, but it takes a sharp manager to understand purchasing and to implement a successful program. Many such systems were sold in California during their energy crisis. This application is dependent on the local rate structure and state laws.

As a side note, many of the systems that perform real-time pricing (RTP) incorporate a demand management application. This can produce significant additional direct savings if properly managed.

Energy/Cost Reduction Dependencies

Actual savings depend on three things: (1) the ease of tying into the existing systems; (2) the current level of management oversight and an understanding of the state of the applications prior to tie in; and (3) the amount of continuous commissioning that is being performed.

Control System Tie-In Costs

Costs to control existing systems depend on how much equipment exists that can potentially be tied into the EIS, and what it will actually cost to do so. Return on investment values must be considered, so predicted savings must be included in the analysis. As a gauge, savings can be predicted based on the amount of equipment that can be controlled by an EIS tie-in. For example, a DDC has the capability to control the majority of energy use in a facility, whereas an EMCS may control multiple buildings or facilities. The number of buildings and the percentage of the overall energy use of a campus that is controlled by the DDCs and EMCS affect the overall ability to achieve savings.

The ability to tie the EIS into secondary controls systems that control major energy uses is also critical. The existence of other types of controls must be investigated. If they are found, a determination must be made regarding their ability to tie into the software. If possible, return on investment analyses must be performed to determine the value of tying into them. For example, a typical commercial building's energy is shown in Table 9-2 below.

Table 9-2. Typical Commercial Building Energy Use[11]

Energy End Use	*Use (percent)*
HVAC	40
Lights	29
Domestic Hot Water (DHW)	9
Miscellaneous	22

With the HVAC load imposing the largest impact on the overall load, tying the EIS into the chiller's DDC alone will provide information about almost half of the load. Obviously, tying into a central chiller or boiler plant for a building (or group of buildings) becomes a high priority because the ability to monitor and manage these plants affects a major portion of total energy use in a facility. Similarly, ties to large fans and pumps can affect major portions of the building load. If large loads can be quantified and tied into the EIS, the remaining loads can be estimated in the equations based on a stable state with reasonably accurate results.

Management Oversight

As discussed previously, there are several shortcomings of traditional maintenance programs. One is related to the simple accomplishment of required maintenance—either half of scheduled maintenance is not performed or it is performed incorrectly. Adding to this maintenance that should have been scheduled but was not reduces the expected maintenance actual completion rate to below 35 percent. In addition, many applications were not commissioned originally or were not reconfigured to accommodate changes in occupancy. In the field of controls applications alone, the combination of these negative synergies results in realizing only 10 to 25 percent of the applications being "on" and utilized.[7,8] The managers responsible for control systems should develop expertise on control applications in order to have a positive impact on the overall savings potential.

Continuous Commissioning

The controls must be recommissioned before they can be baselined or before any strategic program can be reliably established. Once commissioned, the system must be monitored to alarm for any changes made through equipment related programs or through manual changes (putting in manual/hand mode). Regardless of management procedures put in place to minimize these changes, the effects of "average" maintenance will cause degradation on settings and/or operations.

Submeters

The existence and quantity of energy meters and submeters is also important. Submeters provide a simple means to monitor and manage consumption. Accurate and timely consumption data allows a manager to make comparisons, through benchmarking, and to develop an accurate picture of the overall process. A comprehensive submetering package, though not a requirement, helps energy managers achieve maximum effectiveness, thus ensuring better management of accounts and overall use.

Connectivity Costs

Some EISs require the connection of redundant components and hardware to integrate disparate EMCS and DDC legacy systems. This usually results in high capital costs. Fortunately, some EISs can be overlaid upon existing meters, EMCS, and DDCs. By connecting into these data networks and accessing existing input/output sensor and control points, hardware requirements are reduced and the installation of redundant components is avoided. However, because most facilities contain EMCS and DDCs that vary by make and model, a large number of communication protocols are often in use at any given site. To be effective, the EIS may be required to accommodate and interpret data from an array of legacy systems.

The cost to tie into the various DDCs, meters, and other controls can vary drastically. This "connectivity" cost will drive the largest portion of the budget. Connectivity could be twice the cost of the software itself, and does not include surveying, analysis, etc. When costs are totaled, a complete and usable product may well be three times more than the software itself. This point is very important because many purchasers spend more time worrying about and comparing the base software cost, which may be a relatively small percentage of the total project cost. A more meaningful evaluation factor is total system cost, not software cost. This total system cost could be substantially less for software companies that have already developed gateways and have agreements that allow them access to the various equipment and controls protocols.

Another aspect that can end in unexpected cost overruns is programming and connectivity costs that exceed expectations. Even if the software company has confidence in their core software, and has a track record of successful system protocol and other interfaces, it is almost impossible for them, in advance, to determine precisely what they will be tying into. As a result, unexpected or modified systems are discovered and may require custom programming, or unexpected and complex rates are discovered and must be modeled. Unexpected programming and modeling costs add to the client's bottom line costs, while the software company retains and may use that knowledge on future ventures.

Savings Summary

Tables 9-3 and 9-4 present the potential savings and highlight the relationships among the categories. These tables are based on "average" users that do not have superior maintenance programs, do not practice continuous commissioning, and are not applying demand management or real time pricing strategies.

If demand management or real-time pricing functions are being performed, extrapolate or drop to the next lower savings potential category.

Once the savings projection is determined, a simple payback can be determined from Table 9-4. These data are relatively conservative and should be easy to achieve or exceed with typical EIS systems. Note that the annual expenditure should be the total budget for the campus or group of campuses to be controlled by the EIS, and it assumes full enterprise management with all applications/packages discussed above.

Additional indirect benefits include:

Table 9-3. Savings Projections Based on Extent of Planned Applications (12)

Overall Energy Savings (percent)	*Demand Component*	*Metering*	*DDCs and EMCS (percent)*	*Real-Time Pricing*
20 +	Demand	Full Metering	100	Yes
10	Some Demand	Major Energy Users	75	No
5	No	Some submeters	50	No
3	No	Main only	35 to 50	No

Table 9-4. Simple Payback in Years based on Savings Versus Annual Utility Costs[12]

Overall Energy Savings (percent)	*$70M Energy Bill*	*$25M Energy Bill*	*$10M Energy Bill*	*$5M*
20	0.9	1.2	1.4	1.6
10	1.4	2.0	2.3	2.8
5	2.2	3.5	4.1	5.1
3	3.4	5.5	6.6	8.2

- monitoring of, and accountability for, energy use and expense;
- documentation to validate rate, compare against other rate options, and improve negotiation leverage for future rate contracts;
- the ability to validate utility bills and usage
- validation of sub usage and the ability to invoice metered tenants directly and non-metered tenants by computer generated estimates
- a powerful tool to identify and quantify viable energy-saving projects
- the streamlining and automation of energy management reporting including benchmarking, budget and cost analysis, peak load analysis, energy reporting, and energy project analysis
- a reduction in the time and effort spent identifying and addressing inefficient equipment, processes, and systems
- improved data visibility to support energy management applications, processes, and goals across the campus
- the ability to baseline each facility or group of facilities
- the ability to identify higher than expected energy costs in like facilities, and target these exceptions for appropriate corrective actions
- an independent source of data to quantify savings from existing ESPC projects and programs
- access to an archive of operating parameters, energy usage data, and other information that survives personnel changes
- streamlining and automation of energy management reporting
- reductions in equipment downtime, unscheduled equipment replacements, maintenance costs, and critical equipment replacement crises through faster identification of deterioration and abnormal usage; and
- assistance with maintenance implementation, validation, and scheduling.

CONCLUSION

An EIS can be an invaluable tool for the energy manager and the energy program overall, but only if it is properly applied and managed. Key elements of proper EIS management include a proactive team that is willing to change the existing management paradigm and ensure the proper distribution of information to all personnel levels and personnel with the skills to provide expert analysis of the system-provided information. Use of a web-based system for disseminating specific energy information targeted to the appropriate personnel increases the availability of such information by making it easier to review. With the right information, analysis, oversight, and supportive personnel, the team can drastically improve maintenance and increase efficiency, while simultaneously reducing energy expenditures. Combining this with a vigorous program of continuous commissioning promotes energy reduction sustainability along with increased performance and lifespan for systems and equipment. As high quality energy reduction projects continue to dwindle, an EIS can also provide direct savings and a powerful tool to identify and prioritize those projects that remain.

References

[1] Barksdale, George G. Background for Federal ESPC Initiative forum 19 April 2003
[2] Townsend, Terry E., Vice President ASHRAE, Presentation for Federal Sustainability Panel 26 March 2003
[3] Ball, Tim, Associate Director Facilities Management, California State University Long Beach, Presentation at California State—Long Beach on EEM system results for Cal State Long Beach system, Jan 2003
[4] California State and Consumer Services Agency/California Energy Commission. The Summer 2001 Conservation Report. February 2002.
[5] Barksdale, George, Results of Mechanical Equipment Maintenance Evaluation Team reports for 20 bases in Strategic Air Command, 1984/1985
[6] Sharma, Satish, Chief Utilities Privatization and Energy Team, Assistant Chief of Staff for Installation Management, Department of the Army; Briefing for Energy 2002, 13 June 2002
[7] Barksdale, George G., Surveys of DOD Installations by SAIC in 2002/2003
[8] Dixon, Doug, PNL study of FORSCOM installations 2001/2002
[9] Brown, Dirks, Hunt PNNL: Economic Energy Savings Potential in Federal Buildings, prepared for DoE/FEMP Sept 2000
[10] Shaw, Gilbert, Director-Federal Sales; Silicon Energy Experience for installed systems 2002
[11] EIA 1995 CBECS data table EU-1.
[12] Barksdale, George G.; Analysis of EIS customers and systems by SAIC 2002

Section Four

Data Collection and Data Input for EIS and ECS Systems

Chapter 10

The Case for Energy Information

Jim Lewis, Obvius Corporation

THIS CHAPTER DISCUSSES the benefits to a facility of gathering energy use information and explains some of the costs involved in collecting the information. It includes a discussion of the importance of submetering and describes submeter applications including cost allocation, accountability metering, efficiency monitoring, and measurement and verification. Examples of applications covered in the chapter are: accountability metering, tenant submetering, load curtailment/demand response, and web display of existing meter data.

INTRODUCTION

There's an old acronym in the business world that in order to ensure that goals are met, they must be SMART, meaning that they need to have:

- **S**implicity—the goals and tasks to achieve the goals must be easy to understand
- **M**easurability—if you can't measure it, how will you know if you are successful
- **A**uthority—control to affect the outcomes and achieve the goals
- **R**esponsibility—goes hand in hand with authority and makes someone accountable for the success or failure
- **T**imeliness—access to timely and accurate information is crucial to making changes to ensure success

Clearly, these same parameters can be applied to many energy management strategies, but in many cases key elements are left out of the process. Many building owners and managers embark on expensive and time consuming projects that are doomed to failure because some or all of the SMART requirements are either not met or are left to the last minute. How many energy managers have implemented energy projects based on elaborate (or back of the napkin) calculations for projected savings, only to see utility bills that are the same or higher than before the project began? The problem, of course, is that relying solely on utility bill analysis means the process:

- Is Not **S**imple—complex calculations must be applied to the data to try to massage meaningful information
- Is Not **M**easurable—most projects impact only part of the buildings systems while utility bills measure the entire building
- Lacks **A**uthority —total building energy consumption is impacted by almost every user to some extent
- Lacks **R**esponsibility—when everyone is responsible, no one is responsible
- Is Not **T**imely—typically analyzed weeks or months after the usage has occurred

So, why are so many energy managers spending thousands (if not millions) of dollars on projects without any means of measuring and verifying the results? For once, energy managers and energy service providers (ESPs) are all singing from the same hymnal: *measurement and verification of energy savings is too expensive*! The view espoused by many in the industry is that installing and managing hardware and software for data acquisition is simply an added expense and wastes dollars that could otherwise be used to buy more cool energy savings toys.

On the surface, this sounds like a reasonable response, but let's look at this argument from a slightly different perspective: imagine that a Fortune 500 company plans to invest $100 million in research and development of new products. The plan is to put $50 million into two separate technologies with good opportunities for growth, but some degree of uncertainty as to whether the technologies will work and be accepted by the company's customers. In other words, the outcome looks promising for both based on the projections and calculations of the business development people, but the outcomes are not certain. Now imagine that the same company commits to the projects, but decides not to have the accounting department track the progress of the projects to determine whether there is any return on investment. They are content to assume that the business development people did their calculations correctly, that the engineering team met the cost requirements, that manufacturing doesn't encounter any problems in building the new products and that the sales force meets the target sales plans.

Company executives might argue that measuring each project would require hiring some new accountants and maybe investing in some new software or hardware and if the projects meet the projections, these additional investments would be a waste of resources that could be spent on future R&D.

If you were on the board of directors of this company, how much money would you give them? How about when they come back next year with requests for two more projects, but can't give you any results from the prior years' investment?

More importantly, how much of your 401(K) would you like to risk on the management team of this company? Carrying it one step further, how would you feel if the manager of your retirement fund told you that she was investing in a number of different companies, but wouldn't be able to tell you how each was doing and that at the end of the year, hopefully you'd have more money. Unfortunately, if you did lose money, there wouldn't be any way to figure out which investments were good and which were bad so you'd just have to guess and hope things went better next year.

Providing timely and accurate feedback to the success of an investment is such an integral part of our personal and professional lives that it seems very inconsistent to accept anything less in energy information and projects. Near real time information about energy usage "behind the meter" serves to provide accountability for operations and allow for corrections to be made where appropriate.

Another major benefit to gathering energy information is that making someone accountable for energy usage almost always results in a reduction in that usage. If an employee or tenant feels that no one is watching or cares about energy use, the tendency is for that employee to become lax about turning off lights, shutting down computers, etc. Many studies have shown that energy consumption reductions of 5 to 10% commonly occur simply through submetering energy usage and making individuals responsible and accountable for energy.

If we agree on the value of timely and accurate information, we are still confronted with the issue of the cost of acquiring, storing and reporting the information. Historically, this process has been expensive primarily because there has not been hardware and software specifically designed to perform the data acquisition functions, so each installation has required a great deal of specialized integration of sensors and meters to accomplish the task of getting data.

In many cases, energy managers and providers have had to rely on building automation systems (BAS) or programmable logic controllers (PLC) to gather interval energy information. While these systems are capable of gathering the data, their primary focus is to accept inputs from a variety of sensors, calculate appropriate output parameters and implement those commands in a timely manner. Gathering and trending of status points (e.g., electrical meters) is, at best, a secondary function for these systems and consequently the cost of setting up these trending programs is high and the timeliness of the information is suspect.

Recently, a number of companies (including Obvius LLC in Portland, OR) have begun to develop solutions that are focused on the data acquisition needs of commercial and industrial customers. The AcquiSuite from Obvius is a stand-alone server that is preconfigured for many common inputs such as Modbus, plus analog and pulse inputs. This pre-configuration of drivers means that the system recognizes and configures many popular Modbus meters automatically with the only user input required being the naming of the point (i.e., no scaling or multipliers is required). This reduces the installation from hours to minutes and provides a much higher level of accuracy as input errors are minimized.

The second cost driver for energy information is the cost of communications from the local data acquisition server (DAS) to a host server. Since the DAS is a stand-alone web server, all the setup is accomplished by filling out forms using a standard web browser. There is no need for expensive software to configure the system or to set up or change communication parameters (upload times, etc.). The DAS comes with both a modem and an ethernet port and is capable of connection to a remote server via LANs or phone lines with options for direct dialing or connection to an ISP.

The last cost component for energy information is converting the raw data from the DAS into useful information. In the past, getting this information required the purchase, installation and support of proprietary software on a new or existing server. Sharing the costs of supporting a data warehouse was not possible and as a result, the cost of getting useful information was prohibitive. Several companies (including Obvius) offer web based services that provide access to energy information reports using standard web browsers for less than a dollar a day per building.

SUB-METERING FOR ENERGY MANAGERS

Managers of almost any kind of commercial or industrial facility have as all or part of their goals the implementation of an energy strategy. While these strategies take a variety of forms depending on the organization, the overriding purpose is to maximize the efficiency of operations so that the most productivity and profit is realized for the least amount of energy input. For many (if not most) managers, the first challenge in developing an energy policy is to determine how and where energy is being used within the building as most facilities simply rely on monthly energy bills as the measure of performance. The information gained from a monthly bill is generally not timely, nor does it provide indications of where within the building energy is being used and thus it is difficult if not impossible to even know where to start.

Historically, submetering of energy usage within the building has been an expensive and time consuming process due to the lack of hardware and software targeted to this application, so most users have simply avoided looking "behind the meter." Fortunately, recent developments in the energy information field have greatly reduced the time and expense required to gather and analyze data from one or many locations. This chapter will examine some of the ways managers with responsibility for energy can leverage existing systems and technologies to provide cost effective measurement of energy usage.

WHY SHOULD WE SUBMETER AT ALL?

The traditional view of submetering of energy has been that it is, at best, a necessary evil that accompanies investments in energy saving products and services. Today's energy managers realize that simply monitoring energy usage within the building and reporting that usage to the occupants can provide a significant return on investment. Typical submeter applications include:

Cost Allocation

Submeters are installed to provide more accurate information for charging tenants or departments for their energy usage. Cost allocation scenarios can range from very simple programs (e.g., dividing the total bill based on consumption) to very complex programs calculating coincident demand and power factor penalties in addition to consumption.

"Accountability Metering"

Similar to cost allocation applications, accountability meters are used not for billing or accounting purposes, but to identify and impact behaviors of the building occupants in meeting goals for energy usage

Efficiency Monitoring

Submeters and other sensors are installed to monitor the efficiency of specific systems such as industrial processes in plants or chiller plants in commercial buildings

Measurement and Verification

These applications are generally associated with energy improvement strategies and submeters are used to benchmark usage before changes are made and then to measure and verify the actual savings

COST ALLOCATION

In general, cost allocation can be used where a single meter serves a campus or large facility and there is a benefit to assigning the actual costs to tenants or departments or product lines. Typical customer uses include:

Industrial/Manufacturing

Many plants produce a number of different products and one key component of the cost of goods sold is energy used to produce the specific products. Cost of sales numbers and gross profit can be greatly distorted if one or more products require heavy energy input and the costs are simply divided by square footage or sales volume. In order to allocate energy costs accurately and make better operational decisions, manufacturers can install submeters for specific lines and provide a more accurate cost accounting

Retail

Many retail centers have multiple tenants that share common spaces or centralized services such as HVAC. Some users, such as restaurants and bars may have different operational hours and energy needs that are not adequately reflected in the base rental agreement and submetering can be a very effective tool for measuring energy use and for allocating the costs of shared services such as HVAC.

Hospitality

Hospitality providers (including hotels, theme parks, etc.) frequently have space that is rented out to independent businesses such as restaurants or retailers. In many cases, the rental agreement provides a fixed cost per square foot for energy that may be either too high or too low. If the cost is too low, the owner will be subsidizing the costs for the renters and potentially overcharging other tenants. In this case, submetering not only produces a more accurate means of allocating costs, but also provides significant information on the profit provided from these non-owned users.

Commercial

Most commercial real estate properties provide utilities and conditioned spaces during fixed operating hours and have no adequate means of allocating these costs to specific tenants. This means that the building owner has little ability to provide flexible scheduling for tenants who have either long term or short term needs for changes in schedule. It also means that all tenants are treated as having the same energy density and the operations of some tenants (e.g., restaurants, bars and data centers) may be subsidized by other tenants, resulting in a long term competitive disadvantage in the market for the owner of the building.

Campuses/Schools

Most colleges and universities have a single primary feed meter that supplies power to all the buildings on campus. Campuses typically also have one or more central plants providing heating and cooling to multiple buildings via underground lines. This type of installation makes it difficult to assign costs of operations to specific departments or to outside groups using the facilities on a rental basis. Installing electrical submeters and Btu meters on the individual buildings allows for costs to be allocated to groups using the campus facilities

ACCOUNTABILITY METERING

In accountability metering, the submeters are used not for accounting purposes, but to reinforce desired behaviors by providing "report cards" on the activities of the users of the facility. If there is no means of measuring the impact of energy savings activities, it is likely that employees will be less conscientious about managing energy wisely. A large hospitality group found that energy costs were reduced significantly (>5%) simply by implementing a program to submeter the rides and operational areas of their parks so that they could, for instance, determine whether lights and motors were being turned off when the park was not in operation. The results for various areas of the park are published monthly to allow managers to see which areas of their operations are meeting their objectives and which are not. The system also provides the capability to "drill down" into functional areas to isolate particular offenders.

EFFICIENCY MONITORING

Monitoring the efficiency of energy consuming systems such as chiller plants can not only provide valuable feedback about energy conservation and operational practices, but can also be used as an early warning system for maintenance and repairs. For example, if the tubes on a HVAC chiller plant start to plug, the amount of energy required to produce the same amount of cooling (typically measured in kW/ton) will go up. The system will continue to produce chilled water at the desired temperature, but because the efficiency of the heat exchanger is reduced due to buildup in the tubes, the compressor will have to run longer to produce the desired amount of cooling.

Monitoring large motors with current sensors can provide valuable information about bearings going out, insulation breakdowns or other mechanical deficiencies which result in more current being required for motor operation. In addition, a simple current switch can be installed on one leg of the motor to monitor runtime on the motor to determine maintenance schedules, filter changes, etc.

MEASUREMENT AND VERIFICATION

The majority of energy saving retrofit projects are, quite reasonably, implemented based on engineering calculations of the projected return on investment. As with any projections of ROI, much of what goes into these calculations are assumptions and estimates that ultimately form the basis for implementation. As the folks at IBM used to say, "garbage in—garbage out," which in the case of energy retrofits means that if any of the assumptions about parameters (run times, setpoints, etc.) are wrong, the expected payback can be dramatically in error. The establishment of good baselines (measures of current operations) is the best way to determine the actual payback from investments in energy and submetering is a key element in a baselining program.

Just as important as building an accurate picture of the current operation is measuring the actual savings realized from an investment. If there is no effective means of isolating the energy used by the modified systems, it may be impossible to determine the value of the investment made. Using monthly utility bills for this analysis is problematic at best since actual savings which may be achieved can be masked by excess consumption in non-modified systems.

Consider for example, a commercial office building with a central chiller plant with an aging mechanical and control structure that provides limited capability for adjusting chiller water temperature. To improve efficiency, the building owner plans to retrofit the system to provide variable speed drives on pumps for the chilled water and condenser water systems along with control upgrades to allow for chilled water setpoint changes based on building loads. In the absence of baseline information, all calculations for savings are based on "snapshots" of the system operation and require a variety of assumptions. Once the retrofit is completed, the same process of gathering snapshot data is repeated and hopefully the savings projected are actually realized. If the building tenants either add loads or increase operational hours, it is difficult if not impossible to use utility bills to evaluate the actual savings.

In contrast, the same project could be evaluated with a high degree of accuracy by installing cost-effective monitoring equipment prior to the retrofit to establish a baseline and measure the actual savings. While each installation is necessarily unique, building a good monitoring system would typically require:

- *Data acquisition server* (DAS) such as the AcquiSuite from Obvius to collect the data, store them and communicate them to a remote or local host.

- *Electrical submeter(s)*—the number of meters would vary depending on the electrical wiring configuration, but could be as simple as a single submeter (e.g., Enercept meter from Veris Industries) installed on the primary feeds to the chiller plant. If desired, the individual feeds to the cooling tower, compressors, chilled water pumps, etc. could be monitored to provide an even better picture of system performance and payback.

- *Temperature sensors (optional)*—in most installations, this could be accomplished by the installation of two sensors, one for chilled water supply and the other for chilled water return. These sensors do not provide measurement of energy usage, but instead are primarily designed to provide feedback on system performance and efficiency.

- *Flow meter (optional)*—a new or existing meter can be used to measure the GPM and calculate chiller efficiency.

This benefits of a system for actually measuring the savings from a retrofit project (as opposed to calculated or stipulated savings) are many:

- Establishing a baseline over a period of time (as opposed to "snapshots") provides a far more accurate picture of system operation over time and help to focus the project.

- Once the baseline is established, ongoing measurement provides a highly accurate picture of the savings under a variety of conditions and the return on investment can be produced regardless of other ancillary operations in the building.

- The presence of monitoring equipment not only provides a better picture of ROI, but also provides ongoing feedback on the system operation and will provide for greater savings as efficiency can be fine-tuned.

VIEWING AND USING THE DATA

Historically, much of the expense of gathering and using submetered data has been in the hardware and software required and the ongoing cost of labor to produce useful reports. Many companies (such as Obvius) are leveraging existing technologies and systems to dramatically reduce the cost of gathering, displaying and analyzing data from commercial and industrial buildings. Using a combination of application specific hardware and software, the AcquiSuite data acquisition server provides user interface using only a standard web browser such as Microsoft Internet Explorer.

The AcquiSuite DAS automatically recognizes devices such as meters from Power Measurement Ltd and Veris Industries, which makes installation cost effective. The installer simply plugs the meters into the DAS and all configuration and setup is done automatically with the only input required being the name of the device and the location of the remote server. Data from the meters is gathered on user-selected intervals (e.g., 15 minutes) and transmitted via phone line or LAN connection to a remote host where it is stored in a database for access via the internet.

To view the data from one or more buildings, the user simply logs onto a web site (e.g. www.obvius.com) and selects the data to view (see Figure 10-1).

INSTALLATION AND COSTS

It is, of course, difficult to generalize on the costs of submetering as factors such as the amperage of the service and wiring runs will vary greatly from building to building. A couple of examples may provide some rough estimates of typical installations.

The example illustrated below is for a single submeter on a 400 amp service panel.

The installation outlined in Figure 10-2 is for a project with these costs:

- One Obvius AcquiSuite data acquisition server (DAS)
- One 400 amp submeter (Veris model H8035)
- Labor to install the meter and DAS
- Wiring labor to connect the DAS to the internet

Assuming no extraordinary costs of installation, this project could be completed for less than $2,000.

If the data are sent to a remote server via the internet for display, the annual cost for the single AcquiSuite would be approximately $240 (note: the annual cost listed is per AcquiSuite, not per meter, so up to 32 meters can be monitored for the same $240 annual cost).

EXAMPLE ONE: ACCOUNTABILITY METERING

Accountability metering is a term used to describe the use of meters, sensors and software in commercial and industrial facilities to ensure that users are using energy efficiently. This process can be used to compare total usage for a number of different facilities or to mea-

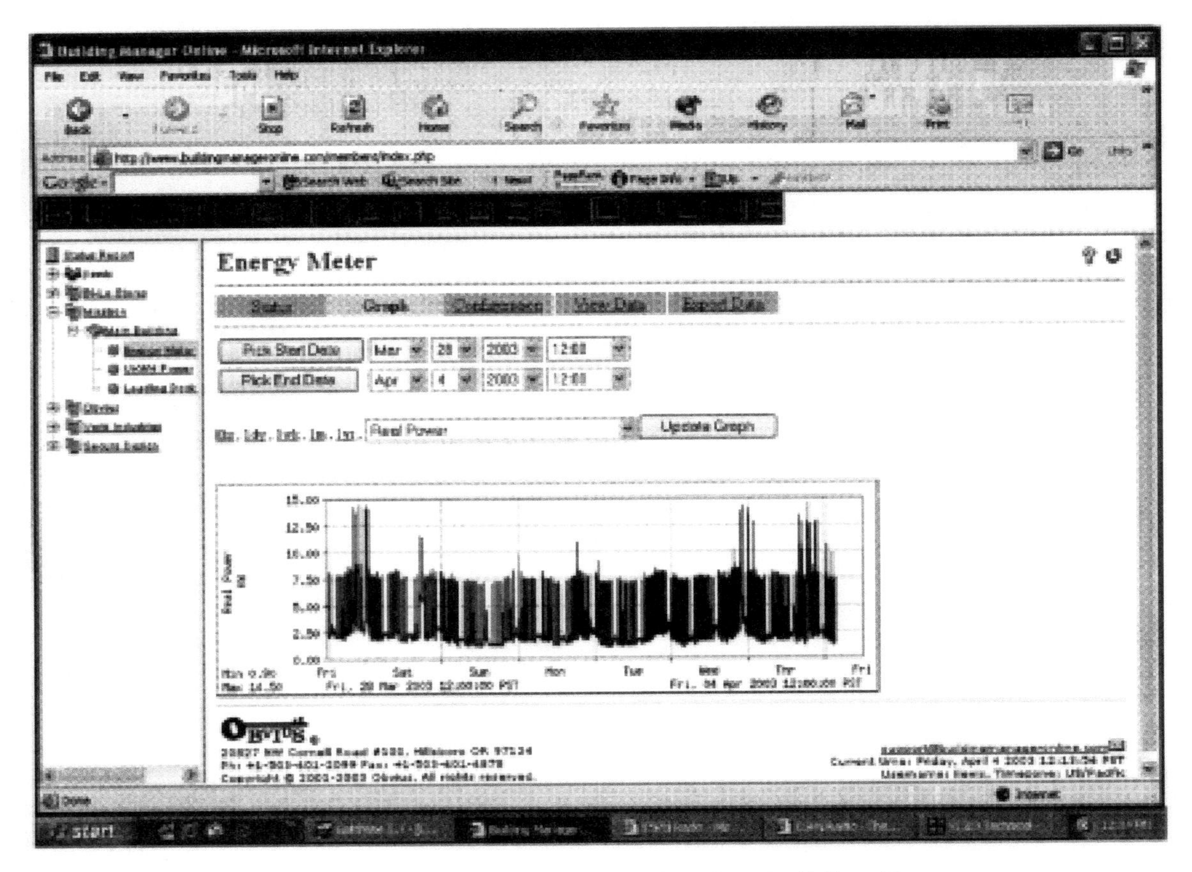

Figure 10-1. Sample Web Page Showing Building Data

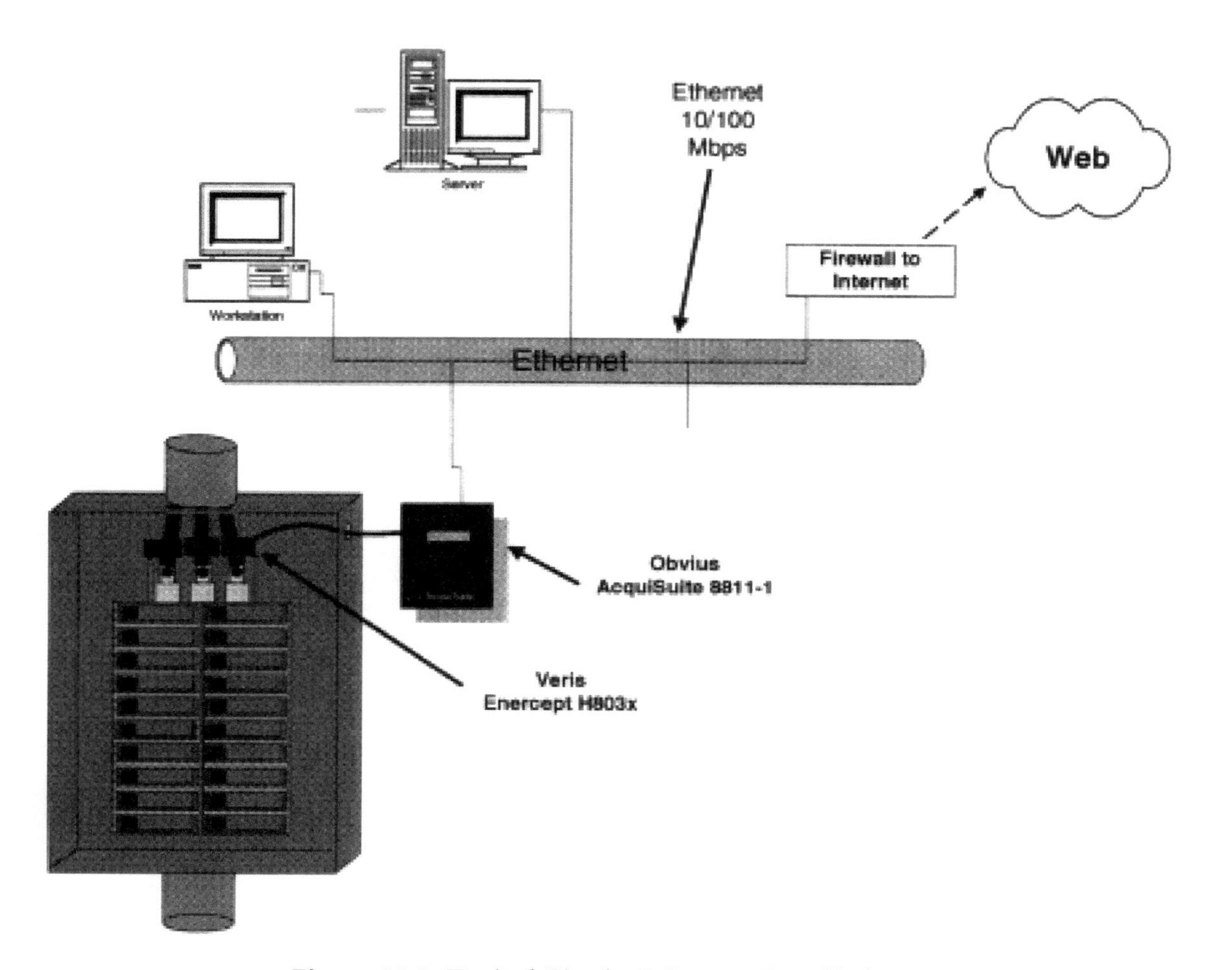

Figure 10-2. Typical Single Submeter Installation

sure energy usage within a single facility (a.k.a. behind-the-meter monitoring) or both. The purpose is to make users accountable for conforming to best practices in energy usage.

Background

Most building owners and managers have established procedures for employees, tenants and other users that are designed to minimize the energy usage in a building (e.g., turn off lights in unoccupied spaces, keep thermostats set to reasonable levels), but lack the ability to monitor whether these procedures are followed. In most cases, users who are not held accountable (financially or otherwise) will, over time, fail to follow through on the procedures and energy will be wasted.

How Does it Work?

The building owner or manager installs equipment (meters, sensors, etc.) that monitors energy usage either for the entire building or parts of the building. The equipment gathers the data on a regular basis and uploads the information to a local or remote database server. This database server can then produce regular or custom reports that provide feedback on the performance of the building occupants in meeting energy goals and following procedures.

Benefits

Many studies have shown that simply metering and sub-metering the energy use and providing accountability for the users will produce savings of 5% to 10% annually. Once building users are aware that the owner has the means to verify that procedures are being documented, they will adjust their behavior to conform to the energy saving goals of the owner. In many cases it has been demonstrated that not only will the users change their behavior to meet expectations, but in fact will actively seek opportunities for additional savings, particularly when these activities are rewarded with incentives for the additional savings (reduced rent, bonuses, etc.).

Drawbacks

Accountability metering requires some investment in hardware, software and time. It may also be viewed as intrusive by some employees or tenants as it may be perceived as "Big Brother" invading their privacy.

Installation Requirements

As with most energy information applications, the specific hardware and software required for any project will vary depending on the systems to be monitored and the level of detail required from the software. At a minimum, the installation in each building to be monitored will be:

- *AcquiSuite data acquisition server (DAS)*—a stand-alone web server located on the building site that communicates with the sensor(s), stores interval information and communicates with the remote server
- *Pulse output* from an existing primary energy meter (electric, gas, water or steam) or a "shadow" meter installed behind the primary meter(s) to provide data to the DAS
- *Sub-meters (optional)* to monitor energy usage for physical areas of the building (departments, operations, HVAC, lighting, etc.)
- *Sensors (optional)* to monitor runtime or energy consumption for specific equipment
- *Phone line or local area network (LAN) connection* for communication with the remote server
- *Software or services* to provide standard and customized reports

Reports

In general, the reports for accountability metering fall into two categories:

On-line Customized Reports

On-line customized reports provide daily information on consumption and demand to users and managers. The purpose of these reports is so that those responsible can monitor their performance in near real-time and be able to fine tune operations. A typical example is shown in Figure 10-3 which shows an actual internet page depicting a week's demand profile for one meter.

This page can be viewed with any web browser (such as Internet Explorer) and shows the peak demand for one week. This information can be used to determine what operations within the facility are contributing to

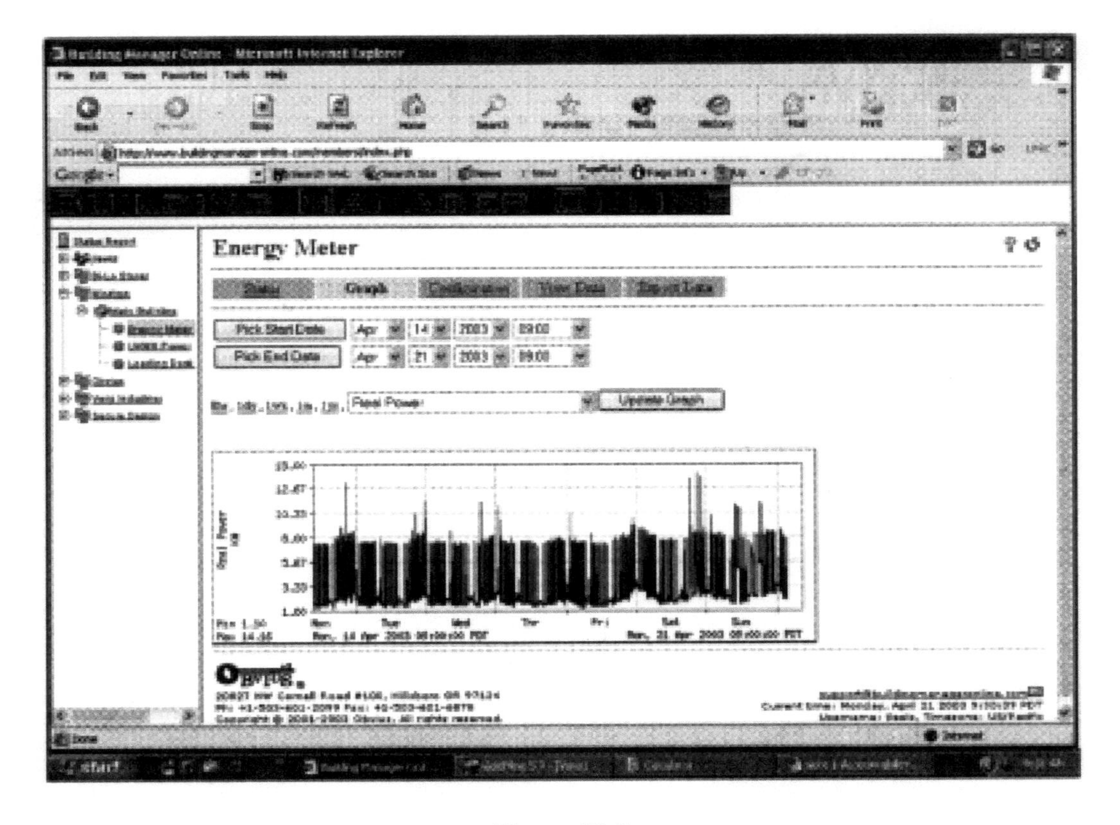

Figure 10-3

the overall energy cost. In the case of buildingmanageronline.com, clicking on the graph "zooms" in to allow the user to determine precisely what time critical power consumption occurred and relate it directly to activities and systems in the building (see Figure 10-4).

Clicking on a point in the graph zooms in to a single day.

Regular Monthly Reports

Regular monthly reports provide a more standardized mechanism for reporting usage and the success (or lack thereof) of individual users or facilities at meeting stated objectives. These monthly reports might be published as a web page or distributed via email to provide accountability. For example, assume that XYZ company has established a corporate goal of 5% reduction in energy this year. A typical monthly report to company managers is shown in Figure 10-5.

Actions

On receiving the report displayed in Figure 10-3, the western regional manager has a great deal of information on which to act. First, the manager knows that while her region is ahead of the 5.0 % goal for the year at 5.26%, the month was only slightly over 3.0% and a repeat for another month will likely push her below plan (and cost her bonus money). More importantly, she can instantly determine that the cause of the excess energy usage for July 2002 is a significant increase in Seattle, a location that has performed well year-to-date. This could signal a change in operating hours, significantly higher temperatures, or failure in the control or mechanical system. Clicking on the hyperlink to the Seattle branch would provide additional insight on run times for equipment and average outside air temperature for the month, potentially valuable data for further analysis. Either way, a call to the manager of the Seattle location is probably in order.

Over the longer term, the regional manager (in conjunction with the corporate energy manager), will not only be able to change the behavior of the employees, but will also be able to identify targets for energy studies and investment. If a location has a consistently higher cost per square foot than other locations, it is a likely candidate for further study and potential installation of energy saving equipment.

Costs

As indicated above, the costs for this service will vary widely, depending both on the nature of the instal-

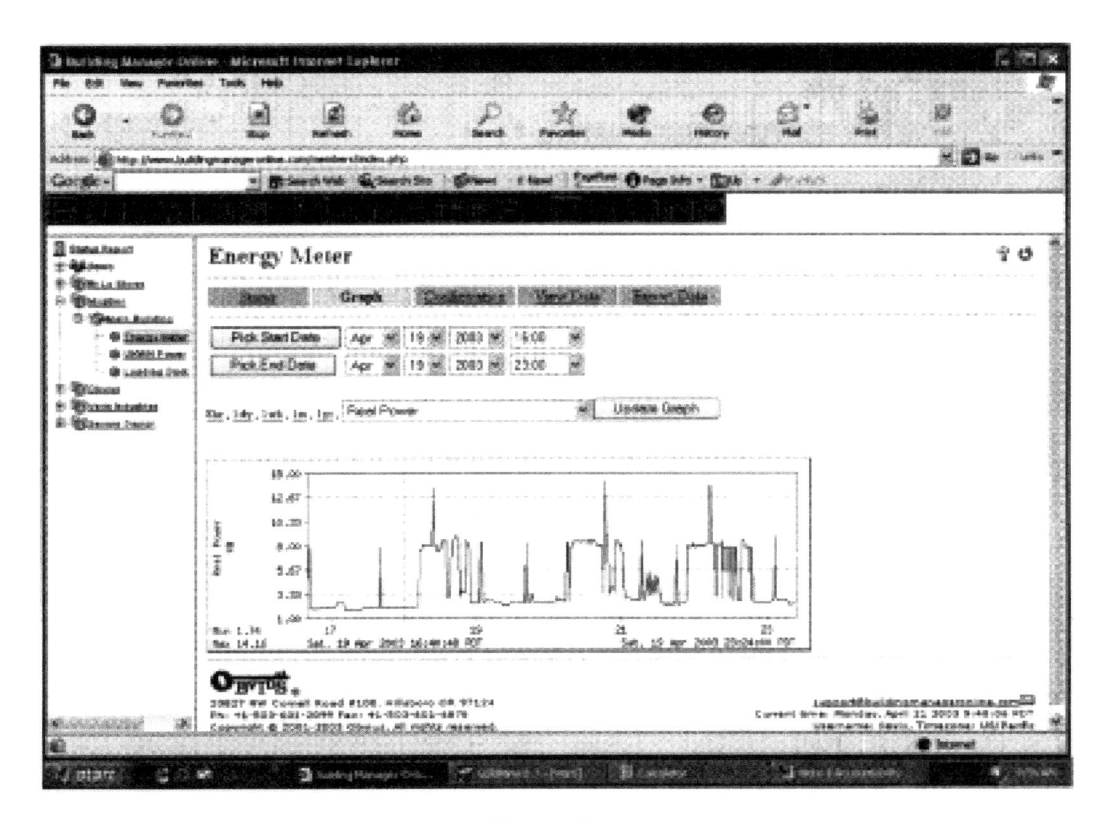

Figure 10-4

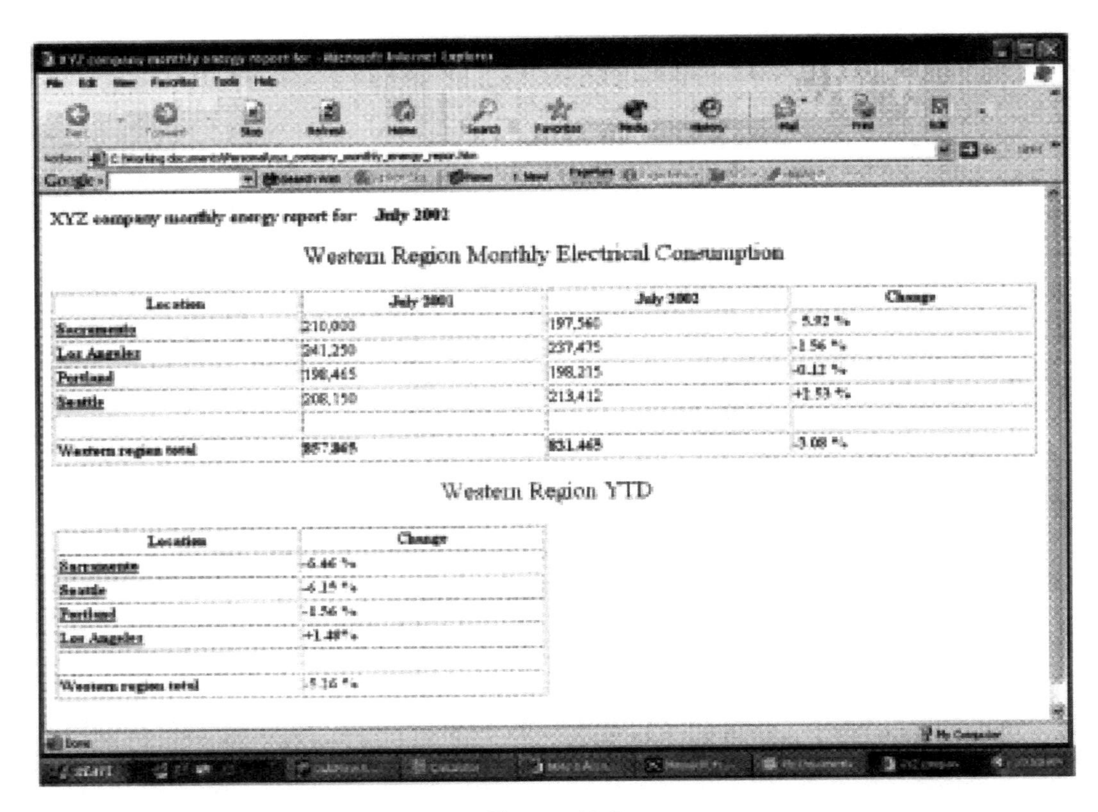

XYZ company monthly energy report for: **July 2002**

Western Region Monthly Electrical Consumption

Location	July 2001	July 2002	Change
Sacramento	210,000	197,560	-5.92 %
Los Angeles	241,250	237,475	-1.56 %
Portland	198,465	198,215	-0.12 %
Seattle	208,150	213,412	+2.53 %
Western region total	857,865	831,465	-3.08 %

Western Region YTD

Location	Change
Sacramento	-6.46 %
Seattle	-6.15 %
Portland	-1.56 %
Los Angeles	+1.48 %
Western region total	-5.16 %

Figure 10-5

lation and the level of detail desired. If we consider a simple installation:

- AcquiSuite server for data acquisition
- Connection to an existing utility pulse output for electricity
- Connection to an existing utility pulse output for gas
- Connection to an existing utility pulse output for water
- One electrical sub-meter
- One temperature sensor

The cost of hardware and installation labor would be less than $2,000 and the ongoing monitoring cost would be approximately $20 per month.

Notes/miscellaneous

The cost of extensive sub-metering (typically several hundred dollars per electric meter) is often prohibitive, particularly for energy managers with dozens or hundreds of facilities. For these applications, it can be cost effective to deploy a two-tiered approach with primary metering in all facilities and extensive metering of sub-systems (e.g., HVAC, lighting) in selected buildings. For example, a retail chain with 150 stores nationwide would likely find the most cost-effective solution to be primary metering of gas and electric in all stores, with extensive sub-metering in perhaps 5 or 10 stores that serve as "models" with similar systems and operations.

Information gathered from the primary meters would be useful in determining how particular buildings perform (for example, energy density and cost per square foot) and for benchmarking usage for later programs. The model stores would provide valuable insight into not just how much was used, but where. Using these model stores, the energy manager can determine what percent of the energy use is heating and air conditioning, lighting, operations, etc. and this information can form the basis for allocating retrofit dollars. If the manager finds that, say, 45% of the electrical energy consumption is in lighting systems, he or she would be wise to devote time and energy to lighting technologies and less on energy efficient motors, fans etc.

Summary

An effective energy management plan begins with an assessment of the current energy situation and the tools are readily available and cost effective in today's world. Using these tools, it is possible to make users accountable for implementing energy strategies and to raise the level of awareness of energy consumption as it relates to operational activities. Accountability metering is the vital first step in the development of an effective energy management program.

EXAMPLE TWO: TENANT SUB-METERING

Tenant sub-metering is a broad term applied to the use of hardware and software to bill tenants in commercial facilities for their actual usage of energy. The goals of tenant sub-metering are: 1) to ensure that the owner recovers the cost of energy from tenants, and 2) to make sure that tenants with high energy usage are not subsidized by those with lower usage.

Background

Many buildings are equipped with only a primary metering system for measuring and billing energy consumption for the entire building. In buildings with this configuration, the tenants are typically billed for energy usage on either a fixed rate (cost per square foot) that is built into the lease, or the bill is allocated to tenants based on their square footage. Each of these methods have inherent flaws, but both share the common problem that energy costs are unlikely to be accurately charged to the tenants. Under a cost per square foot arrangement, the owner will almost certainly collect more or less than the actual bill, and the discrepancy will be even greater during times of energy volatility or low occupancy rates. If the bill is simply divided amongst the tenants on a per square foot basis, tenants with lower energy density (Btu per square foot) will subsidize the space costs for those with higher energy density (e.g., data centers).

These errors become particularly acute when there is a wide variance in occupancy schedules (retail vs. office space) and the building provides central services such as chilled water or conditioned air. Additional complexity is added when the building owner must make decisions in advance on the cost to add when the rate structures for commercial buildings are taken into account.

Residential electrical customers typically pay a flat rate for electrical consumption in the form of cost per kilowatt-hour ($/kWh) which makes calculating a bill relatively simple: read the kWh from the meter and multiply by the $/kWh to get a cost. Commercial rate structures are much more complex. Commercial rate structures typically have the following components:

- Consumption (kWh) charge—this part of the bill is basically the same as the residential charge, but usually has multiple tiers so there is not a single fixed cost. The owner pays at different rates for the amount consumed (e.g., the first 100,000 kWh is billed at $0.07, the next 100,000 kWh is billed at $0.065). These costs will also generally vary by season, so there will commonly be a winter rate and summer rate depending on the supply and demand for electricity that the utility experiences. For purposes of sub-metering, these costs can generally be blended into an average cost per kWh.

- Demand (kW) charge—since the utility has to be certain that adequate supplies of power are available, large customers such as commercial properties are billed not only for the total energy consumed during the month, but also for the maximum power used during a short interval (typically 15 minutes). The demand charge is used to help pay for the costs associated with having generating capacity to meet the highest period of demand (hot summer days, for example) that is not required during lower periods of demand. The demand charge is also applied because the utility will typically have to bring on less-efficient generating plants to meet the peak loads and thus the cost of generating goes up.

- Power factor charge—In a perfect world, the electrical energy provided to a device (e.g., a motor) would be converted to mechanical energy with 100% efficiency. Unfortunately, with very rare exceptions, this is not the case and the inefficiencies associjated with this energy transfer mean that the utility must provide more power than it can actually bill for (the actual math is quite complicated and beyond the scope of this chapter, but it's true). This inefficiency is usually just bundled into the other charges along with things like line loss, etc., but many utilities will bill customers with a power factor penalty if the power factor (the measure of efficiency) falls below a certain level (commonly 92% to 95%).

- Other charges—In addition to things like taxes and surcharges, most utilities are studying or have implemented rate structures that are intended to more directly reflect the actual cost of generation and have large users bear more of the burden during peak times. Things such as time of use metering, load curtailment penalties, etc. will only serve to complicate the average commercial electrical bill even more as time goes on.

How Does it Work?

In the simplest sense, the owner installs meters to monitor the consumption of electricity, gas, water and steam by individual tenants. These meters are connected to a data acquisition server (DAS) like the AcquiSuite from Obvius. The DAS gathers data from the meters on the same schedule as the utility supplying the building and then communicates this data to a local or remote database such as *http://www.buildingmanageronline.com* (BMO). The tenant is then billed at the end of the month at the same rate the building owner pays for the building as a whole and thus the owner recovers the cost of energy from each tenant.

While this process seems very straightforward, as with most things, the devil is in the details. Using the Obvius hardware and BMO service, the gathering, storing and reporting of the data is relatively simple and in most cases the information can be available with just a few hours of installation time. Analyzing the data and producing accurate billings for tenants can be considerably more challenging due to the different rate structures and billing components in a typical commercial setting as outline above. Options for allocating these costs will be considered in the "Actions" section below.

Benefits

The most obvious benefit is that tenants pay their fair share of the energy costs and the owner does not get stuck with unrecoverable costs. An often overlooked, but extremely important benefit is that the building owner is not placed at a competitive disadvantage in the marketplace. If the cost of serving high energy density tenants is spread over all the tenants, the total cost per square foot of leased space goes up and the owner may lose new or existing tenants to lower cost competition.

Drawbacks

There are two key issues to consider before implementing a tenant sub-metering program:

- **Costs**—depending on the layout of the services in the building, the cost of sub-metering may be high, particularly for utilities like water and gas where pipe cutting and threading may be involved.

- **Regulatory agencies**—in many jurisdictions, the state Public Utilities Commission (PUC) regulates the ability of building owners to charge tenants for energy consumption to prevent the owners from overcharging "captive" users. It may be difficult for owners to implement tenant sub-metering programs and to recover the costs of setting up and managing these programs.

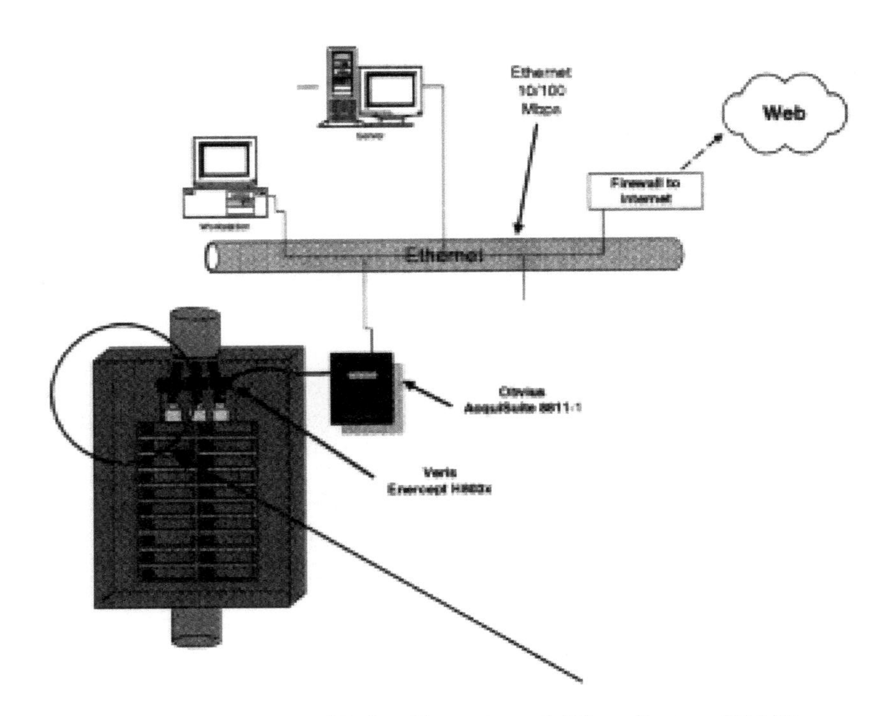

Figure 10-6. AcquiSuite Data Acquisition Server (DAS)

Installation Requirements

Details and costs of installation will naturally be heavily dependent on the layout of the building and utility services, but in general most applications can be met with the following hardware:

- *AcquiSuite data acquisition server (DAS)*—a stand-alone web server located on the building site that communicates with the sensor(s), stores interval information and communicates with the remote server.

- *Electrical sub-meter(s)*—several companies (including Power Measurement, Ltd. and Veris Industries) produce electrical meters designed for sub-metering applications. These meters can be simple pulse output devices or can provide information using serial communications to provide additional information such as power factor, current and harmonics.
- *Flow meter(s)*—flow meters are used to measure the volume of flow of gas, water and steam. There are a variety of technologies that can be employed, but typically these meters produce either pulse or analog signals that can be read by the AcquiSuite and converted to billable units of measure (gallons, therms, etc.)
- *Btu meter(s)*—Btu meters combine flow meters with temperature sensors to measure the actual energy usage for chilled or hot water. These meters can be useful in cases where a central chilled or hot water plant serves multiple tenants as the tenant is billed not only for the water consumed, but also for the input energy required to produce the conditioned water. These meters can provide simple pulse or analog output to the AcquiSuite or can provide more sophisticated analysis via serial connections

Reports

The level of complexity of reports depends on the method chosen for the tenant billing. The data from the BMO site provides all the information necessary for calculating tenant bills except for the rate structure information. In the simplest scenario, the owner simply downloads the consumption data from BMO for each tenant and allocates the total bill cost to each tenant based on his or her proportionate consumption (kWh). For most applications, this simple process provides the most cost-effective solution that distributes the cost fairly without creating a complex and expensive process for administration.

The standard reports from BMO provide all the information necessary to do a more thorough allocation that incorporates the demand (kW) charge and power factor penalty (if applicable). The back end processing required to accurately allocate demand charges can be significant as the owner and tenants (and potentially the PUC and utility) must agree on the mechanism used for allocating demand costs. While it is relatively simple to determine the peak interval for the billing period and compare the demand for each tenant for that same interval, the actual allocation of this cost (known as coincident demand) can be very difficult and time-consuming (see "Analysis/Actions" section below).

Analysis/Actions

In the case of simple allocation based on consumption outlined above, the owner imports data from BMO

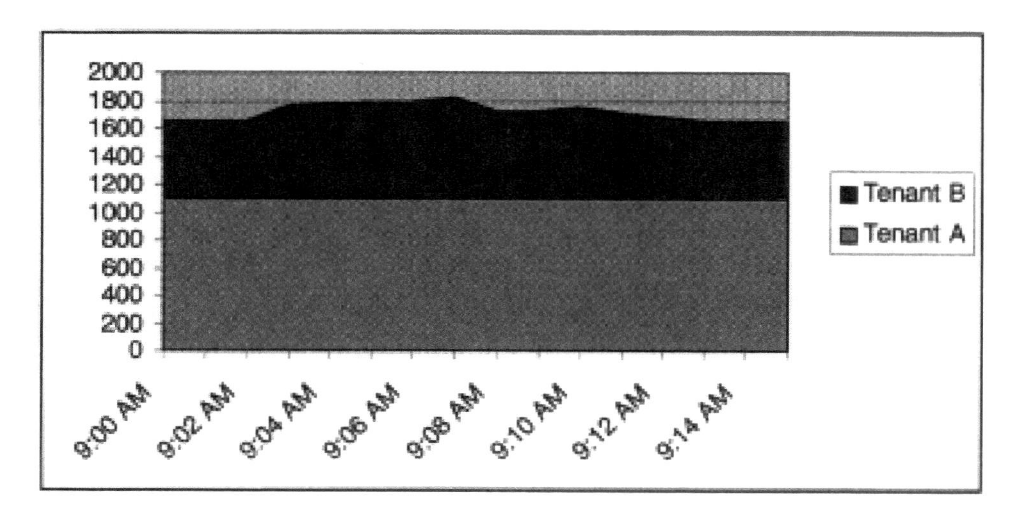

Figure 10-7

into a spreadsheet (or other cost allocation software) and the software generates a bill for the tenant that is added to the monthly rent.

Billing for coincident demand and time or use charges becomes more complex because there are judgment issues involved as well as simple quantitative analysis. Does the tenant with a flat constant demand (e.g., data center) have to absorb the additional penalties for tenants with highly variable rates? Does the tenant whose use is relatively low, but has incremental demand that pushes the total building into a higher demand charge have to absorb all the additional costs or are those costs spread among all tenants?

Notes/Miscellaneous

Simple tenant sub-metering is a relatively easy process that can be implemented by the building owner with the assistance of the providers of the hardware and software. More complex analysis is best left to consultants and resellers who specialize in rate engines and billing analysis.

Summary

Tenant sub-metering is a valuable tool for commercial property owners who want to accurately allocate the costs of energy to tenants and occupants, but it is extremely important to define the scope of the program up-front and to do the homework necessary to ensure compliance with leases and regulatory agencies such as the PUC.

EXAMPLE THREE: LOAD CURTAILMENT/ DEMAND RESPONSE

Load curtailment (or demand response) programs offered by utilities provide commercial and industrial building owners with reduced electrical rates in exchange for an agreement to curtail energy use at the request of the utility. Typically, these requests come during periods of high load such as hot summer afternoons. Building owners or managers who have the ability to reduce loads by turning off equipment or using alternative sources of energy can realize significant savings under these programs.

Background

Utility rate structures typically provide residential, commercial and industrial customers with fixed rates for energy regardless of the generating cost. Not surprisingly, these utilities use the most efficient (lowest cost) generating plants (e.g., nuclear and hydroelectric plants) for the bulk of their load and only bring on less efficient generation (e.g., older coal and gas-fired plants) as load requirements increase. Because of the essentially fixed price of energy to the customer, using less efficient resources has a negative impact on the utility's earnings and they would like to have alternatives.

For the utility, the best option at certain cost levels is to not bring on additional inefficient generating capacity and many utilities find that it is more cost effective to pay customers to curtail loads. If enough customers reduce their usage, the utility does not have to add generation (or purchase additional supplies on the spot market). This compensation can take several forms, but it generally is reflected in a lower overall rate schedule for the owner throughout the year.

How Does it Work?

Load curtailment can take a variety of forms depending on the severity of the shortfall in supply and the type of agreement between the utility and the end user and the equipment in place to implement the reductions. In the simplest form, the utility notifies the owner of a curtailment request (typically a day in advance) and it is up to the customer to voluntarily meet the requested load reduction. Options for the end user range from adjusting temperatures to shutting off lights to closing facilities to meet the requested reduction levels.

The future of demand response is likely to contain more options for automatic, real-time reductions in load, triggered directly by the utility with little involvement of the owner. This option allows for the matching of loads much more closely to actual demand levels in real time, but obviously requires much higher levels of automation and investment. In this scenario, the owner and utility agree in advance what steps can be taken to lower the energy usage in the facility and the utility can initiate load reduction measures remotely using the customer's control system or additional controls installed in the building.

Benefits

For the utility, the primary benefits include:

- Eliminating the cost of bringing another plant on line
- Providing more cost-effective generating sources (i.e., more profit)
- Minimizing the environmental impact of generating plants with poor emissions records (fossil fuel plants)

For the customer, the clear benefit is a reduced cost of energy in the near term and avoiding cost increases in the future since the utility is at least theoretically operating more efficiently. There may also be an additional benefit from the installation of controls and equipment in facilities that provide the user with more information and control over the operation of the facilities during periods when load curtailment is not in effect.

Drawbacks

From a financial perspective, both the utility and customer are likely to incur costs to add or retrofit controls and equipment in the customer's facility. Both must also commit ongoing resources to track and manage the operation of the load curtailment and to provide reports. The customer is also likely to experience inconvenience in the form of less comfortable space temperatures (i.e., higher in summer, lower in winter) than desired if HVAC equipment is shut off, or reduced lighting levels. This kind of program will clearly impact the customer in some respects and these effects need to be maintained within acceptable limits.

Installation Requirements

Most facilities with installed building automation systems (BAS) already have the equipment in place to meet day-ahead requests by using the BAS to initiate new or pre-programmed operational strategies to limit energy use. The primary installation of new equipment could include additional metering and also likely some form of remote monitoring equipment to allow the utility to monitor the success of the program at reducing loads in the building.

Customers without existing BAS systems or those participating in real time demand response programs will have to make additional investment in monitoring and control systems (for example, remote setpoint thermostats). In most cases, the most cost effective way to implement a real time program is to use the internet and web-enabled data acquisition servers (DAS) like the AcquiSuite from Obvius to provide real time feedback to the owner and the utility of the load before and after curtailment. The DAS can also function as the conduit for the utility to provide a supervisory signal to the BAS or to the systems directly.

Reports

The information needed to implement and evaluate the effectiveness of a load curtailment program is near real-time interval data (kW). This report (either for a single facility/meter or an aggregated load from multiple locations) might look like the following figure (Figure 10-8).

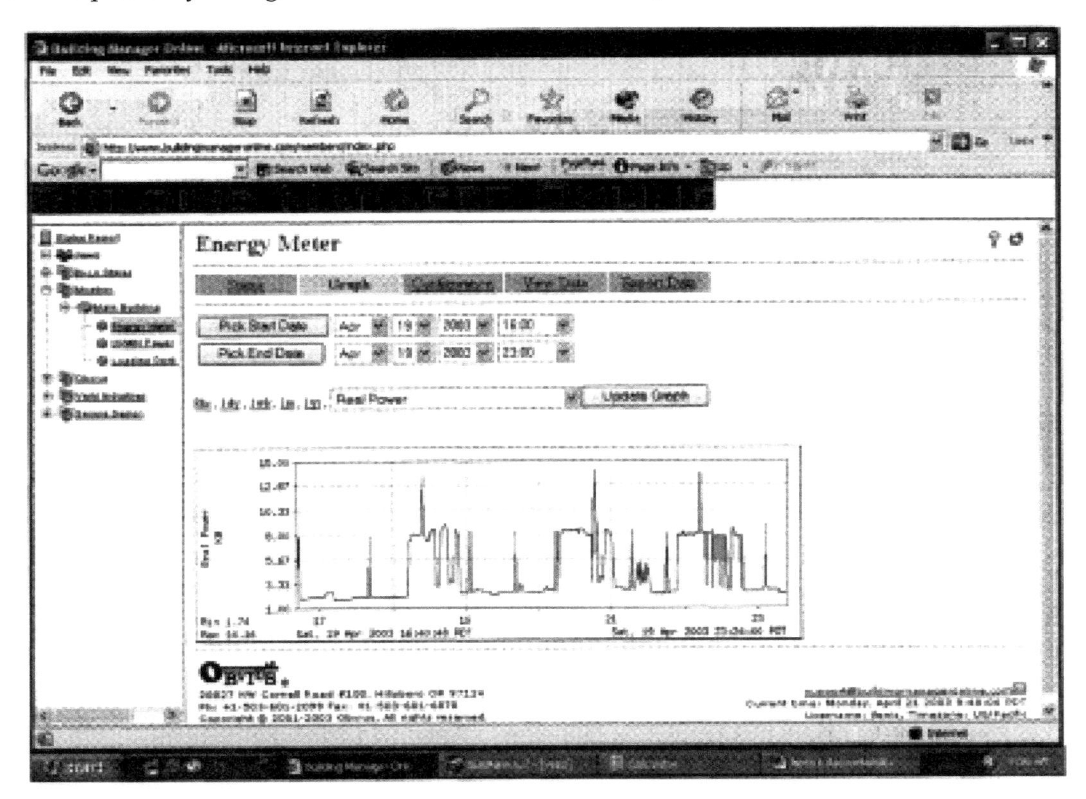

Figure 10-8. Near Real-time Interval Data (kW)

Analysis/Actions

The utility determines that there will be a shortage of available power (or that there will be a need to bring additional generating capacity on line) and informs users with demand response contracts that they will be expected to reduce their demand to the contracted levels. Users are expected to meet these requirements through some combination of automatic or manual shutdown of equipment, temperature adjustment or closing down some or all of their operations. The DAS is used to verify that the user has met the required load curtailment and that the utility has achieved its objectives for taking load off the grid.

Costs

As with all the application notes in this series, it is very difficult to estimate costs due to a variety of factors (wiring distances, communications issues, scheduled shutdowns, etc.), but some general guidelines for costs (hardware and installation) are:

- AcquiSuite™ data acquisition server—$1,200 to $1,800
- Electrical sub-meter (3 phase)—$600 to $1,000
- Data storage and reports—$20 per month per AcquiSuite

Notes/Miscellaneous

To date, implementation of demand response programs has been limited for a variety of reasons:

- Costs to monitor and control energy consuming equipment in buildings are high
- Implementation of these types of programs can be complicated, particularly if the utility desires some form of automated control via the internet
- Occupants have to be willing to except some inconvenience (e.g., higher temperatures) in order to meet curtailment needs
- Options for on-site generation (cogeneration, microturbines, fuel cells, etc.) are in the early stages and not cost effective for many owners
- Low overall energy costs provide limited incentive for negotiating curtailment contracts for many customers

All this notwithstanding, the future for load curtailment contracts is very promising. Improvements in building equipment (e.g., variable speed drives and onsite generating systems) combined with more cost effective internet based data acquisition and control hardware and software greatly reduce the cost and impact of implementing demand response programs. DR programs are likely to become more prevalent and building owners would do well to stay abreast of developments in this area.

EXAMPLE FOUR: WEB DISPLAY OF EXISTING METER DATA

Using either the AcquiSuite™ or AcquiLite™ data acquisition servers (DAS) from Obvius to connect existing electrical, gas or flow meters to the web. Once web-enabled, meter data are available for viewing from any web browser at www.buildingmanageronline.com.

Background

Many owners of commercial and industrial (C&I) buildings have meters installed in their buildings that either provide or could provide outputs (either pulse or serial) that will allow users to see energy information on the web. Many of these meters were installed by the utility for primary metering or as submeters within the building to monitor usage. Examples of meters that are compatible with the DAS from Obvius include:

- Pulse output from any utility meter with a pulse output sub-base;
- Pulse output from submeter;
- E-MON—any submeter with a pulse;
- Veris Industries
  - — H8053
  - — H8035
  - — H8036
  - — H8163 (with comms board)
  - — H8075
  - — H8076
  - — H8238
  - — H663
  - — H704
  - — H8126
- Power Measurement Ltd.
  - — ION 6200
  - — ION 7300
  - — ION 7330
  - — ION 7350
  - — ION 7500
  - — ION 7600
- Siemens
  - — 9200
  - — 9300

— 9330
— 9350
— 9500
— 9600
- Square D
— H8163-CB
— H8076
— H8075

Other meters using Modbus RTU may be compatible with the DAS, contact the factory for information or questions.

How Does it Work?

Each AcquiSuite DAS can support up to 32 Modbus meters and 128 pulse meters and each AcquiLite can support up to 4 pulse meters. There are two basic connection approaches, depending on the type of output available:

Pulse Output Meters

The two wires from the pulse output from the meter (electric, gas, water, steam) are connected to one of the four pulse inputs on the AcquiSuite (A8811-1). The installer uses a web browser to name the meter and add the appropriate multiplier to convert each pulse to valid engineering units. The DAS is then connected to the internet (either via phone line or LAN connection) and data are pushed to the web site for viewing. (See Figure 10-9.)

Modbus RTU Meters

Devices from the list above with a serial output will be automatically recognized by the AcquiSuite as soon as they are connected. The DAS has drivers for these devices that will recognize and configure the device in unit, so the installer only needs to give the meter a name and setup the parameters for reading interval data from the meters and uploading to the remote server.

Regardless of whether the input is pulse or Modbus, the DAS gathers data on user-selected intervals (from 1 to 60 minutes) and stores the data until it is uploaded to the BMO server (typically daily). Once uploaded, the data from all the meters is available for viewing from any web browser (see "Reports" below for sample reports).

Benefits

Many C&I building owners have installed submeters with local display options that are also capable of providing pulse or serial outputs, but the meters have never been connected to a local or remote server due to cost or other constraints. Having only local display means that someone must physically read the meter, record the values and input this information into a spreadsheet or database for calculation. This approach is not only inefficient, but is also prone to error. It is virtually impossible to synchronize readings with the utility bills, which means that accurate accounting is unlikely.

Because the AcquiSuite automatically recognizes supported Modbus meters, installation can be done by any electrician or local building personnel without the need for expensive software and integration.

Using a DAS provides many benefits, including:

- Continuous interval reading makes synchronizing to utility bills simple
- Data from multiple, geographically dispersed buildings are automatic as all data come to a single site for viewing
- Information can be viewed and downloaded from any web browser in spreadsheet and database compatible file formats
- All data are stored at a secure site and record-keeping is minimized
- The DAS can be programmed to call out on alarms via email or pager in the event of a problem

Drawbacks

The major obstacle to this approach is that it requires some investment of time and materials to connect the meter(s) to the DAS and to provide phone line or LAN connection for communications.

Installation Requirements

The requirements for installation depend on the type of installation and whether new meters are being installed. Generally the only requirements for connection to one or more existing meters are the following:

- *AcquiSuite DAS*—used for Modbus or pulse meters, can support up to 32 Modbus meters or up to 128 pulse meters
- *AcquiLite DAS*—used for pulse meters only, supports up to 4 pulse inputs
- *Phone line (can be shared) or LAN connection* for communications

Reports

Once the data from the various buildings are uploaded to the BMO web site (*http://*

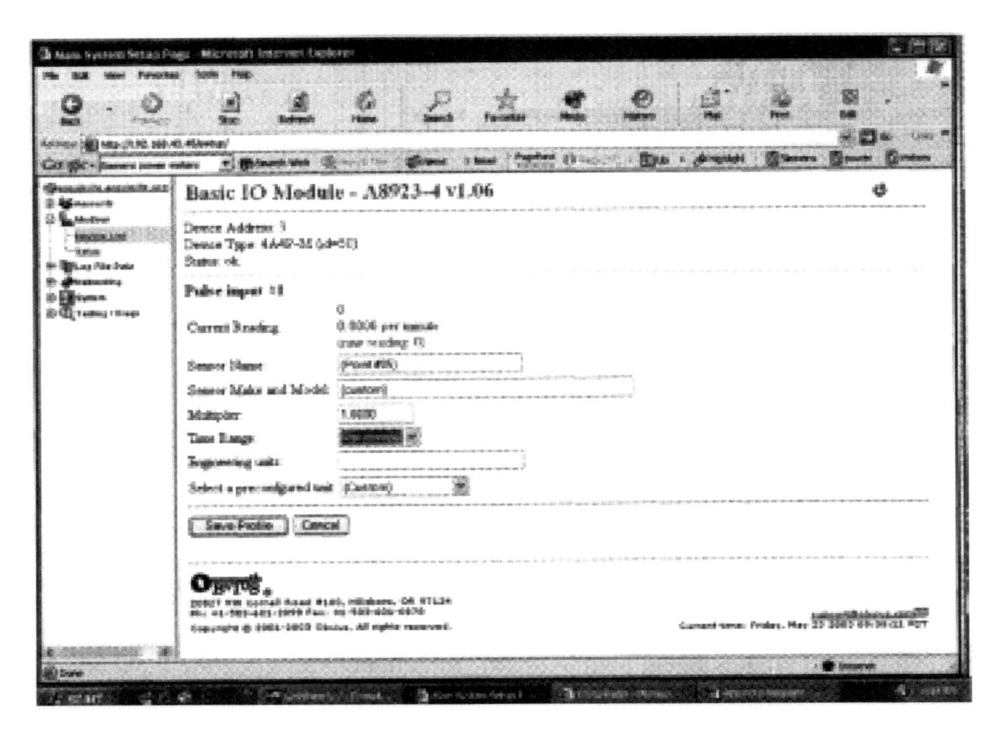

Figure 10-9. Pulse Input Setup Screen

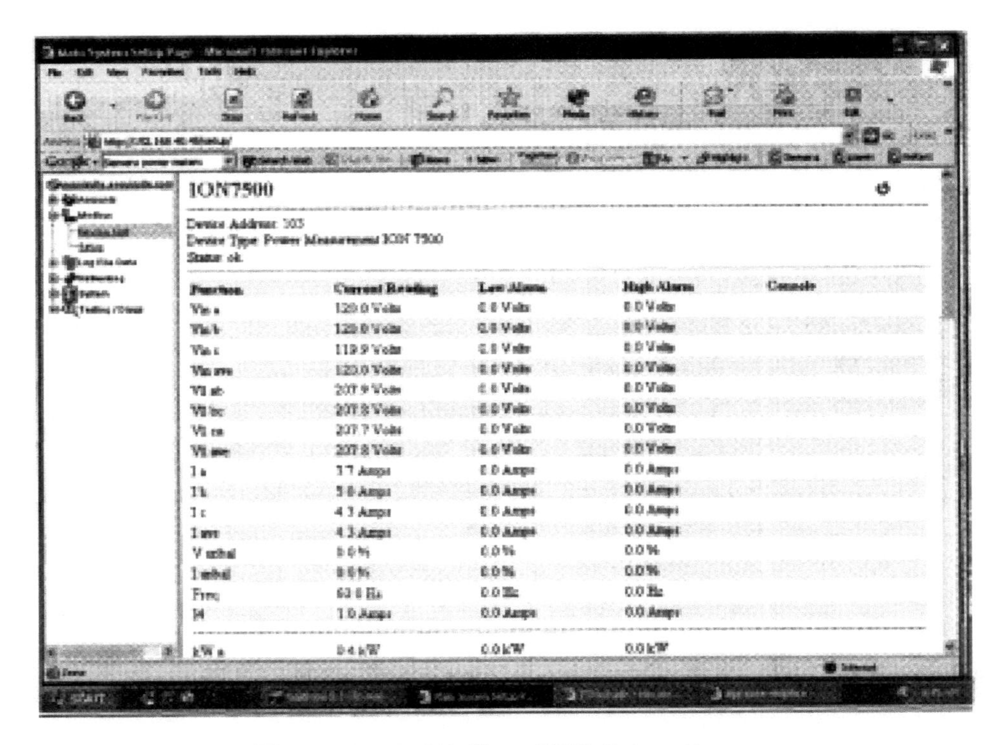

Figure 10-10. Modbus RTU Setup Screen

www.buildingmanageronline.com), they can be viewed using any standard web browser.

In addition to viewing the data from a web browser, users can also download the data in a file format compatible with spreadsheets or databases:

Analysis/Actions

Once the data are exported from BMO to a local spreadsheet, the submeter energy usage can be allocated to the tenant or department.

Costs

Typical installed costs will vary depending on the specific requirements of the job (wiring runs, number of meters, etc.), but in general the installed cost for the DAS will be in the following range:

- AcquiLite™ DAS—$500 to $600
- AcquiSuite™ DAS—$1,200 to $1,800
- Data storage and reports—$20 per month per AcquiSuite or AcquiLite™ (***NOTE:*** *the cost is the same no matter how many meters are connected to a DAS*)

Notes/Miscellaneous

As this chapter shows, it is both practical and economical to add web display capability to existing meters from both local and remote sites. It is important to note that the building owner or manager who wants to gather data from existing meters can also add new submeters to existing buildings at the same time and spread the cost of the installation over more points.

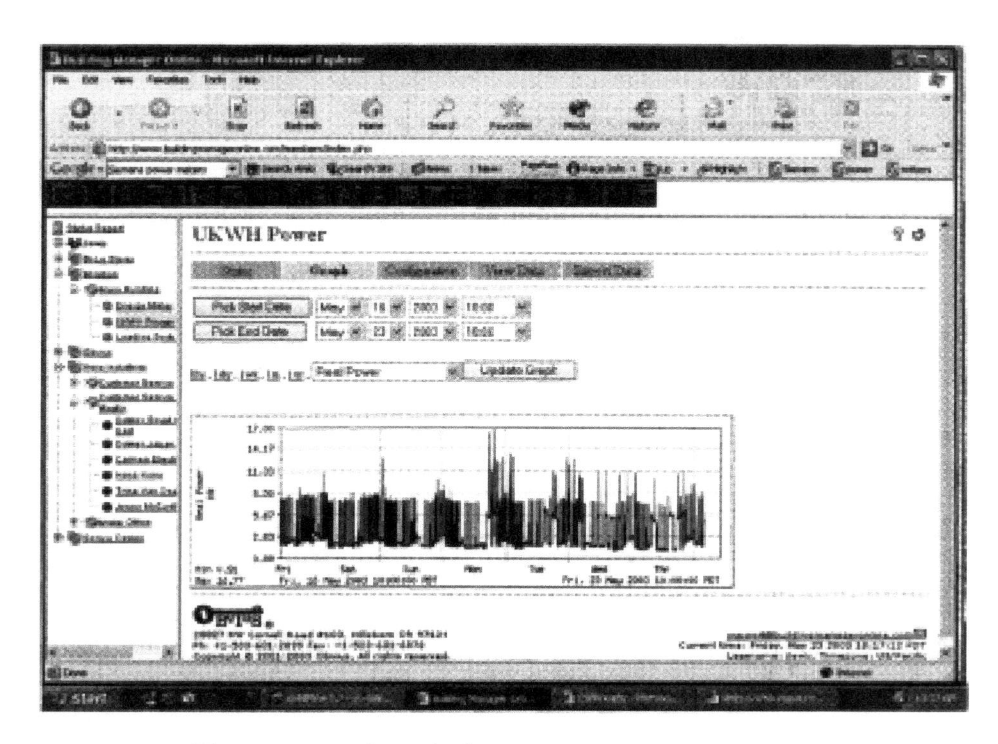

Figure 10-11. Sample kW Report from BMO Site

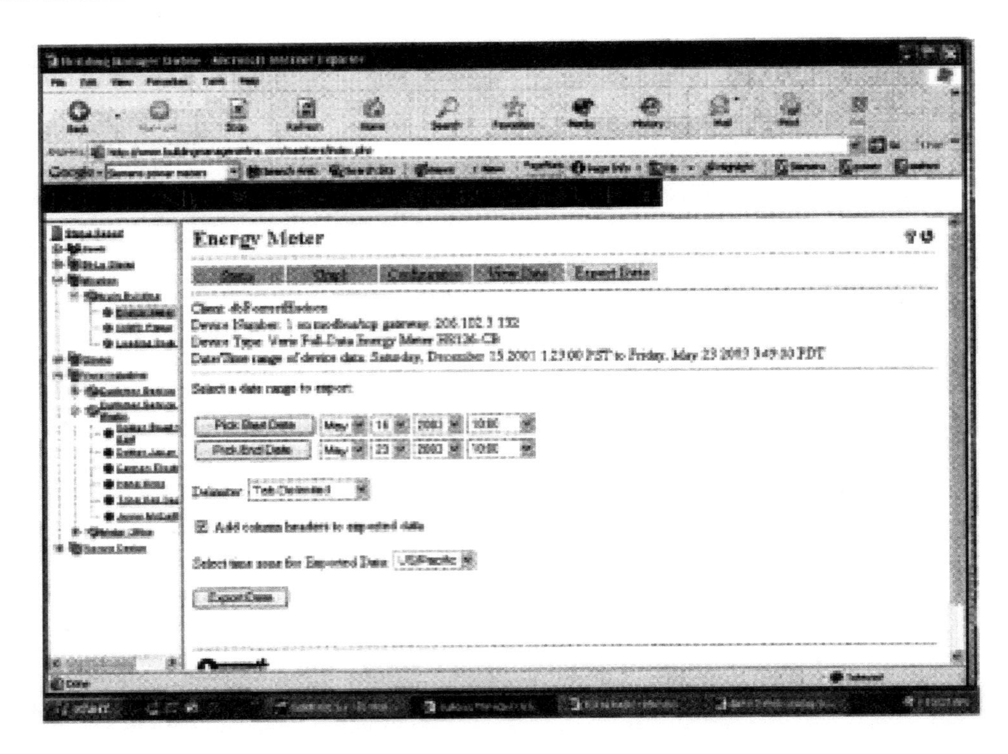

Figure 10-12. Export setup page from BMO site

Chapter 11

Using the Web for Cost-effective Energy Information

Jim Lewis and Steve Herzog, Obvius LLC

THIS CHAPTER ADDRESSES two key technological advances that have significantly reduced the cost and complexity of accessing energy information from remote facilities. The first key is the development of new hardware and software by Obvius, LLC, that provides "plug and play" functionality from the PC world to the field of energy data acquisition. This functionality is achieved through the development of a Linux-based data acquisition server (DAS) that automatically detects Modbus meters from major manufacturers such as Power Measurement, Ltd., and Veris Industries. This feature dramatically reduces the time and costs required to set up and configure software for collecting information from devices in the building.

The second key technology advance described in this chapter relates to the use of the web as the communications channel for data transfer and presentation. Using the existing infrastructure of the web and common tools such as Internet Explorer, users can now cost effectively gather and view data from multiple remote sites on a daily or even hourly basis. The chapter explores how this technology is employed and provides sample reports and outlines of the installation requirements and setup of the systems.

INTRODUCTION

Most of us remember the good old days of personal computers when adding peripherals like printers or video cards was a real challenge, virtually guaranteeing a fun-filled weekend of reboots, reloads and foul language. Standards like USB and "plug and play" have taken much of the joy and challenge out of the task, making it possible for even the novice to simply roll up to CompUSA, select the device that meets their requirements, take it home, plug it in and configure it.

In the old days, the computer user wishing to add a new video card had to either:

1. Pay an exorbitant fee to a local computer shop, or
2. Face the task of doing the installation alone.

Doing the installation alone meant accomplishing the following endeavors (in many cases repeated several times):

- Select an appropriate card by reading the fine print on the side of the box listing the "Minimum system requirements"
- Find an available (and compatible) slot in the PC
- Set the jumpers on the board for an unused IRQ
- Install the board and reboot the computer
- Try to figure out why the computer refuses to reboot and fool with a variety of IRQ's and memory settings
- Once the computer finally boots up, load the appropriate software from the floppy disk using archaic command lines and reboot the computer
- Try to figure out why the computer refuses to reboot and fool with a variety of software settings
- Remove the card from the system and start over again

Sound familiar? It probably does if you have contemplated embarking on an energy information project in the past few years. The options available are basically to select a vendor to supply a total system solution of hardware and software or to cobble together a solution on your own. Vendor solutions generally work, but are typically expensive and any changes or adjustments require hiring the same vendor to come to the site and add hardware or software pieces. The systems integrator selects the hardware (meters, sensors and controllers) and also provides the software and the labor to produce a working solution.

Choosing to do it alone provides a daunting challenge as it requires the user to learn about everything from serial protocols to databases with a healthy dose of hardware knowledge thrown in for good measure. The successful installer will emerge with a vocabulary expanded to include baud rates, CRC's, DHCP, gateways, multipliers, form C's, CSV's, SQL queries and countless other arcane terms sure to provide riveting conversation at cocktail parties.

The optimal solution to this dilemma is for the energy information industry to follow the example of the PC world and provide the equivalent of "plug and play" solutions for energy and operational information. In this scenario, the end user or local contractor would be able to select off-the-shelf hardware for monitoring energy usage, temperatures, flows, etc. and all of the devices would be automatically recognized and configured by a local data acquisition server. This would provide a simple installation where the installer would only have to provide logical names for the connected points and set communications parameters. Data from many buildings would be automatically uploaded to a remote server and published on the internet for ready access without the need for any additional software or plug-ins.

Seems perfectly reasonable, but the EIS world has not evolved to this level due to a number of obstacles:

1. A lack of industry standards, particularly serial communication protocols
2. A wide variety of connection options, including RS485, wireless, power line carrier and ethernet
3. Concerns over security of data on local area networks (LAN) if data are moved to the internet
4. Complexity and expense of systems integration, particularly mapping and configuring points in hardware and software
5. Problems with installing and operating software on a variety of platforms and operating systems
6. Difficulty and expense of handling large quantities of data

Despite these challenges, a number of companies have initiated development projects for hardware and software focusing on the gathering of energy information for commercial and industrial users. The approaches taken by these companies varies, with two fundamental development approaches: 1) Create flexible hardware and software platforms which can be adapted at the site to connect to a variety of sensors and automation systems, or 2) build application specific products which allow some flexibility in configuration, but focus on a narrower range of connection options.

The major advantage to the flexible option is that the hardware and software can generally be configured to meet almost any installation by providing a wide variety of communications interfaces and drivers. While this approach makes the selection of systems components much easier, it also drives up the cost of the installation significantly as the labor required to configure and map the existing sensors and systems increases dramatically. This cost trade-off may be acceptable for installations where a key requirement is the ability to communicate with existing building automation systems (BAS) or other control systems in large facilities.

Several companies, including Obvius (Portland, OR) have opted for a strategic focus that leverages existing technologies to provide low cost, do-it-yourself installations for building owners and contractors. These products provide the industry a package of hardware and services to allow the building owner to see near real time energy and operational information from one or many different locations using only a standard web browser such as Microsoft Internet Explorer™ or Netscape™. This approach not only allows access to data from any internet workstation, but also eliminates the cost and hassle of installing and maintaining software on different hardware and software platforms.

The biggest drawback to the simplified approach is that it does not allow as much flexibility for on-site systems integration and thus may not provide the installer with as much latitude for sharing data with existing control systems (e.g., BAS).

If the user chooses to implement a "plug and play" solution, there are two components that go into providing a simple, cost effective energy information system:

1. Hardware for on-site installation that makes configuration and implementation simple, and
2. Software which provides a clear and simple means to view data

Each of these components will be examined with a focus on the technologies employed by Obvius in commercial and industrial facilities with particular emphasis on the use of existing technologies to provide energy information.

On-site Hardware and Installation

Successful EIS begins with the installation of hardware on the building site to allow connection with existing and new meters and sensors. The AcquiSuite system from Obvius is typical of the emerging solutions and is a Linux based web server which provides three basic functions:

- Communications with existing meters and sensors to allow for data collection on user-selected intervals
- Non-volatile storage of collected information for several weeks
- Communication with external server(s) via phone or internet to allow conversion of raw data into graphical information

The backbone of the system is a specially designed web server. A typical server has the following components:

- **Processor**—386 based microprocessor
- **Memory**—8 MB non-volatile flash memory
- **Operating system**—Linux
- **LAN connection**—Ethernet RJ45
- **Display**—2 × 16 LCD
- **Analog inputs**—4 (any combination of 4 to 20 mA or 0 to 10 Vdc)
- **Digital inputs**—4 (connection to any mechanical or electronic pulse output)
- **Serial connection**—up to 32 Modbus RTU devices via RS 485 port

The data acquisition server (DAS) provides connectivity to new and existing devices either via the on-board analog and digital inputs or the RS 485 port. The analog inputs permit connection to industry standard sensors for temperature, humidity, pressure, etc. and the digital inputs provide the ability to connect utility meters with pulse outputs. The serial port communicates with Modbus RTU devices such as electrical meters from Veris, Square D and Power Measurement Ltd.

The installation ease of these types of servers is accomplished by a set of drivers stored in the server for commonly found meters and sensors. To achieve the required performance, a common protocol (e.g., Modbus RTU) is implemented on the serial port and meters or other devices using Modbus RTU can be profiled into the server. When a device is connected to the server, the server automatically detects the type of meter connected and loads the appropriate driver for communication. Using this design approach, meters and sensors from a variety of vendors can be supported on the serial port without the need for point mapping of each device.

Because the device is a web server, setup can be accomplished without the need for additional software or plug-ins as all configuration and setup is accomplished with a web browser (e.g., Internet Explorer or Netscape). The server can be located in a stand-alone environment with communication via a modem or can be installed and connected to an existing LAN using dynamic host configuration protocol (DHCP) or a static IP address assigned by the network administrator. The LCD provides a convenient means of configuring the network address and verifying the status of the network and serial devices connected to the DAS.

Once the DAS is configured with an IP address, the start-up and configuration is accomplished using a computer either connected directly to the DAS via an ethernet cable or through the LAN. Configuration of the system is done through a series of menu selections that allow the installer to select from configuration choices to customize the installation using a web browser. The DAS serves HTML form pages for the user to complete the configuration of the system. The user can make selections including:

Figure 11-1. Obvius Data Acquisition Server (A8811-1)

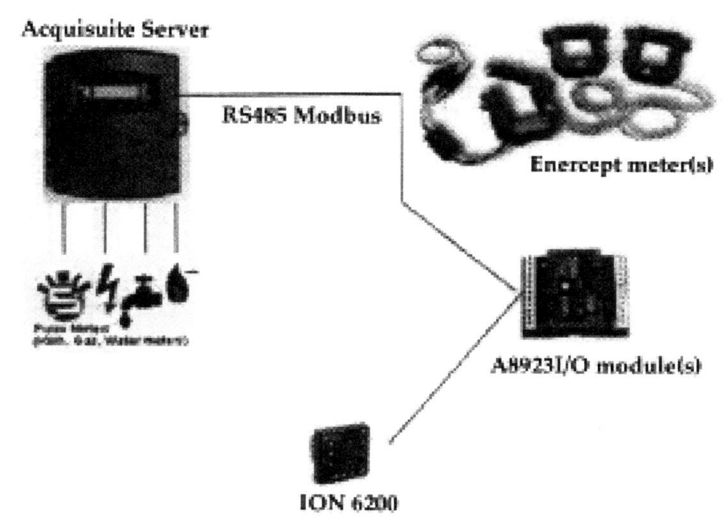

Figure 11-2. Typical AcquiSuite Installation Using RS 485 Modbus RTU Devices

- Interval sample times for the connected devices and sensors (typically from 1 to 60 minutes)
- Configuration parameters for devices or sensors that are not automatically recognized by the system
- Alarm setpoints for connected devices
- Upload schedules
- ISP information (if the modem connection is used for web access)

BASIC FUNCTIONS OF THE DAS (DATA ACQUISITION SERVER)

Regardless of the physical installation, the DAS typically performs 3 functions:

1. Reading data from connected devices—based on user-selected intervals from 1 to 60 minutes, the DAS will poll all connected devices and log the data retrieved into a file. If the device is a profiled Modbus RTU meter, for example, the DAS can log not only basic kWh information, but also other data points such as power factor, current, reactive power, etc. Each of the registers on the device is logged as a separate value for upload to a host server later. Raw data (for example, pulses) is converted to meaningful engineering units by the DAS and stored as a floating-point number.

2. Non-volatile storage of data—to insure that no data are lost due to power failures or other interruptions, the DAS provides non-volatile storage of data, typically using flash memory (e.g., disk on chip or compact flash).

3. Communications with a host computer—the gathered data are not particularly useful unless it can be viewed and analyzed by users. A key component of any DAS is the ability to make a connection with a remote database server to provide storage of historical information from one or more DAS's. In addition to uploading data, the DAS and remote server can use a communication session for critical

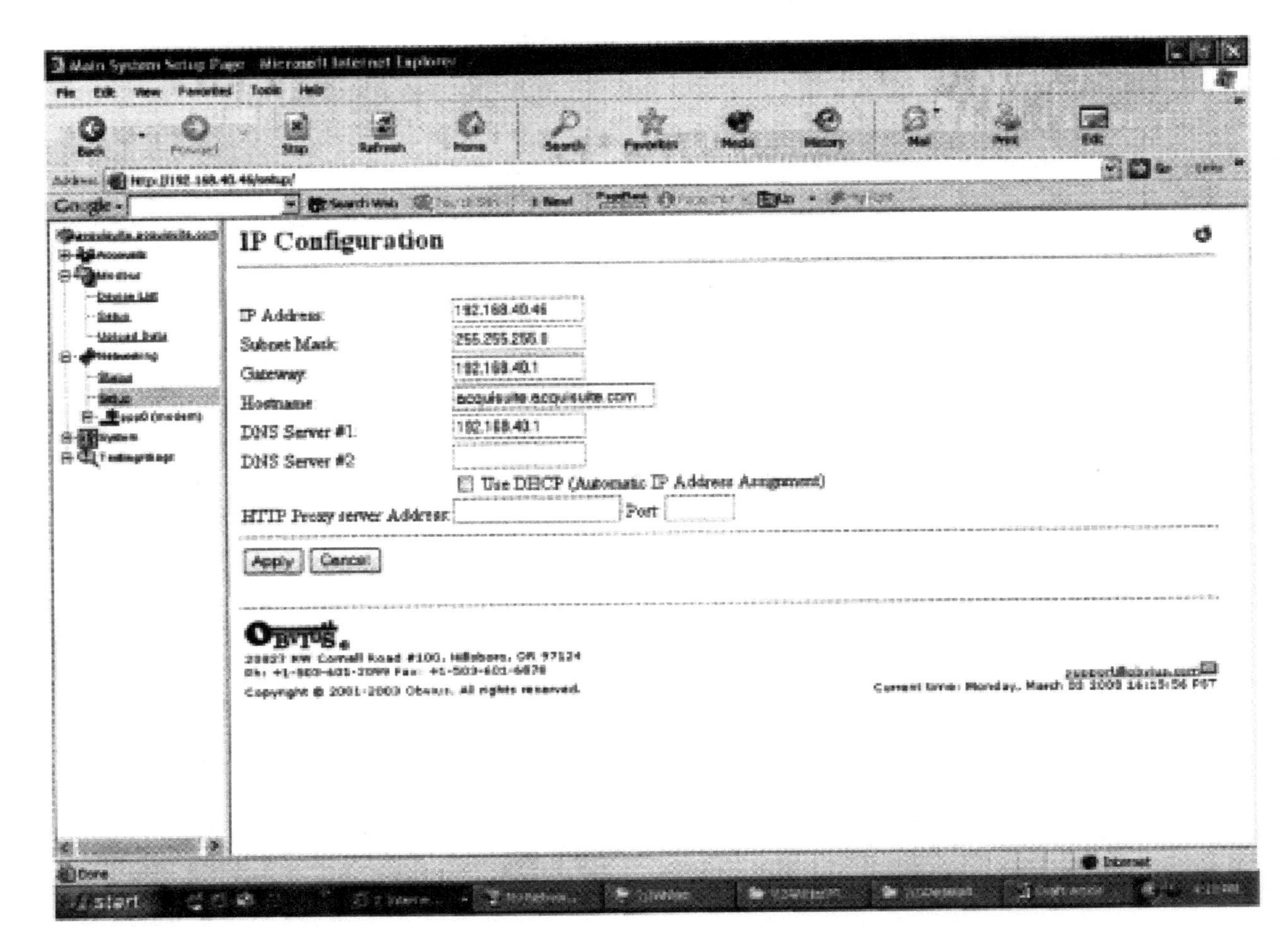

Figure 11-3. Typical Setup of the DAS Serving Web Pages

housekeeping duties such as synchronizing the internal clock on the DAS with the remote server. Many systems (including the AcquiSuite) are also capable of downloading firmware updates via the internet allowing for all the deployed systems to maintain the most current firmware and to add drivers for new devices. Firmware in the DAS functions similarly to the software loaded on a PC and the downlaod feature allows the user to maintain the latest version of the firmware without having to connect to the DAS or buy upgrades. In addition, the DAS can also have parameters (alarm limits, for instance) changed by the remote host based on inputs from users.

LOCAL AREA NETWORK INSTALLATION AND OPERATION

In many campus environments, there is an existing network infrastructure to provide users with access to internal computing resources and to allow users to browse the web and send email. The most economical means of communicating with the DAS is to utilize the existing network, providing local and external channels for uploading data from the DAS to a remote server for display and analysis. Historically, it has been difficult to convince network administrators to allow connection of third party devices to the LAN, particularly when the server is intended to provide communications with servers outside the network. As one would expect, IT administrators are reluctant to permit unregulated devices on their networks, both because of concerns over security and a fear that these devices may absorb significant bandwidth and cause traffic problems for regular users of the systems. From the perspective of the EIS user, the typical historical method of data transfer would be to use the file transfer protocol (FTP) to upload data from the DAS to the remote server. Unfortunately, the prospect of having a server on the private LAN sending large volumes of unregulated files through the firewall strikes justifiable fear into the hearts of network managers, so there are two fundamental approaches being taken to address the concerns of the network administrators:

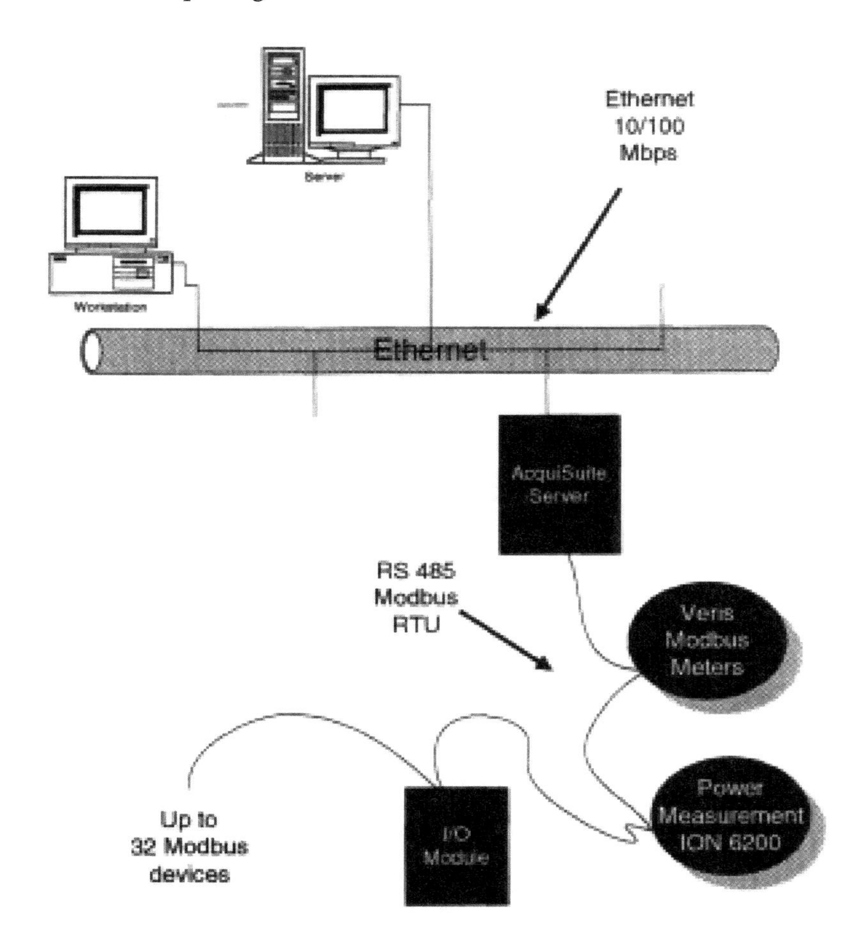

Figure 11-4. AcquiSuite Server on an Existing LAN

- **Email of collected data**—bundling the data into an email virtually eliminates the security questions as the outbound emails pass through the networks existing mail server and can be queued with other emails, eliminating the issue of heavy data traffic on the system. Using this method does have drawbacks, however, notably the lack of feedback from the receiving server to the DAS. Since email is inherently a one-way communication, failure of the email to reach the remote server and be successfully interpreted can't be communicated back to the DAS. In many cases, because the DAS does not get feedback on successful uploads, all the data in the system must be sent each time.

- **Using HTTP commands for communication sessions**—Using this approach, the DAS initiates a call out to the remote server in just the same manner as any web browser on the network. Since the session appears to the LAN and firewall like any other web session, the network administrator can provide the same safeguards (e.g., firewalls) to protect the system and regulate the flow of traffic and access to other connected computers on the LAN. The primary drawback to this approach is that unless the network administrator provides access to the DAS from outside the LAN, all communication sessions must be initiated by the DAS and remote connections directly into the DAS from outside the LAN are not possible. It is highly unlikely that any administrator concerned about data security and system performance would allow access from outside the LAN. The major advantage to the HTTP approach is that the DAS makes a two-way communications link to the remote server and thus can confirm that the data was uploaded successfully.

In a typical LAN installation, the DAS is programmed to initiate an HTTP call to the remote server on a pre-determined schedule (generally once a day). Once the connection to the remote server is made, the DAS sends login information to the remote server that tells the server which DAS is calling. This information can be either coded into the DAS when it is programmed or can be based on an inherent identifier (e.g., serial number) which is unique to each DAS. Once the DAS has been identified and authenticated by the remote server the data from the stored interval samples is uploaded to the appropriate database and appended to the table(s) for that particular DAS. At this point, the data are available for display and analysis from any web browser based on a user ID and password provided to the user.

In a typical installation, the data upload for a day's worth of data takes approximately 30 seconds if the DAS is connected to high speed internet connection, and 3 to 5 minutes if a dial-up connection is used. Upload times will vary depending on the number of connected devices, the interval selected and the time between uploads. In the event a DAS is unable to call the host server (or the connection is lost during the transfer of data) it is programmed to try to make the connection again and will not clear its memory until the host server acknowledges a successful upload. If the DAS is not able to make a connection to the host for a period of time (36 hours in the case of Buildingmanageronline) the host server will initiate an alarm and send an email or page to the designated user of the DAS, alerting him or her that the device has not "phoned home" for some time.

One key feature of the DAS is that it can not only communicate on pre-defined schedules, but can also ini-

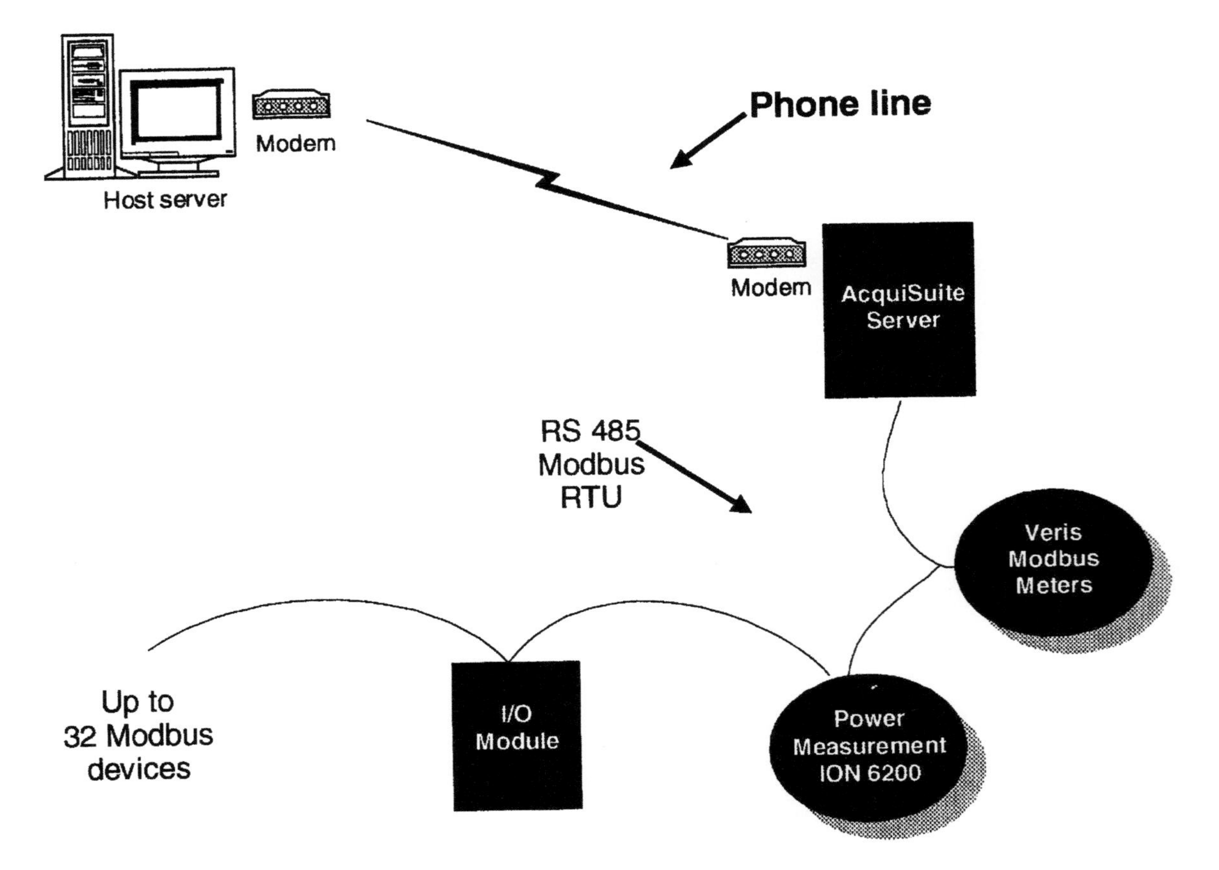

Figure 11-5. Acquisuite Connection via Phone Line

tiate contact with the host server based on alarm conditions. An authorized user can set high and low alarms for any measured value and the DAS will call out to the host server and identify the alarm condition. If the user chooses, he or she can be notified via email or page when critical alarm limits are exceeded and take corrective action if necessary.

NON-LAN BASED INSTALLATIONS

In many cases, it is either impractical or undesirable to connect to an existing LAN and use the network infrastructure for communications to the remote computer. In these instances, it is necessary to use a telephone line for communications and most DAS have a modem which is either built-in or can be added. If a phone line is used, there are basically three ways in which the DAS can utilize the POTS (Plain Old Telephone Service) for communication:

1. *Dial-out directly to the remote server*—the DAS uses an existing dedicated or shared phone line and simply initiates a call directly to the remote server. Once a connection is established, the DAS will exchange information using software built into the DAS and the remote server. The primary advantage to using this method of communications is that the DAS can use a shared phone line (e.g., a fax line) by grabbing the line during off-hours for a daily upload of data. This approach greatly reduces the cost and time required for installation and also minimizes the ongoing cost of operation. There are three primary drawbacks to this approach: 1) the phone line may be busy during normal operating hours, limiting communications in the event of an alarm; 2) it is generally not practical for the remote server to call the DAS due to the lack of a dedicated line, making it impossible to change settings or get updated information on demand; and 3) if the remote server is busy, it may be difficult for the DAS to call in and upload data.

2. *Dial-in directly from the desktop to the DAS*—in this configuration, the DAS has a dedicated phone line and is designed to receive calls from the remote computer. The remote server calls the DAS at a scheduled time and the communication between the DAS and remote is accomplished in the same way as the dial out mode outlined above. The key advantages to the dial-in mode are that the computer can adjust schedules to manage traffic on its modem bank (limiting the potential for a DAS not being able to communicate) and that the remote server is able to initiate a non-scheduled call to the DAS for near real-time updated information. The biggest issues with this approach are that does not scale well to large numbers of DAS and requires long distance charges for each connection.

3. *Dial-out using an internet service provider (ISP)*—this installation is very similar to the dial-out direct outlined above, but instead of calling the remote server directly, the DAS connects to a local (or national) ISP in the same way a computer user connects on a dial-up connection. To accomplish this, the DAS must be capable of using Point-to-Point Protocol (PPP) and the user must provide an ISP account with user name and password. Once connected, the exchange of information is very similar to the LAN connection covered previously, only at the slower speed of the dial-up connection to the ISP. The primary advantages to using an ISP are that the call to upload data is local (i.e., free) and that the remote server does not have to support and manage a bank of modems (minimizing the potential for one or more DAS being unable to upload data in a timely manner). The only negative to using an ISP is that the user must provide an ISP account which may add some time to the configuration and setup process. In many cases, this ISP access can be accomplished using an existing account as the volume of traffic and number of calls will generally be limited.

INTERNET BASED ENERGY INFORMATION SITES

As outlined above, it is now possible to gather and transmit data from a variety of building energy information devices in a cost-effective manner. The raw data sent to the remote server is only useful if the data can be converted to timely information on building operations and performance. There are a variety of services providing this information, ranging from very simple reporting sites to those that provide significant value added services such as energy analysis and rate engine comparisons (the term rate engine refers to database software that allows for interactive, on-line comparison of different rate structures for electricity).

The common element in restoring and displaying

information is the database used for storing the raw data. Once the connection is made from the DAS to the remote host, the data are uploaded into a database such as Microsoft SQL Server or MySQL. Storing the data in a database provides the flexibility for users to access customized or standard reports relating to only the data that is relevant to their needs without the need for exhaustive searches of raw data.

There are three primary means of accessing data from the database:

1. "Shrink-wrapped" software written for a specific PC platform and operating system
2. Web browser interfaces that load Java applets or similar plug-ins to access and display information on a browser
3. Web browser interfaces that do not require any plug-ins or proprietary software on the user machine to display information

Proprietary software generally requires that the user be on a computer that has a physical or network connection to the server hosting the database. The user executes a program (typically written in some version of C or Visual Basic) which loads the tools for graphical and tabular display of information directly to the users machine in a client-server environment. One advantage to this approach are that the data are secure (i.e., hosted in a controlled environment behind firewalls) and access can be easily controlled by the LAN administrator, limiting the potential for unauthorized parties to access the data. Another advantage is that the reports can be highly customized and tailored to the audience as the number of users is tightly controlled.

Key drawbacks to the proprietary software approach are:

- Client software must be compatible with the operating system and hardware of the user
- Updates to the software have to be distributed to all users
- Providing new users with access can be difficult as it requires installation of the software on specific computers
- Software conflicts can occur with other database drivers loaded on the same server

Applets or plug-ins allow users to access data from a web-based remote server using standard browsers such as Microsoft Explorer or Netscape by downloading software from the remote server that loads on the local computer. This scheme allows the user to access the data from a variety of different computers using standard web software and provides the means for updating local software (via new plug-ins), but also carries many of the problems associated with proprietary software. The software must be updated on all computers if changes are made and there may be problems with compatibility of the software with a variety of operating systems, web browsers and hardware. Requests for data are transferred from the plug-in to the remote server which responds by sending raw data to the client computer for display using the plug-in. There are usually performance issues (e.g., slow speed) depending on the structure of the software as this will typically require download of large quantities of data from the remote server to the toolset on the client computer.

Web browser interfaces which do not require plug-ins or applets are becoming more and more common as they provide the benefits of access to data from any web browser without the need to download and run software on the client computer. All requests for information are processed at the remote server greatly limiting the amount of data which must be exchanged between the client and the server. Information returned to the client is in the form of HTML (hyper text markup language) pages, which are very small in size and require little transmission time because the data remains resident on the remote server and results are sent from the server in graphical or textual form, not as data. This system allows the user to access information from any computer without the need to download any software (authentication of the user is provided using a user name and password built into the server). This approach is easy for users to access and provides faster response times, but the major advantage is that all software updates can be accomplished at the remote server and thus the most current software is always available every time the user logs in.

In summary, the advantages to web-based products include:

- Users can access information from anywhere web access is available
- User accounts can be associated with a wide variety of DAS servers located anywhere in the world. In other words, the data available to users can be associated geographically, operationally or by system type. Users could be allowed to see all data from their facility, some data from all facilities, or all data for specific systems from a variety of facilities.

- Software updates (new drivers, bug fixes, etc.) are instantly available to all users as soon as they are loaded on the server
- Many different users can access the data simultaneously without a drop in performance

HOW DOES A TYPICAL TRANSACTION OCCUR?

In the web-based system, a typical transaction consists of the following steps:

1. The user accesses a web site (e.g., www.obvius.com) using a standard browser
2. Upon accessing the log in screen, the user enters a user name and password
3. After authentication, the server looks up the user record and displays the accounts associated with that user on a welcome screen

As shown in Figure 11- 6, the user is greeted with a tree listing all the available accounts for which the user has access (each account will likely have multiple DAS) and a summary of points in alarm. The welcome screen provides immediate access to points in alarm so that the user can "drill down" by clicking on the alarm point to determine what the problem is. The tree of account information is the primary means of accessing detailed information. Clicking on one of the accounts listed in the tree produces a new screen (Figure 11-7) providing information on the DAS associated with the account (in this case, three AcquiSuites associated with the Obvius account):

OTHER OPTIONS FOR VIEWING AND RETRIEVING DATA

In addition to the standard charts and tables shown in Figures 11-6 and 11-7, most providers of web-based

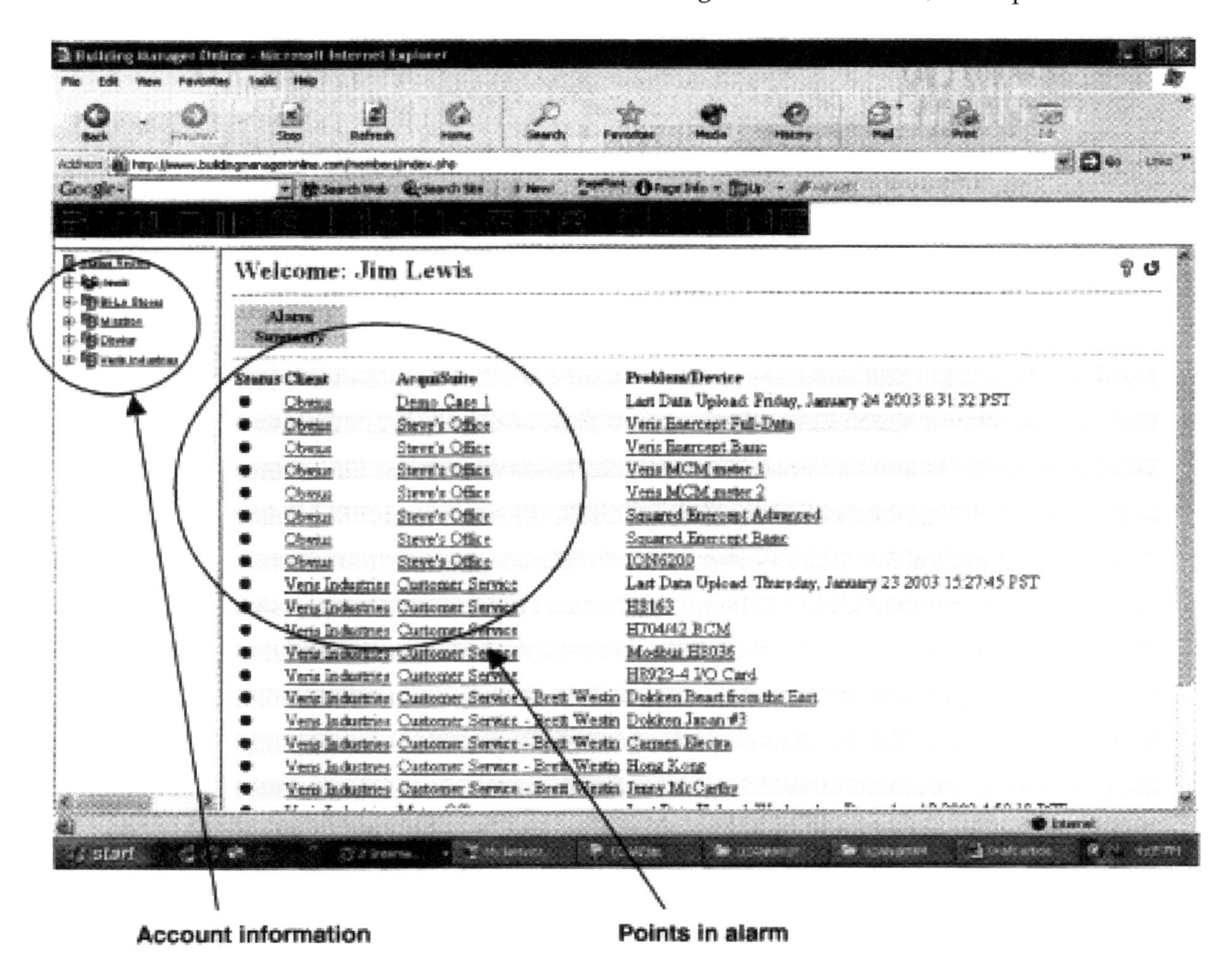

Figure 11-6. Typical Welcome Screen

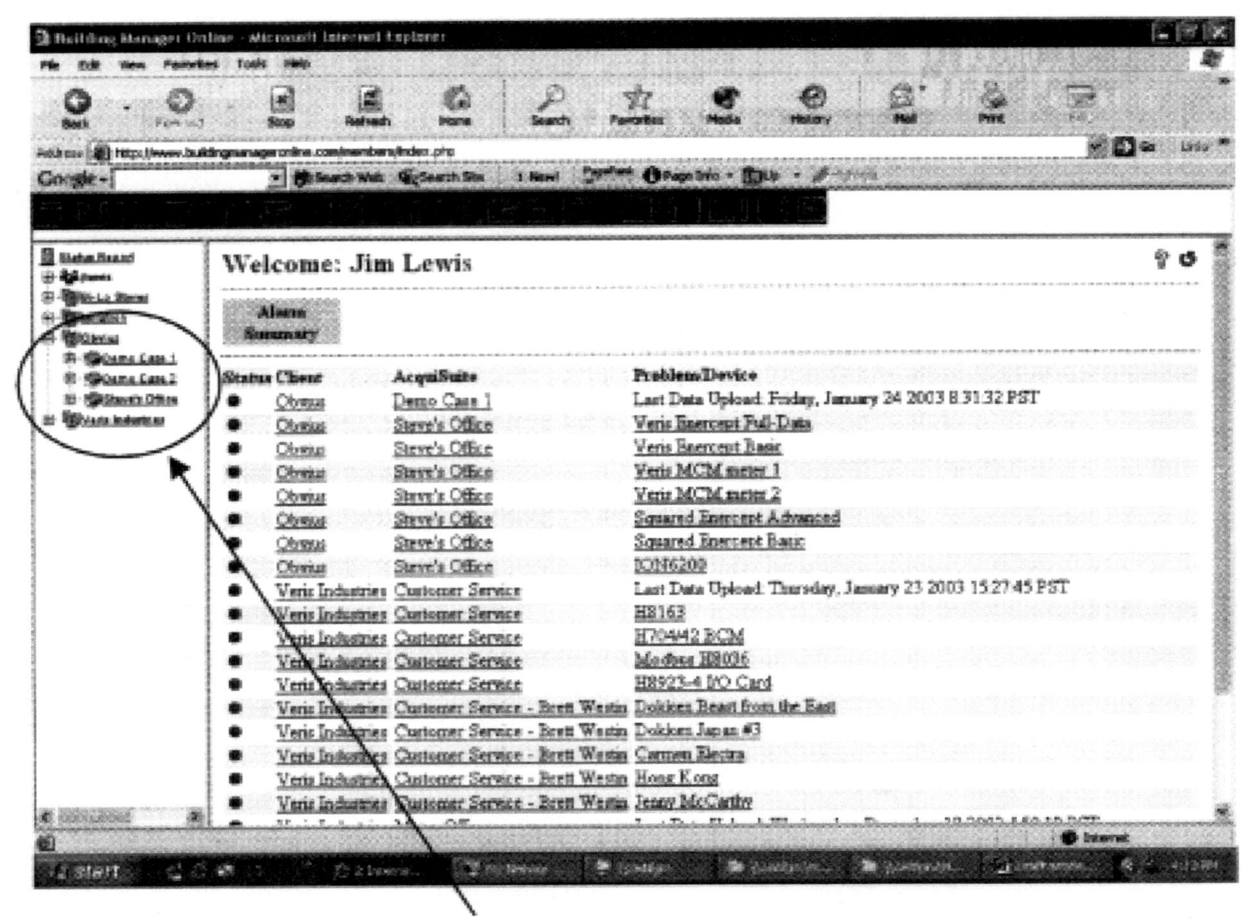

Figure 11-7. Drilling Down into the Obvius Account

energy information provide for the export of selected data to a local workstation. This can take the form of comma- or tab-separated value (CSV/TSV) files or can be accomplished using the file transfer protocol (FTP). See Figure 11-8. If the file is downloaded using text (e.g., CSV) it is generally pre-formatted for popular spreadsheet and database programs such as Microsoft Excel or Access. Once the file is opened using one of these programs the data can be sorted and displayed to meet user-specific requirements.

ALARM NOTIFICATION

In addition to viewing and downloading data, the ASP (application service provider) software provides the capability for the user to receive immediate notification of points in alarm. If a point goes into alarm, the DAS uploads the alarm information to the ASP site and this alarm information is sent to the user via email or page (see Figure 11-9). For example, if a user requests notification on a high temp alarm and the alarm limit is exceeded, the DAS uploads the information to the ASP and an email or page is sent to the user giving the name of the point in alarm, the alarm setting, the current reading and the time the point went into alarm.

INSTALLATION AND MONITORING COSTS

The specific requirements and challenges of individual sites make it difficult to generalize on costs of installing and configuring the DAS, but a look at a "typical" installation may provide some reference points. If we consider a building with the following setup:

- One AcquiSuite DAS
- One pulse input from an existing utility meter
- One additional submeter (400 amp)
- One temperature sensor

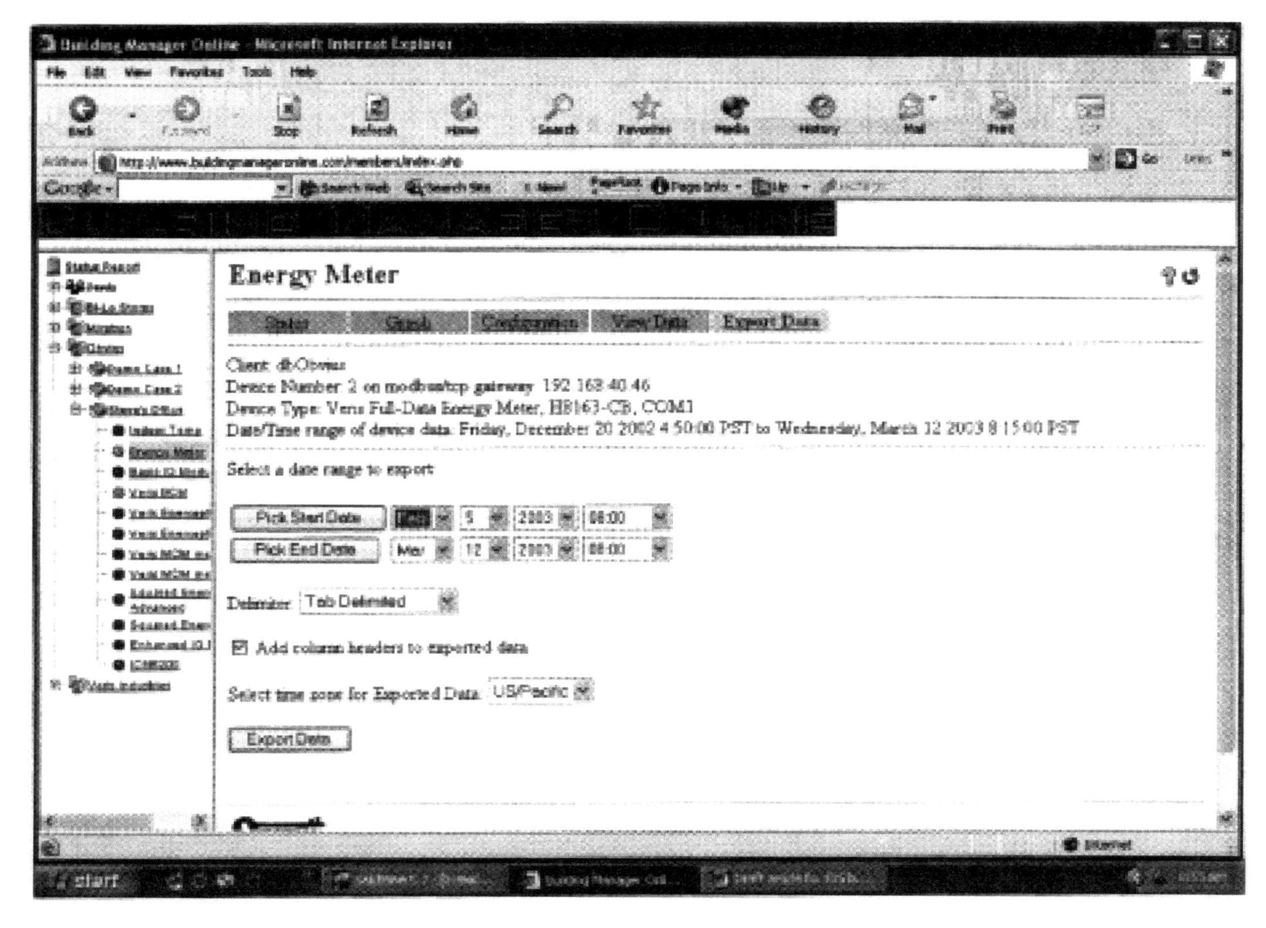

Figure 11-8. Setup for Download of Tab Delimited Data into Spreadsheet or Database

- LAN connection available
- Wiring runs under 50 feet

The total cost for installing this system would be under $2,000 in most buildings and the ongoing monitoring cost would be $20 per month, resulting in a total first year cost of about $2,240 and an ongoing annual cost of $240.

WHAT DOES THE FUTURE HOLD?

Based on the advances in EIS technology made thus far and the proliferation of hardware and software for web-based communications, it is reasonable to assume that some of the major trends in energy information in the future include:

1. Lower hardware and configuration costs as more devices and protocols are included in the "plug and play" lists of major DAS manufacturers
2. More flexibility in the capabilities for monitoring other key parameters (e.g., indoor air quality or processes)
3. A wider range of software applications for aggregation of utility bills, cost allocation and operations
4. More options for using DAS systems to meet requirements of curtailment contracts with utilities by using on-site generation
5. Cost savings for building owners who use energy information for peak shaving, operational savings and lower utility rates

Overall, we are very close to realizing the ideal world of true "do-it-yourself" energy information where the energy manager is able to add and remove devices to the system without the need for costly integration and software. Tailoring the hardware and software to meet specific needs will become routine and will dramatically impact the ability of the energy manager to control costs and minimize downtime.

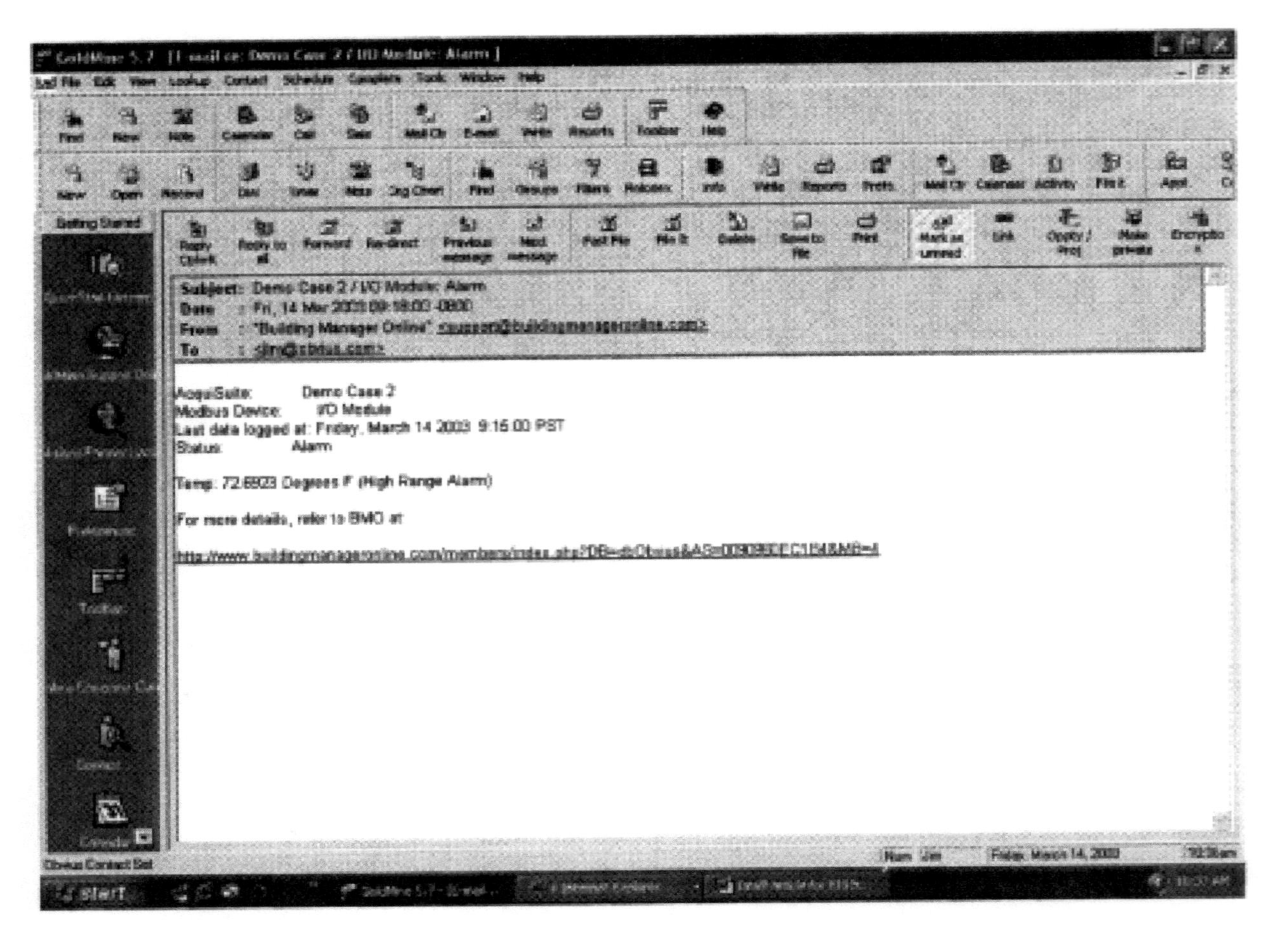

Figure 11-9. Email Alarm Notification

CONCLUSION

Historically, the cost of hardware, software and installation of energy information systems has been prohibitively expensive and has limited implementation to those commercial and industrial facilities that could afford to pay for custom systems integration services. These costs have fallen dramatically as companies such as Obvius leverage the enormous investment in the internet to provide tools to the building owner that make do-it-yourself data acquisition a cost effective reality. Hardware and software designed specifically for data acquisition and using available tools such as TCP/IP, HTTP and Modbus puts valuable energy information literally at the fingertips of today's facility owners and managers and provides the first vital step to a sound energy management program.

Chapter 12

The Importance of a Strategic Metering Plan for Utility Cost Management

Martel Chen
Siemens Building Technologies
Jim McNally PE
Siemens Building Technologies

CONFRONTED WITH INCREASING COST control pressure in higher education today, the facility managers have exhausted "low-hanging fruits" type of energy conservation measures like lighting retrofits, thermal insulation, constant volume/speed systems to variable volume/drive conversion, etc. So when management asks for more contribution from your operation, where do you go for answers? The authors have worked with various universities and colleges and have successfully managed both the supply-side and the demand-side in an integrated fashion and identified what a campus-wide metering and measurement system can do to effect significant savings. This chapter translates that experience into a general program that can benefit any large facility.

INTRODUCTION

To every energy or facility manager, energy and utility cost management seems like a never-ending demand from the bosses. It only gets more intense these days as companies struggle to balance their budgets in the midst of slumping profits while general costs are rising.

We all know that most of the low-hanging fruits of energy conservation measures such as lighting retrofits, thermal insulation, constant volume/speed systems to variable volume/speed conversion have been pretty well picked off. Where do we go to find the next round of cost saving opportunities?

First we need to go beyond the good old "energy conservation measures" and think about the bigger picture of energy management. What about the supply-side management opportunities to reduce the cost of procuring electricity, natural gas and oil for the central plant? Have you ever checked the utility bills for errors before you paid them? What about negotiating a good energy supply deal for a lot of savings by agreeing to manage the load to a certain shape on the demand-side? Have you considered on-site distributed generation possibilities? As soon as we consider "energy management by dollars and sense," instead of just "conservation," many more practical utility cost management options become open to us.

As a result of the Enron implosion and the notoriety of the California energy crisis, the deregulation of retail electricity has been slowed down dramatically. The investment climate in the once high-flying energy sector is extremely pessimistic due to credit crunch. However, we should realize that energy is a very limited resource; as more major players turn away from investing in future energy, we are just sowing seeds for the next round of energy crisis. Now is the best time to take inventory of your energy resources and prepare a comprehensive energy management plan to deal with the uncertain and volatile energy market that will surely come back before long.

As a side note, even though "customer choice" on the electric side has slowed down significantly, the opportunities to buy natural gas competitively are still abundant around the country. The Federal Energy Regulatory Commission (FERC) is pushing hard to restructure the wholesale energy market so that the rules will be more uniform around the country and will encourage fair competition. Several major banks including Bank of America are setting up trading desk operations for energy and starting to fill in the void left open by the demise of yesterday's energy giants. These are all positive signs that a more healthy competitive energy market will re-emerge. Therefore, preparing a total energy cost management program now is important. In this chapter, we will make a case for including a strategic metering plan to support this long term program.

WHY METERING? WHO NEEDS THE INFORMATION?

We have all heard the total quality management guru say: *"You cannot manage what you don't measure."* Beyond what's already proven to save energy and money, finding and proving the next layer of energy cost savings projects will require hard proof via factual data, especially metered data for all important incoming and outgoing energy lines. Load profiling information for meters throughout the facility will give quick insights into problem areas—and hence help identify cost saving solutions. Any proposed demand-side energy conservation project needs measurement and verification of the proposed return-on-investment. Any energy savings report needs to identify and isolate other factors due to changes in weather, seasonal production, and other known energy load changes—again, a difficult energy accounting task without historical interval meter data. Responsibility-based energy budgeting cannot be implemented without the support of factual energy information. Continuous improvement of energy cost savings cannot be accomplished without metered data for performance measurement. In other words, without an adequate amount of accurately metered information, the quality of complex energy decisions deteriorates rather quickly. It pays to think through a systematic approach to measuring utility cost management results including a sensible, progressive metering system.

Besides the energy manager, there are plenty of eager users of the metering information:

- *Finance*: Responsibility-based budgeting using actual energy usage information; utility bill audit and verification for payment; cost allocation to each production unit, department, or tenant.
- *Purchasing*: Load profile data to support competitive energy procurement.
- *Engineering*: Energy system planning and upgrades; ad hoc engineering analysis projects; energy cost control.
- *Operations*: Monitoring of key equipment and system for proper operations including incoming power feeder lines, distribution substations, the central plant, individual building's demand and consumption, individual building's chillers and air handling systems, etc.

ISSUES TO TACKLE BEFORE LAUNCHING THE PROGRAM

Metering is a fairly capital-intensive investment, especially the central database server, either locally or remotely located, and the instrumentation grade meters in steam and chilled/hot water systems. Metering is a means to an end (energy cost savings), and not an end by itself. Therefore, the champion of the metering program needs to find a management sponsor and convince him/her why metering is a necessary piece in order to implement the utility cost management program successfully. The ideal sponsor is the person who has specific cost saving goals to achieve for each business year. It's also important that the \program be phased in over a few years such that "quick-wins" and high-impact projects can be carried out early to prove the value and build up project momentum. Convince the boss that a certain percentage of verified savings should be put back into the pool to fund future metering and conservation projects.

If a rigorous ROI analysis needs to be performed to justify the initial metering project, the project champion should bundle a viable conservation project along with the required verification metering sub-system in order to meet the hurdle rate and prove the case. After the first couple of projects, the boss may very well start to believe in the importance of the strategic metering program that supports the achievement of utility cost control goals of the facility.

The Strategic Metering Process

1. Identify Reasons to Meter
2. Assemble Team of Stakeholders
3. Obtain Senior Management Support
4. Conduct System Surveys
5. Develop Solution Options with Budgets
6. Prepare to Make Revisions

Clearly define the strategic objectives of senior management.

- — Define stake holders, key requirements, and balance team vs. individual goals
- — Identify project costs, phases and timeline
- — Define project deliverables of meter information and management reports including front-end operator interface, facility networking and security requirements
- — Take inventory and check out the existing meters out there for possible re-use, upgrade, or replacement
- — Gap analysis, cost estimate, return-on-investment and funding options
- — Annual budgeting and stepwise implementation
- — Project/fiscal budgeting tool to decide how to prioritize/phase-in the project.
- — For consideration? Establish a utility bill/energy information database as the first step prior to doing load profiling? Also saves money by doing bill auditing; benchmarking; energy accounting; etc.

Consultants

The role of the consultant is more than just to provide a technical solution and a budget. The consultant is a vehicle to confirm your views, to deliver bad news, to build consensus, to educate, to help guide the decision-making of the stakeholders. Do not underestimate the importance of consensus-building in the early stages of the project. Without consensus, the project will surely die.

For the consultants: Provide options, options, options! People like choices; provide plenty of choices. However, don't give a client a choice, you wouldn't want to have to deliver!

Meter Equipment Overview

The following table is a brief summary of the types of meters applicable to university utility systems as well as other large facilities.

COMMUNICATIONS OPTIONS

Dedicated Network

Most networked metering systems began with RS485 two-wire serial communications. The protocols developed were custom tailored to the needs of the hardware on the network.

Advantages

- All features and functions of the meter are available through the native protocol of the dedicated network.

Table 12-1. Summary of Information about Various Types of Utility Meters and Their Uses

Utility	*Meter Type*	*Uses/Comments*	*Maintenance*	*Relative Cost*
Electricity	Pulse signal from another meter	Monitor the electric utility company's revenue meter. When a signal from the utility company's revenue meter is available, use it.	None	1/2$
	Watt meter w/pulse output	Monitor a low-cost kWh-only submeter. Use when a low-cost submeter solution is needed. Use a pre-configured submeter with a kWh output.	None	$
	PQ meter	Most stand-alone wattmeters will be in this category. About 25 data values will be available for every point monitored. These include kW, kVa, kvar, and Amps per phase, Volts per phase. When a stand-alone wattmeter is needed, this level of metering will commonly be selected.	None	$- to- $$
	Power Quality/ Waveform Meter	In addition to the values available in the PQ meter, this meter type also offers harmonic measurement and waveform capture. This is a high end product for the sophisticated user. Use to monitor critical/ sensitive environments.	None	$$$$

(Continued)

Table 12-1. (*Continued*)

Utility	*Meter Type*	*Uses/Comments*	*Maintenance*	*Relative Cost*
Chilled Water	Insertion Turbine with Btu Calculator	Requires one opening in the pipe. [Weld-o-Let]. Hot tap installations possible. High turndown ratio. May come with integral temperature sensor. One, separate thermal well required for Btu measurement.	Inspect annually. Replace rotor every 3-5 years.	$$$$$
	Ultrasonic	Accurate flow measurement. No moving parts to clog or wear. Btu calculator built-in. No pipe penetrations required. Communicates with monitoring system via pulse and/or analog output. Generally recommend thermal wells for RTDs versus clamp-on to avoid errors due to low temperature differences. Quick installation.	Inspect annually. Reset flow-sensing heads every 3-5 years.	$$$$
Utility	*Meter Type*	*Uses/Comments*	*Maintenance*	*Relative Cost*
Steam				
	Orifice Plate	Refurbish Existing No moving parts Higher pressure drop than insertion types means higher cost to have in the line. Pressure compensation not performed. Use to refurbish existing orifice plate station: replace the Orifice plate, replace the DP sensor, add a signal transmitter Error: 1 to 5% electronics annually.	Inspect electronics annually.	$$$
	Insertion Turbine with Btu Calculator	Requires one opening in the pipe. [Weld-o-Let]. Hot tap installations possible. High turndown ratio. Pressure compensation. Steam tables built into energy processor. Can include pressure compensation with built-in temperature sensor and dedicated energy calculating processor. Use where revenue quality measurements are needed. Error: 0.5%	Inspect annually. Replace rotor every 2-4 years. Temperature sensors may need replacing every 6-8 years.	$$$$$
Utility	*Meter Type*	*Uses/Comments*	*Maintenance*	*Relative Cost*
High Temperature Hot Water (HTHW)	Insertion Turbine with Btu Calculator	Requires one opening in the pipe. [Weld-o-Let]. Hot tap installations possible. High turndown ratio. Includes one built-in temperature sensor and dedicated energy calculating processor. Error: 0.5%	Inspect annually. Replace rotor every 2-4 years. Temperature sensors may need replacing every 6-8 years.	$$$$$
	Strap-On Ultrasonic	Accurate flow measurement. No moving parts to wear out. No pipe penetrations required. Requires periodic inspection and head repositioning. Must know precise pipe size and wall thickness. Use for water flow in pipes. High temperature strap-on thermal sensors required above 160°F. Error: 0.5-2%	Inspect annually. Reset flow-sensing heads every 2 years. Temperature sensors may need replacing every 6-8 years.	$$$$
	Insertion Vortex	Requires pipe penetration and welding. Impervious to high temperature. Lower maintenance than turbine meters. No moving parts. Accuracy falls off at low flow rates. Error: 1-2%	Inspect and clean every 2 years. Temperature sensors may need replacing every 6-8 years.	$$-to-$$$$$

(*Continued*)

Table 12-1. (*Continued*)

Utility	*Meter Type*	*Uses/Comments*	*Maintenance*	*Relative Cost*
Natural Gas	Pulse from Gas Utility meter	Most common technique to monitor building-level gas use. Use whenever possible. It is generally least expensive.	None	$ to $$$
	Insertion Turbine with therm calculator	Requires one opening in the pipe. [Weld-o-Let]. Line must be cleared before welding. Comes with integral pressure sensor for pressure adjusted energy measurement. High turndown ratio.	Inspect annually. Replace rotor every 5-7 years.	$$$$$
	Gas Use data from Gas Utility	Gas utility sometimes offers to send gas use data. Generally sent to neutral FTP site for pick-up. Must have willing utility and software to fetch data (FTP pull). Also must have automatic gap detection and filling capabilities.	N/A	$ to $$

- The dedicated network is usually included in the cost of the metering device.

Disadvantage

- The dedicated network is generally proprietary. It cannot be plugged into another manufacturer's network and be expected to operate.

Ethernet

Many metering devices today are able to communicate via the customer's ethernet computer network. Ethernet has many advantages. Because it is also a direct path to the internet, it is very popular.

Advantage

- For facilities with extensive ethernet networks, the incremental cost can be quite small if the roadblocks below are not present. See below.

Consider:

- Meter ethernet entry points require fixed I/P addresses. On some networks these may be in short supply.

- Using ethernet communications requires coordination and support of the I.T. Department which has nothing to gain by cooperating and (they perceive) everything to lose if there is a security breach.

Disadvantage

- Ethernet interfacing hardware/firmware can add $300-$600 to the cost of a metering device.

Use Building Automation System Infrastructure to Communicate Data Histories to Meter Front End System

In recent years, building automation system (BAS) companies have developed "gateways" to communicate with devices and systems made by others. Systems Integration of this type has enabled a certain amount of "mixing and matching" of different manufacturers' products. Cross system communication has been most successfully used to obtain "Status." "Status" is the immediate (or present) condition of a monitored or controlled point. "What's the temperature NOW?" That's Status. The significant limitation of this approach is that historic data arrays cannot be communicated. "What was the electricity use profile last week?" is a question this communication type is not designed to process.

Security

Metering data collection systems need to be clean, continuous, accurate, and secure. This is so because conclusions drawn from bad data will be inaccurate, as will any monetary transactions that are based on the data.

COMMUNICATIONS PROTOCOLS AVAILABLE

Native (Proprietary) Protocols

Native protocols refer to that protocol the metering device was originally designed to use. The native protocol supports the complete set of data acquisition capabilities offered by the metering device. These protocols are often proprietary just as the metering device is proprietary intellectual property. As relates to metering devices, *native protocols are able to communicate energy use histories, waveforms, and setup data residing in the meter. This is a distinct advantage over other communication types.*

MODBUS

MODBUS is a popular, non-proprietary protocol among both meter manufacturers and building automation manufacturers. It is used to communicate status. When MODBUS is used, the calling device constructs data histories by making data requests from the meter. If the calling device calls the various meters at different times (which it does) the resulting data histories may have timestamps that are not aligned. If the calling device is not connected to the meter, is off-line, or is turned off, there will be a *gap in the meter data history record.*

BACNET™

BACNET—much like MODBUS—is a point/status type protocol. It is promoted by ASHRAE (American Society of Heating, Refrigeration, and Air-conditioning Engineers) as a way to have inter-connectivity of various building automation systems. It is a new protocol, which is gaining acceptance among BAS providers. It is also a non-proprietary and open protocol. It is supported by many building automation system companies in addition to their native protocol. *BACNET protocol is not oriented to collecting history files from metering devices.*

Advantage

- Facilitates inter-connectivity between disparate systems.

Disadvantages

- It is not used to communicate meter data histories.
- The protocol is a subset of the capabilities of native protocols.
- BACNET gateways can be fairly expensive.

LonWorks™

LonWorks is a proprietary protocol that has seen some success among second and third tier building automation system manufacturers and specialty system manufacturers. First tier BAS companies also offer LON as an optional communication protocol for the device level network. Component manufacturers embed LonWorks into their product allowing it to be used by any system having a LonWorks network.

Advantage

- A broad selection of devices is available when configuring a system.

Disadvantages

- It is not used to communicate energy use histories.
- LonWorks systems tend to be more expensive due to the cost of licensing the protocol and coordinating it. This issue tends to disappear when Lon is the native protocol.

Front End Software Features and Functions

The front end of a metering system has three basic functions. First it is the window into the system of meters allowing the user to see the status of the various devices being monitored (load flow monitoring). Typically a one-line diagram is superimposed on the screen. At the various meter points, the status of the distribution system is displayed. Secondly, it is the location of long-term historic data collection. Applications such as historic energy use visualization, "shadow billing," cost allocation, energy and cost savings calculations, are commonly available. The third major functional area is alarm notification and a GUI that allows visual review of in-bounds vs. out-of-bounds parameters.

Development of Metering Front End Systems

Historically, multiple device systems had a single front-end computer. To monitor the system with its full array of visual status screens, it was necessary to go to that computer. With the advent of ethernet networks and the internet, front-end systems adopted a terminal server approach to front end visualization. This allowed anyone on a company intranet or internet to gain access to the system. The "terminal server" approach adds users by adding user sessions at a central terminal server. It is scalable by adding hardware. The next development in front end software is to provide access across the internet.

Front end systems should offer user-configurable visual status displays and access to long-term historic data in graphic and tabular formats. These displays should be accessible to users via the intranet, internet, or terminal server.

User Access

Browser-based historic data and status visualizations accessible to anyone on web having credentials to access the system are commonly available in front-end systems today.

Meter system front-end programs themselves have levels of accessibility defined for each user and user group. They are protected by sign-on credentials (Each user has his/her own a user name and password).

Interfacing the Meter System with Other Systems in the Facility

Building automation systems (BAS) can sometimes make use of the power use levels in a facility. Some facilities are equipped with automation systems that automatically reduce power use levels when they reach critical highs (also known as peak-shaving). The useful data for the BAS is the Kilowatt value at the main incoming electric service.

Two commonly used means of communicating this information to the BAS are as follows.

1. Via hardwired conventional, dedicated analog, status, or pulsing points.
2. Via "gateways." Gateways join a building automation system to another network (and its devices).

The hardwired approach is old technology. It is reliable. One data type per pair of wires is handled. Hardwired technology is economic when a few data points are to be communicated

The gateways are like tunnels in a mountain between countries. They connect and translate between two systems. They are digital. They are new technology. They are economically viable if more than 20 or so data points are communicated.

Recommendation: Unless there will be many data points to be communicated between the metering system and the building automation system, the interface should be hardwired.

Long-term, Historic Data Collection

The collection and long-term storage of historic meter data is an important function of the meter system front-end software. The front-end system should be equipped with a SQL-Server or SQL-Server-like database. The system should be scaleable in terms of data table size, volumes of hard drives, number of server processors, and number of servers. It should have a means to be backed up and a means of archiving older data. There should not be significant performance degradation as the system data quantities grow.

Application Reports

Metering systems should have a means of building reports based on historic data. The following are commonly available from system providers.

- Load Profiles (Able to display from 1 day to 1 month)
- Scatter Plot capability
- Historic comparisons
- Power Quality analysis
- Shadow Billing
- Cost Allocation

Weather Data

Local Weather Station

Local weather data are very useful. Data to collect: Dry bulb temperature, relative humidity, and solar gain. The importance of these data is that they are on the site. Also, it will produce a time stamped data stream that can be aligned with all utility data in the monitoring system. This can be very useful for *ad hoc* energy analyses. Some individuals are also interested in wind speed and precipitation. However, these parameters are not generally used in utility analysis.

National Weather Service Data

Weather from the National Weather Service has distinct advantages. Data to collect: Dry bulb temperature, Seven-day forecast of temperature and conditions. The weather forecast data are very useful in forecasting gas and electricity use. However, these forecasts are not load profiles; they are simply the high and low of the day.

TIPS AND "TAKE-AWAYS"

Project Development

- Know clearly why your facility needs a metering system and what strategic goals it will help accomplish.
- Involve stakeholders in the decision-making process.
- Get buy-in early on from senior management.
- Use your consultant to help you "sell" the project to your peers and management as well as developing it technically and providing a budget.
- Put several, viable options and alternatives on the table for stakeholders to discuss.

Meters

- Use pre-wired wattmeters when possible for lowest installed cost.
- Flow meter selection depends on the variations of flow expected. Large inaccuracies may occur if

much of the time the flow is below that which can be seen by the flow meter.

- Inaccuracies in gas and steam may occur if the pressure varies and the meter does not sense and compensate for it. Use pressure-compensated steam and gas meters.

Communications and Front End Systems

- Energy use histories must be clean, continuous, secure, and aligned to the top of the hour
- Native protocols are preferred to "gateways" and "generic protocols" which transfer point data only.
- Data logging front ends should have relational databases such as SQL Server (or better).
- Expect meters to communicate via their own private network(s) and/or the customer's ethernet.
- Expect to have web viewing capability.

CONCLUSION

Facilities that use a metering and measurement system can be successful in managing their energy from both the supply side and the demand-side in an integrated fashion and can achieve significant savings. Metering helps identify problem areas that will benefit from energy efficiency measures. Metering also allows the verification of savings that is necessary in convincing management of the ultimate success of any energy cost control program.

Section Five

Energy and Facility Data Processing, Analysis and Decision Making

Chapter 13

Optimizing The Value of Web-based Energy Information

Gerald R. Mimno, Advanced AMR Technologies

REAL-TIME ENERGY INFORMATION collected and distributed over the web has a pyramid of value. At the bottom of the pyramid is automated meter reading [AMR], which has a value of about one dollar per month per meter. At the top of the pyramid is risk management where knowledge of real-time intervals can create $20,000 value in minutes in the management of energy and futures contracts. Between the top and bottom of the pyramid is data used for load management, managing demand response and time-of-use tariffs, operating digital controls, and aggregating supplies. Facilities managers and energy service providers can generate financial returns by building energy information systems reaching higher up the pyramid of value.

INTRODUCTION

The most formidable dilemma confronting energy information is that no one knows what it is worth. Therefore, when users are asked what they will pay for energy information, they answer "Little or nothing." The task before the energy information professional is to find what energy data are worth, to whom, and why. Then you will know what to charge and how to get other people to recognize this value, what investments to make, and how to get people to pay for it.

The purpose of this chapter is to give you a framework on which you can establish value and pricing and thereby justify the capital expenditure necessary to build and operate an appropriate energy information system (EIS). This chapter will emphasize several themes. First, compared to real-time, on-line expectations in banking, retail, or investments, the state of energy information is primitive. A primary reason is the barrier erected by local telecommunications companies in getting energy data across the so-called "local loop." The answer, for many overlapping reasons, is to embrace internet technologies and build real-time information systems. Web technology will enable you to share data with other co-operating parties in the energy supply and demand chain. It will also enable you to add low cost microprocessor-based controls to energy circuits, not only monitoring but controlling cost, often from remote places. The results will be measured in deciles: up to ten percent savings in monthly energy cost by real-time monitoring and managing of energy, and another ten percent by adding logic and controls to reduce waste. Lighting controls offer especially large returns for minimum cost.

EVOLVING VALUE OF ENERGY INFORMATION

The Meter

The lowest form of energy information is the monthly bill from a utility. This type of energy information has been collected and distributed manually for one hundred years and still applies to 75% of all meters. Higher levels of energy information convey data more often, from more points, and represent more types of data—not only kWh, but kW and kVa indicating how much power you use, when, and whether you have an efficient power factor. But the lowly meter is the foundation of energy information and it is helpful to trace the ways the meter has developed and become automated.

At each building, a meter measures the kilowatt-hours of energy consumed at the electric service entrance. A rotor in the meter increments a counter, which displays a running total on dials or an odometer. Meter grade data are typically within one to two percent depending on the standards of the State Public Utility Commission. Accuracy degrades over time, requiring the meter to be replaced or recalibrated about every seven years. A walk-around meter reader enters the reading in a notebook once a month. These data are then keyed into a billing system which creates a bill mailed to the customer. The entire metering, maintenance, data processing, billing, and mailing for each account costs five to seven dollars per month. A walk-around meter reader can read a meter for a dollar, but this is only a small part of the entire process of sending a bill.

Over the years, various upgrades have been made to this antiquated system. Automated meter reading, AMR, is a broad term encompassing the automaton of any or all steps in the collection and distribution of information. Meters with digital output are replacing the electromechanical meter. A hand-held recording device carried around by a meter reader is used to eliminate the paper notebook and the manual key entry at the utility. The next step is a short hop wireless system which communicates data once a month to a utility van or receiver mounted on a pole in each neighborhood and wired into a utility server. A utility van can read 3000 meters an hour traveling past meters at 30 MPH. But the future of AMR is to collect energy information daily or more frequently over the internet using a DSL, cable, power line carrier, or wireless connection to the internet. California has studied statewide AMR. In their studies the Public Utilities Commission calculated the Net Present Value of an on-line system providing yesterday's hourly meter data delivered once a day after midnight and available over the internet. 43% of the cost was the meter and its installation, 13% of the cost was in communications infrastructure, 39% was in networking and data processing services, and 5% was in integration and maintenance.[1]

In the past, the utility was the main user of energy information. The utility needed data for billing purposes. The customer found billing data of little use as the bill was several weeks old and the single monthly figure or demand charge did not indicate when the highest use occurred or suggest what could be done about it. Seasonal weather changes also obscured meaningful trends. Today, with AMR, energy information is collected more often and used by more types of users. These include the utility, the facility manager, bill payer, energy supplier, remote energy manager, and perhaps risk manager. The same data are manipulated in a server and sent to different users in the form they need. Optimizing the value of energy data is a process of using internet technology to send the same readings to different people all of whom harvest value from the information.

In the future, AMR will be linked to Electronic Bill Presentation and Payment[EBPP]. Electronic bill presentation and payment converts AMR to an internet-based function, automating the entire billing process from reading to payment. Internet bill presentment begins with creating electronic statements and ends with processing payment transactions. It allows consumers or businesses to retrieve their bills through standard web browsers and securely pay them over the internet with a credit card or from their bank account. EBPP can save between $.30 and $.85 per bill with modest investment in information technology. Electronic Bill Presentment and Payment also provides utilities complete interactivity for marketing directly to their customer base.

Return on Investment (ROI)

Data have both values and costs. The return on investment (ROI) is a measure comparing the two. A short ROI means the savings using energy information will exceed its cost in a matter of months. A high ROI ratio means that compared to any other expenditure of capital funds, an investment in energy information will provide the better return for the dollars invested. Sometimes ROI can be calculated to the penny. Other times a relative estimate is good enough. The message of this chapter is that value can be represented by a pyramid. An ROI based on value at the bottom of the pyramid will not look good. As each level of the pyramid is added to the calculation, the returns increase. An ROI addressing the value at all levels of the pyramid—load management, demand response, purchasing, and risk management—will be strongly positive.

Submetering

Historically the metering point is the main electric service entrance. Today, submetering is growing. Mall

Figure 13-1. The value of energy information is a pyramid. The base of the pyramid, interval data, provides a low value to many persons. At each level, the value increase, but to fewer people.

managers used to be satisfied with one bill and would then allocate the bill based on the square feet used by each tenant in the mall. Today, facilities managers want submeters on every store in order to accurately allocate costs and provide incentives for conservation. The mall still has one revenue meter, but the mall uses submeters to allocate the monthly bill among different stores. In the future, more and more points within a facility will be metered, measured, and monitored.

Not all meters or submeters come in meter sockets. Privately owned meters often use current transformers and a transducer to measure load. A current transformer clipped around a wire in the distribution panel can provide an accuracy as fine as 1/10 of a percent. Good CT's cost $50 and up. A hobbyist transducer (actually an analog-to-digital converter) costs $100 and a professional transducer several hundred. A three-phase submetering system with matched CT's is available at around $500. A licensed electrician can install the submeter in an hour. The Square D "Hawkeye" H 8050 Series CT is a very convenient submetering device. In this model, the transducer is built into the CT clips. It installs quickly and the only wire coming out of the panel is a five volt pulse signal any technician can safely work with.[2]

Imagine a contact closes every time a meter turns producing a pulse. This is called a "KYZ pulse" representing a single-pole, double-throw, or Form C, relay. In the field, the technician wires to the KY or the ZY terminal where Y is common. The pulse signifies a certain amount of kWh consumed. The speed of pulses represents the rate or kW being consumed at any given moment. You may own the service entrance, but the utility owns the meters and will likely charge you a fee varying from $200 to $1,000 to get access to this meter data. An annual "maintenance fee" of $100 or more may also be charged. In managing energy, try to get access to the revenue meter pulse as then there can be no argument about accuracy. Alternatively, use your own CT's and calibrate your metering to the utility meter. There is a lot of politics played around meters and meter pulses. Historically, the meter was the utility's cash register and they did not want anyone else near the data. Moving from this mindset to providing open and free access to your own meter data is a long journey. You will find utilities at different stages on this journey. The problems you may encounter are long delays and high prices in getting access to your meter data. Common problems getting a KYZ pulse are described in an article by Kathleen Burns, President of Highland Consulting, "Getting the Data—It's Simple But It's Not Easy."[3]

Local Loop

The local loop is the system of copper wires installed and maintained over the years by the local telephone company. An energy information system may want to "call in" its data 720 times a month for hourly data and 2880 times a month for 15-minute data. Telephone companies want to charge for every call. They maintain an investment of $1000's per customer in the local loop and they aim to get monthly fees of at least $30 per customer from this investment. It is difficult or impossible for any other carrier using cellular, paging, private wireless, cable, or power line carrier to breach the local loop without making an investment in premises equipment, networking, servers, 24/7 maintenance, accounting, customer service, and marketing for less than several $1000's per subscriber. As a result the local loop is a serious and well-defended barrier to real-time energy communications. The penetration of the local loop, or lack thereof, has held back any significant deployment of real-time energy data using public telecommunications. In evaluating any technology, ask yourself, "How does this technology cross the local loop." You will find this the best way to sort out technology, capital cost, and operating cost. For example, your EIS could use a telephone line, a cellular connection, a cable modem, or a direct internet connection over the company LAN. Each has a different capital cost and monthly service fee.

Period

Historically, the period for measuring energy was a month. However, there is a lot of value in measuring energy more often. A reduced period provides a finer grain of detail and may be necessary to respond more quickly to what's happening in the field. The result is a tension between the cost of crossing the local loop and the desire to measure energy more frequently. Compromises are common. Hourly, or fifteen-minute data may be saved in the meter and delivered once a month or once a day by telephone or cellular. Most cellular companies will provide once a day service for a price between $5 and $15 per month. Note the difference between the interval period and the delivery period. We can expect that the growing deployment of "always on" internet service will ultimately allow real-time energy data to penetrate the local loop. Most likely the practical technology will take the form of a short wireless hop or power line carrier to an internet gateway. In homes, this may be a cable modem. In business it will be into a local area network already at the business. A rapidly emerging technology, WiFi, may also provide low cost access across "the last mile" for inserting energy data into the internet. WiFi (IEEE 802.11 unlicensed radio) is one of those technologies exploding from below. While the range is only 300 feet, consumers are building out WiFi hot spots by the tens of thousands.

Two-Way

Historically energy information flowed one way, from meter to utility. Contemporary systems are increasingly two-way. The utility or facility manager can contact the meter and have it execute a command. "Give me a reading now." "Turn off service to this deadbeat customer." In the future the message will be "Critical Peak Pricing has occurred. Turn off circuit X for 30 minutes."

Poll -Push

Energy information systems have evolved out of telephone based technology. Standards, such as the RS-232 protocol for serial data communications between energy devices, were engineered for copper wire decades ago. These circuit switched systems continue their legacy in telephone poll systems such as the widely used Itron MV-90 AMR. A server calls a modem attached to a meter, creates a session, and downloads data. In contrast, the internet uses a packet switched technology in which a microprocessor with logic and memory attached to the meter "pushes" a packet of information, each packet individually addressed, onto the internet on a regular schedule. Routers and switches on the internet read the address and shunt it where it needs to go. The internet is inexpensive because thousands of devices can use the same "wires." Polling systems, when they get large, are administratively unwieldy and expensive. People move, they change phone numbers, the line is misused. A digitally based, internet system has its own issues, notably getting an internet connection and internet security, but it is permanently attached to a meter, not a customer, and shows the direction the field is going. Don't install telephone-type systems in an internet age. Rather use your time to overcome barriers to the internet.

Information Systems and Control Systems

In the past there has been a recognizable separation between an energy information system (EIS) and an energy control system (ECS). The information system reported periodically on the performance of circuits. A control system operated switches on those circuits. However, as communications systems link devices to reporting systems in real-time, the information system and the control system are merging.

Gateway

The combined EIS and ECS is often described today as a "gateway." It sits at the electric service entrance to the building and performs a combination of functions. It receives data on the cost of energy, it measures time of day and coincident demand, it can operate an energy saving routine sending messages to inputs on the building management system (BMS) or to specific lighting or HVAC circuits. Summarizing 17 studies on the effect of information on energy usage, the California Public Utilities Commission noted, "Findings show that feedback alone and in conjunction with other factors can be effective in reducing electricity consumption... The bulk of the literature provides evidence that information feedback can play a role in reducing electricity consumption on the order of from 5% to 20%."[4]

Energy Waste

The reasons energy information and control systems can return double digit savings are that US energy practices are very wasteful. Compared to Europe, the United States uses one third more energy for similar functions. Also, many energy controls still use 1930's type electro mechanical technology deployed in "stove pipes" where one mechanical unit has no knowledge of any others in the facility. Harvesting waste is a matter of adding digital logic and communications. We are moving toward the decade when every appliance will have its own subnet address and will be managed and optimized remotely. Until that decade, the energy practitioner will be heavily involved in retrofitting digital controls and communications to existing systems.

Finding energy waste is easy. It is everywhere in schools, homes, and commercial facilities. For the last 30 years, there has been a widespread practice of retrofitting lighting circuits with energy efficient HID's and fluorescents. The next stage will be controlling that fixture in real time through "day light harvesting" information and controls. Last year, industry purchased 7 billion eight-bit microprocessors. The most popular is the 8051, which originally powered the IBM PC and now comes in dozens of enhanced varieties for a couple of bucks. It is all the brains you need for both logic and communications. If you want to see the future, take an 8051 programmer to lunch. He will show you designs that have a real-time clock, monitor coincident demand and ambient lighting, and dim or cut lighting as it follows the sun around a building. Efficiency is good, but nothing saves more money than turning a light off when it isn't needed.

Energy Software

Historically, billing has been the first focus of energy information and energy software. Utilities used to have huge mainframes on raised floors devoted to billing. The truth was they could process a million bills, all the same way. Unusual bills were done by hand. Servers are much smaller and software is more flexible now. Third-party billing packages offer customer information systems (CIS) and customer relationship management (CRM), which manage billing and any other contacts with customers.

There are many types of energy software on the market. Some consolidate monthly bills from multiple sites and look for errors and trends in the data. Others monitor HVAC building controls, refrigeration, or lighting. Others help in preparing bids for purchasing power in deregulated states. Many of the large energy service companies (ESCOs) operate network operating centers (NOCs) where they monitor and manage power for their customers. Another type of software manages demand response for independent system operators. On the horizon are energy XML packages which will manage secure and verified financial transactions and settlement between energy buyers and sellers. In evaluating a software package, make sure it operates on an internet engine, not just an executable program that can be displayed on the internet. Look for interoperablility such as Java. Look for ANSI, EU, IEEE and CE standards in preference to proprietary systems. Look for open source, Linux, for economy at the server and Microsoft for ubiquity at the browser.

Remote Management

So there are energy micro-controls with access to LANs and WANs in your future. As a consequence it doesn't matter where the management of these controls lies. They may be in your building, or across the world. The advantages of a remote energy manager are the usual—economies of scale and specialization. Even a good building management system is often poorly programmed. The building was recently sold. The people who know anything are gone. The building is programmed to reduce complaints, maintain comfortable temperatures, but it knows nothing about price or the humidity coming up from the south. Remote managers can compare buildings and systems. It is their only business and they are specialists. They can also anticipate conditions and write algorithms which anticipate weather, rather than chase weather. If they don't like what they see, they can reprogram the micros remotely. This is where the history of energy information has taken us.

TURNING DATA INTO VALUE

Energy information systems produce floods of data. Even energy bills, with perhaps fourteen different lines on each bill, can overwhelm. The way to extract information from the flood is to direct the flood into an Excel spreadsheet or database. You will then find it useful to:

1) Visualize the data.
2) Benchmark the data.
3) Find exceptions.
4) Create alarms.
5) Automate controls.

Visualizing Data

Good energy data are fine-grained data. As we said above, you should always push for more frequent intervals and the technology to deliver it. Fine grained data, at least 15-minute intervals, give you a better picture of what is happening in some remote place, and give you more clues about what can be better controlled to halt waste. But you can also drown in data. Often, the first look at a load profile is revealing. But then one profile begins to look like more of the same and customers question why they are paying a monthly fee of $30 to $50 for a load profile service. The answer is to help customers visualize data so they can scan them for understanding and operational needs. For making sense of masses of data, spreadsheets or data base programs are essential. Microsoft Excel is a defacto standard in energy display and charting. It is particularly suitable when sharing energy data with others such as power marketers or energy service companies.

More powerful visualization comes from directing the energy data stream into a data base using structured query language (SQL) such as Oracle SQL, Microsoft SQL, or the open source MySQL. One of these relational database engines will underlie commercial energy management software programs. They work in similar ways. An incoming data stream is directed to rows and columns representing ID, time, and consumption. An administration program relates the data to a customer, a facility, a place, and a time. To present data, the user interface uses structured query language (SQL) to construct queries which find data, perform mathematical operations, and present results in charts or reports. Most queries are canned and always perform the same operation. In data warehousing, middleware programs have the ability to create unique queries on the fly so a user can slice and compare data in original ways without being a programmer. Charts and aggregations are especially meaningful. Energy data are most operational when charted against other variables, e.g., results showing the effects of weather, the savings realized, or the times and places where peak demand occurred.

Benchmark the Data

A key question is always, "I am using more or less energy compared to what?" To answer this question benchmarks should be defined and comparisons made. Is this elementary school using more energy than that elementary school? What is my consumption per square foot? What is the envelope insulation R value of the building. How does consumption compare? Many benchmarks will require additional research and calculation. Excel spreadsheets are also available on the web for facilities by size and type and by class load for common residential and commercial uses. An excellent source of benchmark data is the Energy Information Administration.[5] Many utilities also publish class load statistics. These are excellent benchmarks, especially to tell you how much waste you can expect to find.

Find Exceptions

Once you understand the general picture, you can manage by exception. In this sea of data, find what sticks out. A database allows you to scan data quickly and find the unusual. The database can also be programmed to spot the unusual automatically and flag it for your attention. The exception may be temperature extremes, higher or lower peaks than common, excessive use or cost. Frequently, exceptions will spot some maintenance issue—filters clogged, controls in override, a stuck damper.

Create Alarms

Once you set parameters, create alarms. An alarm combines an exception with an administrative program and communications. The parameter triggers the alarm. The alarm automation program defines the parameter and who to call, how to call, and what to say. Alarms can be current or prospective: "Warning—coincident demand projected to reach monthly peak within twenty minutes." Alarms can be broadcast by email, cell phone, pager, or phone call. Windows and Linux servers come with built-in functions needed to recognize alarms and make calls.

Automate Controls

The highest level of energy information is to respond to an alarm with an automated control. These may be very simple. Where the daily kW peak occurs from 3:00 to 5:00 pm, the automated response is to sequence cooling equipment through this period so that it is not all on at the same time thus limiting coincident demand. The watch words of automatic controls are "Save me money and do it for me."

THE VALUE PYRAMID OF INTERVAL DATA POSTED ON THE WEB

Optimizing the value of energy information also involves optimizing value beyond a single facility, owner, or utility. The greatest value occurs when data are shared over the internet among all the parties in your supply and demand chain. The supply chain includes generators, transmission operators, independent system operators, traders, marketers, and distributors. The demand chain includes energy managers, financial controllers, energy service companies, consumers, and the public. The internet makes it possible and economical to collect, process, and deliver interval data on a real-time basis to the entire supply and demand chain. Real-time, or more accurately "near real-time" interval data are meter data encapsulating a 15-minute period, which is accessible on a network before the close of the next 15-minute interval.

Interval data may relate to a single meter, an aggregation of meters, or a statistical sample of meters representing some aggregated or regional load. Sharing contemporaneous data among the parties in a supply chain adds value to the entire chain and improves the profit of each party. This is the highest objective in optimizing energy information. The members of your supply or demand chain are not strangers. They are contracting parties who sell you power, who give you credits for demand response, who monitor the efficiency and cost of your facilities. According to Richards Energy Group, Inc., one of the largest electric power aggregators in Pennsylvania, power consumers with high load factors will definitely get better pricing from suppliers if interval data are provided. This better information can shave perhaps 2 mills from the price over the entire term of the contract. This is the difference from say 4.0 cents/kWh to 3.8 cents/kWh, enough to make or break a deal. Frank Richards says, "It boils down to the fact that the better the data, the better the deal, since it lowers the supplier's risk. And additional price improvement may result when interval data can be provided not only historically, but also in real-time, which will reduce the supplier's risk further." Without this data, a power marketer can only look at old bills and determine your class load, which may be very different from your actual load. If you are a small user, aggregating your peak demand with others will improve your demand profile and may improve the load factor of the portfolio to everyone's benefit. By aggregating loads, small commercial users in Pennsylvania have saved five percent on their cost of power and ten percent on their demand charges. Aggregating demand response also has significant value to the company, the aggregation portfolio, and the utility supplier. In critical peak pricing incidents, a one percent reduction in demand has dropped prices by ten times as much.

Here is where a number of our themes come together. There is a persistent drive to measure and monitor more points, more often, for more types of data. Buildings and appliances will be retrofitted with microprocessors to manage energy. Energy data will migrate across the local loop to the internet which is the most practical means to both collect and distribute energy information. Energy data will be widely networked on LANs and WANs and processed in servers and databases. Remote energy managers, mostly silicon but some human, will optimize the financial aspects of power. Demand response and distributed generation will proliferate. Energy waste, representing five to twenty-five percent of our current energy bill, will be squeezed out. ROI will be measured in a few weeks or months.

The Bottom of the Pyramid—
Automated Meter Reading

Your first battle will be getting your meter data onto the internet. To get started, survey your options. Interval data comes from three sources: your utility, an energy service company (ESCO), or a system you build yourself. It has taken 50 years for utilities to automate 25% of their meters. Another quarter will get connected within the next five years as AMR becomes commonplace. Utilities already have a record of your monthly use. Find out if this is available on the web or can be provided for a fee. Many utilities are now moving to a once-a-day read by telephone or cellular link. ABB, the big utility supplier, sells a system "ABB On-Line" used by many utilities. The utilities resell this service for $30 to $50 month. It provides a 15-minute load profile over the last 24 hours delivered once a night. Another approach is to buy a special meter from your utility. The meter will cost $200 to $1500 depending on the tariffs set by the Public Utility Commission. The lower cost meter will provide a pulse. The more expensive meters will have a jack for a telephone, RS-232 serial data, or CAT 5 ethernet service. Any new meter will be electronic and will have an RS-232 option. You can buy a serial to TCP/IP converter now for $200 and this offers a practical way to get meter data onto your own ethernet LAN where it can be directed to the internet.

Another service, Enerlink.net, uses the internet and requires a meter with a network interface card, plugs into your LAN, and provides your data on an energy management site. NSTAR in Boston offers monthly service for $95. The service provides 15-minute data, 96 times a day with a latency of about 15 minutes. General packet radio service (GPRS) is a new type of packet data cellular which charges by the packet, not the call. GPRS is available in Europe where 15-minute data, 2880 meter readings a month, cost 30 Euros. Expect GPRS service in the US as well.

Verizon has just announced free WiFi service to persons already buying their DSL service. While WiFi has a range of 300 feet, buildings in high value commercial areas will very likely have WiFi hot spots you may find available to connect meters to the internet.

ESCOs will solve all these problems by providing an energy information system (EIS). Interval data can be provided without analysis, with billing cost analysis, with load management analysis, or with remote monitoring and control. According to a study by E Source, one fifth of 800 businesses surveyed thought cost analysis would be very useful and one in ten thought remote monitoring would be very useful.[6] These are very low market expectations, but they are growing. Respondents generally thought interval data without analysis should be free, but that data provided with analysis would be worth a monthly fee.

Don't think you have to have data from the meter. If your building has an energy management system, it already has a pulse available and you can use an optical splitter to send pulse data to the internet. You can install your own meter-grade pulse meter in your service entrance. Many manufacturers provide private metering equipment using CTs and transducers. Prices vary from $500 to $1500 depending on quality and output options. The advantage of a private meter is you only pay once and there is no recurring fee. You can get a private meter certified by running it against the utility meter if you are buying power from a retail supplier or participating in a demand response program. As you have noticed, AMR seems a costly venture just to put meter data on line. This cost becomes insignificant when you add energy applications higher up the pyramid.

Load Management

The next step up the value pyramid is using near real-time data to manage a load profile and especially coincident demand or peak load. One municipal electric executive has noted that a customer who has never practiced peak demand management can save ten percent in demand charges with the implementation of the very first steps in a demand management program. Since demand charges typically represent about one-third of the cost of power, managing demand in real-time promotes substantial savings. We have been trained since infancy to think that power costs the same day or night and that the way to save money is to turn out the light. This training is wrong. Peak load power always costs 30% more than base load, sometimes costs three times base load, and at Critical Peak Pricing incidents (threatening a brown out or rolling blackout) prices can reach thirty times base load. Utility regulators used to protect consumers from the time-varying cost of power. This is changing. Commercial customers are now very likely to be exposed to the time value of energy, and residential customers will also see the real costs of energy in their future.

There are many financial reasons to be concerned with your load profile and to spend the money to display it on the web where you and others can see it, make changes, and save money. The waste in your monthly utility bill is somewhere between five and twenty-five percent. For a company earning a ten percent profit, finding $1,000 annual savings in the utility bill is the same as expanding sales by $10,000 every year. The waste is caused by inefficiency, poor controls, and poor procedures. Without information, you will be near the high end. With information, you can begin to reduce waste to zero. Many people, schools especially, think they have conquered waste because they have a building management system and a designated energy manager. Without objective measurement, neither a device nor a person really knows whether they are saving or not. People familiar with controls will tell you, "A BMS starts to deteriorate the day it is installed." If this is true of a well-run operation, you can imagine waste in a facility with no more information than a monthly bill. Typically 30 % of energy consumption is fixed and 70% varies with the season and the weather. In order to tell one from the other, and moreover find waste, you need both good data and benchmarks.

The level of detail needed is at least 15 minutes, though some systems are now moving to five minutes. Your load profile will tell your daily use pattern, your load factor, and your coincident demand. Each of these factors will also be reflected in tariffs with users penalized for anything not optimized. Even residential users who assume they are on a flat tariff will pay eventually for poor load factors or spikes in coincident demand because the whole system is charged wholesale power fees which reward or penalize good or bad performance. It may be especially galling that you are paying for the high coincident demand of your neighbor's air conditioner. Allocating costs fairly is another reason regulators are beginning to expose consumers to the variable cost of energy.

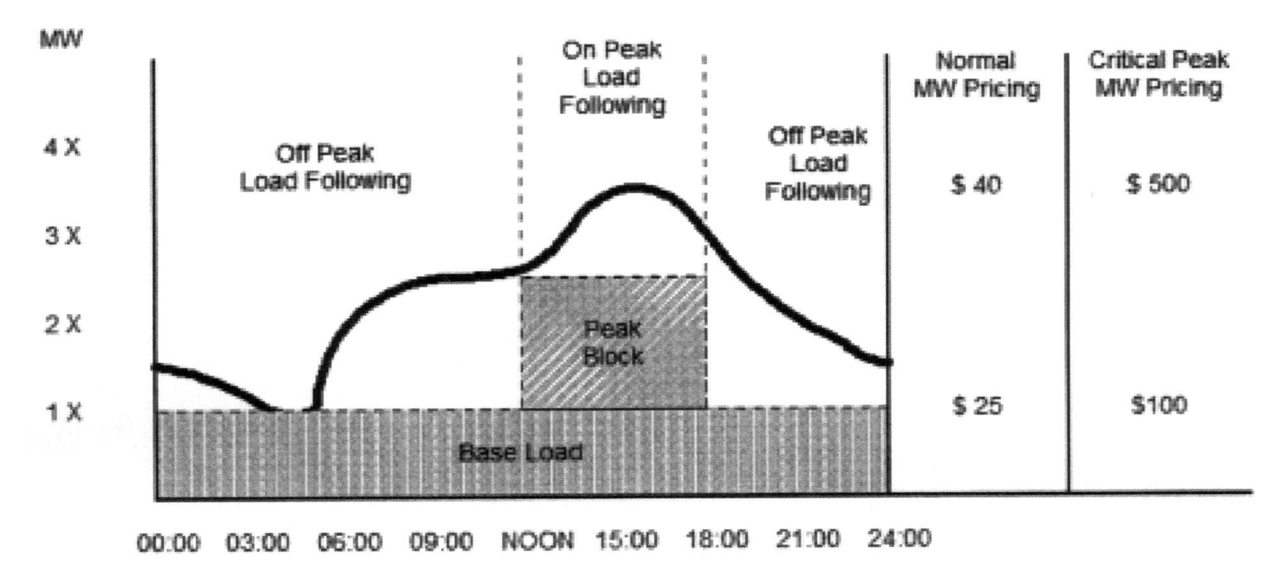

Figure 13-2. During the 24 hour cycle the amount of power used peaks at about three times the base load. During normal times, the peak price will be about 1.5 times the base load price. However, price is more volatile than consumption. On a hot day, critical peak prices can reach many times normal pricing.

Schoolteachers, store managers, office workers can be very helpful in conserving energy when they are presented with a load profile and benchmarking. The web provides a convenient way to enlist these users in an energy conservation program. The first presentation of a load profile can be described as the "Discovery" stage. Users discover what energy they use and when, and they often discover they can reduce peak consumption by five percent. This happened at Puget Sound Energy when over 300,000 customers were showed the difference between using energy on peak and off peak. Experiments like this call into question rigid thinking about elasticity of demand curves. Given access to information, both residential and commercial customers show they can change consumption patterns by five percent with ease.

An expert can read a whole story in a load profile but what is really needed is benchmarking. When a load profile is plotted against degree days, the data are "normalized for weather." You or your software can begin to puzzle out what changes are weather-related and what result from poor efficiency, poor controls, or poor management. There is no end to the kind of benchmark data that can be collected or calculated. Government agencies, universities, trade associations, and professional associations publish weather data, class load profiles, insulation value of different types of construction, and engineering specifications for mechanical systems. Unit calculations such as kWh consumption per square foot, per class room, per apartment, per pupil are useful. In benchmarking, the first data may be the most useful. Once a relative pattern or trend is revealed, you will know where to begin a load management plan.

Demand Response and Time of Use

The next step up the pyramid of value is demand response (DR). DR programs were popular in the energy cost spikes of the 70's and 80's but faded as energy prices fell in the 90's. Now they are back with considerable regulatory backing from the Federal Energy Regulatory Commission (FERC). The objective of demand response is to bring market discipline to the unregulated wholesale energy market. In other words, to curb gaming the market and price spikes. Supply side load balancing is relatively inelastic. When customers want power, the grid has to deliver or the grid will go down. This "gotta have it" situation opens the door to profit-seeking generators who can spike the summer price from $30 megawatt to $3,000. This results in a critical peak pricing incident. But what the market causes it can also fix if the rules are structured to give customers the same high prices in payment if they reduce demand. A

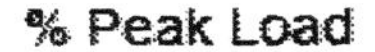

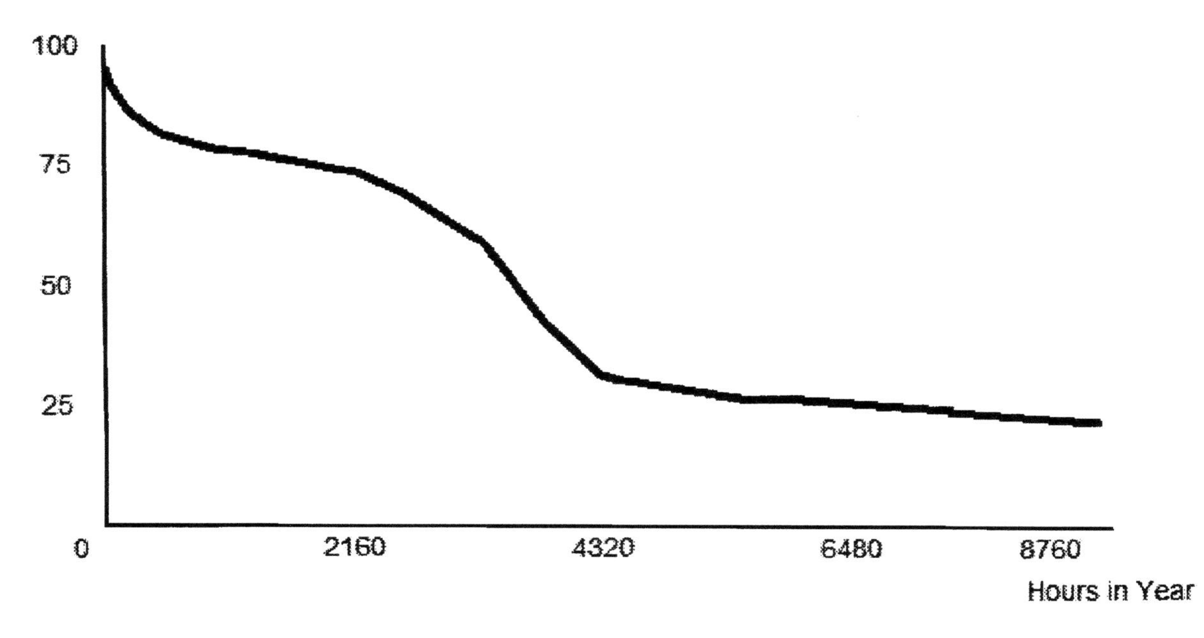

Figure 13-3. This load curve shows capacity utilization over the course of a year. Critical Peak Pricing incidents may occur 7 days a year, but it is of immense value to the system operator to enlist your cooperation in reducing demand during these days. Your demand reduction will be rewarded ten to thirty times the normal cost of power during these incidents.

quarry operator in PA usually pays $120 for the energy needed in one afternoon. By shutting down the crusher on a hot afternoon, the operator can get paid $4,000 when the price spikes to $1,000 MWh. By scheduling the crusher and stockpiling stone, he can optimize his demand response.

The quarry noted earlier participates in the PJM (Penn, Jersey, Maryland) Interconnect curtailment program. FERC authorized a similar program at the New England Independent System Operator (ISO). The NE Demand Management Plan operates a Day Ahead program where businesses can bid from $50 to $500 per MWh for curtailing load. If they want to seize an opportunity, they can get paid up to $100 MWh any day power exceeds $100 or up to $150

MWh for a minimum of 2 hours if they can react within thirty minutes. ISO-NE has a price cap of $1000 or roughly 33 times the normal price. These numbers suggest the direction demand response will take in other parts of the US, particularly as FERC implements region by region a Wholesale Power Market Platform. FERC objectives have some very relevant features to our discussion:

1. FERC wants demand response to be the functional equivalent of generation. If the grid needs megawatts, they may be supplied either by adding a generator or reducing demand. The system operators will accept bids for either one.

2. The rules are designed to aggregate demand response by adding up the combined response of hundreds or thousands of small changes, rather than ask a single business to cut 500 MW and effectively stop operations.

3. Demand response should be automatic and sustainable. Sustainable DR may be cycling air conditioners, setting back thermostats, coasting coolers, or dimming lights.

4. ESCO's may find a new business aggregating demand response.

None of this would be possible without internet-based energy information systems. But note that the system you build for managing your load on a day-to-day basis also enables your participation in demand response programs. Should you participate? You may want to consider that demand response opportunities generally represent one to two percent of the hours in the year, but during those hours, you can get a power credit of ten to 30 times the normal price. It is your opportunity to make a lot of money very quickly by managing your energy in a manner that will not even be noticed by your customers, clients, students, or renters.

Another example of DR is distributed generation in a congested service area such as the Chicago region on a hot summer afternoon. Typically an important building will have stand by generation for 25% of its load for operation of elevators and key systems. The ability to deploy stand-by generation has substantial value. In the Midwest, a typical demand charge is $10 per kW. A typical contract price for interruptible load is $12 to $22 per kW for load shed. The spot market price for a MWh will vary from $30 to $600 and has hit $3,000 and more. Assuming favorable environmental review for operating distributed generation, if this load could be coordinated on a hot day, the value of stand by generation is worth twenty times its regular price, whereas under present interruptible tariffs, it is typically sold for twice its regular price. Your real-time interval data allow an auction market for stand-by generation and load shedding that can unlock this value.

Many large utilities offer time-of-use tariffs. You can find this on their FERC Form 1 which they have to file every year.[7] It lists their tariffs and the number of customers on each tariff. Some have residential TOU tariffs. Should you be afraid of TOU? Many customers do not want to be exposed to a utility rate that changes depending on the time of day or time of year. Nor do utilities want "free riders" who happen to use most of their power off peak. The optimum combination is to create some flexibility so that more power can be consumed when rates are low and less when peak prices occur.

Time of use depends on elasticity of demand. If a business operating under a TOU tariff cuts consumption by 10% in response to a price rise of 100%, the business has an elasticity of demand of minus 0.10 which is the ratio of demand change to price change. According to the California Public Utility Commission, "Customers, on average, reduce electricity demand in response to higher electricity prices and in response to having more information about their energy usage."[8] Many studies of TOU cited by the CAPUC show the price response of residential users was 3% of total consumption and 4% of peak consumption. The use of radio-operated demand response systems on air conditioning, water heaters, and pool pumps can double these averages. This may not sound like a lot, but it can have a profound effect on pricing and system reliability. In critical peak pricing, when a transmission system is facing a brownout and prices have shot up to $1,000 a MW, a 5% reduction in demand will drop prices by ten times as much, i.e. prices will fall 50%. This is the reason both PUCs and FERC intend to expose customers to real-time pricing.

Until recently it was thought that commercial properties could not respond as effectively to TOU tariffs and critical peak pricing incidents as residential users. This is an error. In fact there is more waste in commercial properties, especially small and mid size ones, than in residences. The problem and opportunity is to build the energy information systems to find this waste and cut it down. Double-digit savings are your incentive to install a web-based EIS and digital controls. Many ESCO's employ performance contracting to harvest efficiency. The ESCO typically charges clients an up-front fee and 33% of proven savings, or no up-front fee and 50% of proven savings.

Automated Digital Controls

The next step up the value pyramid is remote monitoring of energy consumption at a central monitoring station coupled with the remote operation of the facility to achieve efficiencies. A formal name for this is an enterprise energy management system (EEMS) able to control set points and thermostat settings remotely, to disable or reduce chiller operation, and to dim interior lights. Most larger facilities have on-site building management systems. However, these systems have become too complicated for on-site personnel who are reluctant to optimize a system or who implement very limited adjustments. The systems also tend to be programmed using tight constraints for comfortable temperatures in a building and loose constraints on cost. A skilled off-site energy manager can optimize both comfort and cost by anticipating operational requirements from a very small data set trending ambient temperature, humidity, chilled water temp, flow rate, and specific loads. Based on this trending, the remote manager can benchmark a facility and operate optimum on-site energy management scripts. Energy service companies can point to operating savings of ten percent in office buildings from efficient mechanical management without loss of comfort or significant new investment. Since utilities typically cost $1.75 sq.ft., saving ten percent in a 100,000 square foot office could add $17,500 to the bottom line of a Real Estate Investment Trust (REIT) and leverage the equity value by ten times that amount. A real-time remote monitoring system makes possible continuous energy optimization, potentially saving customers thousands of dollars a month.

Another ubiquitous example is a soft drink vending machine. An average high school may have twenty scattered around the halls. A 1500-watt unit with power cost of $.15 and a 40% duty cycle will consume $68 in energy per month. A control program addressing kW and kWh would coast the compressor (it would still vend) through the peak demand period every day and turn it off every night from midnight to 5:00 am. Savings are $15 month in peak demand (1.5 kW) and $9 month in kWh. The digital controls require a timer and relays. The savings occur every month without inconveniencing anyone and without any thought or action required.

The cost benefit of adding automated digital controls to your facility is a payback measured in months. Someone has described this as saying "I do not want to save a million dollars, but I want to save a dollar a million times." Small scale, digital, automated controls are the way to do this. The best targets are in order:

1) Refrigeration controls on reach-in coolers, walk-in cold storage, vendors, or any refrigeration
2) Lighting controls, especially dimmers and occupancy sensors
3) HVAC set backs, especially coordinating operation of multiple units
4) Industrial processes, especially running secondary functions off peak

Aggregation

Continuing up the value chain, the next step up the pyramid is using interval data to manage aggregation and energy user groups. Market experience in Pennsylvania shows that a user group can save 5% on kWh charge through bulk purchasing and 10% on the demand charge by smoothing the load factor. It is expected that gas and electric deregulation will eventually lead to significant bidding for aggregated loads. Anticipated targets are business, schools, and institutions. Massachusetts suppliers have announced numerous aggregations including all state and municipally-owned buildings, health and hospital facilities, colleges and education facilities, companies in the High Tech Council, companies in traditional industries, non-profit organizations, the Building Owners and Managers Association, and more. As a rule of thumb, eighty percent of power is sold to the twenty percent of accounts representing large users. This aggregation potential represents well over half the power sold in a region. Networking these meters will accelerate the benefits of deregulation. Utility deregulation, despite its rocky start and roll back in California, is proceeding in half the US states. Initially, the information available to alternative energy suppliers was minimal. Regulations specified that the incumbent utility had to provide customers with historical data. This data was monthly billing data, it was not normalized for weather conditions, and it was often months out of date. The energy marketing industry managed to get by with this static monthly data. Settlement between the generator and retail energy seller was often based on class load rather then the particular use and demand of a specific business. The wholesale supply contracts were "take or pay" so that if power was not used, the contract had to be paid for anyway. There was no way to adjust either the wholesale or retail contract to real-time conditions. Some of this is now changing.

The deregulation process in PA and MA illustrates what will likely happen elsewhere. The incumbent utilities have been given a grace period of five to six years to recoup their stranded costs. This period is now ending and businesses are getting exposed to real-time market prices. In Massachusetts this spring, businesses on the default rate saw regulated commodity prices rise from $0.05 to $0.08 per kWh raising their commodity cost by 50%. Each year, the formula for deregulation will create a more favorable climate for alternate power marketers. Additionally the FERC's push toward a wholesale power market platform, regional transmission organizations, and locational marginal pricing will all drive the need for real-time load management by your electricity supplier. For the first time in Massachusetts, electric bills indicate what region the customer is in such as "Boston NE." Power costs will now differ by location. If you want to go to sleep, your cost will rise. If you want to cooperate with your energy supplier, then you need to share information from your EIS in real-time. If your supplier knows you have contracted for more power than is being used, your marketer may sell the excess in the spot market. Power you purchased for $.08 may be worth $2.40 kWh (30 times as much) in the spot market. Even if your state has never touched deregulation, your wholesale market is deregulated. Eventually this will filter through to the retail market. You can sit still if you want, but the market is sending you a powerful message—if you want to control cost, you need information and you need an EIS.

Risk Management

The highest value in using real-time meter interval data on the web is risk management or the process of using contemporaneous data to adjust positions in the energy forward market, energy derivatives, or spot market. Energy trading may be conducted at a power exchange, through bi-lateral agreements between parties in the energy supply chain, or in the futures and options markets. On average, energy or power is traded five times before it is delivered. A single trading position may be planned a year ahead of delivery and adjusted six months out, three months out, and then at decreasing intervals of 1 month, 2 weeks, six days, three days, and one day ahead.

Today's traders have a lot of detailed contemporaneous data on supply and market pricing but little data about current demand. What data are available on specific high voltage transmission and distribution links is dis-aggregated and not keyed to customers. There is very little contemporaneous data available at the customer level. Most customer data are collected for monthly billing and is likely a month or more out of date. The internet offers the opportunity to access current demand data in a data warehouse giving traders insight into the unpredictable effects of climate, local weather, business operations, and demographic variation on the demand side.

Some traders want to avoid risk by locking in future prices; other traders seek profit in the spread between current and future prices. Energy risk managers, who do not expect to take physical delivery of the product, use financial derivatives such as futures and options to manage price exposure in the market place. These financial instruments are based on different estimates of the future, and the derivatives manager who has current demand data will have significant advantage over colleagues who are trading while looking through the rear view mirror.

Energy supply is no longer the simple process of delivering energy but an increasingly complex process of filtering energy through the financial markets and requiring historical and real-time data to adjust trading and marketing positions.

CONCLUSION

It is quite a puzzle why good managers have not tried, or have not succeeded, in wringing the waste out of monthly utility bills. Some people think of utilities in the same way as death and taxes—not much you can do about them. Others would rather not spend a dollar now to fix something even though that might bring two dollars back in the future. The future is a long way off. And a third group thinks, "My control systems aren't broken. Why should I fix them?" We hope you can now see the problem is a lack of information. With a proper web-based energy information system, you or your customers will know when facilities are wasting money, that the bottom line savings are significant, and that the key to savings is digital information and controls.[9]

It is true that the first hurdle, getting real-time energy data into the web is formidable. The cost is tangible and significant. The benefits seem diffuse and ephemeral. The intent of this chapter has been to show you value is a pyramid. The bottom of this pyramid is real-time data. In and of itself, it is expensive, nice to look at, but hard to value. It is the layers you erect on top of the base that create significant, tangible value. With each step up the pyramid, the value increases, but to a smaller group of people. Anyone can do load management, most can do demand response, many can automate their controls, some can do aggregation, and a few can do risk management. But all use the same data. So as an energy professional, your task is first to get facilities on the web, and then to connect them to higher and higher levels of value. We know from our experience, that you can take your facilities, or your clients on this climb up the pyramid and, as you get higher and higher, someone will thank you for it.

References

[1] California Public Utilities Commission. "Rulemaking 02-06-001, Proposed Pilot Projects and Market Research to Assess the Potential for Deployment of Dynamic Tariffs for Residential and Small Commercial Customers." Section 4.3 Advanced Metering System Deployment Cost Estimates. http://www.cpuc.ca.gov

[2] Square D H8050 Series. Self contained pulse transducer in a split core current transformer. http://www.veris.com/products/pwr/805x.html

[3] Burns, Kathleen, "Getting the Data—It's Simple But It's Not Easy," Energy Pulse, November 12, 2002. http://www.energypulse.net/centers/article/article_display.cfm?a_id=71

[4] California Public Utilities Commission. "Rulemaking 02-06-001, Report of Working Group 3." Section 2.4.2 Information Results: Effect of Information on Energy Usage. http://www.cpuc.ca.gov

[5] U.S. Energy Information Administration. http://www.eia.doe.gov

[6] Sunshine, Gary, "Estimating Markets for Energy Information Services," E Source, Platts Research & Consulting, 2002. http://www.esource.com

[7] Form 1, U.S. Federal Energy Regulatory Commission. http://www.ferc.gov/electric/f1/f1-viewer.htm

[8] California Public Utilities Commission. "Rulemaking 02-06-001, Report of Working Group 3." Section 2.4 Information Results: Customer Demand Response. http://www.cpuc.ca.gov

[9] Motegi et al, "Web-based Energy Information Systems For Large Commercial Buildings," Lawrence Berkeley National Laboratory, May 2002. http://eetd.lbl.gov/btp/buildings/hpcbs/Pubs.html

Chapter 14

Guide to Analysis Applications In Energy Information Systems*

Naoya Motegi
Mary Ann Piette
Satkartar Kinney
Karen Herter
Lawrence Berkeley National Laboratory

WHILE DATA STORAGE and visualization functions are increasingly offered by EMCS vendors, these features are often limited or underutilized. Energy information systems (EIS) are equipped with a range of features invaluable for understanding and utilizing the data collected by EMCS and secondary data acquisition systems. This chapter reviews several common EIS analysis functions and their roles in building performance analysis. Some additional analysis methods not currently widespread in EIS are also included. Lastly, the role of EIS in demand response programs is reviewed.

INTRODUCTION

The use and availability of web-based energy information systems (EIS) for buildings has been increasing in recent years; however, the range of EIS features and application options have not been clearly identified for consumers. The limited analysis features available in control systems generally go unused by building operators, so often they have little experience with large amounts of building data[1].

This chapter summarizes key application features available in today's EIS products and is taken from a more extensive categorization and review of EIS application capabilities conducted in 2002[2]. This project involved an extensive review of research and trade literature to understand the motivation for EIS technology development and information. This is not an exhaustive review of all EIS products or features. The following are the areas of applications discussed:

- Data analysis, including graphing, forecasting, benchmarking and financial analysis
- Fault detection and diagnostics
- Demand response evaluation and implementation

DATA ANALYSIS

Most EIS products have data visualization capabilities, some more complex than others. The most common data analysis features of EIS are designed to facilitate the analysis of whole building energy data. Data analysis functions range from simple plotting features to complex analytical routines. Diagnostic functions typically utilize HVAC component data to detect and diagnose performance problems. Financial analysis and whole-building energy benchmarking tools are also useful for building operators and energy managers to assess overall performance and conduct long-term planning.

Data Visualization

Visual presentation of data, in the form of graphs and charts, provides a simple way for operators and analysts to quickly summarize data and assess building performance. Here we discuss some of the graphing capabilities of EIS and their use in evaluating building performance. Some of the most common features are described below, and examples of data visualization are shown in 0.

- **Summary**—The data summary feature is used to aggregate data by day, week, month, or other se-

*This chapter is part of a report for the California Energy Commission, Public Interest Research Program, that was originally published by the Commercial Building Systems Group, Lawrence Berkeley National Laboratory, as Report No. LBNL—52510 (April 18, 2003)

lected time period. Data typically are displayed in bar charts and/or tabular summary statistics. Summary statistics are useful for analyzing the historical sequence of energy usage (for single buildings) over time, since monthly or yearly summary statistics can be compared.

- **Energy Use Breakdown**—Energy usage breakdowns show energy use for individual or multiple buildings, either by energy source (electricity, gas, oil, etc.) or by end use (lighting, chiller, refrigeration, etc.). Breakdowns by energy source are easier to obtain as data are collected separately. End use breakdowns require meters for every end use that is to be reported. It is common to meter the end uses that use the most energy, such as cooling equipment.

- **Multi-site Comparison**—Multi-site comparison is a crucial feature for EIS users with multiple buildings. The typical method of comparison is to plot whole-building energy consumption for each building in a bar chart. This is helpful in targeting sites for energy saving measures or retrofit.

- **Normalization**—Typically a client's sites differ in many respects. In order to make a fair comparison between buildings, some products utilize normalization techniques. Some normalization factors include building area, number of occupants, outside air temperature (OAT), and cooling or heating degree-days (CDD, HDD). Expressing the data in terms of energy use per square foot normalizes energy use data to account for variations in building size. Weather normalization techniques often involve a more complex method such as a regression model.

- **Load Duration**—A load duration curve indicates the percentage of time the load persisted at the defined levels of magnitude, in kW. If the curve is relatively flat, it indicates the energy consumption is steady. If the curve spikes up at the left side of the graph, it indicates that the peak demand is high and infrequent relative to the baseline, which could result in a high peak demand charge on the next utility bill.

- **X-Y Scatter**—X-Y scatter plots are useful for visualizing correlations between two variables. Some common X-Y plots used in chiller diagnostics are chiller tons vs. kW/cooling-ton and fan CFM vs. power. For whole building analysis, daily or monthly energy may be plotted against outside temperature for buildings with power consumption dependent on weather conditions. This is useful for users interested in managing peak load under high temperature conditions.

Time Series

One of the most common ways that building operators view EMCS data is through simple time-series graphs. With EIS, building operators and remote users have access to more sophisticated time-series plots, such as daily overlays, averages, highs and lows, point overlays, and 3D charts—all of which are combinations of different daily profiles or different layouts of the same time-series data. The main use of time-series visualization is to quickly analyze data trends. Some common time-series visualization features are described below and time-series visualization examples are shown in Figure 14-2.

- **Daily profile**—Time-series daily load profiles are displayed with time, in intervals of an hour or less, along the horizontal axis and load along the vertical axis. This is the most common EIS function for visualizing energy consumption data. It can be used to verify operation schedules, identify peak hours, and develop baseline load profiles. Incorrect scheduling of equipment (air-conditioning, lighting, etc.) can be quickly determined from the daily profile.

- **Day overlay**—Overlays plot multiple daily profiles on a single 24-hour time-series graph. Daily overlays are useful for finding abnormal days that would otherwise be difficult to find in single daily profile, and also can be used to obtain a quick estimate of the average 24-hour load profile of a building.

- **Average**—The Average function calculates average hourly energy consumption values for selected days and displays an average daily profile, which can be used for baseline reference.

- **Highs and lows**—Indicates maximum and minimum hourly consumption values for the day or plots a daily profile of the maximum and minimum day within selected days.

- **Point overlay**—Allows viewing of multiple time-series data points on the same graph. This is useful

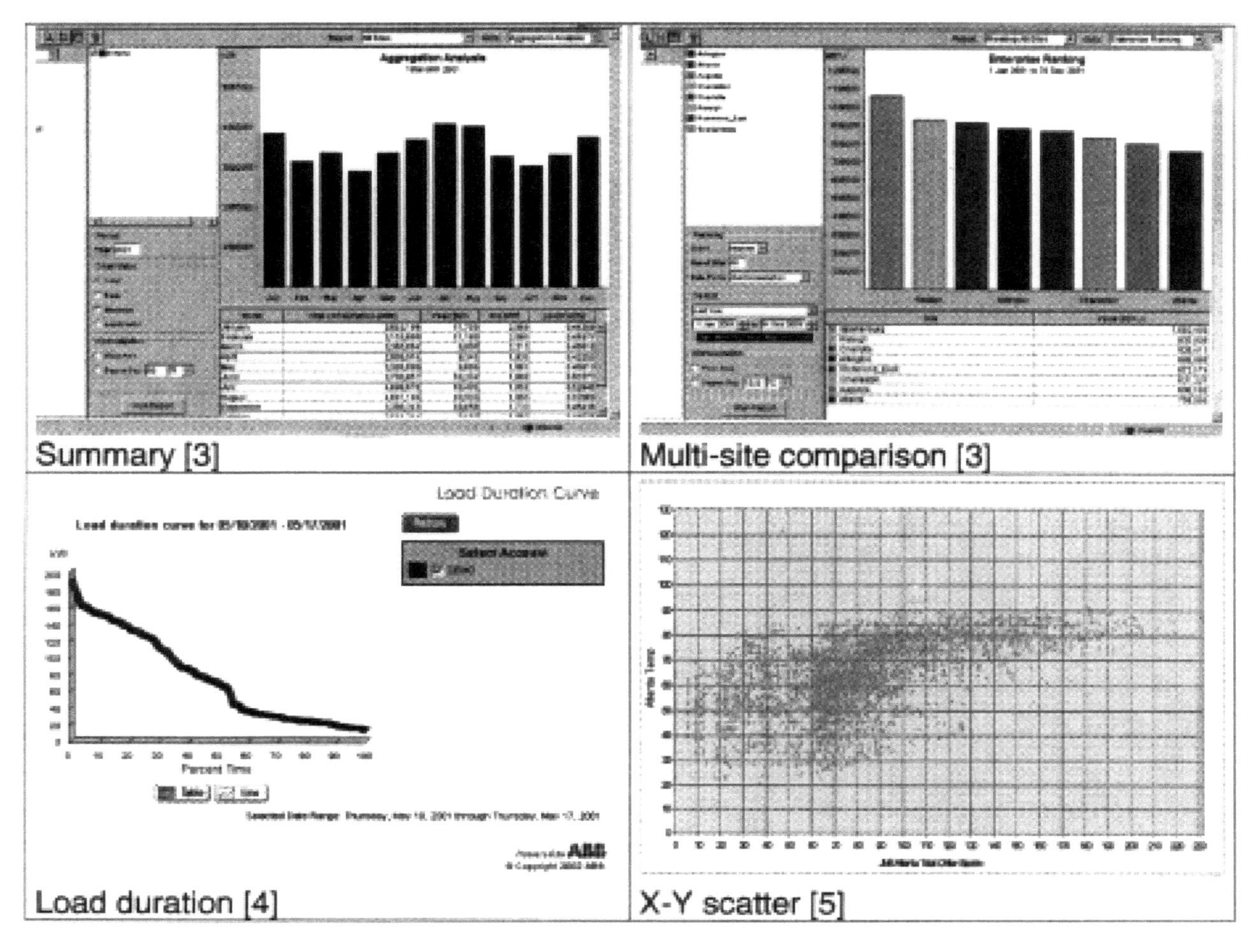

Figure 14-1. Examples of EIS Data Visualization

for finding the most energy-consuming sites (for multiple sites) or components (for sub-meters). It is also useful to overlay highly correlated data, such as power and temperature for office buildings. Some EIS products offer on-line access to real-time and forecasted temperature data.

- **3D chart**—Three-dimensional surface charts often display the time of day, date, and variable for study. These can be used to quickly pinpoint problematic time periods in large amounts of data. A user finding an abnormality in a 3D chart can then study a 2D daily profile in more detail or examine other detail graphs for that day. This is useful for a top-down diagnostic approach.
- **Calendar profile**—View up to an entire month of consumption profiles on a single screen as one long time series. The calendar profile displays the historical sequences of daily profiles and weekly trends.

Forecasting

Forecasting functions use historical time-series data to predict future values. In EIS this function typically consists of 2- to 5-day forecasts of electricity consumption. The methods and the parameters used for the forecast vary in each EIS. The primary use of forecasts is to plan future operational schedules, including DR program participation. When the forecast indicates high peak demand, the users can plan for a DR program curtailment or schedule peak avoidance strategies. A limitation of the forecasting feature in most cases is that human intervention is required to determine and implement operational changes in response to forecasts. Table 14-1 describes the forecasting capabilities of some existing EIS products.

Financial Analysis

Energy managers generally do not have time to incorporate energy usage information into their companies' financial strategies. Financial analysis tools provide

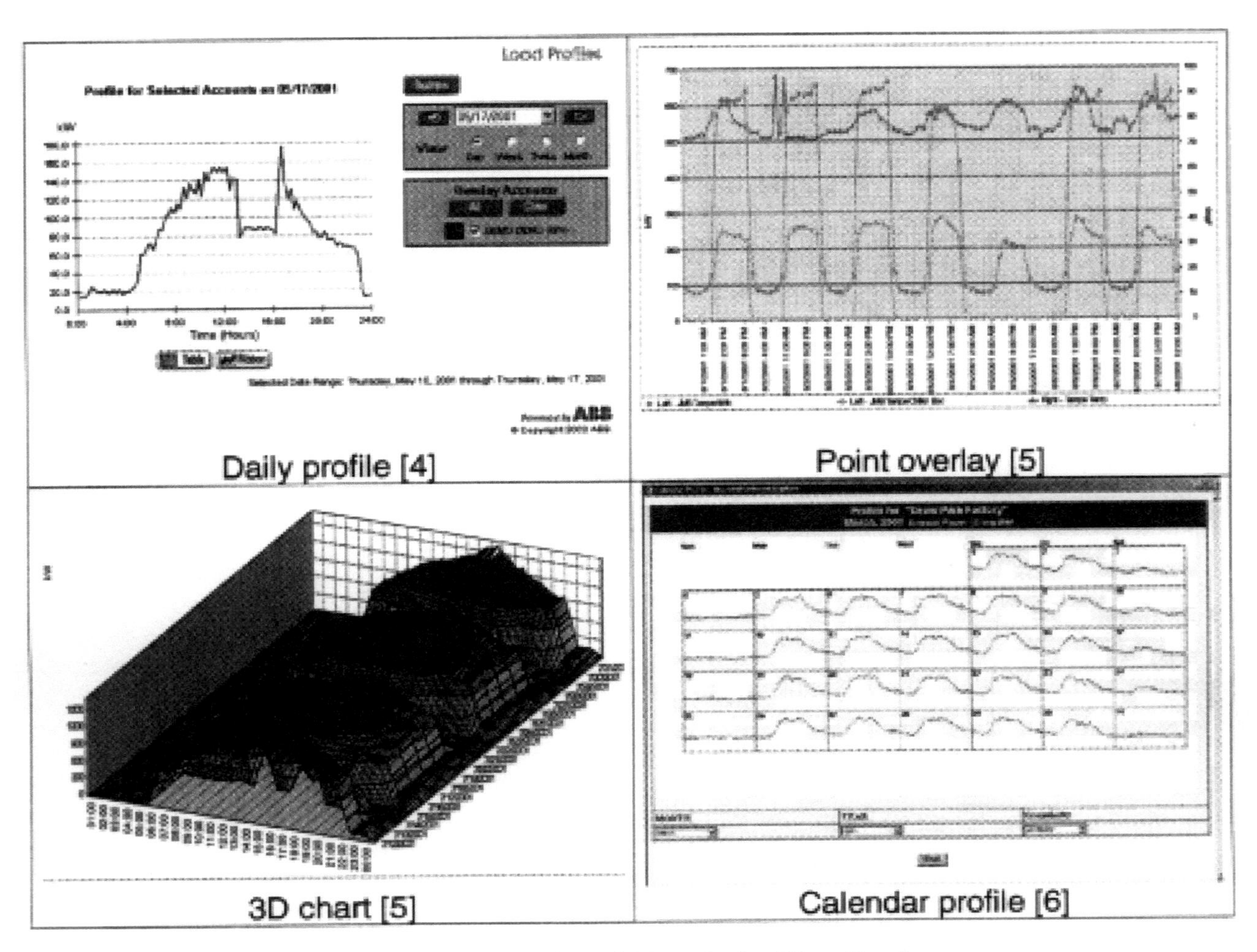

Figure 14-2. Examples of Time Series Visualization

Table 14-1. Forecasting Functions of Existing EIS Products

IUE	IUE automatically forecasts end-use-level, time-series electricity consumption by assessing device consumption data in conjunction with variables that drive energy use and cost, such as: external weather, space occupancy and temperature, real-time prices, and market demand. These forecasts can be viewed at both the building and the portfolio level. IUE's Intelligent Agents utilize data feeds every 2 minutes to learn building consumption patterns and their reactions to change. IUE can predict future energy usage for any period of time—1 hour, 2 days, and more than 30 days. Accuracy of the forecast, however, depends on the accuracy of weather forecasts and other data inputs[7].
Automated Energy	Automated Energy incorporates data from AccuWeather, the internet weather provider, into its system so that weather can be factored into its forecasts[8].
EEM Suite	EEM Suite's Forecasting module normally uses its own proprietary algorithm based on historical load and weather data to forecast 24- to 48-hour energy usage. However, it has the capability to use customer-generated algorithms. Various statistical analysis techniques are used to calculate expected consumption or demand on an hourly or daily basis. Techniques include facility and profile cluster analysis, weather and production normalization, day-type normalization and hybrid statistical-neural network modeling[5].

a powerful way for energy managers to identify costly energy behavior, plan cost-saving procurement, or assess economical profitability of DR events. Common features of financial analysis functions are defined as follows.

- **Rate tariff**—Software contains a series of rate tariffs to fit to clients' utility rates. Users can compare energy procurement alternatives or estimate DR event benefits.
- **On-line rate tariff**—Rate tariff data can be dynamic according to a real-time price structure downloaded from the web.
- **Bill validation**—Utility bills are compared to meter readings to validate the accuracy of bills.
- **Real-time tracking**—Calculates electricity costs every day or hour using real-time meter reading and rates.
- **Breakdown summary**—Breaks down monthly utility bills by daily or hourly profile and identifies the highest cost day or hour. This combines aspects of the real-time tracking and rate tariff features; however, a breakdown summary may also be performed with downloaded or historic data that is not necessarily provided in real-time.
- **End-use allocation**—Estimates end-use energy consumption from whole-building energy use according to user-defined parameters and algorithms. Generally used for cost allocation to building tenants. A common parameter definition is energy consumption per square foot.

Rate tariff features are often included in EIS offerings from utility companies. The breakdown summary shows the product of rate and hourly load profile, which may indicate more striking results than electricity consumption data alone. Final results depend on the hourly rate, demand charge, and other rate features. These charges can comprise a large part of the total utility bill, so the breakdown summary, a cost structure approach, has significant energy cost-saving potential. Although end-use allocation is not really considered an energy-saving measure, it may lead to energy savings. It can be used to allocate the utility bill among tenants on a fair basis, and it can help building managers understand where they use energy. Table 14-2 describes the financial analysis capabilities of some existing products.

Benchmarking

Benchmarking compares the energy use of one building with the energy use of other buildings. Benchmarking can be useful in energy audits and for targeting individual buildings for energy-saving measures in multiple-site audits. Energy service companies and performance contractors may use "best practice" and "typical" benchmarks, such as annual energy use intensities (EUIs), to communicate energy savings potential. On the other hand, control companies and utilities tend to track actual energy usage and compare it to historical data and/or combined data from multiple buildings. Energy managers and building owners have an ongoing interest in comparing energy performance to

Table 14-2. Financial Analysis Functions of Existing EIS Products

EP Web	EP Web has the basic features of financial analysis, including rate tariff, bill validation, real-time tracking, and end-use allocation. These features can be applied to any monitored or logged value[6].
Automated Energy	Automated Energy estimates upcoming bills using load forecasting and "what-if" scenario analysis.[8].
EnterpriseOne	The RateEngine module compares purchasing options from multiple energy providers and verifies cost/benefit analyses by third party energy service companies[9].
EEM Suite	EEM Suite has wide variety of financial analysis tools, including all of the financial analysis functions listed above[5].
UtilityVision	UtilityVision integrates performance contract procedures into the EIS, showing a monthly summary of energy savings with baseline, target savings, utility rate, and estimated cost savings.[10]

other similar buildings. Integration of benchmarking information in EIS provides users with constant information on the performance of their buildings relative to each other and to other similar buildings.

Benchmarking features are becoming increasingly common in EIS. Typical benchmarking features allow comparisons between different buildings and loads connected to the EEM. A few EIS tools are also starting to provide comparisons to public survey data or industry benchmarks.

There are a number of free on-line benchmarking tools that could easily be linked or integrated into any other on-line EIS. For example, Silicon Energy is providing benchmarking information to its California customers by sending queries to LBNL's Cal-Arch benchmarking tool. Most national benchmarking tools we have found are based on DOE's CBECS (Commercial Building Energy Consumption Survey[11]. Two examples of on-line benchmarking tools are given in Table 14-3.*

DIAGNOSTICS

Detecting and diagnosing problems in buildings can lead to improved control and occupant comfort, prolonged equipment life, and lower energy use and maintenance costs. The data visualization and analysis features described in the previous section are useful for analyzing detailed trend data to detect operational problems; however, human expertise is required to determine which data points are to be plotted and what patterns in the data or graphs indicate a problem. Even knowledgeable users may find this to be time-consuming; moreover, user selections will not be consistent among all sites and over time. This section describes some features that provide additional guidance in building performance diagnostics as well as some that run automated diagnostic procedures.

Pre-defined Diagnostic Graphs

Data analysis can be a time-consuming process, even with the data visualization tools described previously. Some EIS products facilitate building diagnostics by providing streamlined access to the most commonly used graphs. Pre-defined diagnostic graphs designed for specific variables may still require that the user interpret the information, but can save valuable time. Commonly viewed plots such as chiller tons vs. tons/kW could be prepared as built-in charts. Figure 14-4 shows an example of a pre-defined chart.

*A list of on-line benchmarking tools is maintained at http://poet.lbl.gov/cal-arch/links/.

Table 14-3. Benchmarking Functions of Existing EIS Products

Cal-Arch	Cal-Arch is an on-line benchmarking tool that provides distributions of actual energy use in California buildings. It is based on California's Commercial End Use Survey, which has data on approximately 2,000 buildings. Users can compare the energy use in a building to other buildings in the database and have the option of comparing by size, climate zone, and building type. The results are displayed as histograms (Figure 14-3) and summary statistics of the energy use for the buildings contained in the database[12].
Energy Star	Energy Star's Portfolio Manager is the most commonly used national tool. Portfolio Manager allows users to track multiple buildings over time and calculate Energy Performance Ratings (EPR). Eligible buildings with an EPR higher than 75 are said to be among the top 25 percent in terms of energy performance and qualify for an Energy Star buildings label. The scores are developed from complex regression models developed using CBECS data. Separate models and scoring systems were developed individually for several different building types[13].

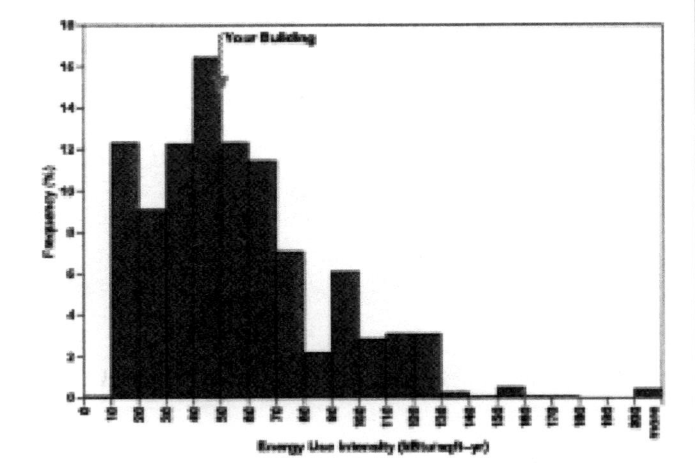

Figure 14-3. Cal-Arch Benchmarking Plot

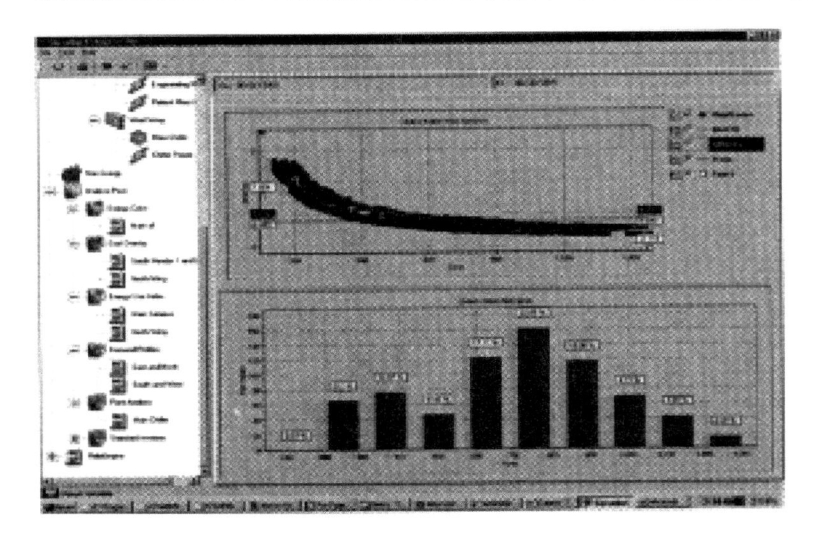

Figure 14-4. Tons vs. Tons/kW Chart and Tons Histogram[9].

Automated Diagnostics

Sophisticated building energy information systems have the potential to deploy automated fault detection and diagnostic analysis. Although many EIS products provide alarm features that produce alarms according to simple algorithms such as maximum/minimum limit, this is not sufficient to detect many HVAC problems. Some EIS, however, do have automated diagnostics functions. These more advanced tools will run algorithms to search for anomalous data representing potential faults and suggest possible causes and solutions. Diagnostic tools integrated with an EIS, EMCS, or other data acquisition system provide a greater opportunity for real-time analysis and a greater level of automation because data can be automatically transferred from the EIS into the diagnostic application. Automated diagnostic procedures can be run continuously or periodically, requiring user action only when a fault is detected. This saves a great deal of time and the embedded expertise allows less-experienced users to perform high-level assessments[14].

One type of automated diagnostics is rule-based diagnostics, in which system failures and operational problems are detected by rule-based algorithms, and the diagnosis presents potential causes and recommendations to alleviate a problem. For example, an algorithm can be programmed to send an alarm if the supply fan air pressure passes a minimum setpoint. This is only a very simple example, but an algorithm can be made very complex by combining multiple subroutines or calculating a parameter from multiple data with mathematical formulas.

While these automated diagnostics features are emerging in EIS, they were originally developed as EMCS data diagnostic tools[14]. Additional discussion on automated diagnostics is found in Xu and Haves[15]. Here are some examples of EIS that have automated diagnostics either in practice or in progress:

Other Diagnostic Tools

Diagnostics in EIS are generally focused on whole-building energy use. There are many other types of diagnostics procedures, with other system components and faults diagnosed using automated methods. A range of

Table 14-4. Automated Diagnostics Functions of Existing EIS Products

EMAC	eMAC has a rule-based diagnostic algorithm in its on-site controller and diagnoses rooftop units in real-time. If the controller detects an abnormality, it immediately sends an alarm with estimated cause via e-mail, pager, and/or some other method[16]
UtilityVision	UtilityVision's FacilityVision module uses PowerNet as its automated diagnostics platform. Professional agents operate FacilityVision, and provide diagnostic solutions to their clients[10]
EEM Suite	EEM Suite includes a rule-based exception analysis and web reporting capability to alert personnel to abnormal equipment, process and/or facility operating performance. Expected energy consumption and/or electrical demand are compared to actual values to identify anomalies within commercial facilities or industrial processes[5]
FX-TEM	Statistical methods are applied to hourly time-series data to detect anomalous values based on daily energy consumption and peak load. Energy consumption for a particular day is compared with recent energy consumption for up to 30 days of the same day type (e.g., Weekdays). Anomalous days are often an indication of faulty equipment operation or scheduling problems[17].

diagnostics tools exist that facilitate data collection, analysis, and visualization, enabling building operators and energy engineers to continuously assess building performance. At a minimum, a diagnostic program for building analysis will process data and provide summaries of relevant performance metrics and common diagnostic plots for manual analysis. Friedman and Piette discuss diagnostics in greater detail and compare several diagnostic tools[14]. Table 14-5 shows examples of diagnostic tools and methods.

DEMAND RESPONSE PROGRAMS

The rise in demand response (DR) program implementation has corresponded to a rise in EIS-type tools that facilitate participation in these programs. In addition, DR features have been added to existing EIS products. The use of these features may also be useful for demand management, regardless of program participation. Some of the analytic functions that are useful for demand response include forecasting, baseline and savings calculations. In addition, web-enabled EIS are used in DR for communication between the program administrator and program participants, and in some cases, for remote implementation of control sequences for demand reduction.

Evaluating Demand Reduction

Energy information systems that are designed for demand reduction program participation typically include a savings analysis feature. The demand reduction is measured by predicting what the load shape would have been in the absence of a demand reduction event, typically referred to as a baseline. The amount of the reduction is estimated to be the difference between the baseline and actual power use as monitored by the electricity provider. The baseline calculation methods used by many DR programs are designed to be simple to understand and implement and may not necessarily reflect accurately the demand reduction achieved. Other forecasting methods can be used to obtain better estimates of a building's load in the absence of a power curtailment. Xenergy has reviewed several demand reduction baseline methods[19].

Implementing Demand Reduction

Implementation of a reduction in peak demand is greatly simplified and more quickly achieved through the use of special control routines. Control routines may

Table 14-5. Other Diagnostic Tools Available in Existing EIS Products

PowerVisor	PowerVisor provides a platform for creating and running diagnostic routines. While some commonly used routines are provided in the software, the user or administrator can also program custom routines with a built-in graphical programming language. When routines trigger an alarm, the alarms can be sent to a web page, pager, or email address along with recommendations and related information[18].
AHS Plots	Annual Hourly Statistical Plots (AHS Plots) were developed to aid in the analysis of data collected over several years in the Energy Edge project carried out by LBNL and Bonneville Power Administration between 1986 and 1993. The plots, as shown in 0, have continued to be useful in commissioning and diagnostics research; however, to our knowledge, no on-line or commercial software automatically generates them. An AHS Plot is actually 12 monthly time-series plots displayed on a single page. Each month is divided into daily hours between 0 and 23, and for each hour, the mean, median, maximum, minimum, and quartiles are calculated and displayed on the graph. Some examples of information that can easily be read off the graphs include scheduling errors, start-up control, and peak usage.

Table 14-6. Savings Estimation Functions of Existing EIS Products

EEM Suite	EEM Suite's Curtailment Manager calculates 10-day baselines for savings verification, which can be customized to adjust to a variety of baseline formats. Pacific Gas & Electric deploys Curtailment Manager, which is customized to use the California ISO baseline method[5].

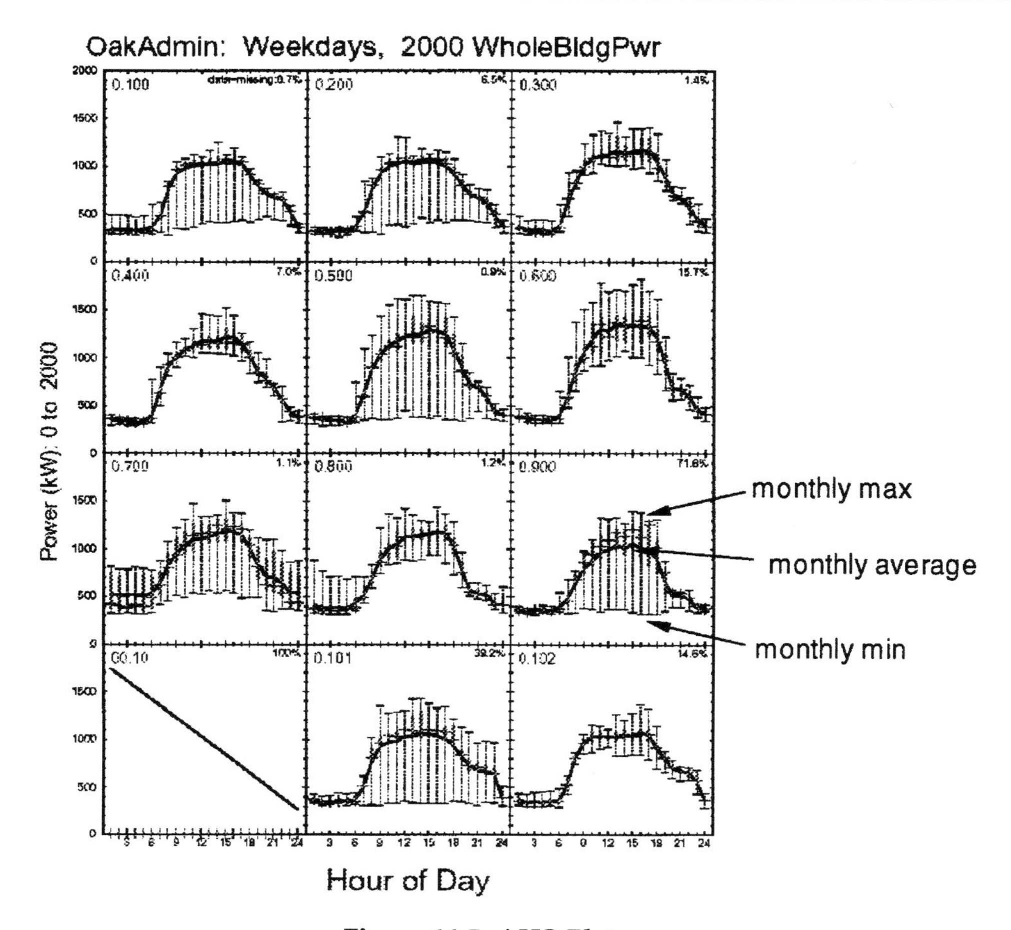

Figure 14-5. AHS Plots

either be implemented manually by an operator or automatically implemented in response to a signal from a DR program administrator. The advantage of using an EIS for control is that a user can operate multiple buildings at once over the internet, allowing energy managers to act promptly on DR events.

Manual control capabilities have various degrees of facilitation. For example, the user can customize a range of demand-shedding strategies for each site and execute the strategies remotely when a curtailment event occurs. An example of a hypothetical demand-shedding strategy is shown below.

Site A	Level-1:	Thermostat setting up to 76°F
	Level-2:	Thermostat setting up to 82°F
Site B	Level-1:	60% load chiller operation
	Level-2:	30% load chiller operation

Popular control strategies are chiller load reduction, load rotation, thermostat reset, supply air temperature reset, and dimmed lighting. The levels to select depend on the level of curtailment needed. These preset control commands facilitate DR event participation and other energy-saving strategies, so that a user can easily execute load shedding across all sites. Table 14-7 is an example of manual control functionality.

Some EIS products function as an EMCS by automatically controlling building facilities in response to monitored building conditions. These automated control technologies may also use real-time control algorithms for DR programs, so that even energy managers do not have to respond to an alert in real-time. Often these automated control capabilities are not utilized, partly because clients are not fully convinced of their usefulness and reliability. Further demonstrations and user education will be required to popularize the use of these capabilities.

SUMMARY AND FUTURE DIRECTIONS

This chapter has reviewed the most common data visualization and analysis features found in energy in-

Table 14-7. EIS Products with Manual Demand Response

UtilityVision	Each site has its own load curtailment settings ranging from level 1 to 5. These curtailment settings include chiller shedding, dimming lights, and changing temperature setpoints. The user sets the start time, duration, and curtailment level. When the start time comes, UtilityVision automatically controls the facilities according to the user's manual settings[10].

Table 14-8. EIS Products with Automated Demand Response

IUE	IUE's Intelligent Agents automatically deploy proprietary, pre-defined intelligent conservation strategies to minimize energy usage. Based on a forecast, energy-saving strategies are aimed at permanent load reduction, peak load management (avoidance and reduction), and voluntary and emergency curtailment. Specific tactics include: set point optimization, supply air temperature optimization, 24-hour-7-day-a-week load shed, and load rotation. A pre-cooling strategy is now in development. All of the strategies that are automatically and manually deployed can be viewed in real-time on-line[7].
eMAC	Unlike most web-EMCS products, eMAC comes with predefined but customizable control strategies. eMAC has several modules specified for rooftop units, lighting and refrigeration. This "component-specific" design reduces the system variation of their products and lowers the production cost[16].
UtilityVision	UtilityVision deploys load curtailment according to a proprietary electric demand forecast system.† UtilityVision automatically starts up a curtailment event when the demand forecast exceeds a user-defined peak demand or when the target demand reduction is not being met by on-site electricity generation. UtilityVision then attempts to meet the requested target demand reduction by choosing the appropriate curtailment level. Once the desired demand reduction is met, UtilityVision processes a series of commands to reverse the curtailment without causing unwanted spikes in the electric demand due to re-energizing equipment within a short period. Currently this automation feature is under field test at some sites with photovoltaic systems.[10].
EEM Suite	Using the Universal Calculation Engine module, a user can create a control algorithm that calculates energy conditions and automatically sends alarms or controls equipment in real-time. In combination with its Forecasting module, it can also implement "forward-basis" control strategies[5].

formation systems today. The market is rapidly changing, however, and thus this review is not intended to be exhaustive nor does it include all products. However, the information should provide the reader with a basic understanding of the types of features available to them in an EIS and some of the ways in which these features can be used to help reduce energy use, peak demand, and energy costs in buildings.

This chapter also discussed the use of EIS applications in demand reduction evaluation and implementation. Chapter 7 discusses the technical issues associated with communication between EIS at participating sites and program administrators for demand response programs. An additional project, currently underway, by LBNL and the California Energy Commission will evaluate automation in demand-responsive buildings using EIS and related technologies, including issues of communication, control strategies, and level of reductions attained. LBNL also has an ongoing interest in commissioning and diagnostics. Related research will continue to explore how to best utilize data of the type collected by EIS to detect and diagnose building performance problems.

References

[1] Piette, M.A., S. Kinney, and P. Haves. "Analysis of an Information Monitoring and Diagnostic System to Improve Building Operations," Energy and Buildings 33 (8) (2001), 783-791.

[2] Motegi, N., M.A. Piette, S. Kinney, and K. Herter, 2003. "Web-based Energy Information Systems for Energy Management and Demand Response in Commercial Buildings." Report to the California Energy Commission.

[3] Tridium Software, 2002. Vykon product information. http://www.tridium.com.

[4] ABB Energy Interactive, 2002. Energy Profiler Online product information. http://www.energyinteractive.com.

[5] SiliconEnergy Inc., 2002. Enterprise Energy Management Suite product information. http://www.siliconenergy.com.

[6] eLutions, 2002. Energy Partner web product information. http://www.elutions.com.

[7] WebGen Systems, 2002. Intelligent Use of Energy product information. http://itswebgen.com.

[8] Automated Energy, 2002. Automated Energy product information. http://www.automatedenergy.com.

[9] Circadian Information Systems, 2002. Enterprise One product information. http://circadianinfosystems.com.

[10] CMS Viron, 2002. Utility Vision product information. http://www.cmsenergy.com/MST/.

[11] Kinney, S. and M.A. Piette, 2002. "Development of a California Commercial Building Energy Benchmarking Database." Proceedings of the 2002 ACEEE Summer Study on Energy Efficiency in Buildings.

[12] Lawrence Berkeley National Laboratory, 2002. Cal-Arch Building Energy Reference Tool. http://poet.lbl.gov/cal-arch/.

[13] U.S. Environmental Protection Agency Energy Star Program, 2002. Benchmarking/Portfolio Manager. http://yosemite1.epa.gov/estar/business.nsf/.

[14] Friedman, Hannah, and M.A. Piette, May 2001. "Comparative Guide to Emerging Diagnostic Tools for Large Commercial HVAC Systems," Final Report for the California Energy Commission, Public Interest Energy Research Program, High Performance Commercial Building Systems Program, LBNL Report #48629.

[15] Xu, Peng, P. Haves, 2002. "Field Testing of Component-Level Model-Based Fault Detection Methods for Mixing Boxes and VAV Fan Systems." Proceedings of the 2002 ACEEE Summer Study on Energy Efficiency in Buildings. LBNL Report #50678.

[16] Pentech Solutions, 2002. EMac product information. http://www.pentechsolutions.com.

[17] Seem, J.E., 2002. "Using Intelligent Data Analysis to Detect Abnormal Energy Consumption in Buildings."

[18] PowerNet Software, 2002. PowerNet product information. http://www.powernetsoftware.com.

[19] Xenergy, Inc., 2002. Protocol Development for Demand Response Calculation—Draft Findings and Recommendations. Consultant Report to the California Energy Commission, Sacramento, CA.

Acknowledgments

The authors are grateful for assistance from Grayson Heffner (LBNL), and our sponsors Martha Brook, California Energy Commission (CEC), and David Hansen, US Department of Energy (DOE). We are also grateful to the numerous EIS and web-EMCS vendors who assisted in this effort. This work is part of LBNL's High Performance Commercial Building's Program, Element 5 (Integrated Commissioning and Diagnostics). This program is supported by the CEC and by the Assistant Secretary for Energy Efficiency and Renewable Energy, Office of Building Technology, State and Community Programs of the U.S. DOE under Contract No. DE-AC03-76SF00098.

Chapter 15

Data Analysis and Decision Making: Using Spreadsheets and "Pivot Tables" To Get A Read On Energy Numbers

Partha Raghunathan
Director, Product Marketing
i2 Technologies, Inc, Dallas, Texas

IN RECENT YEARS, the computing world has witnessed a large-scale adoption of spreadsheet technology by business users worldwide to gather and manage numeric information. Today, spreadsheets have become the de-facto standard for business data analysis. Spreadsheet applications like Microsoft® Excel© are very easy to use, easy to share, have many rich in-built analytic and statistical functions and can handle large volumes of data while still providing sub-second performances and, finally, are easily integrated to web pages on the internet.

This chapter's objective is to take a deeper look at data analysis features in spreadsheets and understand how they can be applied to common tasks performed by an energy manager. Using a simple tutorial, we will attempt to show how spreadsheets can be a very effective, cheap and simple solution for most data analysis needs of an energy manager.

INTRODUCTION

The last decade has seen a massive proliferation of software applications and databases in almost every conceivable business process. From buying a book on the internet to paying bills from a bank account to tracking a FedEx® package, web-based software systems are dramatically changing the very way we live our lives. These systems allow companies to electronically capture vast amounts of information about their customers, their suppliers and their competitors. Consequently, these companies that are now capturing vast amounts of information, are demanding sophisticated analysis tools to deal efficiently with this information and make profitable decisions. Mainly driven by this commercial segment, the data analysis and decision support industry market has seen a huge boom in the past decade—technologies like Data Warehousing, Business Intelligence, On-line Analytical Processing (OLAP), Data Mining etc. have seen massive advances—as early adopters of these technologies attempt to get a competitive advantage through intelligent analysis of data culled from consumers, competitors and business partners.

As a few elite business analysts ride this wave of leading-edge, decision-support technology on the lagging edge, business users worldwide are very quietly beginning to use everyday spreadsheet technology in very imaginative ways, stretching its limits to accomplish the most sophisticated of analyses. Hundreds of software vendors are using more advanced features of spreadsheets such as the open application programmable interfaces (APIs) to develop complex business applications for budgeting, forecasting, manufacturing and even strategic operations planning as well as consumer applications such as personal financial planning.

Today's spreadsheet applications, like Microsoft® Excel©, are very easy to use, easy to share, have many rich, in-built analytical functions and can handle large volumes of data while still providing sub-second performances and, finally, are easily integrated to web pages on the internet.

SPREADSHEET TUTORIAL

This chapter contains a tutorial on using spreadsheets, showing how they can be an effective solution for most data analysis needs of an energy manager.

What You Will Need to Follow Along with the Tutorial

To go through this tutorial on your computer, you will need the following:

- A Personal Computer running on Windows® platform (98, 2000, NT 4.0, XP or higher)
- A licensed copy of Microsoft® Excel© version 9.0 or higher
- Access to the spreadsheet containing the sample input data for the tutorial which is located on the web at: http://www.utilityreporting.com/spreadsheets/IT_2003_Plastico_Sample_Data.xls
- At least 128 MB RAM (256 MB or more preferred)
- Older versions of Windows platforms as well as Microsoft® Excel© may also be used but certain parts of the tutorial may require work-arounds. Other spreadsheet applications like Lotus® 1-2-3® may also be used, but there will be minor variations in the tutorial steps.

CASE STUDY FOR TUTORIAL

This case study shows how an energy manager, Bob Watts, uses spreadsheet technology to gather quick insight into the energy consumption patterns of a small commercial customer. This example will also show how this insight leads to decisions resulting in reduced energy usage and costs. Further, we will illustrate how Bob can easily share his findings with his customer and his peers over the web. In our estimate, this whole analysis would take Bob, a skilled spreadsheet user, less than 2 hours to complete from start to finish.

Meet Bob Watts, an account manager at a leading utility company in Atlanta, Georgia. Bob owns a few large commercial customer accounts and is responsible for providing services such as energy cost reduction programs to his customers, not only to keep customers happy but also to make sure the utility company doesn't run into an energy shortfall.

Bob recently visited Plastico Inc., a manufacturer of small plastic widgets, to assess their energy usage patterns and suggest ways to reduce consumption and costs. Plastico is headquartered in small town, south of Atlanta, in Georgia and has sales offices around the country. The headquarters in Georgia has 3 major locations:

- Office: Offices for sales & marketing staff, executive management staff.
- Warehouse: Holding area for finished product inventory to be loaded onto trucks.
- Plant: Facility for manufacturing, testing and packing operations.

While he was at Plastico, Bob collected information about the various electric and gas equipment in the 3 major areas, along with estimated hours of operation. Table 15-1 shows a sample snapshot of the types of data Bob was collecting. The column "Efficiency" is Bob's qualitative assessment of the relative efficiency of the equipment (for example, a standard incandescent would rate "Low," a standard fluorescent would rate "Medium" and a high-efficiency compact fluorescent lamp would rate "High").

Bob now wants to analyze the data he has gathered. The next section will describe how Bob converts this data into meaningful information. First, he enters all the data into an Excel© spreadsheet. Table 15-1 shows sample data in a spreadsheet. In this tutorial, for the sake of simplicity, we will only consider the information for lighting appliances, although Plastico also has several pieces of production equipment, gas heaters and furnaces, and the HVAC systems. This example can easily be extended to all kinds of energy equipment.

Table 15-1: Sample energy Consumption Data Collected by Bob at Plastico

Location	Area	Category	Sub Category	Quantity	Efficiency	Rating (Watts)	Annual Usage (hours)	Comments
Plant	Shopfloor	Lighting	HPS	316	Low	300	8760	
Warehouse	Working Area	Lighting	Fluorescent	240	Medium	48	8,760	
Office	Cafeteria	Lighting	Fluorescent	156	Medium	48	2,480	
Warehouse	Main Storage Area	Lighting	Metal Halide	150	Low	480	8760	
Warehouse	Cafeteria	Lighting	Fluorescent	128	Medium	48	8760	
Warehouse	Back-end Storage	Lighting	Metal Halide	100	Low	480	3744	
Plant	Cafeteria	Lighting	Fluorescent	96	Medium	48	8760	
Office	Laboratory	Lighting	Fluorescent	92	Medium	28	8760	
Office	Back Office	Lighting	Fluorescent	76	Medium	48	2,480	

Step 1. Preparing the data for Analysis

The first step involves performing all basic computations. In this case, Bob uses "Annual Usage (hours)," "Rating (Watts)" and "Quantity" to compute estimated "Annual Usage (kWh)" and "Demand Usage (kW)" using the formula:

Annual Usage (kWh) = Annual Usage (hours) (Rating (watts) (Quantity/1,000 (Wh/kWh)

Demand Usage (kW) = Rating (watts) (Quantity/1,000 (Wh/kWh)

Table 15-2 shows the computed column along with the collected data. It is possible to have several other computed columns like "Average Hours of Operations Per Day," "kWh consumed per day" etc. using simple spreadsheet formulas.

Step 2: Creating a "Pivot Table"

Once the data are complete and ready for analysis, Bob creates a "pivot table" for his analysis. A "pivot table" is an interactive table that Bob can use to quickly summarize large amounts of data. He can rotate its rows and columns to see different summaries of the source data, filter the data by displaying different pages, or display the details for areas of interest.

> *Tip: For more information on "pivot tables," refer to Microsoft® Excel© Help on PivotTable"*

To create a pivot table, Bob highlights all the data in the spreadsheet and selects from the Excel© Menu, Data–> PivotTable and PivotChart Report.

In the window titled, "PivotTable and PivotChart Wizard - Step 1 of 3," Bob clicks "Finish." This results in the creation of a new worksheet titled "Sheet2" that has a blank table and a floating menu titled "PivotTable."

Step 3: Some basic reports

Once the pivot table has been created, Bob wants to generate some very basic reports to understand where to begin looking for opportunities. There are three reports that Bob wants to create to understand these basic trends in Plastico's energy consumption patterns:

Table 15-2. Fully Computed Data

Location	Area	Category	Sub Category	Quantity	Efficiency	Rating (Watts)	Annual Usage (hours)	Annual Usage (kWh)	Demand Usage (kW)	Comments
Plant	Shopfloor	Lighting	HPS	316	Low	300	8760	830,448	94.8	
Warehouse	Working Area	Lighting	Fluorescent	240	Medium	48	8,760	100,915	11.5	
Office	Cafeteria	Lighting	Fluorescent	156	Medium	48	2,480	18,570	7.5	
Warehouse	Main Storage Area	Lighting	Metal Halide	150	Low	480	8760	630,720	72.0	
Warehouse	Cafeteria	Lighting	Fluorescent	128	Medium	48	8760	53,821	6.1	
Warehouse	Back-end Storage	Lighting	Metal Halide	100	Low	480	3744	179,712	48.0	
Plant	Cafeteria	Lighting	Fluorescent	96	Medium	48	8760	40,366	4.6	
Office	Laboratory	Lighting	Fluorescent	92	Medium	28	8760	22,566	2.6	
Office	Back Office	Lighting	Fluorescent	76	Medium	48	2,480	9,047	3.6	

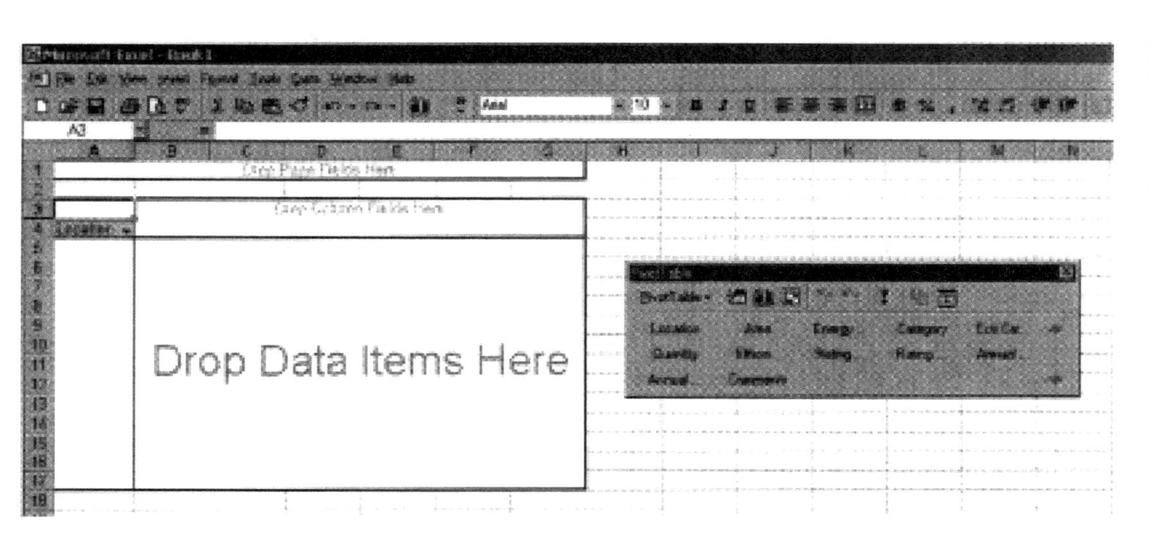

Figure 15-1. A Pivot Table (left) with the Floating Pivot Table Menu (right)

- #1 - Annual Usage (kWh) by Location—i.e. office, warehouse, plant
- #2 - Annual Usage (kWh) and Demand (kW) by sub-category—i.e. fluorescent, incandescent etc.
- #3 - Annual Usage (kWh) and Demand (kW) by efficiency—i.e. high, low, medium

Report #1: Annual Usage (kWh) by Location

To create this report, Bob starts with the empty pivot table created in Step 2.

1. Bob selects the "Location" field from the "pivot table" menu and drags it onto the "Drop Row Fields Here" section in the empty pivot table.
2. Next, he selects the "Annual Usage (kWh)" field from the "pivot table" menu and drags that onto the "Drop Data Fields Here" section in empty pivot table.
3. Bob then double-clicks on the pivot table title "Count of Annual Usage (kWh)." In the "PivotTable Field" window, he selects "Sum" instead of "Count" (Figure 15-2).

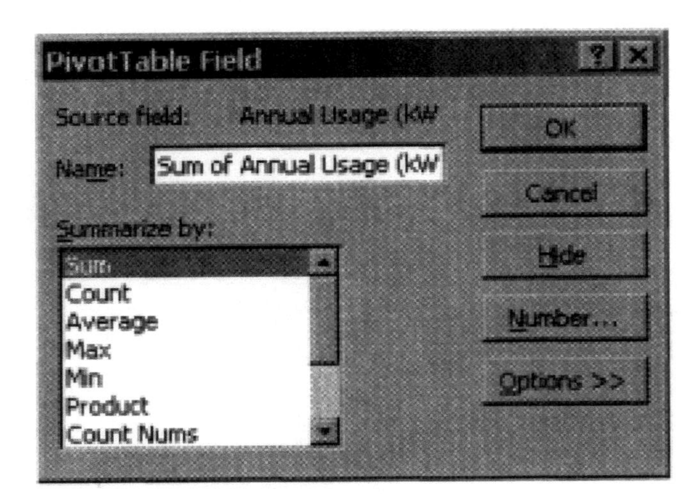

Figure 15-2. Changing the Summarize Option of a Pivot Table Field

4. In the same window, Bob also clicks on the button "Number," formats the numeric display to "0 decimals," and checks the box "Use 1000 separator (,)." He then clicks on "OK" in the PivotTable Field window. Table 15-3 shows a finished report "Annual Usage (kWh) by Location."

(This report has also been formatted. To apply a format to a finished report, you can right-click anywhere on the pivot table and select "Format Report" and choose one of the several standard formats. The format selected below is "Table 7.")

Table 15-3. Report #1—Annual Usage (kWh) by Location

Location	Annual Usage (kWh)
Office	162,183
Outdoor	159,782
Plant	970,377
Warehouse	1,015,070
Grand Total	2,307,413

Tip: Unlike the normal formatting in a spreadsheet, pivot table formats are dynamic. By applying the formats once, the fonts, colors, etc. are preserved automatically as the content in the table changes. For example, if Bob adds more lighting entries in the main spreadsheet, all he has to do is refresh the pivot table and the results are automatically recomputed.

5. This report can easily be converted into a chart by clicking on the "Chart" icon on the Excel© toolbar OR by selecting Insert–>Chart from the menu (Figure 15-3a). Since Bob is going to use the same pivot table to create other reports and continue with his analysis, he now takes a snapshot of this report and copies it over to anotheer file. To do this, he selects the entire results table, copies it (Edit–>Copy) and pastes it into a new spreadsheet file.

Report #1: Analysis

This report gives Bob a good feel for the break-up of energy usage by area. From Figure 3b, since the Plant and Warehouse areas account for 86% of overall kWh consumption, one straightforward conclusion is that Bob should first investigate these two areas before the other two. Lets see what the other reports reveal...

Report #2: Annual Usage (kWh) and Annual Demand (kW) by sub-category—i.e. fluorescent, incandescent etc.

To create this report, Bob starts with Report #1:

1. First, Bob needs to replace "Location" with "Sub-Category" in the Row field. To do this, Bob opens the "Layout" window by right clicking anywhere on the pivot table and selecting "Wizard" (Figure 15-4). From the resulting screen, he selects the "Layout" button.

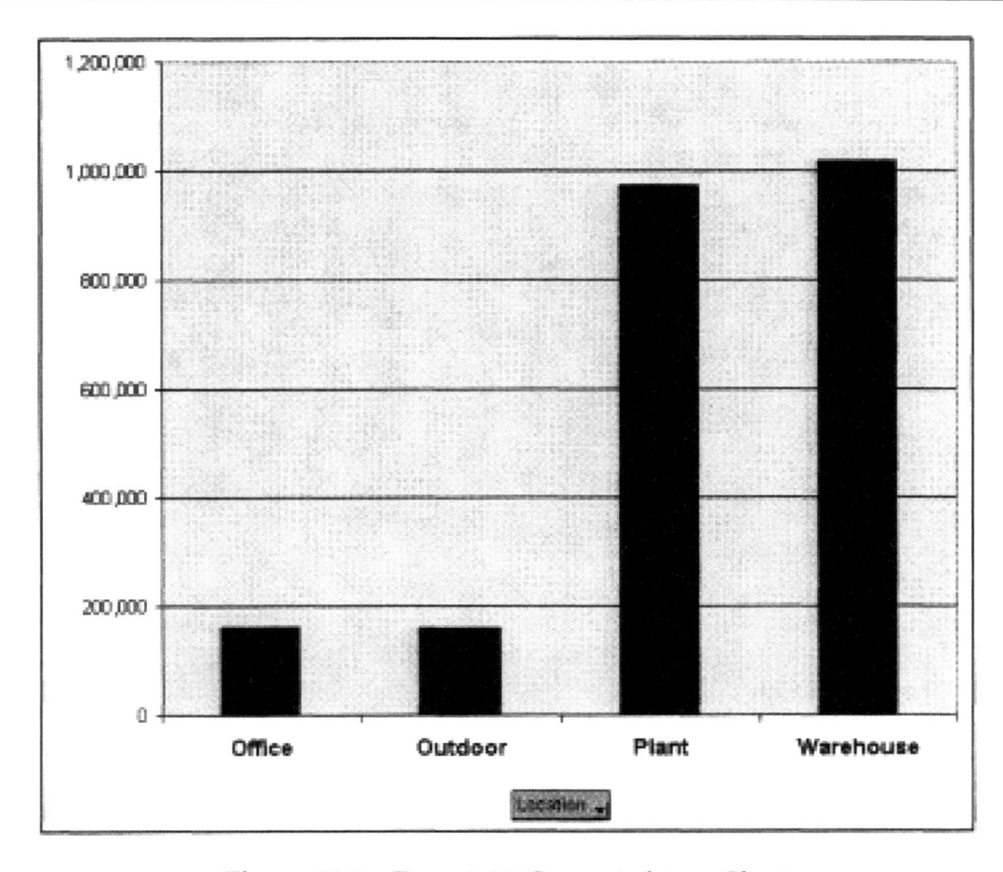

Figure 15-3a. Report #1 Converted to a Chart

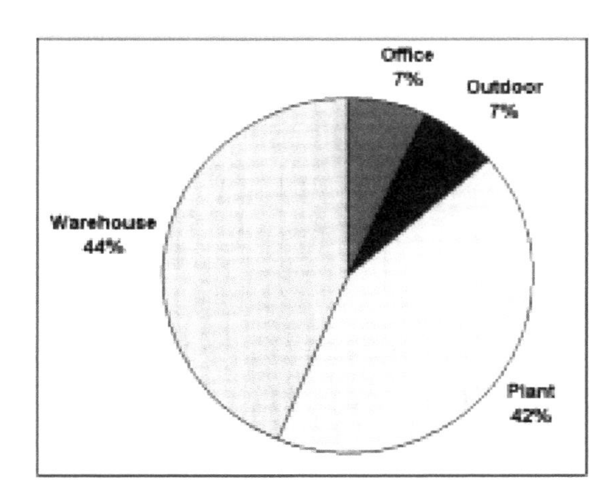

Figure 15-3b. Pie-chart View of Figure 3a

2. In the "Layout" Window, Bob re-arranges the report view. He first drags "Location" from the ROW area onto the PAGE area on the left. Next, he drags the "Sub Category" field from the palette onto the ROW area. He also drags the "Demand Usage (kW)" onto the DATA area along with "Annual Usage (kWh)." Finally, he formats the two data measures to show the "Sum of" values instead of the default "Count of." (See Step 3 of Report #1 above). Figure 15-5 is an illustration of the finished layout.

3. He then clicks "OK" and in the Wizard window, selects "Finish."

4. This results in a report that shows Annual Usage and Demand Usage by lighting sub-category. The resulting report is the completed report as shown in Table 15-4.

Tip: You might have figured out by now that it is possible to create dynamic computations or formulas within pivot tables (right-click on any DATA *measure title and select* Formulas-Calculated Field*). For instance, Bob could just as easily have computed demand usage (kWh) using a pivot table formula instead of computing it in the datasheet. These are two ways to approach the same problem, each with its pros and cons. Computing formulas in pivot tables keeps the original datasheet small and simple but at the same time, it takes up computation time each time you pull up a pivot report as the computations are being done dynamically. As a general guideline, it is better to perform commonly used computations up-front and use pivot tables for formulas that are only used in a few reports.*

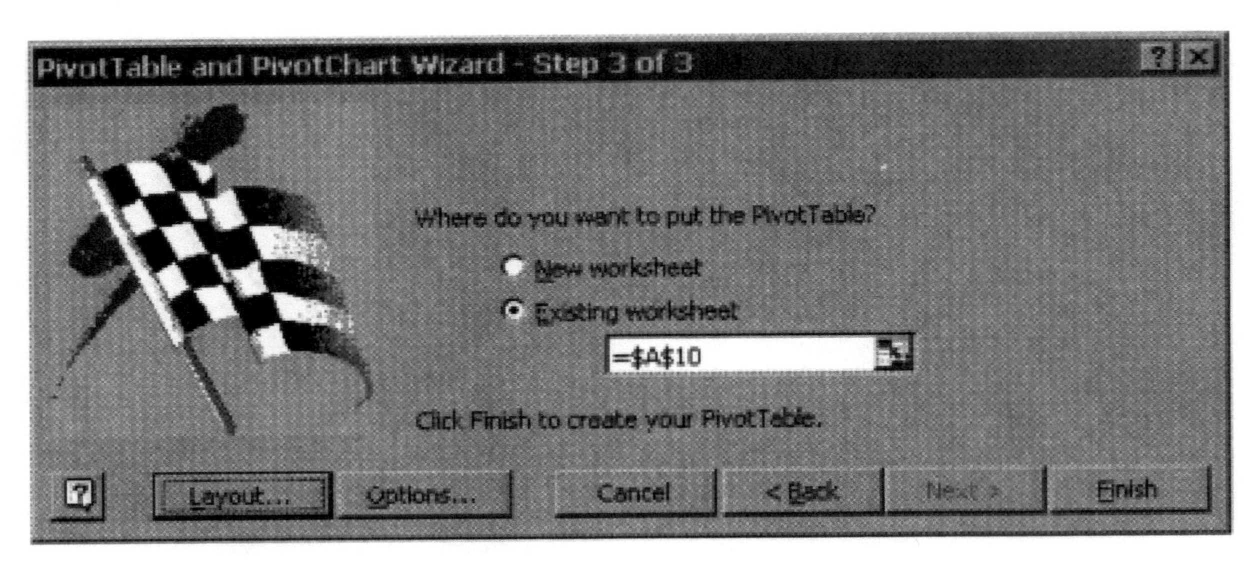

Figure 15-4. Modifying Existing Reports using the PivotTable Wizard

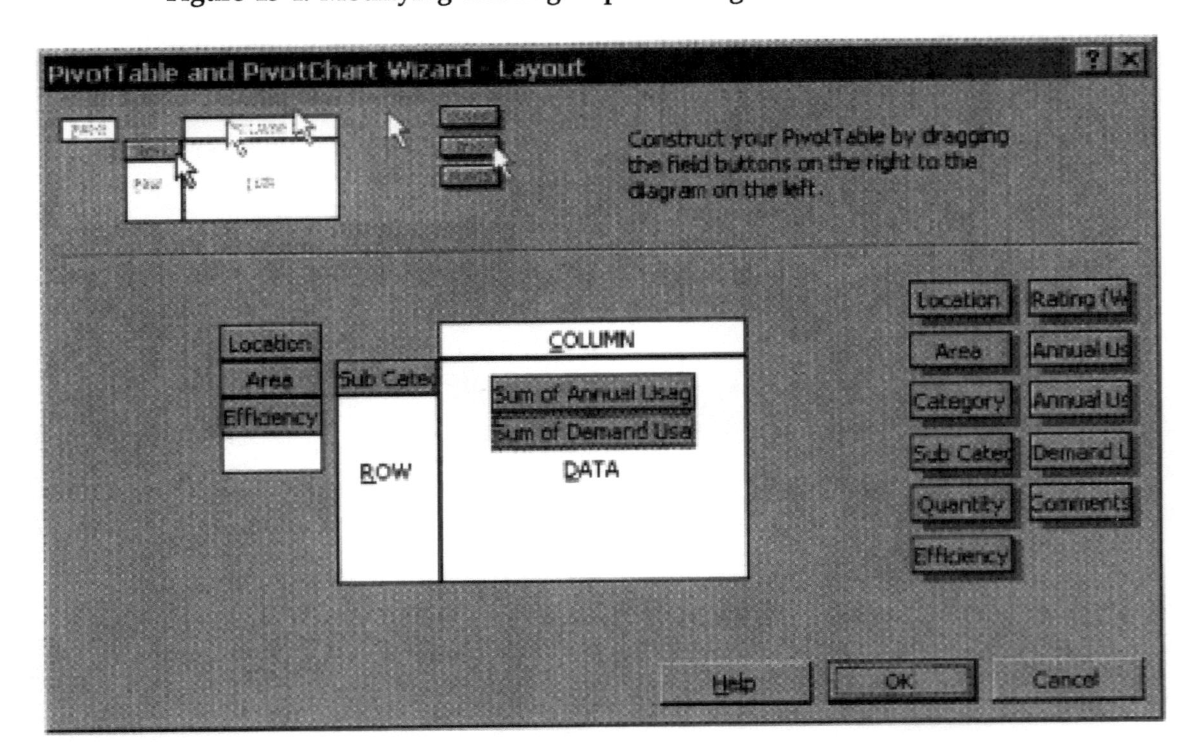

Figure 15-5. PivotTable Layout Tool

Sub Category	Annual Usage (kWh)	Demand Usage (kW)
Fluorescent	366,658	62
HPS	830,448	95
Incandescent	140,093	19
Metal Halide	970,214	156
Grand Total	**2,307,413**	**332**

Table 15-4. Report #2: Annual Usage (kWh) and Demand Usage (kW) by Sub-Category

Report #2: Analysis

Looking at the report #2, it is clear that metal halides and HPS lamps account for a bulk of the kWh usage. Also, metal halides account for nearly half of the total annual demand (kW). If reducing annual kWh usage is a big priority, obviously, these two lamp types are the best targets. To show the power of pivot tables, with a single-click, Bob can drill deeper into metal halide and HPS usage. First from the row drop down, Bob deselects all other lamp types except metal halide and HPS. Then, he drags "Location," "Area" and "Rating (Watts)" fields onto the Row area, which explodes Report #2 into detailed usage. Figure 15-6 shows a chart view of this report (remember: clicking the chart icon automatically creates a chart of the pivot table report).

From Figure 15-6, it is quite clear that the two big areas of kWh usage are the 150 480-Watt MH lamps in the "Main Storage Area" and 316 300-Watt HPS lamps in the Shop floor. Together, they account for 1.5 Million kWh (approx. 75% of the total 2.3 million kWh usage by Plastico). A 10% reduction in this usage either through wattage reduction or usage reduction will result in savings of 150,000 kWh—Bob quickly estimates that to be worth nearly $15,000 at 10 cents/kWh, which does not include any savings from demand (kW) usage reduction. Also, a 20% reduction in metal halide wattages will lead to approximately 30kW monthly demand reduction (actual peak demand reduction is typically less and depends on usage patterns). At $10/kW/month, this represents roughly $3,600 of annual savings. Bob jots this away for later and continues analyzing further trends.

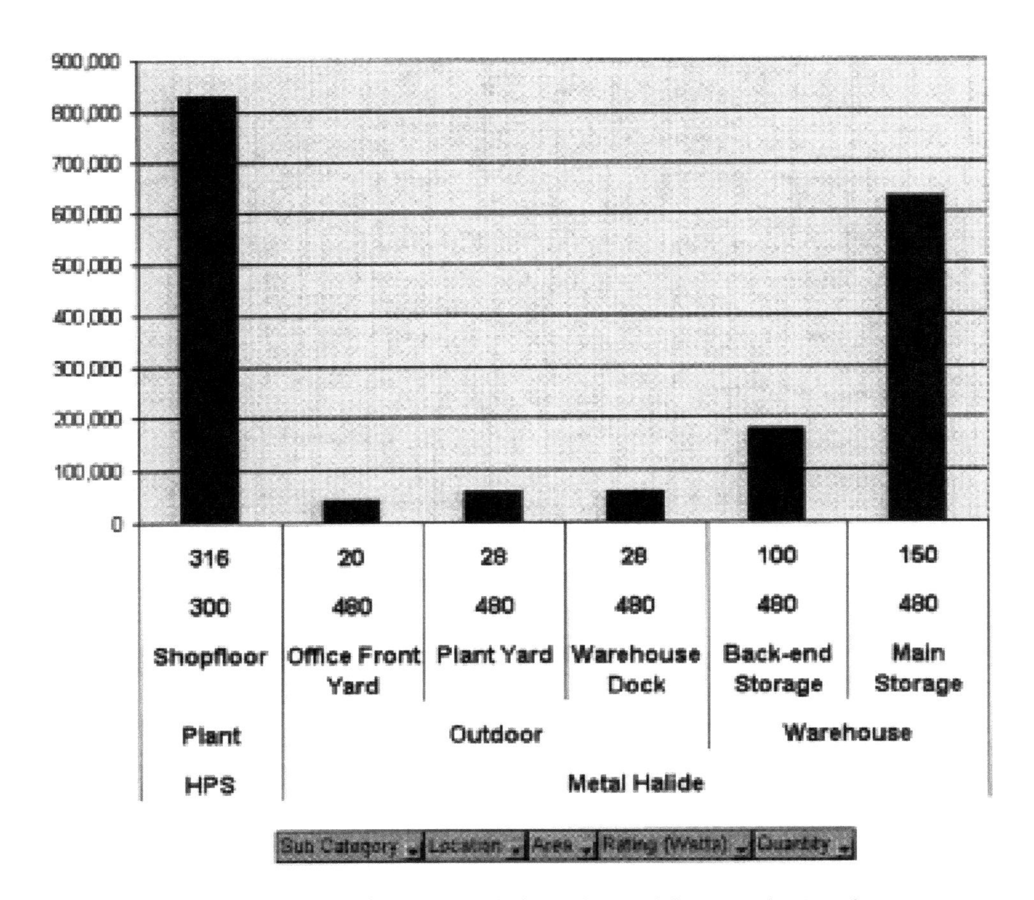

Figure 15-6. Gathering Insight using ad-hoc analysis of Report #2

Table 15-5. Report #3: Annual Usage (kWh) and Demand Usage (kW) by Efficiency

Sub Category	(All)	
Location	(All)	
Area	(All)	
Efficiency	**Annual Usage (kWh)**	**Demand Usage (kW)**
Low	1,940,755	270
Medium	366,658	62
Grand Total	**2,307,413**	**332**

Report #3: Annual Usage (kWh) and Demand Usage (kW) by efficiency—i.e. high, low, medium.

To create this report, Bob starts from scratch.

1. First, from the "Layout" window, he removes all items from the ROW area.
2. Next, he adds "Efficiency" to the ROW area.
3. He makes sure Annual Usage (kWh) and Demand Usage (kW) are in the DATA area and properly formatted.
4. Bob now wants to drill deeper into the Low-efficiency devices, and get a prioritized list of the top 5 energy saving opportunities from the low-efficiency devices. To do this, Bob takes Report #3 and first adjusts the layout.
5. First he drags Area onto the ROW section and moves Efficiency to the PAGE section. After clicking "Finish" in the layout window, from the pivot table, he selects only "Low" from the "Efficiency" drop-down box in the PAGE section of the report. In the resulting report, Bob adds "Rating (Watts)" and "Sub Category" to the ROW section which leads to a listing of all the low-efficiency devices along with their kWh and kW usages. (Table 15-6)
6. The next step is to narrow down this list to the top 5 devices. This task doesn't seem that difficult since there are few rows to begin with but as the number of rows in the spreadsheet increase, an automatic top 5 list can save an energy manager a lot of time. Bob first double-clicks the ROW title "Area" and then selects "Advanced" from the Pivot Table Field Settings Window. This leads him to an advanced settings window. In this window, he first sorts in descending order of Annual Usage (kWh) and further applies a "Top 5" filter as shown in Figure 15-7.
7. As a result, Bob now has a top 5 report as shown in Table 15-7 (chart version of this is shown in Figure 15-8).

Table 15-6. All Low-efficiency Devices

Efficiency	Low				
Area	**Quantity**	**Rating (Watts)**	**Sub Category**	**Annual Usage (kWh)**	**Demand Usage (kW)**
Back-end Storage	100	480	Metal Halide	179,712	48.0
Cardboard Storag	36	90	Incandescent	28,382	3.2
Entryway	8	60	Incandescent	4,205	0.5
Mailroom Entry O	32	60	Incandescent	4,762	1.9
Main Storage Are	150	480	Metal Halide	630,720	72.0
Maintenance Hall	28	60	Incandescent	3,931	1.7
Mens Lockeroom	54	60	Incandescent	28,382	3.2
Mezzanine Office	68	60	Incandescent	35,741	4.1
Office Front Yard	20	480	Metal Halide	42,048	9.6
Paint Office	4	60	Incandescent	2,102	0.2
Paint Room	22	60	Incandescent	11,563	1.3
Plant Yard	28	480	Metal Halide	58,867	13.4
Shopfloor	316	300	HPS	830,448	94.8
Warehouse Dock	28	480	Metal Halide	58,867	13.4
Womens Lockero	40	60	Incandescent	21,024	2.4
Grand Total				**1,940,755**	**269.9**

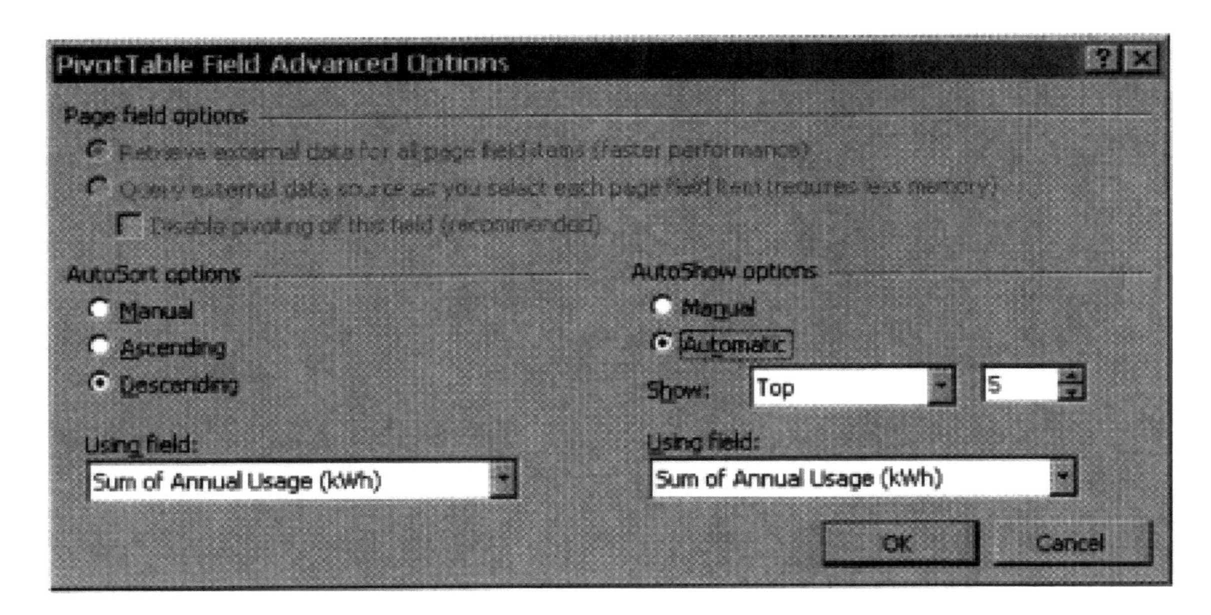

Figure 15-7. Sorting and Filtering Results in a Pivot Table

Table 15-7. Top 5 Low-efficiency Areas (by Annual kWh Usage)

Efficiency	Low			
Area	**Rating (Watts)**	**Sub Category**	**Annual Usage (kWh)**	**Demand Usage (kW)**
Shopfloor	**300**	HPS	830,448	94.8
Main Storage Are	**480**	Metal Halide	630,720	72.0
Back-end Storage	**480**	Metal Halide	179,712	48.0
Warehouse Dock	**480**	Metal Halide	58,867	13.4
Plant Yard	**480**	Metal Halide	58,867	13.4
Grand Total			**1,758,614**	**241.7**

Report #3: Analysis

Report #3 suggests that a high-efficiency lighting upgrade will greatly benefit Plastico. There are primarily 3 opportunities that stand out:

— The 300 watt HPS Lamps in the shop floor

— The 480 watt Metal Halide Lamps in the warehouse (2 storage areas)

— The 60 to 90 watt Incandescent Lamps all over the Offices

Looking at Figure 15-8, the first two account for nearly 80% of all low-efficiency kWh usage. Therefore, these two areas are immediate priority areas. The office upgrades are also important because this has minimal impact on operations.

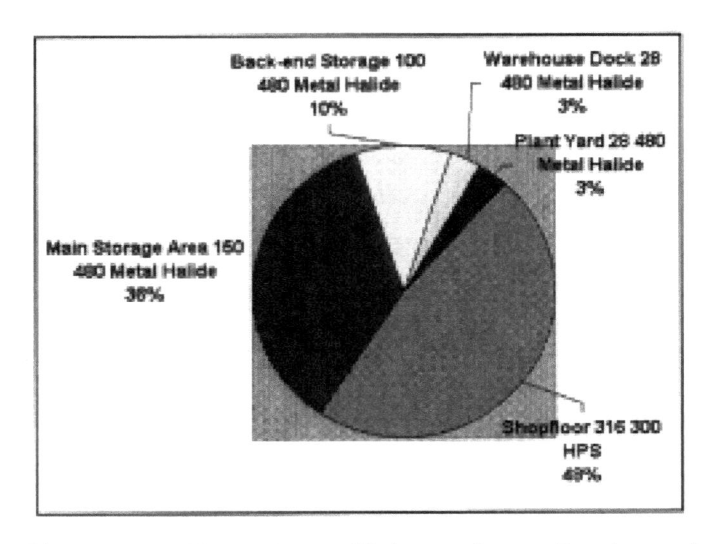

Figure 15-8. Top 5 Low-efficiency Areas (by Annual kWh Usage)

Step 4: Recommendations for Energy Savings

Based on the quick analysis Bob has conducted in his 3 basic reports, he has already come to a conclusion on the key opportunities he wants to pursue. Obviously, this is by no means an exhaustive or final list of recommendations. However, it is a list of a few obvious areas that can result in significantly large energy savings:

1. Replace the 250 480-Watt metal halide lamps in the warehouse (main storage and back-end storage areas) with 400-Watt metal halides
2. Replace the 316 300-Watt HPS lamps in the shop floor with 250 Watt lamps
3. Install timers for outside lighting

To begin building his recommendation, Bob first creates a detailed report of usage as shown in Table 15-8.

Next, he copies it over to another spreadsheet and only retains the 3 areas of interest to him (Table 15-9). There are 2 ways of getting from Table 15-8 to Table 15-9:

Method 1: From the Area dropdown, Bob can deselect all areas of no interest

Method 2: This is a lot easier (and is obviously the one Bob prefers!) than Method 1. In the pivot table, Bob deletes the rows that contain areas of no interest to him. This automatically deselects these areas from the Area drop-down.

Now, Bob goes about adding columns to help estimate the savings from making the changes. He adds columns for suggested hours (applicable to the outdoor lighting timer recommendation) and suggested wattage (applicable to the shop floor and warehouse lamps). He then computes the estimated annual energy savings (kWh/yr) and estimated demand savings (kW) using simple spreadsheet formulas as follows:

For Warehouse and Plant areas:

Estimated Demand Savings (kW) = Estimated Wattage Reduction (Watts) (0.001 kW/Watt

Estimated Annual Energy Savings (kWh/yr) = Estimated Demand Savings (kW) (Annual Usage (hours/yr)

For Outdoor area:

Estimated Annual Energy Savings (kWh/yr) = Estimated Annual Usage Reduction (hrs/yr) (Rating (Watts) (Quantity (0.001 kW/Watt

Table 15-10 shows the detailed recommendations and Table 15-11 is an executive summary. Overall it appears that these simple measures will yield an annual savings of over $34,000.

Step 5: Reporting and Collaboration

This section will describe how Bob quickly communicates his findings and analyses with his team members in the utility company as well as with his client, Plastico. Bob does not have extensive web-site authoring skills and will again depend on the basic capabilities in MS Excel© spreadsheets to report and collaborate his results.

First, Bob decides what he wants to share with whom. In this case, Bob has a very simple collaboration need:

- Post the final recommendations (Table 15-10 and Table 15-11) as well as Figure 15-6 (Top 5 low-efficiency usage areas) in an internet web site and send a link to Plastico so that they can view the results

Given recent advances in internet publishing, there are several ways of achieving the above two objectives—starting with email (directly to the concerned people) and all the way to the other extreme end which involves creating a sophisticated interactive web site that requires a user-id and password to access the files.

In this exercise, we will show how Bob converts all the charts/tables he wants to share with partners, to "HTML format." In doing so, Bob will have the information in a format that is compatible with every operating system and will be easily rendered in a browser on any machine. It is also possible to link these HTML files within a web site easily using simple "URL tags" (Please talk to your web administrator to help you post your HTML files on your web site or on the company intranet) and a basic editor like Microsoft® FrontPage™.

For the first task, Bob takes snapshots of the three reports in Table 15-10, Table 15-11 and Figure 15-6 and saves them as separate Excel© spreadsheets, say, "Plastico_Detail.xls," "Plastico_Summary.xls" and "Plastico_Usage.xls" respectively.

Next, he opens "Plastico_Detail.xls" and then selects File->Save As. In the resultant window, he selects "Web Page" in the field "Save As type" (Figure 15-9). He leaves the file name as the same. He also clicks on the button "Change Title" and renames the web page as "Plastico Detailed Recommendations" and then clicks "OK" in the "Set Page Title" window (Figure 15-10). He also clicks OK on the "Save As" window.

Table 15-8. Detailed Energy Usage Report

Location	Area	Rating (Watts)	Quantity	Sub Category	Ann	Annual Usage (kWh)	Demand Usage (kW)
Office	Back Office	48	76	Fluorescent	2480	9,047	3.6
	Break Room	48	14	Fluorescent	2480	1,667	0.7
	Cafeteria	48	156	Fluorescent	2480	18,570	7.5
	Closet room	48	4	Fluorescent	2480	476	0.2
	Conference Roor	28	24	Fluorescent	2480	1,667	0.7
	Conference Roor	28	12	Fluorescent	2480	833	0.3
	Conference Roor	48	8	Fluorescent	2480	952	0.4
	Entryway	60	8	Incandescent	8760	4,205	0.5
	H Room	48	24	Fluorescent	3968	4,571	1.2
	Laboratory	28	92	Fluorescent	8760	22,566	2.6
	Literature room	48	12	Fluorescent	2480	1,428	0.6
	Lobby	48	56	Fluorescent	2480	6,666	2.7
	Locker Room	48	28	Fluorescent	2480	3,333	1.3
	Mailroom Entry (	60	32	Incandescent	2480	4,762	1.9
	Main Personel O	28	76	Fluorescent	2480	5,277	2.1
	Mens Lockerroom	60	54	Incandescent	8760	28,382	3.2
	Mens Restroom	48	4	Fluorescent	8760	1,682	0.2
	Mens Restroom I	28	8	Fluorescent	8760	1,962	0.2
	Press Office	109	16	Fluorescent	8760	15,277	1.7
	Reception	48	12	Fluorescent	3968	2,286	0.6
	Snack Room	48	16	Fluorescent	2480	1,905	0.8
	Womens Locker	60	40	Incandescent	8760	21,024	2.4
	Womens Restroo	48	4	Fluorescent	8760	1,682	0.2
	Womens Restroo	28	8	Fluorescent	8760	1,962	0.2
Office Total						167,183	35.8
Outdoor	Office Front Yard	480	20	Metal Halide	4380	42,048	9.6
	Plant Yard	480	28	Metal Halide	4380	58,867	13.4
	Warehouse Dock	480	28	Metal Halide	4380	58,867	13.4
Outdoor Total						159,782	36.5
Plant	Cafeteria	48	96	Fluorescent	8760	40,366	4.6
	Employee Devel	28	44	Fluorescent	2480	3,055	1.2
	Main Maintenan	28	28	Fluorescent	2340	1,835	0.8
	Maintenance Hal	60	28	Incandescent	2340	3,931	1.7
	Maintenance Offi	28	64	Fluorescent	2340	4,193	1.8
	Mezzanine Offic	60	68	Incandescent	8760	35,741	4.1
	MIS Center	28	28	Fluorescent	8760	6,868	0.8
	Offices	48	72	Fluorescent	8760	30,275	3.5
	Paint Office	60	4	Incandescent	8760	2,102	0.2
	Paint Room	60	22	Incandescent	8760	11,563	1.3
	Shopfloor	300	316	HPS	8760	830,448	94.8
Plant Total						970,377	114.8
Warehouse	Back-end Storag	480	100	Metal Halide	3744	179,712	48.0
	Cafeteria	48	128	Fluorescent	8760	53,821	6.1
	Cardboard Stora	90	36	Incandescent	8760	28,382	3.2
	Individual Office	48	60	Fluorescent	3968	11,428	2.9
	Loading Area	48	24	Fluorescent	8760	10,092	1.2
	Main Storage Ar	480	150	Metal Halide	8760	630,720	72.0
	Working Area	48	240	Fluorescent	8760	100,915	11.5
Warehouse Total						1,015,070	144.9
Grand Total						2,307,413	332.0

Table 15-9. Key Areas for Recommendations

Area	Location	Rating (Watts)	Quantity	Sub Category	Annual Usage (hours)	Annual Usage (kWh)	Demand Usage (kW)
Shopfloor	Plant	300	316	HPS	8760	830,448	94.8
Main Storage Area	Warehouse	480	150	Metal Halide	8760	630,720	72.0
Back-end Storage	Warehouse	480	100	Metal Halide	3744	179,712	48.0
Warehouse Dock	Outdoor	480	28	Metal Halide	4380	58,867	13.4
Plant Yard	Outdoor	480	28	Metal Halide	4380	58,867	13.4
Office Front Yard	Outdoor	480	20	Metal Halide	4380	42,048	9.6
Grand Total						1,800,662	251.3

Table 15-10. Energy Recommendation Details

Area	Location	Rating (Watts)	Quantity	Sub Category	Annual Usage (hours)	Annual Usage (kWh)	Demand Usage (kW)	Suggested Hours	Suggested Wattage	Energy Savings	Demand Savings	$ Savings
Shopfloor	Plant	300	316	HPS	8760	830,448	94.8	same	250	138,408	15.8	$ 15,737
Main Storage Area	Warehouse	480	150	Metal Halide	8760	630,720	72.0	same	400	105,120	12.0	$ 11,952
Back-end Storage	Warehouse	480	100	Metal Halide	3744	179,712	48.0	same	400	29,952	8.0	$ 3,955
Warehouse Dock	Outdoor	480	28	Metal Halide	4380	58,867	13.4	3,650	same	9,811	0	$ 981
Plant Yard	Outdoor	480	28	Metal Halide	4380	58,867	13.4	3,650	same	9,811	0	$ 981
Office Front Yard	Outdoor	480	20	Metal Halide	4380	42,048	9.6	3,650	same	7,008	0	$ 701
Grand Total						1,800,662	251.3			300,110	35.8	$ 34,307

This creates a file named "Plastico_Details.htm." This file can now be easily embedded in any web page. Figure 15-11 shows an example of this report that has been linked into a web site.

Similarly, Bob creates HTML versions of the other reports he wants to share and links them up to the same web site to with his peers and with Plastico.

CONCLUSION

This tutorial has explained how to use Excel for easy energy use calculations. The tutorial is available on the web at: http://www.utilityreporting.com/spreadsheets/IT_2003_Plastico_Sample_Data.xls

Using the techniques shown in the tutorial, Bob

Table 15-11. Energy Recommendation Summary

Area	Location	Sub Category	Rating (Watts)	Recommendation	Quantity	Energy Savings	Demand Savings	$ Savings
Shopfloor	Plant	HPS	300	Reduce to 250 Watts	316	138,408	15.8	$ 15,737
Main Storage Area	Warehouse	Metal Halide	480	Reduce to 400 Watts	150	105,120	12.0	$ 11,952
Back-end Storage	Warehouse	Metal Halide	480	Reduce to 400 Watts	100	29,952	8.0	$ 3,955
Warehouse Dock	Outdoor	Metal Halide	480	Install Timer and reduce usage	28	9,811	0	$ 901
Plant Yard	Outdoor	Metal Halide	480	Install Timer and reduce usage	28	9,811	0	$ 981
Office Front Yard	Outdoor	Metal Halide	480	Install Timer and reduce usage	20	7,008	0	$ 701
Grand Total						300,110	35.8	$ 34,307

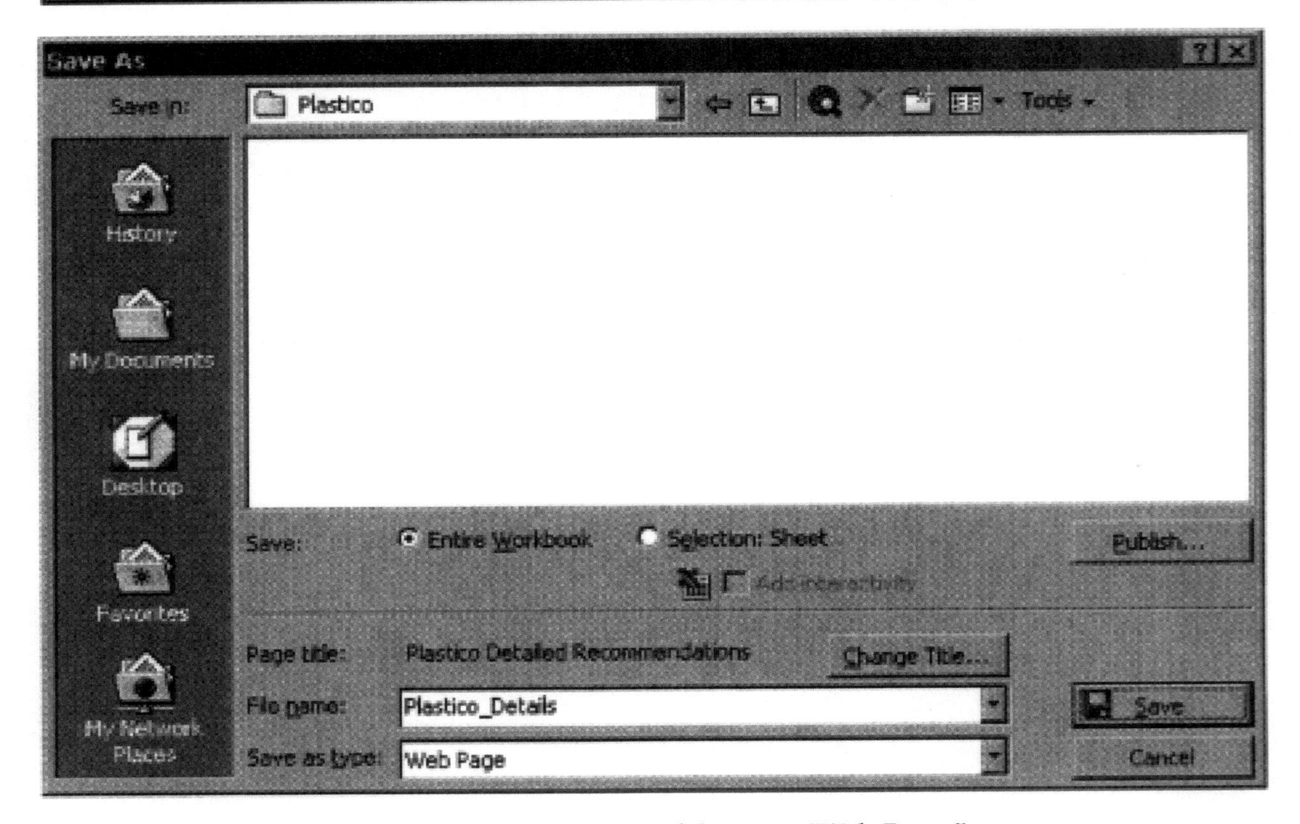

Figure 15-9. Saving Spreadsheets as "Web Pages"

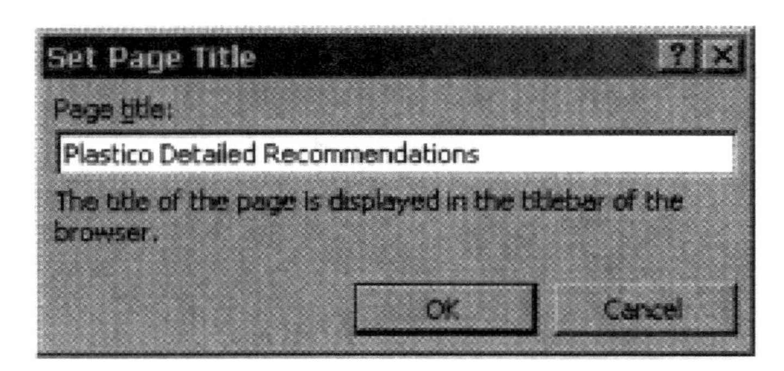

Figure 15-10. Renaming a Web Page Title

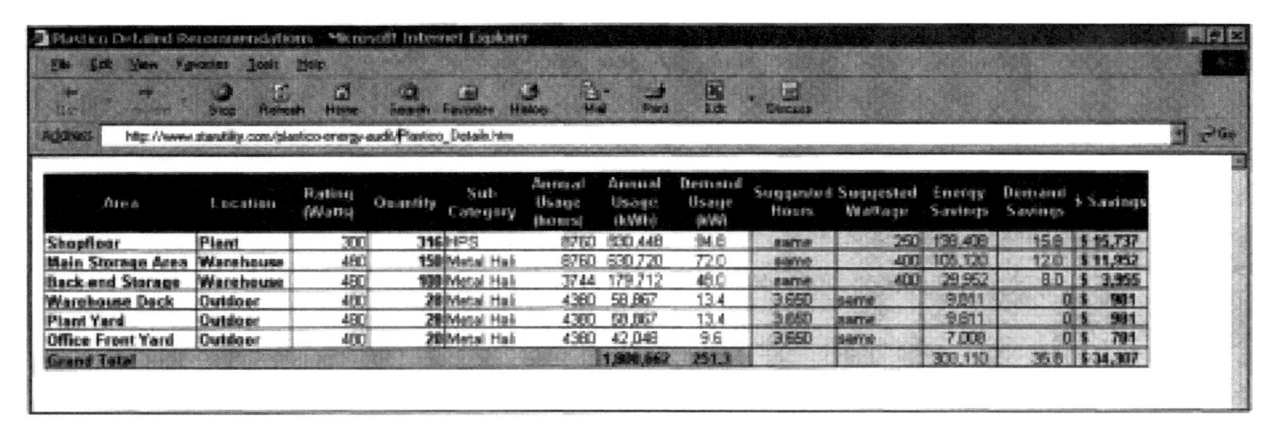

Area	Location	Rating (Watts)	Quantity	Sub Category	Annual Usage (hours)	Annual Usage (kWh)	Demand Usage (kW)	Suggested Hours	Suggested Wattage	Energy Savings	Demand Savings	$ Savings
Shopfloor	Plant	300	316	HPS	8760	830,448	94.8	same	250	138,408	15.8	$ 15,737
Main Storage Area	Warehouse	480	150	Metal Hali	8760	630,720	72.0	same	400	105,120	12.0	$ 11,952
Back and Storage	Warehouse	480	100	Metal Hali	3744	179,712	48.0	same	400	29,952	8.0	$ 3,955
Warehouse Dock	Outdoor	480	28	Metal Hali	4380	58,867	13.4	3,650	same	9,811	0	$ 981
Plant Yard	Outdoor	480	28	Metal Hali	4380	58,867	13.4	3,650	same	9,811	0	$ 981
Office Front Yard	Outdoor	480	20	Metal Hali	4380	42,048	9.6	3,650	same	7,008	0	$ 701
Grand Total						1,800,662	251.3			300,110	35.8	$ 34,307

Figure 15-11. HTML Report Rendered in a Web Site

was able to slice and dice a large spreadsheet of data, perform quick analyses and narrow down to the top energy saving opportunities in two hours. Further he was able to instantly compute the estimated savings and share that via the internet with his customer and his peers. Bob is no technology wizard and yet he was able to use everyday spreadsheet technology to perform such a comprehensive technical analysis.

This oversimplified case study was easily handled by Excel and it is quite possible that a larger or more sophisticated audit would require more advanced decision support tools and maybe a full fledged database to hold the data. However, the features contained in spreadsheets can easily handle a large majority of energy management analysis. Also, as spreadsheets continue to mature and advance, newer versions will allow users to handle larger data volumes, provide more computational algorithms and finally, provide more internet publishing capabilities.

Chapter 16

Knowledge Practice In a Sea of Information

Michael Bobker, MS, CEM
Director of Strategic Planning
Association for Energy Affordability, NYC

As DATA BECOME more available through web-based communications there is an increasing need for a strategic approach to data management. Technical, cognitive, and strategic dimensions are involved. As data acquisition and communications hardware and protocols become standardized, software tools and interfaces that make information intelligible and usable will differentiate systems and products and will provide competitive advantages to organizations. IT strategy should focus on the need and value of such tools and prepare an organization for their use.

INTRODUCTION: HAVE A STRATEGIC VIEW OF IT DEVELOPMENT

The goal of this chapter is to provide a strategic approach to web-based information technology (IT) applications to building controls and building management systems. The prominent change introduced by IT is a great expansion of the availability of data and information. But availability does not automatically translate into meaning and use. So the over-arching theme of this chapter is how the newly expanding potential for data acquisition and broadcasting can be strategically developed to become, first, comprehensible, actionable information and, then, more broadly, organizational knowledge.

Technology Convergence and Data Intensity

The silicon-based technology convergence we see on the web encompasses data collection, communications, information processing, storage and archiving. New devices and media have evolved rapidly and continue to do so. Taken together they comprise a new paradigm for how we live and how we manage our tasks. Massive amounts of data can be relatively inexpensively collected, transferred and shared.

The combination of communications and database technologies is especially potent in leading to new functionalities and levels of data intensity. We are all familiar with its impact through the banking industry on our daily lives: credit cards could not exist without this combination of technology. The banking industry's adoption of massive, near-instantaneous telecommunicated data transfer has completely revolutionized the way we pay for things over the past two decades.

Banking is one, relatively early, example of how IT adoption can affect daily life. There is a broad business and institutional evolution occurring, suggested in Figure 16-1, that both drives and utilizes IT, resulting in increasing data intensity across many fields of endeavor, requiring increasing levels of participation and sophistication by an ever-broader slice of the working population.

In considering the banking revolution, as consumers we use our credit or debit cards as access keys to the secure and invisible back-office processing of financial transactions. The system's intelligence is at the processing center and the end-points are "dumb" terminals. But over time technology and data processing intelligence have been pushed out from closeted back-office IT departments into user hands through distributed network architectures. As building control users we are actively involved "clients" with data management interests and responsibilities.

Yet as more is expected of users at points of distributed intelligence, data can become more overwhelming. Data acquisition and recording in intervals of minutes is becoming commonplace. The amount of data available, paralleling the growth in data speeds and storage capabilities, has grown by orders of magnitude. This growth has resulted in what might be considered a "sea of data," accessible through databases but not necessarily meaningful or even readily comprehensible.

So technology developments and convergence pose a two-fold challenge: large amounts of undigested data and access to it by large numbers of users. The basic challenge of emergent web-based building controls and information systems, if we truly expect to improve building and system performance, is how the new capabilities can be harnessed for use. To accomplish this means tackling the issue of *information intelligibility* and its strategic relation to *organizational knowledge.*

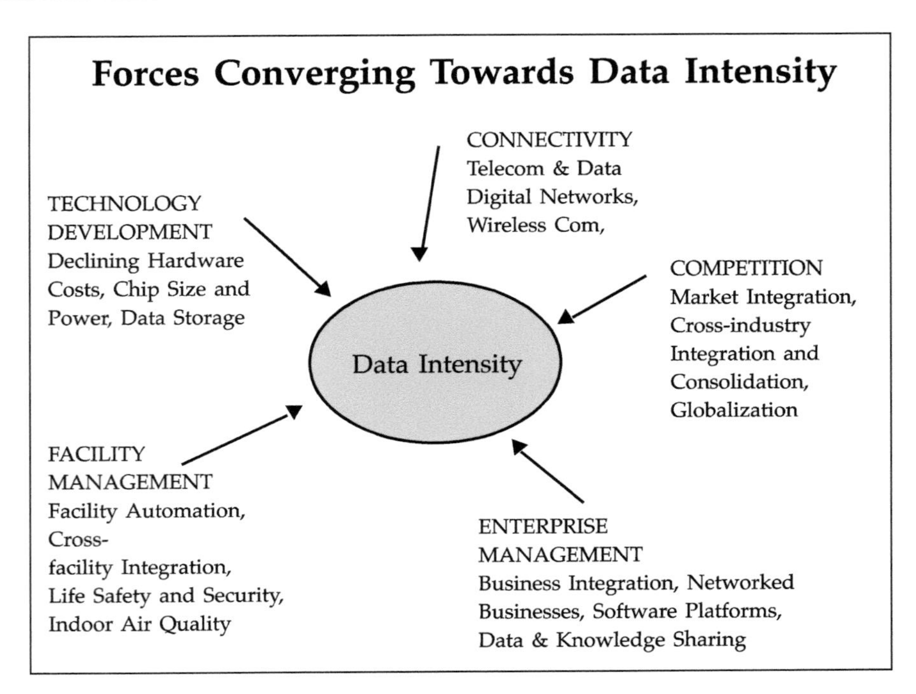

Figure 16-1

Strategic Context of IT Development for Building Controls

Terry Hoffman, global products manager of Johnson Controls, sees the building controls industry adopting an IT model with web-based (TCP/IP) protocol in order to be part of its investment pattern and ability to move products quickly to market.[1] Lawrence Lessig argues that the internet provides a natively open protocol for end-to-end standardized communications; in doing so it pushes "intelligence" out to the periphery where user creativity can flourish without restriction from coding built into the network.[2] Technology marketing theorist Geoffrey Moore describes how the IT industry develops by focusing on the special characteristics and needs of a particular market segment.[3] It may be now, with the collapse of the speculative growth bubble of the 1990's internet start-ups, that the IT industry will turn seriously to the staid and stable building controls segment.[4]

Given the forces at play (see Figure 16-1), it is virtually certain that our facility control systems will be driven to new levels of data intensity and will have new demands for enterprise integration placed upon them. The organization that understands how new data utilization can optimize facility operations gains an advantage in its industry. Facility managers who understand how to accomplish this will contribute more powerfully to their organizations and will become more fully integrated into business planning.

Within the area of facility management we see a convergence of functions and systems made possible: HVAC equipment and temperature control, lighting control, indoor air quality, security, metering, maintenance. Today, separate, specialty, proprietary, and legacy systems still dominate in highly automated facilities. But IT is driving functional integration onto standardized communications platforms. We are seeing the beginnings of this today with large amounts of effort going into the development of gateways to link separate systems into integrated enterprise systems. Barriers between control systems will erode with data accessible not through dedicated controls, panels, and work stations but through standard networking devices. A powerful trend in IT industry evolution is data communications integrated through portable devices such as notebooks, PDA's and data-enabled cell phones.

Data integration techniques extend the possibilities of organizational integration as well as systems integration. Looking inward, facilities and departments are becoming increasingly integrated and trackable. Looking outward, facility transactions with utility suppliers can be integrated to real-time levels and from there to broader markets. The development and increasing emphasis placed on "open" communications protocols and web-based standards such as XML are making broader data sharing and network integrations feasible. The facility sees its discrete boundaries dissolving, its internal functions becoming a web of glowing usage nodes continuously recording bits of data into shared, web-accessible databases.

Even more important than the scope and breadth of data sharing is how much more intensively data are being acquired with respect to time. Control has always occurred in "real time" but control events have usually gone unrecorded. Where control events have been recorded, as by an energy management system (EMS) or a unitary digital control for a piece of central plant equipment, such data has largely remained "islanded," not readily accessible for integration to wider systems. Steadily increasing digital capacity to store data at decreasing cost has made recording and storing of control events feasible. With such media, measurements that were taken only periodically, such as meter readings, can become time-marked "interval" data, recorded at much greater frequencies and coordinated with time-marked events, such as those used in increasingly familiar electricity programs like demand response (curtailment), time-of-use and real-time pricing, and power quality monitoring.

But precisely here emerges the paradox that in its sheer volume greater amounts of raw data becomes less easily intelligible as information.

Enable Information Intelligibility and Sense-making

In recognizing that there is too much data for ready comprehension, we need to look more carefully at what makes data comprehensible. This suggests that we need a cognitive approach in order to fully utilize information technology's new capacities and capabilities.

Cognitive Levels in Information Science

Cognitive science examines how data are processed by the human mind. We will here adapt from this science an over-simplified hierarchy of cognition from data acquisition to knowledge to help us understand cognitive function at the organizational level.

Data are defined here as the recording of discrete events or measured quantities (such as a temperature or a stock price) at a moment in time. In a perhaps confusing phrase, it can be thought of as a "bit of information." Huge amounts of such data can be automatically written to a database. Stock reports are a familiar large database that is continuously updated and published. Where does the electronic data go to after each momentary update? Daily closing reports are published in newspapers but even this volume of data exceeds what the unaided mind can make sense of. The term "data warehouse" calls to mind the vast inventoried storehouse of numbered, boxed items frequently pictured in movies when a piece of evidence is stashed away perhaps never to be seen again.

Information is a next level, both the pattern of data and events that depart from the pattern. Regular, repeating data can be described mathematically, statistically, and graphically and that expression of the pattern is information about the mass of data points. Mathematical formulation compresses large data series into compact formulae.[5] All manner of mathematical functions can be graphed for visual representation: what is intuitively apparent to the gifted mathematician, can be made visually apparent for the common data user. In this way we can expect to literally see the information pattern of the large data sets that will be generated in web-based monitoring of building systems.

However, in a routinely repeating pattern it is the departure from pattern that bears information. Think of

Table 16-1: Cognitive Levels in Information Science

Cognitive Level	*Type of Cognition*	*Organizational Dimensions*
Data	Measurement results	Data acquisition hardware, storage
Information	Patterns in data	Database viewing tools
Meaning	Actionable Information	Job functions, reporting systems
Knowledge	Understanding and skills	Training and education
Strategy	Specialized planning knowledge	Environmental awareness, organizational intelligence

the familiar machine hum lulling the operator to drowsiness; what we think of as normal background noise provides no news. But the sudden clank does and the operator's senses are called to the alert. In information science this is the distinction between statistically repetitive background "noise" and the difference that creates new information, a message.

Meaning introduces the element of the human actor. In contrast to the pattern of information that resides in the data, meaning occurs in a human mind. It is related to how people think about something, and their resulting intentions and actions. With reference to organizations, we might say that something becomes meaningful when it has a place in an organizational function. In this context we can see how "information" interacts with "meaning" to become the much sought-after "actionable information." How this comes about is worth some further examination.

Meaning, Purpose, and Actionable Information

Let's first look at how information becomes "actionable" in a network system such as a LonWorks control ring. In this distributed network architecture, smart devices are continuously sending outputs ("publishing data"), receiving inputs ("reading"), and responding with actions to certain of those data.

- Sensed data about the controlled environment, such as temperature, occupancy, available daylight, etc., and about the status of equipment on the network is continuously published. Sensors are basically "publish-only" devices. Other devices for archiving system data may be "read-only."
- The ethernet (or other media) network continuously makes available current, real-time data to and from all connected devices.
- Devices follow their programming to sort through the data stream for specific types of data as marked by the standard encoding protocols
- Devices find the appropriate data, read the value and, again based on their programming, initiate signals to actuate their controlled equipment.
- Message is published to the network about the control action.

This sequence might be considered a paradigm of what it means for information to become "actionable." The devices read and respond to specific data based on programming. They know exactly what data to attend to and they identify it by its coding. Then, through the if-then statements of programming, they "know" what action is required.[6] But let's look a bit more closely at how this model stands up when we look at it in relation to introduction of a human element.

Humans can be described as acting similarly but we do so with (at least usually and to some degree) the element of consciousness. It is significant therefore to think about how human actors determine what data to attend to and how they will respond to it. At least in our work functions we most often follow established procedures for, say, taking routine actions, completing transactions, recording data, and filing reports. When a company establishes procedures, it is doing something akin to programming. It is focusing employees' attention on certain data or information inputs and specifying certain outputs.

In comparison to sensors, we humans generally have a wider scope of data and control points to which we attend. Where input-output relationships are not so well defined, people look for guidance to what they know or believe about a system's overall operation as well as about the organization's more general goals and strategic objectives. For example, a building operator deciding whether to deliver more heat may consider what he has recently understood about company policy statements about energy conservation and about tenant comfort and complaints. It is in this more generalized kind of response where people distinguish themselves from automata and where organizations gain the ability to learn and improve.

Compared to programmed devices, human beings become aware of a broader view of environmental variables (even if our attention is focused to some degree by procedures, guidelines and goals), interpret directives (programming) in light of circumstances, and take actions with greater flexibility and creativity. For this reason giving improved information tools to process operators is able to lead to improved process performance. Operators will almost always seek to improve their processes and if they can acquire data and information that helps them to do so, they will creatively go beyond their "programming." Superior information tools will enable operators to obtain new kinds of "actionable information." We will return to the implications of this for organizational knowledge-creation at the end of this chapter. We will now turn to considering what makes for superior informational tools

Statistical Tools

Superior information tools are those that help process operators see significant patterns in data generated from their processes. It is important to understand that this goes beyond mere summarizing of data. Summaries, while generally useful and necessary, if poorly constructed can actually hide significant patterns. Superior tools are carefully constructed to avoid this. Superior tools also enable operators to explore data in ways that may not have been pre-programmed; it is most difficult to anticipate where on-going operational involvement with process data may lead.

Statistics is the science of quantitative analysis of large data sets.[7] Concepts of average, range, and standard deviation are basic to looking at data sets. More complex are methods for simultaneously handling and correlating multiple variables, called multivariate analysis. Regression analysis identifies the degree of relationship between occurrences of separate variables and provides a variety of tests to assess the reliability of such correlations.

Mean, median, and mode, all various forms of the "average," express *central tendencies* of a data set. While this characterization is important, it can also be deceptive. For example, an acceptable average of temperatures throughout a building over the course of a day may be composed of unacceptably wide swings and local imbalances. Such temperature variations would lead to occupant discomfort and perhaps complaints, the cause of which would be hidden by the average temperature. Variations at certain times of day or from location to location would similarly be masked by reliance solely on averages.

To counterbalance this effect in the use of statistical central tendencies, analysis also employs measures of *dispersal tendencies*, such as ranges and outliers. These show us the spread of distribution in the occurrences. A narrow range of values shows consistently tight building performance and control. A wider range of values might lead to questions about how often, when and where values were at the outer edges of the range. Thus seeing the statistical dispersal tendencies lead to more detailed questions about how the facility or process is operating.

A certain range of values is to be expected and can usually be accepted; how wide the acceptable range can be depends on the nature of the process. The range simply stated may reflect a small number of occurrences that lie outside of what might be considered a "normal range." Such occurrences are termed "outliers." Outliers may be removed from calculation of central tendencies and also from the normal range. But they cannot be ignored. They show points of erratic system performance or system failure and should be flagged as alarms for specific attention.

A fairly comprehensive selection of univariate statistical functions are provided as formulae for MS Excel spreadsheets. They can be used to build up automatic tracking reports to operate on data that is brought into a spreadsheet environment from a database. A large data set can be boiled down to a small set of statistical indicators, including a list of outlier and dispersal details as noted above.

Multivariate analysis usually proceeds beyond tracking summaries to approach causal relationships through hypotheses and correlations between data sets for at least two variables. Such data sets can also be set up on a spreadsheet and presented graphically but for tests of statistical validity use of a package such as SPSS (Statistical Package for Social Science) would generally be used.

Data Visualization

Data visualization is an essential element for changing raw data to information. Accurate visualization via graphs and charts allows the data to "speak for itself." In large data sets, direct observation of numerical values, as in a table or a spreadsheet, is generally unreliable for finding significant patterns. Edward Tufte suggests the rule of thumb that tables are useful for sets up to some 20 data points. Beyond this rather limited level, the human mind starts to get lost and graphical representations are necessary to reveal a pattern. You can follow your stock prices by reading their daily values but over an extended period of time the only way to track the pattern of their behavior is to graph it.

The pattern becomes meaningful in allowing us to see the structure of normal behavior of a system. The pattern also provides a baseline for understanding departures from normal behavior of the system. In a graph, the outliers, those data points that do not fall within the normal pattern, become obvious.

For our purposes in building controls, the time series is probably the most significant kind of data series to present in a visual format. A time-series, as the name suggests, plots occurrences of a variable across a period of time. More than one variable can be so plotted and their behavior relative to one another can be seen in the shape of the patterns they make over time. A single variable, such as temperature, can be recorded from different parts of a facility and plotted in time-series as different lines. With data richly recorded to a data warehouse, a process operator can call up many different kinds of plots to better understand and document his process.

Why is it that the time-series form is so important? In the excitement over real-time capabilities it is easy to forget that while we *control* processes in real-time, we *understand* them through observing the trends of their behavior over time. The standard graphical user interface for generations of energy management systems was the diagram display of each numbered HVAC unit, showing current data such as damper position and outlet air temperature, typically one screen, one AHU diagram. The real time data would change at some specified time interval, leaving the operator trying to remember the series of data values. Calling up the point's history would bring a screen full of recorded values; gone was any pretense of graphical support for data visualization. Anyone who has ever driven a car with a digital speedometer knows how limiting this kind of display is. For example, just to know if you are speeding up or slowing down requires monitoring a series of flashing numbers. The traditional, analog speedometer shows us this trending in a way that takes advantage of our minds' short-term memory capabilities for data processing through analog visual patterns.[8]

The problem with the present commercial standard format of building automation screens is that it is difficult for the operator to get an overview—of the entire building function, of that point's history, or of that point in relation to other variables. Building control system vendors have begun to build basic graphical support into their products. Going beyond present commercially available capabilities, a prototype "Information Monitoring and Diagnostics System" (IMDS) was developed at Lawrence Berkeley National Labs. Box 1 describes the system requirements, as specified for a pilot installation at a San Francisco office building, beyond standard commercial building automation. The beauty of the system is that the intensive data are simplified in presentation such that building operators could easily interpret patterns that needed corrective attention: "a key point for operators was that if the time-series data appeared to fluctuate by large amounts... something wasn't right" (Smothers and Kinney).

The lesson of this example is clear: data acquisition technology is indeed making large amounts of data collectable but this does not have to make data utilization difficult. As data sets become larger, data visualization becomes increasingly critical. In fact, good data management and presentation tools can make building operations simultaneously more intuitive and more accurate. Operators will need such tool in order to be able to make good use of newly available mass-data. *As web-based system functions become standardized, one of the key differentiating features among IT-based building control products will be the quality of these tools and the intelligence they enable at the end-user level.*

Box 1

Additional Specification Requirements Reflecting IMDS for 160 Sansone St. San Francisco, CA

There are three key elements of additional functionality that are a requirement of this bid: abandonment of the trend log metaphor in favor of a data acquisition metaphor, a time-series graph output in place of a "system graphic," and remote system access and control using any current web browser software.

To implement a data acquisition metaphor, the new EMS must gather and store all digital and analog points at operator selectable data rates. The accumulated data files must be downloadable to an IBM-compatible PC in a format directly accessed by MS Excel (*.csv is acceptable and intended). The system will be capable of storing and accessing a minimum of 18 months of data at any time. KWP's intent is that all data be gathered at one-minute intervals for the main data archive, with the capability of logging any single datum at one second sample rates for a minimum of five hours, or up to four data points at one second sample rates for one hour.

The main operator display wilt be a graph of data, with time as the X-axis and up to eight operator selectable points plotted on the Y-axis. The operator will be able to specify groups of points to be displayed "on-the-fly." An example of the minimum functionality required is the two-dimensional display capability of the software provided as part of the IMDS project. KWP is not looking for pictures of the systems with point values pasted on the picture.
(emphasis added).

Remote access to the system will be provided by web-based software.

Smothers & Kinney 2002

Sense-Making

Following the terrorist attack of 9/11/2001, the need for making sense of large quantities of scattered data was recognized by the national intelligence community. Putting together bits and pieces of fragmentary data into a larger pattern highlighted the need for new methods and tools. Thus, the concept of sense-making was born as a technology application and market segment.

Beyond tracing patterns of a single variable, sense-making looks for multivariate connections, through overlays of multiple variables, connections between separate databases, tracking of events, names, transactions, keywords, and so forth, to try to see multivariate relationships, and trace chains of events and connections. As a methodology it is similar to the process of tracking medical treatment or public health where multiple variables are correlated: symptoms, body chemistry indicators, and drug application (see Tufte 1997 pp 29-37, pp 110-111). While relatively simple time-series can help us track building system functions, more complex multivariate analysis will be necessary to press building science towards cures for "sick building syndrome," bringing with it the potential for enormous gains in health and productivity.

This idea of sense-making by organizations is built on an analogy of how individuals such as detectives put together a story from bits of evidence, circumstance, and inference. It provides a useful transition from the consideration of revealing patterns within data to the use of that information by individuals and organizations.

MANAGE IT STRATEGICALLY

The technical dimensions of IT receive much more attention than the strategic dimensions. IT seems like a prototypical technical subject with a strong "how-to" imperative. But IT strategic opportunities need to be articulated so that they can lead the way in creating, deploying, and managing systems that make information useful. In particular, as we've already suggested, as certain IT functions become standardized, they no longer provide a basis for competition and creativity moves to other functions. Thus, as the web standardizes communications, we will see greater emphasis on creating better end-user functionalities such as reporting and data visualization interfaces.

In thinking about how to manage IT strategically, we need awareness along various dimensions:

- **Competitive advantage endowed by superior performance through IT applications**. Applications that make a difference will make other systems work more efficiently and reliably, and will lower operating and transactions costs. Knowledge-based process improvement through benchmarking is supported by IT, as will be discussed further below.

- **Differentiation between IT product packages based on features and functions to fit customer and client needs**. IT producers will compete on this basis, while users must make procurement decisions about infrastructure that will be with them for years. IT managers must base their efforts on understanding gained from sound relationships with their ultimate end-users in client departments or groups.

- **Coordination and harmonization of IT and client departments**. IT must work well with client departments and departments must have an appropriate level of understanding and skill to work with IT in order to realize new capabilities, application use, and process improvements. Failing in this can create a huge waste of time and investment as well as cause organizational conflict.

- **Contribution to organizational alignment**. IT makes possible a bi-directional dynamic between executive-managerial levels and technical-operational levels. Metrics and reporting allow non-specialist executives and upper level managers to communicate to technical people, such as facility managers, what is strategically important and most valued, and therefore where their focus should be. Statistical and visualization products enhance the ability of technical people to communicate back up the chain what is important and significant in their operations and operational trends. Software tools need to be designed to facilitate this kind of organizational communication.

It is important to note that the broad strategic goals of organizations change over time. The appropriate metrics, technical attention, data acquisition, and reporting then also change over time.

Data Structures and Access

As discussed elsewhere in this volume, data will be stored or warehoused in a database. Microsoft SQL Server is the most common database structure today and might be considered to represent an industry standard. Corporate database management uses a client-server architecture under which data users are the clients of the data storage server. The details of the server are the realm of the IT planner. How the clients will access data is the critical issue for the rest of us.

"Access" often refers to the level of security and permission, controlled by log-in protocols and passwords. Various segmentations and restrictions are possible. For our purposes more importantly, access also refers to how users are able to view data from a database. Users do not look directly at the underlying data in the database. The underlying level is protected for security and data integrity purposes, and in any event is usually coded and organized in ways that would not be readily comprehensible or useful to the client user.

Instead, clients access data through intermediate forms. Different terms are used by different databases but the basic concepts and terms are:

- Queries: requests per protocol to search the database for and return certain data
- Reports: structured set of queries that is standardized as a template for periodic use

Reports commonly go beyond simply listing out database data by performing programmed operations on the data. So, rather than viewing all temperature occurrences over the course of a day, the report might compute the hourly average at various locations, show the mean, median and range and list outlier locations and times. The design of reports, then, is a powerful tool, the exercise of which can greatly differentiate the quality and usefulness of the information system.

Use of the term "report" introduces a useful ambiguity. The basic intent is that of reporting data from the database to a user. In this meaning of report, as a data-viewing tool, it is structured for and/or by a user without necessarily going beyond that user and without any organizational reach. However, "report" also suggests a request for and transfer of data between levels of an organization. This second use of "report" suggests a strategically standardized procedure that structures data according to (at least someone's view of) organizational information needs.

The creation of standardized reports has a strategic value in structuring what data the organization as a whole will see and what client users may be able to view, depending upon their ability to create their own reports (see next section). Therefore the design of reports should not be left to a single party—be it a corporate executive, IT manager, end-user, or vendor. *Designing how and what data will be accessible should be the substance of a cross-organizational task force with strong top level executive leadership.* Work of this task force should not be allowed to delay or bog down the IT deployment process but rather to guide it through a series of decision-making steps.

Usability Tools and Training

In website design "usability" refers to the graphic standards for creating pages, page links and site structure. While such usability will undoubtedly be important, usage here is broader in referring to data access, reports, statistical and graphical tools that are intuitive and easy to adopt by the non-IT person. The development or selection of tools (see below) should incorporate this kind of user-friendliness as a key criterion. User trials are a good idea and IT should be instructed to work closely with a sample of "client" users.

Increasingly IT developers are coming to understand the importance of reflecting and capturing business processes *as practiced* in order to have their new tools readily adopted and smoothly functional. So usability includes a dimension of understanding how work is conducted. For example, sending alarm messages via e-mail to a maintenance manager who is expected to be making rounds much of his time may prove less usable than a pager or cell phone message. Understanding the meeting habits of a work group may be important in determining when and where data and reports should be made available.[9]

Most technical people have by now had experience with the Windows and browser environments that are most commonly used for building up specialized applications. Nevertheless, training in use of the new tools should be a planned part of any new system "roll-out." The design of training goes beyond the scope of this chapter but anyone responsible for this process should be aware that learning of new tools is incremental. Initial introductory training will help people feel comfortable in using and exploring a new system but overload will be counter-productive. Cycles spread over time for going deeper into capabilities is a better approach, especially when combined with on-line support from IT and the development of peer-to-peer user-groups.

It is important to keep in mind that the new system and tools are only as useful as their adoption. When fully adopted, work-process operators will see the tools improving their capabilities and productivity. They will find new, often unanticipated uses for solving problems and making improvements. This is the desired outcome, leading to an exciting "learning organization," rather than slavish acceptance of required practice and reporting.

Partnering and Outsourcing

Getting the right tool set is important but having it in the right time frame is at least as important. The development of an energy information system and associated IT tools is not something that even large organizations any longer take on as an in-house project. It is too complex as a one-off undertaking for in-house use. Needs are sufficiently similar from organization to organization that commercial developers have been able to create suitable products that can come off the shelf ready to be customized for the specific organization and application. This is advantageous in terms of development cost, speed of deployment, product reliability, and specialized support.

IT marketing theorist Geoffrey Moore suggests that IT develops by focusing on a client industry, mastering its needs through engagement with individual customers, and developing new product to suit. In the electric sector with new market requirements of deregulation, this has occurred very strongly, with emphases on building-out utility SCADA systems to the customer level, creating electronic exchanges, and providing C&I (Commercial and Industrial) customers with management tools. The idea of getting closer to the customer with new value propositions has become appealing to utilities and their new competitors.

Software developers have been ready to use IT to support this business evolution and there is now a "market space" for vendors of an array of products with functionalities based around meter data acquisition and data management.[10] They provide standardized reports, in forms such as multi-meter, multi-site aggregations, usage benchmarking, bill simulations, "what-if" modeling, and graphics including time-series and interval data 3-d viewing. Attention has focused heavily on electricity although most vendors claim to handle multiple fuels. Less progress has been made in integrating HVAC and other building systems into these products. However, one developer built an extensive software library of "gateway" interfaces to proprietary building control systems and made this the keystone of its enterprise integration platform.

Vendors in this space also differ in whether they will provide hardware, communications, and service bureau functions in addition to the energy information tools. So comparing vendors in a procurement process is not necessarily straightforward. It behooves the client to understand well the mix of services involved, the options for them, and how they fit into the organization's other IT-based systems and plans.

Under a traditional licensing arrangement, a client buys the right to use the software product, with price related to the number of sites at which it will be used. The term of the licensing agreement may also be limited and subject to renewal. Support and updates are typically included in licensing. However, the buyer's IT department is responsible for maintenance of the software, updating it as new versions are released, and deploying it to new locations.

An interesting alternative is the application service provider (ASP) model in which the software vendor retains responsibility for the tools and the client accesses them for use through the ASP's web site. In this model the vendor maintains the software, assures that it works, updates it, and controls access to it. The client's IT department has less responsibility for upkeep and support of the software, although it must still assure data quality and characteristics such that the software can operate on it. In contrast with a service bureau model, there is no transfer of data to the service provider. Because ASP software use can be structured on a "pay-as-you-use" basis, up-front costs may be lower than outright licensing and therefore the "lock-in" to a platform and product is less. But if a high volume of data processing is expected, ASP costs might exceed licensing fees.

It can be seen that even with a partnering or outsourcing arrangement replacing in-house development, the procurement process still requires a careful assessment of different vendor offers and how best to structure the relationship with a software partner. Finding the right fit of tools and partners is a key strategic step that needs careful execution.

CONCLUSION: DATA SYSTEMS AND KNOWLEDGE CREATION IN ORGANIZATIONS

Going back to the cognitive hierarchy, we might notice that we've put the cart before the horse, treating strategic dimensions in the previous section before getting to knowledge. Strategy, a specialized application of organizational knowledge, is intuitively understood. Knowledge and its organizational creation require a more abstract view and so may be a fitting place to conclude.

If we understand an organization's knowledge as the sum of all that is known by the individuals within it, we can grasp the epistemological problem posed by the title of the knowledge-management classic If Only We

Knew What We Know. To make things harder, this sum might be taken to include both the formal, explicit knowledge articulated in manuals and other documents and the unarticulated tacit "personal knowledge" of skills and how things are done. An important aspect of organizational knowledge creation is the process of bringing tacit knowledge forward into more consciously recognized and embodied form.[11]

The common view of knowledge creation is that it is about research and development for new products or for new break-through processes. Such R&D is typically thought of as compartmentalized and supported with its own separate R&D budgeting. But complementary to this formal R&D level, knowledge creation also exists, or can be made to exist, much more broadly at the incremental level of finding and communicating better ways of doing all manner of organizational tasks.[12] This process has come to be known as "benchmarking" and IT plays a key role in accomplishing it.

The fit and feedback between organizational learning and the data management process is suggested in Figure 16-2.

The IT reporting system enhances the consistent measurement across the organization so that exceptional performers can be discovered and publicized and subsequent progress towards benchmark goals can be monitored. The strategy for charting key performance variables leads the data acquisition that will be put into force. Process operators will creatively use acquired data to improve their process performance. Publishing of results leads to new awareness of process performance patterns (information) that become transformed into new knowledge. For example, the intuitive skills of the best facility operators become apparent, through their

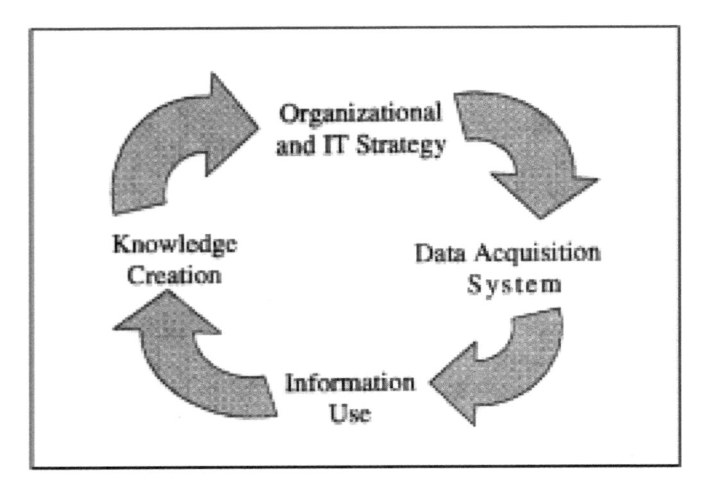

Figure 16-2

reporting; with data collected the dimensions leading to superior performance can be identified and analyzed, thus started on the road towards organizational knowledge available for replication. New knowledge and new practices in turn may point towards new strategies for data collection and reporting.

If, as this chapter has suggested, building controls will be integrated with IT-based data analysis and reporting to the enterprise level, then the organization will have an important platform for learning. Data-enabled facility operators will be able to see more about their plant operations and will be called upon to communicate about it, sharing knowledge with the rest of the organization. With new forms of monitoring made possible, physical plants will operate better and, the ultimate goal, enhance the productivity of all the work that goes on within them.

References

Brown, John S. and Duguid, Paul 2000 The Social Life of Information Harvard Business School Press, Cambridge

Leonard-Barton, Dorothy 1995 Wellsprings of Knowledge: Building and Sustaining the Sources of Innovation Harvard Business School Press, Cambridge

Lessig, Lawrence 2001 The Future of Ideas: The Fate of the Commons in a Connected World Vintage, New York

Moore, Geoffrey 1995 Inside the Tornado: Marketing Strategies from Silicon Valley's Cutting Edge Harper, New York

Nonaka, Ikujiro and Hirotaka Takeuchi 1995 The Knowledge Creating Company Oxford University Press, New York

O'Dell, Carla and C. Jackson Grayson 1998 If Only We Knew What We Know: The Transfer of Internal Knowledge and Best Practice The Free Press, New York

Polanyi, Michael 1958 Personal Knowledge University of Chicago Press, Chicago

Polanyi, Michael 1967 The Tacit Dimension Doubleday, New York

Smothers, Frederic and Kristopher Kinney "Benefits of Enhanced Data Quality and Visualization in a Control System Retrofit" Proceedings of the ACEEE Summer Study on Buildings and Energy 2002

Tufte, Edward 1997 Visual Explanations: Images and Quantities, Evidence and Narrative Graphics Press, Connecticut

Utterback, James 1996 Mastering the Dynamics of Innovation Harvard Business School Cambridge

Endnotes

[1] Presentation at the NY Chapter of the Association of Energy Engineers, March 2003

[2] Lessig 2001

[3] Moore 1995

[4] Economist Joseph Schumpeter, writing of a previous time much like our own: "It is in recession, depression, and revival that the achievements initiated in the prosperity phase mature and fully unfold themselves, thus bringing about a general re-organization of industry and commerce, the full exploitation of the opportunities newly created, and the elimination of obsolete and inadaptable elements" Business Cycles p.183 1964 (1939)

[5] See Gelman The Quark and the Leopard for a good explanation

[6] Using the term "know" is probably inaccurate anthropomorphizing. Machine-evolution, sentient machines, and science fiction aside for the moment, there is no consciousness involved. The process is an automatism, pre-programmed and with limited flexibility. Fuzzy logic, neural nets, and other forms of AI are seeking to rectify this for the machines.

[7] There are thousands of texts and primers on statistics and statistical analysis. One good summary that is available in the public domain is the US Environmental Protection Agency's "Guidance for Data Quality Assessment: Practical Methods for Data Analysis," EPA QA/G-9

[8] I am indebted to Frederic Schnookler for this argument and example, from the presentation of his paper at the ACEEE 2002 Summer Study on Energy in Buildings.

[9] See Brown and Duguid pp. 99-112 for an extended discussion and example of this kind of analysis.

[10] Some of the leading names at present include RETx, Elutions, Excelergy, Silicon Energy (now a division of Itron), Circadian, and Canon-Yukon. These vendors will act as OEMs, providing software licenses for others (such as utilities) to brand under their own names. RETx combines various solutions developed by others into its applications suite.

[11] The formulation of explicit versus tacit knowledge is from Michael Polanyi 1958, 1967, somewhat obscure works in the philosophy of knowledge that were re-discovered in the 1980's as part of more practical studies in business behavior. One such important work that examines the process of turning tacit, skill-based knowledge into articulated, product-embodied knowledge is Nonaka 1995.

[12] The classic work on new product and process development is Utterback Mastering the Dynamics of Change. Among the large number of recent works on the "learning organization," a classic statement of how organizations remain creative and dynamic is Leonard-Barton 1995.

Chapter 17

Some Methods of Statistical Analysis for EIS Information Signals

Valerij Zvaritch, CEM
Ukrainian National Academy of Sciences

AT THE PRESENT TIME, artificial intelligence applications, monitoring systems, expert systems, and information systems in particular, are recognized as providing efficient solutions to a wide range of practical problems. Methods of control and monitoring decision-making may be divided into two main approaches: deterministic and statistical.

In the statistical approach, the solution of the information systems development usually consists of the following stages: elaboration of a mathematical model of information signals; verification that the mathematical model corresponds to experimental data; separation of the most informative diagnostics or control parameters using experimental data; and, construction of decision rules.

This chapter discusses some specific statistical procedures that should be applied in the operation algorithms for energy information systems (EISs) when statistical methods are the basic methods of the information system.

INTRODUCTION

Improving performance is important to every player in competitive energy markets such as the US and Europe. As a result, generators, traders, transmission companies and others are constantly seeking accurate, timely data on plant performance, transmissions information, fuel costs, electricity prices and more.

The increase of information technology capabilities due to the development of the internet and the World Wide Web has made it possible to develop more powerful and often widely dispersed energy information systems. Although these new systems often merely use a new way to implement some well-established information technology (IT), decision-making technology or artificial intelligence technology such as planning, research, and problem solving, they definitely have their own identity. These technologies are tightly linked with the internet, and have opened new fields for application and practical use of energy information technologies during the last decade.

Intelligent agents are computer programs that function autonomously or semi-autonomously in communication with other computational agents, programs, or human agents. Agents typically communicate by exchanging messages written in a standard format using a standard communication language. They can identify, search, and collect resources, and perform independently and rapidly under changing conditions.

Knowledge management is the process of converting knowledge from the sources available to an organization and connecting people with that knowledge. Knowledge management facilitates creation, access and reuse of knowledge, typically using advanced technology, such as the World Wide Web, Lotus notes, the internet, and intranets.

EXAMPLES OF ENERGY MONITORING AND EXPERT SYSTEMS

Energy information systems (EIS) are capable of recommending energy conservation opportunities in areas such as lighting, boilers, motor selection, belt driven systems, insulation of heated surfaces, and air compressor operation.

An *expert system* can be defined as an intelligent system composed of a knowledge base, an inference engine, a working memory, a user interface, and an explanation-based system, capable of solving problems that are generally unstructured and too difficult for humans to solve easily. Most expert systems sold today are personal computer or workstation based[1,2].

Several companies are helping to meet the need in the power industry for accurate, real-time data for reporting and analysis. Some examples of energy monitoring systems and energy expert systems include:

eDNA

Major energy companies such as Southern California Edison, Exelon and Ontario Power Generation use eDNA, an enterprise, real-time data historian from Chicago-Based InStep Software[3]. eDNA, developed about 15 years ago by PG&E in California, is a process historian—i.e. a high-speed compression database that captures, archives, and time stamps historical and real-time data from power plant operations sources. It takes data from any data generation product in the plant—for example, distributed control systems, maintenance systems, and diagnostic systems—and passes it to the historian for storage and long term compression.

The storage data are made accessible across the enterprise via the web or a client server and can be integrated with other systems for preventive maintenance, forecasting and other analytic functions. Users can look at both real time data and historical data to help them analyze plant operations, improve the accuracy of trading bids, and reduce plant-operating costs. eDNA can collect and store values from a 15,000 data point system for decades, and it is highly accurate.

eDNA also includes a resolution agent known as the service directory, which gives end users the ability to maintain the data historian. Server applications also available with eDNA include high end graphical and trending tools, security server, notification and alarm servers. The eDNA web allows the user to gather and display data from eDNA as well as other data capture products that a company may have installed (for example an Oracle system). This way, all pertinent data can be seen together[3].

Webgate

WebGate system from MU Net gives utilities a way of tapping the residential customer market through the internet to supply services and collect data on customers[4]. By using an internet protocol gateway device located within a household's electricity meter, the system establishes a secure two-way communication pathway to the utility that allows the utility to not only monitor and record energy use in the home, but also offer a range of additional services to the homeowner.

MU Net's gateway technology uses the standard TCP/IP protocol over existing networks, making the technology and its applications economically feasible by avoiding the need to install single purpose narrow bandwidth networks. The cost to utilities of obtaining frequent detailed metering information and performing energy management function is therefore lowered.

The WebGate System (IRIS for the residential market and ICIS for the commercial market) consists of three main components: the WebGate gateway, WebBOt central control software and Home-HeartBeat, a tool that allows the utility to deliver information to the consumer.

The WebGate IP gateway is the device that sits in the home and creates a business-quality, utility-owned gateway to the consumer. It uses standards-based open architecture and communications protocols, including HTML and XML and is usually installed under the glass of the electricity meter. The gateway consists of the following elements:

- *Metrology*: MU Net is not a metering company and so it deploys technology from others companies, including Siemens, ABB, Schlumberger and GE.

- *Data Acquisition and Database Management Systems*: The microprocessor-controlled data acquisition system listens for pulses from the electricity meter and passes the information to the database management system. It can also acquire data from gas and water meters or any other devices connected to the gateway via a home network.

- *TCP/IP Internet Server*

- *Broadband Modem*: Information is passed from the database to the internet via a broadband modem and an ethernet connection. The open standards of WebGate mean that any number of broadband modems available on the market can be used, and MU Net is in the process of obtaining approvals for its own modem. Other devices, such as a DSL ladder, a fiber optic media converter or a cable modem going over the cable TV signal, can also be used, depending on the local broadband infrastructure. WebGate is therefore indifferent to the method of internet access, providing that there is a connection to a broadband system rather than a dial-up network[4].

New England ISO System

The New England ISO (Independent System Operator) manages the coordination and dispatching of electric power from over 150 generating units scattered around the Northeastern regions of the United States[5]. These generating units range in size and complexity from nuclear plants to "hog-fuel" fired cogeneration units at paper mills. The NE ISO runs a sophisticated energy demand and pricing model that updates unit commitment, dispatch point, and pricing information for each unit.

The NE ISO needed an automatic, reliable and auditable mechanism for delivering this data to unit owners and for receiving verification from these owners. They selected the Remote Intelligent Gateway (RIG) technology, combined with commercial frame-relay networking, to address this need.

The New England ISO, like other ISO entities formed in the USA, is responsible for coordinating with the participating generator owners (called "Designated Entities" or DE) to purchase power and match generation supply against demand. To accomplish this task, the NE ISO uses an EMS system to model the demand and produce five-minute "dispatch point" updates (including ramp rates, pricing data and operating mode data) for every unit[5].

Turbine Startup Simulator

The Mexican electric utility, Comision Federal de Electricidad (CFE) and the Instituto de Investigaciones Electricas (IIE) developed a turbine startup simulator (TS), which began operation in 1993[6]. The simulator mathematical model reproduces the steam generation in the boiler and its flow through the equipment that integrates the turbine system.

One of the main features of this simulator is its transportability. Normally, it remains in the control room of each thermoelectric power station of CFE for one month. Training is carried out by taking advantage of operation dead times. Equipment damages may be avoided when operators know all the process related with turbine startup. The CFE has estimated that every time the equipment is repaired, an economic loss of about 3 million dollars is incurred.

The development of an intelligent tutoring system (ITS) for turbine startup training (TS-ITS), began in 1998 as an attempt to optimize operators training[6]. To reduce instructor-student dependency, instructional systems in instructional design process, computer-assisted instruction and artificial intelligence are used. This allows trainees to move forward at their own pace, by using appropriate dynamic scenarios and continuously monitoring operator performance.

MEASURING OPERATIONAL SUCCESS

The lifecycle effectiveness of the above systems is the overall measure of their operational success[7]. This measure incorporates several goals, such as lowest lifecycle cost, highest possible efficiency and the ability to meet environmental targets. There are two key categories of measurement: the performance of the equipment and the personnel operating it, and the factors that make up the equipment's operational availability.

Plant measurement systems should preferably employ the newest and most efficient technology available in order to keep up with today's requirements and tomorrow's challenges in monitoring and data collection. Environmental aspects are another important consideration, especially the need to obtain accurate information on vital energy production parameters to keep the plant within environmental limits. To achieve the maximum level of performance, older installations need to consider automation modernization packages that upgrade to the latest technical standards. In many cases this means adopting digital in place of analog systems.

Data acquisition systems with remote connections can be also installed. Companies such as the Finnish company Wärtsilä now offer specialized systems covering data acquisition, fault analysis, planned maintenance and technical information. Monitoring a power plant from a remote location is becoming increasingly popular and can be a key factor when customers are choosing the turbine supplier. Complete updating and monitoring can be supplied either for part of the power plant or for the whole facility, including the mechanical, electrical and automation systems.

Remote monitoring offers fast and easy on-line support while trouble-shooting system faults, resulting in less downtime. It is possible through this system to discuss operational parameters (trends) with staff, making it possible to avoid problems before they occur and to optimize turbine performance. The power plant receives support at all times because experts are able to monitor plant performance without having to take time to visit them. Following major overhauls, the turbines can be closely observed to ensure optimum performance is being achieved.

Power plants, large or small, are a problem when it comes to maintenance and supervision. Not only are there a number of high-tech systems requiring continuous supervision and control, but there are also increasingly high requirements on the optimal operation of these systems. All this, plus the escalating number of measurements and tools for evaluation, is leading to an information overload[7].

ANALYZING THE DATA

The management of a power plant demands a variety of tools for analysis, reporting and general maintenance logging. There are many different systems and tools used for these processes and personnel trying to use these different forms of data encounter hindrances in the efficient operation of a plant.

The flow of data involved in the daily operation and maintenance of a plant is vast, and it has become evident that the average human cannot utilize the given information to its potential. Automated support for the decision-making process is needed.

The information systems data can perhaps best be treated with a statistical approach because information signals are random processes. Many statistical procedures are based on the stationary state of analyzed random processes. But experimental investigations have shown, that the random information signals of an EIS cannot always be assumed to be stationary processes[8]. If this is not taken into account, the non-stationarity of the information signals will reduce the effectiveness of the EIS application and in some cases will cause the system to lose working capability. (For example, statistical spectral analysis and autoregression analysis can be applied only for stationary random information signals[8-11]).

The methods used to estimate the stationary state of random information signals on the basis of certain statistical criteria are explained in depth in the appendix to this chapter. The following examples show how these methods are used to analyze the information signals for the vibration of roller bearings in a piece of rotating machinery.

EXAMPLES OF THE STATISTICAL PROCEDURES APPLICATIONS

A Stationarity Test Application

We consider the test application for investigation of roller bearings vibration signals. To simplify the procedure for estimating vibration stationarity from data corresponding to different test regimes, we selected samples for which the hypothesis H that they belong to a normal process has been accepted. We took respective significance levels of 2% for the criteria used to test to see whether the skewness and kurtosis coefficient are equal to zero. The hypothesis H is accepted if the estimates of the skewness coefficient g_1 and kurtosis b_2 coefficient for $n^* = 1000$ satisfy the inequalities $|g_1|<0.17$, $-0.3 <b_2 <0.39$ and, correspondingly, if for $n^* = 800$ we have $|g_1|<0.202$, $-0.35 < b_2 <0.46$. The critical limits of the inequality were found from [12]. The number of samples selected and the corresponding test regimes are indicated in Table 17-1.

Table 17-1.

Bearing test regime	N	t_2	C_2	t_3	C_3	H
Functioning	8	0.13 39	0.14 02	1.3 8	2.576	H_S
Misaligned	10	0.10 74	0.11 29	1.4	2.576	H_S
Defect on inner race	9	0.12 78	0.12 51	1.5 4	2.576	H_N

We used Eq. (17-3) (see appendix) for computer calculation of the estimates of the mathematical expectation m_i and variance σ_i^2. On the basis of these estimates we used the Cochran test and the Two-sample test to calculate the values of the statistics t_2 and t_3 for each of the test regimes. The results of the analysis are shown in Table 17-1. Here the largest of all the statistics t_3 is calculated for all possible pairs of samples for the test regime is given.

In the next stage, we determine the critical values $C_{i,N,n,\alpha}$. We have to select the value α. For a constant sample size, the smaller α, the lower the probability that a stationary process will be taken to be a non-stationary process, but as α diminishes there is an increase in the probability of taking a non-stationary process to be stationary. A significance level $0.01 \le \alpha \le 0.15$ is often selected. We selected a significance level $\alpha = 0.015$. For the Cochran criteria, $C_{i,N,n,\alpha}$ values for $\alpha = 0.015$ were calculated for each of the test regimes on the basis of the approximation formula proposed in [12]; they are shown in Table 17-1.

Comparing the values given in Table 17-1 for the statistics t_2, t_3 and the corresponding critical values shows that in accordance with the inequality (6a), the hypothesis H_S when the variance and mathematical expectation are constant is accepted with significance level of $\alpha = 0.015$ for each vibration of properly functioning bearings and misaligned bearings when ESh-176 lubricant is used. The hypothesis H_S that the variance is constant, and for the given case the hypothesis that vibrations are stationary for a bearing with a defect on the inner race, ESh-176 lubricant being used, are rejected with significance level $\alpha \geq 0.015$.

Decision Making Rules Application

Considered in [8-11], the methods of vibrodiagnostics contain the following three stages: "teaching," "experiment design," "diagnostics." We shall examine in detail the special features of using the method of decision rules making to diagnose 309ES h_2 bearings in a stand while analyzing rolling bearings vibrations. In the "teaching" regime after estimation the stationarity of data of vibrations of 309ES h_2 bearing and carrying out autoregression analysis of vibration data[9], it is necessary to formulate the teaching sets and estimate the matrices Θ_m, M_m corresponding to different technical states of the examined bearings.

As mentioned earlier, the mathematical basis for these decision rules is given in the Appendix. Equations 17-1 to 17-18 are in this chapter's appendix, and several references are made to these equations. Thus, calculations here are denoted with equation numbers starting with 17-19.

Using the relationships (17-19) (see appendix), we obtain the matrices corresponding to bearings suitable for service:

$$\theta_1 = \begin{vmatrix} \pm 1.0288 \\ 1.244 \end{vmatrix},$$

$$M_1 = \begin{vmatrix} 1.0162 \times 10^{\pm 3} & \pm 0.5207 \times 10^{\pm 3} \\ \pm 0.5207 \times 10^{\pm 3} & 4.6817 \times 10^{\pm 3} \end{vmatrix}$$

and for misaligned bearings

$$\theta_2 = \begin{vmatrix} \pm 1.1087 \\ 1.399 \end{vmatrix},$$

$$M_2 = \begin{vmatrix} 1.4517 \times 10^{\pm 3} & \pm 0.2549 \times 10^{\pm 3} \\ \pm 0.2549 \times 10^{\pm 3} & 1.8445 \times 10^{\pm 3} \end{vmatrix}$$

It is now necessary to find the matrices of orthogonal transformations.

For the example given, these matrices are:

$$B_1 = \begin{vmatrix} 0.1379 & \pm 0.9904 \\ 0.9904 & 0.1379 \end{vmatrix} \quad B_2 = \begin{vmatrix} 0.4384 & \pm 0.8987 \\ 0.8987 & 0.4384 \end{vmatrix}$$

Expanding the matrix relationship $Y_m = B_m \times X$, we obtain the following equations for transition to statistics $\tilde{y}_1 \ldots \tilde{y}_4$:

$$\begin{aligned} \tilde{y}_1 &= 0.1379\tilde{a}_1 \pm 0.9904\tilde{a}_2, \\ \tilde{y}_2 &= 0.9904\tilde{a}_1 + 0.1379\tilde{a}_2 \\ \tilde{y}_3 &= 0.4384\tilde{a}_1 \pm 0.8987\tilde{a}_2 \\ \tilde{y}_4 &= 0.8987\tilde{a}_1 + 0.4384\tilde{a}_2 \end{aligned} \tag{17-19}$$

The transition to normalized statistics is carried out as follows:

$$W_{1(n)} = \begin{vmatrix} \mu_1/\sqrt{\lambda_{11}} \\ \mu_2/\sqrt{\lambda_{21}} \end{vmatrix} = \begin{vmatrix} \pm 1.4234 \\ \pm 0.84048 \end{vmatrix}, \; W_{2(n)} = \begin{vmatrix} \pm 1.7333 \\ \pm 0.38307 \end{vmatrix}$$

$$Y_{1(n)} = \begin{vmatrix} \tilde{y}_1/\sqrt{\lambda_{11}} \\ \tilde{y}_2/\sqrt{\lambda_{21}} \end{vmatrix} = \begin{vmatrix} \tilde{y}_1/\sqrt{4.754 \times 10^{\pm 3}} \\ \tilde{y}_2/\sqrt{0.94367 \times 10^{\pm 3}} \end{vmatrix},$$

$$Y_{2(n)} = \begin{vmatrix} \tilde{y}_3/\sqrt{1.9637 \times 10^{\pm 3}} \\ \tilde{y}_4/\sqrt{1.3303 \times 10^{\pm 3}} \end{vmatrix} \tag{17±20}$$

To verify the hypotheses H_1, H_2, the expression for the elementary ratio of the probability has the form:

$$\begin{aligned} \mu_{1,2n} = {} & \pm 299.103\tilde{y}_1 \pm 287.958\tilde{y}_2 \\ & + 884.198\tilde{y}_3 + 287.958\tilde{y}_4 + 384.45 \end{aligned} \tag{17-21}$$

In the "experiment design" state, it is necessary to estimate the mathematical expectations and variance of the logarithms of the elementary likelihood ratio for the hypotheses H_1, H_2. It is also necessary to determine the required number of observations n and the threshold c for the given values of the probabilities of errors of types I and II. For this example, the mathematical expectations of the elementary likelihood ratio and also its variance for verification hypotheses H_1, H_2 are respectively:

$$M_1(u_{1,2}) = 3.837, \; M_2(u_{1,2}) = -6.371, \; \sigma = 18.402$$

Taking into account the relationships (17-12, 17-13—see appendix) and specifying the probabilities of the $\alpha = 0.05$, $\beta = 0.01$, we can determine the required number of observations n and threshold c

$$N = \frac{(u_\alpha + u_\beta) \times \sigma^2}{[M_1(u_n) \pm M_2(u_n)]^2} = \frac{(1.645 + 2.326)^2 \times 18.402}{(3.837 + 6.371)^2} = 2.87 \approx 3$$

$$c = \frac{[M_1(u_n) + M_2(u_n)]}{2} + \frac{\sqrt{\sigma^2} \times (u_\alpha + u_\beta)}{2\sqrt{N}} = \frac{3.837 \pm 6.371}{2} + \frac{\sqrt{18.402}(2.326 \pm 1.645)}{2\sqrt{N}} = \pm 0.622$$

Consequently, the hypothesis H_1 (bearing is in good working order) is accepted if

$$v_{(1,2)n} \geq c_{1,2} \tag{17-22}$$

where $v_{(1,2)n} = (u_{(1,2)I} + u_{(1,2)II} + u_{(1,2)III})/3$

If the condition (17-22) is not fulfilled, we accept the hypothesis H_2 (bearing is misaligned).

In the "diagnostics" regime, the technical state of the examined bearing is tested as follows. According to the result obtained in the "experiment design" stage, to determine the misalignment of the 309ES h_2 bearing with the errors $\alpha = 0.05$, $\beta = 0.01$ placed on a stand for examining vibrations, it is necessary to record three independent data points of the vibrations and carry out their autoregression analysis.

The first index at the autoregression coefficient denotes the number of data points of the vibration; the second is the order number of the autoregression coefficient. Using equations (17-19), we determine the values of the statistics $\tilde{y}_1 \ldots \tilde{y}_4$

$$\tilde{y}_{11} = \pm 1.4867 \quad \tilde{y}_{12} = \pm 0.87723 \quad \tilde{y}_{13} = \pm 1.6843 \quad \tilde{y}_{14} = \pm 0.3107$$
$$\tilde{y}_{21} = \pm 1.4602 \quad \tilde{y}_{22} = \pm 0.8317 \quad \tilde{y}_{23} = \pm 1.6653 \quad \tilde{y}_{24} = \pm 0.3311$$
$$\tilde{y}_{31} = \pm 1.454 \quad \tilde{y}_{32} = \pm 0.8464 \quad \tilde{y}_{33} = \pm 1.6454 \quad \tilde{y}_{34} = \pm 0.2924$$

Substituting the resulting values of $\tilde{y}_1 \ldots \tilde{y}_4$ into the expressions of the elementary likelihood ratio (17-21) we obtain

$$u_{(1,2)i} = 32.65 \qquad u_{(1,2)ii} = -6.766 \qquad u_{(1,2)iii}) = 23.29$$

$$v_{(1,2)n} = u_{(1,2)I} + u_{(1,2)II} + u_{(1,2)III})/3 =$$
$$= (32.65 - 6.766 + 23.29)/3 = 16.39$$

Since $v_{(1,2)} \geq c_{(1,2)}$ we accept the hypothesis H_1—the bearing is suitable for service.

CONCLUSION

We discussed examples of energy information systems and described methods of information signals statistical analysis that may be applied to the information signals operation algorithms used by EISs through the example of diagnosing vibration data from roller bearings. Application of these methods can improve decisions made based on data from the monitoring systems of these EISs.

APPENDIX: A PROCEDURE FOR TESTING THE STATIONARITY OF EIS INFORMATION SIGNALS

We must start with some definitions. A random process $\{\xi(t), t \in (-\infty,\infty)\}$ is said to be weakly *stationary* or *stationary in the wide sense* if its mathematical expectation $M\xi(t) = m$ is constant and independent of the time t and the correlation function $R(t_1,t_2) = R(t_2,t_1)$ depends only on the difference $|t_2 - t_1|$. If the finite-dimensional distributions of the process of $\xi(t)$ are independent of the time, then $\xi(t)$ is said to be stationary in the strong sense.

A stationary process with discrete time ξ_t, $t \in Z$ (where $Z = \{\ldots,-1,0,1\ldots\}$ is a set of integers) with independent values ξ_{tk} and ξ_{tl}, $l \neq k$, is discrete white noise. It is stationary in the wide sense if its mathematical expectation and variance is constant. The process ξ_t is strongly *stationary* or stationary in the strict *sense* if its one-dimensional distribution function $F(x)$ is independent of time. To be rigorous we shall take the process ξ_t to be ergodic and non-periodically correlated.

The statistical analysis of the process ξ_{kt} is based on samples. If an experiment yields a series of numerical values $x_i = \xi_{ti}$, $i = \overline{1,n}$ of the process ξ_{kt}, then the set $x_1, x_2, \ldots, x_n$ is said to be a sample of size n for a realization of the process ξ_{kt} of the segment $k = \overline{1,n}$. The samples of the process ξ_{kt} should be used to test the hypothesis H_s that the process is a stationary state process against the alternative, hypothesis H_n, that it is a non-stationary state process.

As follows from the definition of a stationary state process in the strong sense, testing of the hypothesis H_s consists in checking to see that the N independent samples obtained at different times, each of size n, have been taken from the same random processes ξ_{kt} with distribution $F(x)$. If $F(x)$ is continuous, then H_s may be tested on the basis of the Kolmogorov-Smirnov criterion. For this purpose we define the statistic

$$D^{i,j}_{n,n} = \max_{\pm\infty < x < \infty} \left| F_{in}(x) \pm F_{jn}(x) \right| \tag{17-1}$$

where $F_{in}(x)$ and $F_{jn}(x)$, i, $j = \overline{1,N}$, $j \neq i$ are empirical distributions of the i-th and j-th samples respectively. If for the pair of samples

$$D^{i,j}_{n,n} > k_{1\pm\alpha} \sqrt{n/2} \tag{17-2}$$

then accept H_s and otherwise accept H_n. The values of quantiles $k_{1-\alpha}$ of the Kolmogorov distribution are found from the tables given in [12]. Parameter α is the level of significance of the test. It is equal to the probability that a stationary process will be taken to be a non-stationary on the basis of two samples, i.e. when $N = 2$. In evaluating the stationary state of information signals, we calculate the statistics (17-1) obtained from the pairs of samples of the information signals. If the calculated statistics satisfy the inequality (17-2), the information signals are a stationary process with significance level less than α.

The procedure for estimating the stationary state of information signals becomes simpler if signal ξ_{kt} is normally distributed. In such a case, estimating its stationary state consists of testing the hypothesis H_s that the variance and mathematical expectation are constant. It is convenient to implement such a test on the basis of the criterion t_1, the Cochran criterion t_2 and the two-sample criterion t_3. The variance is first tested for constancy, and only after this is the mathematical expectation estimated.

Estimating the stationary state of normal information signals ξ_{kt} may be carried out in the following sequence: For each of N samples we calculate the estimates of the mathematical expectation m_i and variance σ_i^2, using the formulas

$$m_i = \frac{1}{n}\sum_{k=1}^{n} x_{ki},$$
$$\sigma_i^2 = \frac{1}{n \pm 1}\sum_{k=1}^{n}\left(x_{ki} \pm m_i\right)^2, \; i = \overline{1,N} \tag{17-3}$$

where x_{ki} is the k-th reading of the i-th sample.

Constant Variance Tests

For sample size $n_i \neq n_j$, $n_i = n_j$ and number of samples $N = 2$ we use following statistic (F-criterion)

$$t_i = \frac{\sigma_1^2}{\sigma_2^2} \tag{17-4}$$

For sample size, $n_i = n_j = n$ and number of samples $N > 2$ we use following statistic (Cochran criterion)

$$t_i = \max \frac{\sigma_i^2}{\sum_{i=1}^{N} \sigma_i^2} \tag{17-5}$$

Constant Mathematical Expectation Test

For sample size $n_i \neq n_j$, $n_i = n_j$, and number of samples $N = 2$ we use following statistic (Two-sample criterion, variance of compared samples assumed equal):

$$t_3 = \frac{\left\lfloor m_i \pm m_j \right\rfloor}{\left(\frac{\left(n_i \pm 1\right)\sigma_i^2 + \left(n_j \pm 1\right)\sigma_j^2}{n_i + n_j \pm 2}\right)^{1/2}} \times \left(\frac{n_i n_j}{n_i + n_j}\right)^{1/2} \tag{17-6}$$

For the criteria chosen on the basis of the calculated m_i and σ_i^2, we use formulas (17-4)-(17-6) to calculated the values of the statistics t_i, $i = 1,2,3$. On the basis of the distribution of the statistics t_i the critical values $C_{i,N,n,\alpha}$ have been calculated and tabulated in [12] for these criteria. The critical values depend on the number of samples N and sample size n, as well as on the significance level α. The hypothesis H_s that the variance or mathematical expectation is constant is accepted, provided

$$t_i < C_{i,N,n,\alpha} \tag{17-6a}$$

for corresponding criterion.

If H_s is accepted for σ_i^2 and for m_i we then make the decision that the normal random information signal being analyzed is a stationary random process.

To test the normality of the information signals the criterion proposed in [12] may be applied. This consists in testing the hypothesis H that the skewness and kurtosis coefficients fall within the critical limits for the process being analyzed. To do this, on the basis of the significance level and the sample size, we use tables[12] to find the critical limits for the estimates of the skewness g_1 and kurtosis b_2 coefficients. If the calculated g_1 and b_2 fall between the limits, we then accept hypothesis H. This means that the analyzed sample has been taken from a normal process, with an accuracy of up to the first four moments. The direct utilization of the samples obtained to estimate information signals stationarity by means of the proposed criteria involves errors, since in the general case the readings obtained following quantification of information signals will be dependent. If we equate the amounts of information contained in samples obtained from the processes with dependent and independent readings, then the sample from the process with independent readings will be smaller in size. We refer to this size as the effective size n^*.

We consider an example of the method of stationary state estimation on the basis of roller bearings vibration signals investigations[8,9]. It follows from the model proposed in [9-11] for the vibration of roller bearings of electrical machines and experimental data that the value of the vibration autocorrelation function is equal to zero for a certain time interval Δt between readings. Readings of stationary normal vibration signals taken at intervals Δt will be independent, i.e. the information signals may be represented as a random process $\xi_{k\Delta t}$. If $\Delta t \geq T_d = 1/f_d$ a sample with independent readings will have the maximum possible effective size, which may be estimated from the following formula:

$$n^* = \mathrm{mod}[n/f_d S_o] \quad (17\text{-}7)$$

where mod[a/b] - integer part of division a/b;

$S_o \leq \Delta t$ is the time interval between the absolute maximum and the moment at which the autocorrelation function first passes through zero;

f_d is the sampling frequency.

For the particular case in which $\Delta t = S_o = T_d$, we obtain $n^* = n$. More detailed information concerning the method of stationary state estimation and application of the method for diagnostics of electrical machine roller bearings has been considered in [8].

A PROCEDURE FOR DECISION-MAKING RULES

Information on the technical state or control state, obtained on the basis of statistical estimates of the informative parameters, is represented in the form of a random vector with coordinates $\Xi = a_1 a_2$.

The solution of the problem is reduced to accepting one of the following hypotheses:

1. H_1 - random vector Ξ has the distribution density of probabilities P_1;
2. H_2 - random vector Ξ has the density of the distribution of probabilities P_2;
3. H_3 - random vector Ξ has the density of distribution of probabilities P_3.

Initially, we verified the hypotheses H_1, H_2 and then H_2 and H_3. Since the components of the vector of diagnostic parameters have a normal (Gaussian) distribution with a probability of 0.95, the combined densities of distribution of diagnostic parameters have the form

$$P = \frac{1}{\sqrt{2\pi |M_m}} \exp\left\{ \pm \left[\left(\Xi \pm \Theta_m\right)^T M_m^{\pm 1} \left(\Xi \pm \Theta_m\right) \right] \right\} \quad (17\text{-}8)$$

where Θ_m is the vector of mathematical expectations $\Theta_m = \{\Theta_{m1}, \Theta_{m2}\}$;
M_m-is the covariance matrix of diagnostic (or control) parameters;
$|M_m|$ is the determinant of the covariance matrix;
Ξ is the vector of diagnostic (control) parameters,
m is the number of hypotheses;
l - is the number of diagnostic parameters.

Determination of Θ_m M_m for hypothesis H_1 and H_2 is carried out using teaching sets corresponding to different technical states (control states) of the diagnosed section (control section) by means of the following expressions

$$\theta_{ml} = \frac{1}{n}\sum_{i=1}^{n} a_{iml}$$
$$M_m = \begin{vmatrix} \gamma_{1,1,m} & \gamma_{1,2,m} \\ \gamma_{2,1,m} & \gamma_{2,2,m} \end{vmatrix}$$
$$\gamma_{1,1,m} = \frac{1}{n \pm 1}\sum_{j=1}^{n} \left(a_{1,j} \pm \theta_1\right)^2$$
$$-\gamma_{2,2,m} = \frac{1}{n \pm 1}\sum_{j=1}^{n} \left(a_{2,j} \pm \theta_2\right)^2 \quad (17\text{-}9)$$
$$\gamma_{2,1,m} = \frac{1}{n \pm 1}\sum_{j=1}^{n} \left(a_{1,j} \pm \theta_1\right)\left(a_{2,j} \pm \theta_2\right)$$
$$\gamma_{1,2,m} = \gamma_{2,1,m}$$

where n is the volume of the teaching set;
Θ_{ml} are the mathematical expectations of the diagnostic (control) parameters;
$\gamma_{1,1,m}$, $\gamma_{2,2,m}$, are the variance of the diagnostic parameters;
$\gamma_{1,2,m}$ is the mixed central moment of the diagnostic parameters.

A similar procedure can be used to determine the elements of diagnostic spaces in a different dimension. The Neiman-Pearson procedure is based on the analysis of the logarithm of the relationship of probability v and is then reduced to the optimum selection of some threshold c which separates the set of permissible values of u into two non-intersecting subsets, i.e. on the set of permissible values of v, it is necessary to select threshold c for which it may be concluded, at the given value of the error of the first kind α, the fixed volume of the sample n and the lowest value of the error of the second type β, that the hypothesis H_1 holds if $v \geq c$ and H_2 if $v < c$.

The logarithm of the ratio of probability for the hypothesis H_1, H_2 is

$$u_N = \ln\sqrt{\frac{|M_2|}{|M_1|}} \pm \frac{(X \pm \Theta_1)^T M_1^{\pm 1}(X \pm \Theta_1)}{2} + \\ + \frac{\left[(X \pm \Theta_2)^T M_2^{\pm 1}(X \pm \Theta_2)\right]}{2} \quad (17\text{-}10)$$

where Θ_m, M_m are respectively the vectors of mathematical expectations and the covariant matrices of diagnostic (control) parameters for the hypothesis H_m; X is the realization of the observation vector Ξ;$(.)^T$ - sign of conjugation; $(.)^{-1}$ sign of matrix inversion.

It is assumed that the analyzed data of the information signals are independent. If we carried out registration and analysis of N independent data from information (control) signals of sections, the logarithm of the ratio of the probability is $v = u_1 + u_2 + \ldots + u_n$. To determine threshold c and the required number of observations n, it is necessary to know the distribution of the logarithm of the elementary probability u. This problem can be solved by several methods. We shall test some of them.

The distribution of the logarithm of the elementary ratio of probability can be estimated using the Monte Carlo method (i.e. by carrying out statistical simulations). However, this method has a significant disadvantage—for every test condition of the diagnosed (control) section it is necessary to carried out a relatively laborious process of statistical simulations of the logarithm of the elementary likelihood ratio and estimation of its distribution and know in advance several parameters of such models.

There is another method of solving problems of this type. Using the transformation of the parameters of the logarithm of the elementary likelihood ratio Θ_m, M_m and also realizations of the observation vector X. u can be reduced to the distribution of a known type, for example, normal, χ^2. Transformations can be both linear and non-linear. To use these transformations of the parameter of the logarithm of the likelihood ratio does not require the laborious process of statistical simulation.

Initially, it is convenient to reduce the covariance matrixes M_m to the diagonal type. For this purpose, it is necessary to determine the corresponding matrixes B_m for which the following relationships hold:

$$B_m^T \times M_m \times B_m = A_m$$

$$A = \begin{bmatrix} \lambda_{1m} & 0 \\ 0 & \lambda_{2m} \end{bmatrix} \quad (17\text{-}11)$$

Consequently $Y_m = B_m \times X$, $W_m = B_m \times \Theta_m$, where $\gamma_{1m}\ \gamma_{2m}$ are the eigen numbers of the covariance matrix; B_m are the matrixes of orthogonal transformations.

If we normalize the vector Y_m, W_m the matrixes $A_1^{\pm 1}$, $A_2^{\pm 1}$ in the expression for the logarithm of the elementary likelihood ratio are converted to unit matrices.

Now, carrying out the transformation described previously, we arrive at the case of recognition of images of two classes of two—dimensional random quantities having normal distributions, identical diagonal matrices and different vectors of mathematical expectations. Therefore, the logarithm of the elementary likelihood ratio has a normal distribution. According to the classic Neuman-Pearson procedure, specifying the probabilities of errors α and β, we can determine the required number of observations N and the threshold c using the expressions

$$N = \frac{(u_\alpha + u_\beta)^2 \times \sigma^2}{\left[M_1(u_n) \pm M_2(u_n)\right]^2} \quad (17\text{-}12)$$

$$c = \frac{\left|M_1(u_n) + M_2(u_n)\right| \times \sigma^2}{2} + \frac{\sqrt{\sigma} \times (u_\alpha u_\beta)}{2\sqrt{N}} \quad (17\text{-}13)$$

where u_α, u_β are the quantiles of the normal distribution;

$M_1\ (u_n)$ is the value of the mathematical expectation of the logarithm of the elementary likelihood ratio for the hypothesis H_1;

$M_2\ (u_n)$ is the value of the mathematical expectation of the elementary likelihood ratio for hypothesis H_2.

Consequently, the decision-making rule has the following form:

if $v \geq c$, then the hypothesis H_1 is accepted with the probabilities of the error of type I α and type II β;

if $v < c$, then the hypothesis H_2 is accepted with the probabilities α of errors of type I and β of errors of type II.

Another linear transformation can be applied. If we use a preliminary centering of diagnostic (control) parameters and further we carry out orthogonal transformation of correlation matrixes of the parameters, then the elementary likelihood ratio u has a non-central χ^2 distribution with the number of degrees of freedom equals to the number of diagnostic parameters (dimension of diagnostic or control space).

SOME BASIS OF STATISTICAL SPLINE APPROXIMATION

We consider the main aspects of spline function prediction method in a brief form.

Let y be the functional dependence given on the interval $[b,d]$ with distortions by some stochastic process. It is defined as

$$y = f(t,A),\ t \in [a,b] \tag{17-14}$$

where A is a determined vector of unknown real number parameters.

These parameters are linear entries into the y and the parameters are independent of t. Then stochastic values of the function $\left\{y_i, i=\overline{1,n}\right\}$ are observed at the given points $t_i \in [b,d], i=1,n.$. On the assumption that the stochastic values are uncorrelated values and they have a Gaussian distribution then

$$\begin{aligned} \mathbf{M}y_i &= x_{i0}a_0 + x_{i1}a_1 + \ldots x_{ir}a_r \\ \mathbf{D}y_i &= \sigma^2, \qquad i=\overline{1,n}, \end{aligned} \tag{17-15}$$

where $A = (a_1, a_2, \ldots, a_r)$ and σ are unknown parameters,

$\mathbf{X} = \left\{x_{ji}, i=\overline{1,n}, j=\overline{0,r}\right\}$ is a rectangular matrix of determined coefficients.

The coefficients are functionally dependent on t_i. This dependence on t_i is not necessarily linear. The symbols Y and A will denote the column-vector of the corresponding values of the function $y_i, i=\overline{1,n}$ and parameters $a_i, i=\overline{0,r}$ respectively. Therefore, equations (15) can be written in the matrix form

$$\begin{aligned} \mathbf{MY} &= \mathbf{XA}, \\ \mathbf{DY} &= \sigma^2\mathbf{I}, \end{aligned} \tag{17-16}$$

where I is an identity diagonal matrix of n-th order.

We have to construct the interval statistical estimation of the prediction value of function (14) with the application of estimations of unknown A vector parameters from the results of observation Y. The arbitrary selection of plan matrix elements X is assumed. The given problem is traditionally determined by solving a polynomial class of functions. However, spline method is a more suitable method of approximation from the computing point of view. Spline function $S_m(x) \in C^m[b,d]$ is a function that consists of "pieces" of different functions.

The functions are connected by an appointed plan. $C^m[b,d]$ is the set of continuous functions on the interval $[b,d]$. The set has continuous m-th derivative. It should be find the lattice $\{t_j\}_+^r$ on the interval $[b,d]$ when r is given. The lattice must satisfy the following properties: a spline $S_{k+1}(x) \in C^k[b,d]$, k = 0,1,..., is constructed on the lattice Δr to provide the minimal statistical estimation $\hat{\mathbf{A}}$ of the vector A in the mean square deviation sense. The values of vector $\hat{\mathbf{A}}$ are determined by the solution of the normal equation system

$$\hat{\mathbf{A}} = \left(\mathbf{X}^T\mathbf{X}\right)^{\pm 1}\mathbf{X}^T\mathbf{Y}, \tag{17-17}$$

where $(.)^T$ is designated the transposition matrix. If we introduce the designations $\mathbf{C} = \mathbf{X}^T\mathbf{X}$ and $\mathbf{H} = \mathbf{X}^T\mathbf{Y}$ then

$$\hat{a}_i = \sum_{j=0}^{r} \left\{C^{\pm 1}\right\}_{ji} h_j, i=\overline{0,r}, \tag{17-18}$$

where $h_j, j=\overline{1,n}$ are the elements of the matrix $\boldsymbol{H}$.

The desired lattice $\left\{\hat{t}_i\right\}_+^r$ is found by a sorting method. The statistical estimation of vector $\boldsymbol{A}$ is minimal in the mean square deviation sense. A system of equations (17-17) always has a solution. The statistical estimate of $\hat{\mathbf{A}}$ is unbiased estimate. Peculiarities of the solution of the equation system (17-17) and an optimal selection of plan matrix $\mathbf{X}$ has been discussed in [12,13]. A calculation of the confidence interval with the given confidence level is realized with the application of values of Student statistic t that are given, e.g., in [5].

The software for this method has been developed by scientists from the Institute of Electrodynamics, Ukrainian National Academy of Science.

Bibliography

[1] Capehart B.L., Turner W.C., Kennedy W.J. *Guide to Energy Management. Second Edition,* The Fairmont Press, Inc., Lilburn, GA 1997.

[2] Gopalakrishnan B., Plummer R.W., Nagarajan S. Energy: Expert Systems for Industrial Energy Conservation and Management. *Energy Engineering,* vol. 94, No. 2, 1997, pp. 58-79.

[3] Green S. Blueprint for Success. *Power Engineering International.* April 2002, pp. 37-39.

[4] Green S. Service Reaches New Heights. Power Engineering International. October 2001, pp. 41-45.

[5] Shaw W.T. Intelligent Technology. *Power Engineering International*. July 2001, pp. 35-37.
[6] Lopez M.A., Flores C.H., Garcia E.G. An intelligent tutoring system for turbine startup training of electrical power plant operators. *Expert Systems with Applications*. vol. 24, 2003, pp. 95-101.
[7] Pettersson D. A look at the lifecycle effectiveness. *Power Engineering International*. March 2002, pp. 34-36.
[8] Zvaritch V.N., Martchenko B.G., Protsenko L.D. Vibration Analysis and Detection of Electric-Machine Malfunctions. *Power Engineering* (USSR Academy of Sciences), Allerton Press, vol. 23, 1985, N4, pp. 25-31.
[9] Zvaritch V., Martchenko B. Linear autoregressive processes in the vibration diagnostic problems of electrical machine parts. Technical Diagnosis and Non-destructive Testing, Riecansky Science Publishing Co. 1996, No. 1, pp. 38-44.
[10] Martchenko B.G., Myslovitch M.V., *Vibration Diagnostic of Rolling Bearing Units of Electrical Machines*, Naukova Dumka, Kiev, 1992, (in Russian).
[11] Martchenko B., Myslovitch M., Zvaritch V. Vibration Signal Expert System for Fault Detection of Power Equipment Rolling Bearings. Proceeding of IFAC 14 World Congress, Beijing, China, July 5-9, 1999, vol. P. pp. 181-186.
[12] Bolshov L.N., Smirnov N.V., Mathematical Statistics Tables. Nauka, Moskow, 1983 (in Russian).
[13] Bendat J.S., Piersol A.G. *Measurement and Analysis of Random Data*, John Wiley & Sons, Inc. 1968.
[14] Shutko (1981). Application of the Spline-functions in the Problems of Statistical Analysis of Information Signals. Znanie UkSSR, Kiev (in Russian).
[15] Denisuk V.P and B.G. Martchenko (1995) Splines and their Application in the Problems of Simulation and Acquisition of Measurement's Signals. National States Technical University (KPI), Kiev, (in Russian).
[16] Zvaritch V.N., A.P. Malyarenko, M.V. Myslovitch and B.G. Martchenko (1994). Application of the statistical splines for prediction of radionuclide accumulation in living organisms. Fresenius Envir. Bull, 3, 563-568
[17] Deverdzic V. A Survey of Modern Knowledge Modeling Techniques. Expert systems with Applications. vol. 17, No. 4,, 1999, pp. 275-294.

Acknowledgment

The author gratefully acknowledges Prof. Dr. B. Capehart for his kindness in making it possible for me to provide a chapter for this book.

Section Six

Network Security for EIS and ECS Systems

Chapter 18

Introduction to Network Security

Curt Phillips, Senior Consultant
erg/Technicon, Smithfield, NC

THE THREATS TO INTRANETS are real, and the damage that can result when they are compromised is substantial. However, with the application of state-of-the-art network security technologies and strategies, network security can reduce these threats and provide for confidence in network operations.

INTRODUCTION

The ready and easy access to energy data via intranets and the internet opens new avenues for controlling and managing energy. However, the increasing use of intranets and the internet also opens new avenues of access into sensitive business information and even control of vital operations to those who may have malicious intent regarding your organization.

Many people assume that intranets are more secure than public networks such as the internet. But, any network is vulnerable, even internal networks operated by companies who sell secure on-line transaction systems.

It Can Happen to Anyone

Niku Corporation of Redwood City, California, is such a company. They develop commercial exchange software, but in mid-2002 they discovered that someone had used pilfered passwords to sneak into their network more than 6,000 times over approximately a nine month period, downloading some 1,000 documents.

In the lawsuit against arch rival Business Engine Software Corporation, whom they alleged was responsible for the actions, Niku contended that the stolen files were the crown jewels of their software company, including upcoming features, lists of potential customers, pricing, and customizations for clients. The downloaded items reportedly included the detailed plan for a sales demonstration to a major potential client, a potential client that Business Engine then approached using information that could have been obtained only from the confidential Niku documents.

The Niku-Business Engine lawsuit was settled in late 2002, with Business Engine reportedly to pay Niku $5 million and provide assurances that Business Engine product releases would not incorporate Niku trade secrets. But if this can happen to a company like Niku, whose business specialty is to develop secure on-line systems, what other less computer savvy companies might suffer from such trespassing and never realize what had damaged their competitiveness.

The total array of issues and solutions for internet/intranet security are beyond the scope of this chapter. The goal of this chapter is to introduce you to the issues in sufficient depth that you will be aware of the need to take network security seriously, and either do the work or commission someone to do the work required to implement sufficient security for your network.

THE THREATS

Why the Concern for Network Security?

Although the need for security is self-evident to most computer users, occasionally people will take the attitude that their network has little information of value, or that no one would be interested in the data on their network. This attitude leads them to think that there is little need for network security at their organization beyond the most nominal (and low cost) solutions. This attitude tends to increase proportionally with increases in estimates of costs for implementing more effective network security.

But implementing only the most minimal network security makes no more sense than expecting the physical assets of the organization to be adequately protected by using the most simple and least expensive padlocks available at a discount store.

The concept of "security through obscurity" is a myth. Some organizations think that they have a low enough profile, and/or their system is so unique and unusual that no one could or would break into it. But some hackers target smaller or lesser known organizations for their network busting efforts because they perceive them to be less secure. And the same open protocols and ready availability of tutorial demos that are used to sell networking equipment and network enabled controllers, allow the development of needed specialty skills by would-be unauthorized users.

What Has Your Organization Got to Lose?

A typical organization's intranet holds such sensitive information as employee records, customer and partner data, financial information, including marketing and production data, as well as proprietary engineering and technical documents. Virtually everyday, as users increasingly appreciate the power and utility of intranet accessible information, more data of value is placed in this vulnerable position.

This book, by emphasizing the advantages of using intranets for a new purpose, adds to the value of intranet access both for legitimate users and for those who would trespass into it. By enhancing the value of the network with energy data and control functions, its value is enhanced to outsiders as well.

What Are the Chances of a Security Violation?

The need for network security is underscored by the data contained in the *2003 CSI/FBI Computer Crime and Security Survey*. This edition of the yearly survey states in summary that, "...the most important conclusion one must draw from the survey remains that the risk of cyber attacks continues to be high. Even organizations that have deployed a wide range of security technologies can fall victim to significant losses. Furthermore, the percentage of these incidents that are reported to law enforcement agencies remains low. So attackers may reasonably infer that the odds against their being caught and prosecuted remain strongly in their favor."

Data compiled from this survey shows that:

- 56% of respondents reported unauthorized access
- 42% reported denial of service
- 21% reported theft of proprietary information
- 21% reported sabotage

The total annual losses reported in the 2003 survey were $201,797,340. It is important to remember that this figure is simply the total losses reported by a specific number of organizations (251 of them). As in prior years, theft of proprietary information caused the greatest financial loss ($70,195,900 was lost, with the average reported loss being approximately $2.7 million). The second most expensive computer crime among survey respondents was denial of service, with a cost of $65,643,300.

Although the theft of proprietary information represents a self-evident financial loss for an organization, the financial loss implications of denial of service may not be as familiar to those new to computer networks.

Denial of Service Attacks

A denial of service attack works by flooding a web server or internet I/P address with hundreds or thousands of false requests for information. The attack overwhelms the system's ability to process web traffic and eventually crashes it. In May 2001, vandals used such an attack to swamp www.whitehouse.gov, the public relations site of the President of the United States, essentially removing it from the internet. According to researchers at the University of California at San Diego, on-line vandals launch denial-of-service attacks more than 4,000 times per week against a variety of targets, ranging from Amazon.com and America Online to small businesses and individual users.

For a non-computer analogy of how a denial of service attack operates, imagine a teacher in a classroom with 30 students, taking questions and handling problems for each of the students. The teacher is busy, but capable of taking care of the needs of this number of students. Imagine now that someone hostile to the education of the 30 students in this classroom constructs 3000 robots that look just like the children, and flood the classroom with these robots. The robots are virtually indistinguishable from the real students, so much so that it takes the teacher a minute or so each time to tell a robot from a real student, if the teacher can tell the difference at all.

The robots, out-numbering the real students by a 100 to 1 ratio and all calling for attention from the teacher at once, would take up so much of the teacher's time and attention that the real students would be lucky to get any attention from the teacher at all. Thus the builder of the robots, hostile to the purposes of the classroom, would have successfully prevented the teacher from accomplishing the goals for which they were assigned to the classroom.

This is the essence of a denial of service attack. The attack is based on the standard network procedure for establishing a typical connection. A user device sends a message, asking a server to authenticate it. The server returns the authentication approval to the user. The user acknowledges the approval, and then is allowed on the server.

In a denial of service attack, an interloper seeking to inhibit the operation of the computer system sends many (hundreds or more) authentication requests to the server, filling it up. All the requests have false return addresses, so the server can't find the user when it wants to send the authentication approval. The server waits, sometimes more than a minute, before determining that there is a problem with a given authentication requests. Finally the server closes the connection. As soon as it closes the connection, the interloper sends a new batch of forged requests, and the process begins again. By continually sending false authentication requests, all the available connections, lines, registers, and memory of the server are kept occupied, so it cannot service any legitimate requests. Thus the money spent to purchase, maintain and operate the server is wasted for the period of time that the server is incapacitated.

So, Who Might Wish to Disrupt Your Organization's Computer Operations?

Corporate Spies

The first group is unethical competitors. Although the stereotype of the greatest threat to computer network security is bored hackers out for a thrill, these "kids" aren't the major threat. According to Richard Power, the editorial director for the Computer Security Institute (CSI), it's corporate espionage that provides the major threat to on-line data safety. "More and more hacking for the sake of trade secret theft is going on," said Power, who wrote the cybercrime book "Tangled Web." "But it's very difficult to prove. No one wants to talk about them."

William Malik, an analyst at Gartner Group, says he has consulted in two cases of electronic espionage during the past few years that cost companies over $500 million. In one, two "heavy manufacturing" firms were bidding on a $900 million contract; one outbid the other by a fraction of a percent.

Based upon his investigation, Malik said it wasn't just a case of "bad luck." The losing company happened to be testing network monitoring software during the bidding process and later discovered that someone had broken into the company's computer network and accessed files that contained bidding strategy information. Although Malik said that there wasn't enough evidence to go to court, the company called in his company to make sure that it never happened again.

An average of approximately 45% of information security professionals cited corporate competitors as likely sources of cyber attack over the past five years of the anonymous CSI/FBI security survey. The previously mentioned Niku-Business Engine case is a rare example that makes the business press.

In addition to pricing and cost structure information and marketing strategies, among the treasures that unethical competitors can reap from raiding your organization's network are current manufacturing capacity and information of general sensitivity (and possible embarrassment) to your organization such as product design and manufacturing problems or labor problems.

Computer espionage doesn't just occur between competitors in the "dog-eat-dog" world of business. In 2002, an associate dean at Princeton University was removed from his post after admitting he had used easily guessed passwords to gain unauthorized access to a student admissions site set up by another Ivy League university

Despite the reticence to discuss computer security breaches by those whose systems have been compromised, browsing through the DOJ's case list (at www.usdoj.gov/criminal/cybercrime/cccases.html) can provide many more examples of how this type of activity is not rare.

Foreign Governments or Foreign Corporations

In the 2003 CSI/FBI survey, 28% of the respondents listed foreign governments as the likely source of attacks and 25% of the respondents listed foreign corporations as the likely source of attacks. Foreign corporations have the same incentive as domestic companies to engage in corporate espionage. A foreign government may be engaged in such activity as a means of assisting the industry in their country, or as hostile act against the United States and the U.S. economy. With computer networks, distance is no barrier.

One computer security consultant reports of chasing a Chinese national for six to seven years who regularly hires U.S. teenagers to hunt down documents. In one case, he said a 17-year-old U.S. hacker was paid $1,000—and promised $10,000 more—for stealing design documents for kitchen appliances from U.S. firms.

Disgruntled or Ex-employees

Another group of potential interlopers is disgruntled employees or ex-employees. The range information sought by these people is similar to that of unethical competitors. Although they are generally most interested in sensitive and embarrassing information, they may initiate contact with competitors to seek to do harm by divulging confidential data.

Consider the case of Richard Glenn Dopps, who plead guilty to one felony count of "obtaining information from a protected computer." Here's the summary from the Department of Justice's web page on computer crime cases:

> *Until February 2001, Dopps was employed by The Bergman Companies (TBC), a contracting firm based in Chino. After leaving TBC to go work for a competitor, Dopps used his internet connection to gain access to TBC's computer systems on more than 20 occasions. Once Dopps was inside the TBC systems, he read e-mail messages of TBC executives to stay informed of TBC's ongoing business and to obtain a commercial advantage for his new employer. Dopps' unauthorized access into TBC's computer system caused approximately $21,636 in damages and costs to TBC.*

A similar group of would-be trespassers are people with personal grievances against individual employees or members of your organization. Since individuals are their target, the data they would seek to obtain would tend toward items like salary information and confidential personnel records. Although this information might not harm the organization as an entity initially, the injured party may take legal action if there is a belief that the organization did not take reasonable care to maintain the privacy of their personal information.

Employees who turn against the organization are one of the most dangerous threats a computer network will face. Because they have been (or still are) on the "inside," they can have knowledge of all the nuances of the system operation and the strengths and weaknesses of the any implemented security features.

Political or Ideological Opponents

A serious threat to your organizations network may be posed by those with a political or ideological agenda. Your organization may consider itself to absolutely apolitical, but in today's world, any group may take offense to the most seemingly trivial issues.

Who could be more inoffensive than the manufacturer of cookies, a treat to both young and old? In the spring of 2003, someone became offended by one. A major cookie maker faced a lawsuit from an individual who felt their cookie recipe was unhealthy. So anything can be a cause for grievance by someone, and one or more of those who perceive themselves to be aggrieved may seek recourse by breaking into the computer systems of their "opponents."

Hackers

A final group who will want to access your network and data without permission is the group loosely known as "hackers." For the most part, these are people who break into computer systems and networks just for the challenge of doing so, and the thrill of exploration. They may disrupt systems operations inadvertently or they may alter files and data in a form of computer graffiti, to mark where they have been. Occasionally they will attempt sell information obtained in such pursuits to help finance their activities.

The total number of "expert" hackers who can break into "any" system or network is relatively small. Ira Winkler, president of the Internet Security Advisors Group, estimates that there are probably only a couple of hundred "genius" hackers who can reliably exploit network and operating system security holes. He adds that there are perhaps five times that number of programmers who can write tools to make use of that knowledge.

But the exploits of these few are leveraged by a large number of hacker wannabes, who obtain information via the internet from the highly skilled group and promote mischief with it. The large portion of this subset are those known as "script kiddies." While the "expert" hackers take pride (perhaps inordinate pride) in their knowledge of computer systems, network weaknesses, cracking techniques and social engineering, script kiddies are those who lack the inclination or ability to learn to break into systems themselves. But on the internet and from other sources, there exist "scripts" detailing the typical ways to break into common systems and networks, using known weaknesses of commonly implemented software programs. Using these cookbook-type instructions like a large ring of master keys, the script kiddies test system after system and break into those that remain vulnerable to the known weaknesses.

Since script kiddies are not as computer savvy as the more traditional hackers, they are more likely to damage files and systems by accident through their awkward exploration of systems to which they have gained access. And the sheer number of "cookbook" hackers like script kiddies can help mask the actions of the experts who have deliberate, malicious intent as they break into computer networks.

SECURITY SOLUTIONS

Although other people in your organization may have the primary responsibility for the security of your computer systems, when you develop new resources for intranet access and invite new users to access these resources, you become a co-owner of the security issues.

As you begin to investigate solutions to computer security vulnerabilities, first, determine if there is a person in your organization's IT (information technology) staff responsible for IT security. Second, determine if your organization has an existing security policy. A security policy is a statement of procedures and requirements. A typical policy decision will state what software applications are supported; for example, the Internet Explorer browser and the specific security settings are selected for it.

If a security policy exists, coordinate your security efforts with the existing rules and policies and under the direction of the person responsible for IT security, if one has been designated. Regardless of the security policies and procedures implemented at any organization, an intranet system is only as secure as the people with access to the system. The most sophisticated security system can be compromised by the actions of a single individual. Keeping a computer system secure depends upon the staff overseeing it.

Start with a Risk Assessment

A computer network security strategy should start with a comprehensive risk assessment. A possible approach to developing this policy is to examine periodically the following:

- "What are the security vulnerabilities in the organization and its computers and networks
- What resources is the organization trying to protect?
- Which people does the organization need to protect the resources from?
- How likely are the threats?
- How important are the resources?
- What measures can the organization implement to protect the assets in a cost-effective and timely manner?
- Do the security measures in place meet the security benchmark determined to be appropriate"

Once you've quantified all of the risks that face your intranet, the next step is to define your security policy if one has not been previously defined. A security policy takes the concepts developed in the development of a security strategy and codifies them into a set of procedures and information. It defines the rights and responsibilities of system users. It can also be a guide for purchasing decisions at your organization because it defines standard classes of protection for various pieces of computer hardware and software.

Incident Response Action Plan

An important, and often overlooked, step in defining security policy is to create an incident response action plan. A good policy defines preventive measures to reduce threats, but it's also important to have an incident response plan outlining how to react to threats as or after they occur. When there's a security breach, your organization needs to be able to respond in real time.

Other important phases of the risk assessment cycle include implementing your security policy, promoting awareness among your employees, and measuring the policy's effectiveness.

It cannot be overemphasized: an intranet is only as secure as its weakest link, which is usually the people managing and using the intranet. System users should be taught the importance of their taking personal responsibility for maintaining intranet security, and the information and techniques required to assist them in doing so. For instance, do the system users know that they shouldn't open an email attachment with a .vbs or .exe file extension and why? Regular training should keep the users apprised of common risks and appropriate responses.

Finally, also have a policy in place to check for creeping links to unsecured locations, back doors from the internet to your intranet, proper use of passwords and other security mechanisms provided by your intranet setup.

Basic Security Techniques

After completing a risk assessment and crafting a security policy, it's time to put a plan into action. There is no definitive blueprint for designing a secure network; network security has to be custom tailored to fit the needs of your organization's network. The following actions are some basic ways to begin to provide security for a network.

Access Control and Permissions

The first principle of a well-designed system of access control is to ensure that the appropriate people have access to the information and resources they're authorized to use, and nothing more. This means identifying and authenticating network users, and having the right tools to manage access control.

User authentication is the first step in controlling access to sensitive information. Only after a computer system identifies a user can it decide what the user is allowed to do, and what the user is not allowed to do. To prove their identity to a computer system, the user must supply it with a set of credentials. There are three broad types of credentials that can be employed to verify a user's identity to a computer system:

- something only the authorized user should knows (for example, a password),
- something only the authorized user should possess (for example, a smart card or some other physical token),
- something only the authorized user could have (for example, DNA, biometric identification).

A strong system of authentication, like that used in ATM cash machines, attempts to verify credentials from at least two of these categories, typically a card and password or personal identification number (PIN) code. Highly secure systems, like those used in the military, authenticate users based on all three. The more types of credentials a system requires, the more trust you can place in its authentication abilities.

Currently, password protection is the most common form of access control, but as security receives more emphasis and technology costs decrease, the use of ID cards and biometrics will become increasingly common.

Enhancing Password Security

Passwords are one of the most basic elements of network security, yet are one of the most ineffectively utilized. Whether are not you are responsible for intranet security, you can work to ensure better password management within your organization. The concepts are readily implemented; they simply require that users follow rules for effective password management.

Three rules for effective password management are, 1) to require frequent changes to passwords (usually, at least every 90 days), 2) to require that passwords be complex and not "dictionary" words, and 3) to ensure that passwords are not written in readily accessible places (or better yet, not written down at all.)

The need for effective password management has existed for decades, but it remains a weak point in many computer networks. A excellent example of how NOT to practice effective password usage can be seen in a movie which predates the widespread access to the internet, 1983's "WarGames."

In a scene where Matthew Broderick's character is sent to the principal's office to be reprimanded, he slips behind the desk of the absent secretary. Beside her computer terminal, he opens a desk panel and written on it is the word, "pencil." By its location and context, it is obviously her password into the school computer. The Matthew Broderick character will later use the password to illicitly enter the school's computer system and change his grades.

Thus the secretary in the movie violated at least two of the three rules. Since "pencil" is at the top of a list of crossed-out words, it appears that she may have changed her password frequently, so perhaps at least that aspect of effective password management was followed. However, writing the password down adjacent to the computer terminal negated the benefit of frequent changes, and the password "pencil" was a word found in virtually any dictionary, decreasing its usefulness for security.

The insecure nature of "dictionary" passwords gained widespread notice in Clifford Stoll's true story of tracking international hackers in governmental computer networks in the book *The Cuckoo's Egg*. As implemented in most computer programs, user's passwords are stored in an encrypted manner. The manner of algorithms used for encryption are typically irreversible, and the analogy of a "sausage grinder" can be used to explain the process.

In a sausage grinder, meat put into it comes out as sausage, but a sausage grinder cannot operate in reverse. That is to say, although it can turn pieces of meat into ground sausage, it cannot turn ground sausage into pieces of meat.

In a mathematical manner, a computer would encrypt the password "pencil" and store it as something that looks like meaningless gibberish, perhaps "kxf#q+m." But if someone should find in a stored file that "kxf#q+m" was the password of a person's account, the program would not reverse the encryption and reveal that the actual password was "pencil." Sometime in the 1980s, someone decided that this security feature could be circumvented by running every word in a dictionary through the encryption algorithm of the program to which illicit access was desired. In this way, a cross-reference could be created, and in this cross-reference it would show that the password the correlates to "kxf#q+m" is "pencil," giving a would-be interloper a legitimate password and unauthorized access to the program or system.

So it is now well known that passwords that are single words or names are not very secure. Worse are passwords using a child's name or a pet's name or any other word that can be easily identified with a user. These types of passwords can be readily guessed without even using the brute force dictionary creation method as outlined above. For a movie example of such password breaking, see "Clear and Present Danger" where the password Jack Ryan (Harrison Ford) uses for his ATM is quickly guessed by a computer analyst working for him.

A good password is a combination of letters, numbers and keyboard symbols of at least eight characters in length. So a good password would be something like "k@x48$eo." But since almost no one could remember that password, they would be likely to write it down and thus negate most of the benefit of this complex password.

One complaint that many users have with frequently changing passwords is that they have difficulty in coming up with new passwords on a regular basis. Here is a method that provides a relatively easy way to create complex passwords, and even a method to help remember them without unduly compromising password secrecy.

Instant Super Passwords

Each person will need to create some lists that mean something to them. Something like a favorite TV show, some names distantly memorable to the person (NOT parents, children or siblings) and some numbers of distant memory and association.

For this example, we will use the TV show, "Friends." The names can be something like the names of childhood neighbors and the numbers can be the last four telephone digits of the first company you worked for, your teenage best friend, your first girlfriend (or boyfriend) and the number of your favorite professional athlete. The greater the number of lists and the longer the lists, the better for creating more and diverse passwords.

In the TV show "Friends," the character names in credits order are Rachel, Monica, Phoebe, Joey, Chandler and Ross. Let's suppose that neighbors you remember from your childhood are Ed (dad), Ann (mom), Jean (daughter) and Jerry (son). The last four digits of your first employer's phone number (US Steel, to be abbreviated USS) was 7483, you best friend Jack's phone number ended with 2230, your first girlfriend Beth's phone number ended with 1906 and your favorite athlete is Michael Jordan, whose number was 23. Both the list of childhood neighbors and numbers are so deeply rooted in your past and so arbitrary that no one you did not tell would guess their relationship to you.

Using these lists to create a password, for instance, take the first four letters of the third character (in credits order) from "Friends" ("phoe" for Phoebe) and the digits from Jack's old phone number, 2230. Separate them with a keyboard symbol like '&' and you have the unique password "phoe&2230" which contains all the elements of a good password and is something that would not occur in any dictionary. Although you would *still* want to keep any password reminder well hidden, a reminder written as "[FR3]3[jack]" would be meaningless to anyone who encountered it, but could remind you that the password consisted of the third character in "Friends," the keyboard character which is yielded when you shift the number "7" and the old phone number of your high school friend Jack.

Another password can be quickly created and denoted as "[beth]8[NM]," with the "NM" standing for "Neighbor-Mom" and the password being "1906*ann." The concept of using only the first four letters of any name having more than four letters is a rule that you can use or discard at your choosing.

As you can see, with only a few lists and a list of numbers of distant meaning to you, you can create a virtually endless string of unique, varying length, diverse passwords that follow the guidelines for secure passwords, using elements that cannot be readily guessed at, while being at least reasonably easy to remember, and capable of being cryptically noted should your memory require a little boost on occasion. Teaching the people using your computer network to use more secure passwords is a small but important step to better network security.

But there is one other aspect of password use for access control with an intranet. There are two main forms of HTTP authentication: basic and digest. Basic authentication transmits unencrypted user names and passwords and provides no real security. This is probably acceptable if the passwords aren't traveling outside of a trusted network of the organization, but in most other cases digest authentication will be needed to provide better security. Digest authentication typically uses a sophisticated hashing algorithm to obfuscate user credentials before they're sent from client to server. Many types of server software allow for a choice of authentication methods.

BASIC INTER-NETWORK SECURITY

Firewalls

Firewalls are an essential part of network security. A network firewall is more than just a piece of hardware, like a router. It is an integrated system of hardware and software, part of an overall security plan that also includes protection from internet and corporate intranet links. They are used to protect your intranet from external threats that can compromise your system. A firewall is usually deployed so that everyone entering or leaving your network must pass through it, somewhat like the drawbridge of a medieval castle surrounded by a moat.

As with most aspects of computer security, a firewall is only as good as those who operate it. If a firewall is not set up properly it may let interlopers in. Furthermore, as the defenders of ancient Troy discovered, attacks often come from within. But despite its limitations, a firewall is still a basic and effective way to protect a network that must be connected to the internet.

Firewalls work by examining network data packets as they attempt to move from inside to outside your intranet, and vice versa, and then applying a set of rules to:

1. Protect the internal network from attack by external sources.
2. Prohibit access from the internal network to outside sources; in effect, forcing compliance with your organization's Security Policy.

In addition, firewalls can perform the following functions depending on their type:

- Provide network address translation (NAT); convert internal IP addresses on an internal network into globally unique public IP addresses for use on the internet
- Keep a record of network traffic that has passed through the firewall that can be used for other purposes.
- Divide a network into segments according to differing security needs.

Types of Firewalls

There are several ways to classify firewalls. Some useful sets of classifications are defined by the firewall's intended use, its system architecture, and the way it uses filtering technology.

If the purpose of the firewall being installed is to protect a network, it is then known as a *"network firewall."* A network firewall should be installed at a point where one network meets another in order to act as a choke point for network traffic. While the most common use for a network firewall is to protect an inside network from the internet, a network firewall can also be used to protect or segregate networks whose security requirements differ from each other. Most of the firewalls discussed in this section are network firewalls.

If a firewall is used to protect a single PC, then it is called a *"personal firewall."* Personal firewalls have become much more common since the advent of broadband internet access for home users. Personal firewalls share many of the same traits as network firewalls, such as protecting an internal "network" (or single computer) from the internet. Some personal firewalls limit access to the internet, as well.

Firewall Classifications

A firewall may also be classified by the way it is designed to use its hardware and software. In terms of software, firewalls fall into two distinct categories: firewalls that have an operating system separate from the firewall software and those that don't. Firewalls that don't require a separate operating system are sometimes referred to as "firewall appliances" or "single-box firewalls." The great advantage of purchasing a firewall appliance is that it does not require the user or administrator to worry about installing an operating system, making sure it is secure and then maintaining it. In this case the firewall software is also its operating system. However, the administrator of a firewall appliance has to learn the appliance's proprietary operating system at the same time they learn to administer the firewall's filtering system. These devices are generally more expensive than firewalls based on a standard, separate operating system.

Regardless of whether the firewall requires a separate operating system or not, firewalls can also be classified by the number of network interfaces they employ. A firewall usually has two or more network interface cards that connect it to two or more networks. Network administrators often deliberately move servers that create more risk than others, such as web servers or e-mail servers, onto a third network that is set apart from the internal network but is still not completely on the outside. In this way, more access can be granted to them while still affording them some protection. This type of firewall network architecture is known as a "multi-homed firewall with a perimeter network."

Filtering Technology

The most common way firewalls are classified is by the way the firewall actually works. Depending on how fine a distinction is made in the ways that firewalls work, they can be grouped into at least two categories:

Packet Filtering

The simplest function performed by a firewall is packet filtering. A packet filtering firewall describes a firewall that examines the basic building blocks of network transmissions, called packets, compares them with filtering rules set by the network administrator, and then decides whether to forward the packets into the network. In total, these filtering rules are known as a ruleset. The firewall examines portions of the packet known as its headers, specifically the packet's originating IP address, destination IP address, TCP/IP source port, and TCP/IP destination port.

Modern firewalls further improve packet filters by adding a feature called *stateful packet inspection*. Using stateful packet inspection means that the firewall not only inspects packets as they cross it, but that it also looks at the characteristics of the network transmission, known as its state, which gives it important information about whether or not the packet is legitimate.

Stateful packet filters can record session-specific information about the network connection, including which ports are in use at the client and server. Information on network connection ports is important because, although most internet services run on well-known ports, internet clients may be using any port above 1024. If a stateless packet filter is used, it can't tell which port is being used, so in order to let web servers respond to browsers at the higher port numbers, it leaves them ALL open. Stateful packet filters, on the other hand, can make intelligent decisions about which ports to open for return communications and which to keep shut. Stateful packet inspection also enhances security by allowing the filter to distinguish on which side of the firewall a connection was initiated.

Be aware that because packet filtering firewalls only look at a very small amount of information contained in the packet's headers, they must examine and decide upon thousands of packets in a very short time, so that the overhead imposed on the transmission time is minimized. However, because they basically allow direct connections without examining much more than basic information, they are open to particular types of network attacks where packets are deliberately altered. Furthermore, since the actual data portion of the packet is not considered at all, higher-level manipulation of the packet, such as screening for content, cannot be performed.

Application Proxy

A proxy is someone who acts on your behalf; in the computer network world, an application proxy firewall never allows direct connections from one network to another. Instead, the firewall acts on behalf of the network and intercepts each packet, examines it and compares it with the filtering rules set by the network administrator and then (if the packet is to be allowed on to its destination) creates a mirror of the application that sent the packet in the first place, copies the data into it, and then sends the packet on.

Application proxies are considered to be very secure because actual packets never make it through the firewall. Application proxies can also filter content. In addition, they can provide for user authentication and web caching to speed up perceived network speed. However, because this operation is complicated and resource intensive, it can be very slow, especially if it is used on a network with a lot of traffic. Furthermore, if it sees packets and information that the application proxy doesn't understand or can't handle, they may not be forwarded properly.

Another powerful security feature offered by most firewalls is network address translation (NAT). This allows the network to hide the IP addresses of the intranet's hosts from the external world, making it a powerful technique for masking the structure of your intranet from those who might be seeking information to be used maliciously. It does this by converting the private IP addresses on the intranet into unique IP addresses for use on the internet. To perform NAT, a firewall maintains a list of internal sockets and matching external sockets. When an intranet client attempts to establish a connection with an external host, the firewall mates the internal socket with one of its external sockets. The same process, working in reverse, lets external hosts send data back to intranet hosts without knowing the actual IP address of the internal host.

A useful additional benefit of NAT is the conservation of IP addresses. Many smaller organizations get only 8 or 16 IP addresses from their ISP and the number of hosts on the intranet can rapidly exceed that number. Using NAT, only the border devices at the perimeter of your network need to expose legal IP addresses. Internal hosts can be assigned nonroutable IP addresses to conserve the number of legal ones.

Hybrid Firewalls

Many commercial firewall products offer a hybrid of firewall types and functions. A basic packet-filtering firewall can be created using a router. The router could be programmed to drop packets attempting to enter the intranet if they are deliberately masquerading as packets coming from within the intranet (known as IP spoofing.) Putting this type of router between your network and the internet would have it function as a border router. If such a device were combined with another type of firewall, like an application proxy, the intranet would have even more protection. Providing unauthorized users with multiple firewalls that are attempting to constrain their activities might cause them to seek other networks to hack.

What Firewalls Cannot Do

Firewalls are important for securing the border of your trusted network, but don't let your firewall lull you into a false sense of security. Firewalls cannot protect your network from attacks coming from within the network, nor can they protect attacks arriving via connections that don't go through it, such as modem connections. Many firewalls can be administered remotely by telnet. You don't want to leave any holes open in your system that would let a cracker telnet into your firewall and reconfigure it.

Firewalls aren't very good at protecting a network from viruses, and they aren't very good at detecting new types of attacks, either. To maintain its effectiveness, a firewall needs someone who is trained and knowledgeable to keep them properly configured. Given that some threats may have an insider providing assistance, all firewall equipment should be placed in a locked equipment room or cabinet.

SECURE COMPUTER PLATFORMS

Because of the importance of preventing the theft of proprietary data, the concept of using trusted computer platforms is making a comeback. One way of thinking about this general trend is that it focuses on adding security to the end-user desktop computer (though, of course, the same tools will doubtless be adopted on server equipment). At the desktop, there is currently no effective way to tell at a distance (from the perspective of the application server, for example) whether someone or some rogue process has tampered with the software or data running on the desktop.

Recently chip makers and Microsoft began to tackle this problem both at the hardware and operating system levels. The basic idea is to embed a tamper-resistant security chip into computer systems, providing a location to store information about what the software on the system is supposed to "look" like. Low-level routines in the operating system then use that information to verify the trustworthiness of the system before it is allowed to run software.

This idea of trusted systems isn't new—there was considerable interest in the idea in the 1980s—but it's new to desktop computers. Early production of the security chips is well in the works. IBM and Microsoft are already working on compatible hardware and software combinations. Given the level of losses due to the theft of proprietary data reported in the 2003 CSI/FBI survey, many other options are likely to be tested in the near future.

IPSec

The IP Security Protocol Working Group of the Internet Engineering Task Force (IETF) is working to develop the IP Security (IPSec) Protocol, a standards-based method of providing privacy, integrity, and authenticity to information transferred across IP networks. The goal of IPSec is to address all of these threats in the network infrastructure itself, without requiring expensive host and application modifications.

IPSEC, defined by IETF, consists of a set of open standards to provide security equivalent to a private network in the shared infrastructure, the internet. Unlike SSL, which is implemented in the application layer, IPSEC provides security at the network layer. It provides security to all the packets of applications belonging to a security policy. There is no requirement to change the applications to support security.

IPSEC makes use of different cryptography technologies to provide confidentiality, integrity and authenticity. Bulk encryption algorithms provide data confidentiality. Keyed hashed algorithms provide per-packet authentication and data integrity. Digital certificates, based on RSA and DSA, provide identity verification and access control. An algorithm called the Diffie-Hellman algorithm provides key material for encryption and authentication.

IPSEC mainly consists of two components—IPSEC Packet Processing, and Internet Key Exchange (IKE). The IPSEC packet processing component secures the packets by encrypting and authenticating them. The internet key exchange component negotiates security proposals between two entities and generates the key material. The IKE component uses digital certificates and pre-shared keys to authenticate the peers, and Diffie-Hellman algorithm to create shared keys.

IPSec is a relatively new protocol, and there may be some incompatibilities between it and older network equipment. But because of the level of technology and sophistication being developed by IPSec to provide network security, it should be reviewed by anyone involved with implementing network security. Detailed information on IPSec can be found at the IETF web site, listed on the resources page at the end of the chapter.

WIRELESS NETWORK ACCESS SECURITY

The advent of wireless access has added a powerful new capability to intranets and the internet, but it also adds a multitude of new ways that unauthorized users can make mischief with an organization's intranet and connected computer systems and equipment.

Although a first impulse may be simply to decree that wireless access with not be used with the organization's intranet, this is not practical. Wireless offers too many advantages to increase your organization's flexibility, productivity and competitiveness to ignore it. Plus, the low cost, ready availability and easy installation of such devices means that there is a high likelihood that one of your organization's users has or soon will have installed one for their personal use and convenience, without the sanction of IT management. These rogue APs (Access Points) must be eliminated or incorporated into the intranet under the appropriate control and security management, and vigilance must be maintained for the possibility of new rogue APs appearing.

The first important concept to be made clear about wireless networks and security for them is the fact that they are **wireless**. Radio transmissions back in Marconi's day were known as wireless; wireless networks today use radio transmissions to allow the computers to communicate with each other. Although this should be self evident, for a long time, many people seemed to be unaware that their cordless phones or cellular phones were indeed, *radios*. They are, and the equipment that connects computers together wirelessly is first and foremost, a radio. So the communication between the computers is sent out "over the air," through the ether (as it used to be known) and is therefore susceptible to being intercepted by anyone with a compatible radio receiver, over a distance that probably exceeds the property lines of your organization's facility. In this case, a compatible "radio receiver" is any wireless network device compatible with your system. It can be purchased from virtually any computer sales outlet. So, via wireless, a person located outside the control of your organization with equipment easily purchased from computer stores can attempt to connect to your intranet.

Fortunately, the need for security was evident to the first designers of wireless network access. Unfortunately, many of their attempts at security were far from sufficient.

The original IEEE standard for wireless networking, 802.11, uses authentication and encryption for its basic security. When shared-key authentication is enabled, stations can associate with the AP only if they have a 40- or 128-bit key known to both stations. If Wired Equivalent Privacy (WEP) is enabled, the same key is fed into a cipher called RC4 to encrypt all the data frames. Only stations that possess the shared key can join the wireless network, but the same key decrypts frames transmitted by other stations.

Although the security features of 802.11 are a useful first line of defense, these are just a start. WEP is a wimp, and can be cracked by a determined "script kiddie" using tools readily available on the internet.

The 802.1X standard offers an "extensible authentication protocol" which provides an interface for vendors to implement more robust authentication, and 802.11I provides for a temporal key integrity protocol (TKIP) to implement more robust encryption.

Wi-Fi is the brand name of 802.11 products certified by the Wi-Fi Alliance, a consortium organized to promote 802.11 products and interoperability among them. Wi-Fi Protected Access (WPA) is a security enhancement for current-generation wireless network hardware. WPA incorporates just the stable parts of the 802.11i advanced security standard, which is still a work in progress. WPA products can interoperate with the older WEP products.

WPA defines TKIP, which derives keys by mixing a base key with the transmitter's unique media access control (MAC) address. An initialization vector is mixed with that key to generate per-packet keys. This stops WEP-crackers from comparing frames encrypted with the same key. WPA also includes a message integrity check (MIC) to prevent data forgery.

For a higher level of security, organizations should use WPA with 802.1X for key delivery and refresh. Those using WEP should apply certified WPA firmware as soon as upgrades become available. The final 802.11i standard will add more robust security using next-generation hardware, but that will be a major upgrade rather than simply a firmware upgrade.

Remember that none of these security features are "bullet-proof." There are numerous third-party hardware and software available to boost the security of wireless networks. Just as with security for the intranet, these should be considered based upon the level of threats to your intranet and the potential losses that could be incurred.

As will be covered below, it is always a mistake to leave equipment in it's default state when it is installed, but the complications of making this mistake are increased by several orders of magnitude when it is made in conjunction with the application of wireless networking equipment. All security features should be enabled (most often the default is disabled), and all passwords and other means of access should be changed.

Intranet Strategy for Wireless Connections

There are a couple of strategies that should definitely be implemented within the intranet if wireless access is allowed. Wireless APs should be treated as "untrusted entities" and should only be sited outside the firewall—NEVER inside the firewall.

After entering the wired network, traffic from the wireless access points should be segregated so that different policies can be applied. Intranet servers, edge routers and bandwidth managers can be updated to filter on addresses to your wireless APs. Even when addresses are hidden behind network address translation (NAT), Virtual LAN (VLAN) tags can be employed to avoid having wireless traffic from traveling unfettered though your intranet.

Wireless Security—Offense and Defense

With wireless access being such a new area, everyday there are reports of new ways in which wireless networks have been attacked, and everyday new practices and software are being developed to counter the threats. Using wireless access obligates those in charge of network security to educate themselves continuously.

Keep all software updated with the latest patches, and stay apprised of threats both to wireless access points in general and to the specific software and hardware used for wireless access by your organization.

Wireless Mobile Devices

One advantage of wireless networking is that it allows for the easy use of wireless mobile computing devices like hand-held computers and personal digital assistants (PDAs). But the use of these devices adds another layer of security precautions that must be considered.

The problem of security with portable computing devices (including PDAs) when your users take them off premises may not seem directly relevant to an attack on your intranet. After all, even though these devices are vulnerable to being compromised, especially if wireless connections are made in their use, hacking into them will not allow for a direct attack on the servers and connected equipment on your intranet. However, the information gleaned from unauthorized access of this type could assist an interloper in an attack on the physical intranet of your organization. So you must stress the need for security and secure operating procedures for your organization's computing devices, both on and off-site, to maintain a secure intranet and a secure computing environment.

SECURITY FOLLIES: CHECK AND DOUBLE CHECK

Whether the elements necessary to provide for intranet security have been long implemented on your organization's networks or are just underway, there are some special points to check and double check, to be absolutely sure that your network avoids these common follies.

Access Control Follies and Password Follies

Make sure that users only have access to files and areas that they have a definite need to access. Broad network access should not be given upon request, even to managers and supervisors unless they need that access as a part of their job. The broader the access that is granted, the greater the chance that interlopers will illicitly enter the network in a manner that will allow them maximize the damage they do.

Certain key files should only be accessed by specific individuals. For example, password or system configuration files should only be accessed by network administrators.

Many organizations allow access from any user to any resource on the intranet. Dividing your intranet into segments—such as human resources, engineering, manufacturing, sales, and others—protects assets and information from unauthorized users. This is readily accomplished by using firewall technology to control traffic and access between work groups and departments.

For example, compartmentalization, which provides extra measures of security between external servers such as web and commerce servers and the internal servers and databases, is a good idea. A secure operating environment and strict networking and access controls would be appropriate for any server that is exposed to public use and has access to internal databases.

The importance of following password management techniques should be constantly reinforced to all users. Although passwords are just a first line level of defense, they are a very important part of total network security, and their importance should not be ignored.

Helpful User Follies

Most users like to be helpful. So when a fellow employee asks them for their password, so they can access network areas not normally available to them or for whatever reason, they will often accommodate them. Sometimes they will give out passwords or network information to people who request it over the phone or e-mail, especially if the person seems to be someone who *should* have that information. The concept of an interloper pretending to be such a person and getting inside data from legitimate users is part of what is known as "social engineering," and is a basic technique that hackers use obtain the information they need to compromise intranets. Noted hacker Kevin Mitnick is teaching courses on "Social Engineering" around the country, presumably to instruct organizations how to protect themselves from those seeking to gain access to network information illicitly. But simply emphasizing and re-emphasing to users the importance of maintaining the confidentiality of network information can go a long way towards accomplishing this goal.

Encryption Follies

The need for encryption as data are transmitted across networks has been mentioned previously and is emphasized in almost all texts on network security. However, files containing the most sensitive data—including salary information, strategic plans, and intellectual property—require extra protection. In addition to restricting access to these files to the bare minimum of users who have a legitimate need for it, these files should be individually encrypted.

Although both Microsoft Word and Excel offer password protection encryption for their files, you should be aware that programs to defeat this encryption can be purchased over the internet for less than $50. So while they will stop illicit access by the most casual users, anyone seriously seeking to access files encrypted in this manner can do so easily.

More sophisticated encryption methods are available, and should be obtained and used for highly sensitive data. Some advanced operating environments provide multiple levels of file protection and logging utilities to track users who access, or attempt to access, the data.

Remote Access Follies

Perhaps the biggest single weak point in intranet security today occurs in remote access implementations. That is, banks of dial-in modems ready to provide access to corporate network resources and information.

One problem is what are known as "wargame dialers," named after the previously mentioned 1983 move of that name which popularized them. These freely available programs will automatically dial hundreds of telephone numbers at random, which means they may dial your modem phone numbers even if those phone numbers are not publicized anywhere. Even if the system is not fully accessed through this means, many remote access servers reveal host names or LAN router prompts, information useful to people trying to gain illicit access to the network.

Complicating the problem of remote access systems is that many of them transmit users names and password prompts "in the clear," meaning that anyone listening—"snooping"—to the activity on a remote access port can see and record user information along with the data. Programs to do this are also freely available.

Often when a user is logged into the remote access system via modem, they have access to any system resource (*see Access Control follies*), sometimes because the modem access capability was installed prior to implementation of network security procedures. The danger here is that once a hacker has broken through the remote access system, they also have free access to any system resource.

All systems of remote access should be checked and re-checked to assure the adequate security features are installed and enabled with them.

Wireless Problems

Default Settings Follies

Most pieces of security-related hardware and software come out of the box with default security settings in place. Many of them have default settings like Username: Admin and Password: Admin to allow the new user to start them up.

All to often, these defaults remain in place long after the hardware and software has been installed and is in use. All default passwords should be changed or disabled, and other default settings should be changed unless (as in the case of some port settings) they must remain in order for the system to properly function. This includes servers, firewalls and operating system software.

Outside Software Follies

Many experienced users will want to bring in third party software and install it on one of the organization's computers. However, this opens the computer system to Trojan horses, backdoors, sniffers and the like that can be used to assist unauthorized users in breaking into the system.

Although users should not be allowed to add software to the system just because they don't like the organization's choice of some basic program (e.g. a standard browser), power users should not be prohibited from adding software that will truly add to their productively. A strong system of acceptance testing for all software should be implemented that will allow for new software to be installed, but only after its utility and security have been demonstrated and tested.

Application Design Follies

As new uses and applications are developed for the computer systems and networks, new software will be developed to implement them. Custom software developed in-house or by contractors should have to pass at least the same security tests as any other software in use on the system. Since such special software has a higher likelihood of involving confidential and competitively sensitive data and process, the developers should make a special effort to make it secure and resistant to those who would seek unauthorized access.

Security Audit Follies

The need for security in an intranet is on-going. The threats to the system, internal and external, and weaknesses of the network are ever evolving. Therefore, security audits of the network should take place on a regular basis; not just acceptance testing of a new piece of hardware or software.

Ideally, the security audit would be conducted by an outside organization, which would have a strong incentive to find any weaknesses and no incentive to cover up existing weaknesses. But from whatever source, regular testing, documentation of problems, resolution of problems and retesting is required to provide the intranet with security that will have a high probability of resisting attacks.

CONCLUSION

As mentioned early in this chapter, the issue of computer system and network security is much larger than can be handled in one chapter. Entire books have been devoted to single aspects of these issues. Be sure to take advantage of whatever internal resources your organization currently has to ensure than adequate security has been implemented. Do not hesitate to hire outside assistance if your internal resources lack either the knowledge or time to implement them. Saving a few dollars in this area now can easily cost you a thousand times more later.

Resources

The Computer Security Institute—www.gocsi.com
The Network Security Library—secinf.net
The SANS (SysAdmin, Audit, Network, Security) Institute—www.sans.org
National Institute of Standards and Technology (NIST), Computer Security Resource Center (CSRC) csrc.nist.gov—Guidelines for the Security Certification and Accreditation of Federal Information Technology Systems
The Internet Engineering Task Force (IETF), IP Security Protocol (IPSec)—www.ietf.org/html.charters/ipsec-charter.html
Information Security magazine—http://www.infosecuritymag.com/index.shtml
IEEE 802.11 Standard—http://grouper.ieee.org/groups/802/11/main.html
http://www.isaca.org/articles.

Bibliography

Cooper, Frederic J.; Goggans, Chris,; Halvey, John K.; Hughes, Larry; Morgan, Lisa; Siyan, Karanjit; Stallings, William; Stephenson, Peter, "Implementing Internet Security," Indianapolis: New Riders Publishing, 1995.
Hinrichs, Randy J., "Intranets: What's The Bottom Line?," SunSoft/Prentice Hall, 1997.
Hughes, Larry J., "Actually Useful Internet Security Techniques," Indianapolis: New Riders Publishing, 1995.
McCarthy, Linda, "Intranet Security: Stories from the Trenches," Sun Microsystems Press, 1998.
Siyan, Karanjit and Hare, Chris, "Internet Firewalls and Network Security," Indianapolis: New Riders Publishing, 1995.
Intranet Organization: Strategies for Management Change, by Steven L. Telleen, Ph.D., Iorg.com, 1996.
Internet Security Policy: A Technical Guide, Gaithersburg: National Institute of Standards and Technology, 1997.
Internet/Intranet Firewall Security for TCP/IP and IPX Networks: A Manageable, Integrated Solution, Ukiah Software, Inc., 1997.
Microcomputer Network Controls and the Intranet, by Frederick Gallegos and Steven R. Powell, Auerbach, 1997.
Sholtz, Paul; "Internal Security: Rules and Risks" (Web

Techniques, July 2001)
"Secure Remote Access With IPSEc VPNs, White Paper," Intoto, Inc., 2002.
"Security Out of Thin Air," WatchGuard Technologies, November 2002.
"Mobile Insecurity: A Practical Guide To Threats and Vulnerabilities," BlueFire Security Technologies, January 2003.
Arnold, Steve; "Ten Intranet Security Pitfalls and How to Avoid Them" (Intranet Professional, January/ February 2003).
Esposito, Leonardo; "Don't underestimate passwords in Web apps," (Builder.com, 26 March 2003)
Phifer, Lisa; "Air Safety," (Information Security, April 2003)
Landgrave, Tim; "Designing secure intranet applications," (Builder.com, 29 May 2003)

Web Address Shortcut

Typing complex web site addresses into a browser can be tedious and time-consuming, and they must be typed precisely. To allow for readers of this chapter to simply cut and paste the web site addresses into their browsers, a listing of all of the web site addresses listed herein can be obtained in electronic format by sending a request to the e-mail address, energybookinfo_w4cp@yahoo.com.

Chapter 19

Computer Network Security: An Overview

Robert E. Johnson, P.E.
Network Security Coordinator
College of Engineering
University of Florida

ANY SYSTEM CONNECTED to the internet will be exposed to what to date has been a steadily increasing level of probes and attacks. Any such system must be managed in a manner that protects it from these attacks. This chapter introduces the basic principles necessary to understand the proper use of protective measures, and provides specific recommendations for network and system configuration.

INTRODUCTION

Any computer connected to the internet will be subjected to probes and attacks from a variety of sources. The motivation of any individual attacker can vary considerably, but the most common attacks are aimed not at obtaining access to information stored on the computer, but at gaining control of the computer to use it as a base from which to mount further attacks. Attackers may, for instance, use a system they have broken in to as a cover for an attack against the real target in order to make it difficult to track their identity. They may also use it as one of many computers involved in an effort to overwhelm the target by the sheer number of systems simultaneously attacking it (a DDoS, or distributed denial of service attack). Attackers may also gain control of a system in order to set up an FTP or web server to distribute pirated software or video, or illegal pornography. Because of this risk, every computer on a network, regardless of whether it stores sensitive data, should be protected from attack via its network connection.

BASIC NETWORKING

To understand the basic defenses against network attacks, one must understand the basic operation of the network. The internet is based on a variety of data communication protocols that act in concert to provide a variety of services and capabilities. Of these protocols, the two that are most important from a security perspective are UDP (user datagram protocol)* and TCP (transmission control protocol).[1,2] The common internet services such as www, Email, FTP, streaming video, and similar services are based on these two protocols.

UDP is a relatively simple protocol in which each packet of data is transmitted from the sending station to the receiving station without any provision for detecting loss of data. In other words, the sending station will not know if some of the data packets get lost and never reach the receiving station. UDP is used when speed is more important than absolute data integrity, or when the software using the data has another means to detect and handle loss of data or data packets received out of sequence. In particular, streaming audio and video are often transmitted using UDP.

TCP is a more complicated protocol in which the receiving station sends acknowledgments back to the transmitter to let it know which data packets have been correctly received, so the transmitter can automatically repeat data that is lost. The overhead involved in this process slows data transfer, but it guarantees that data will reach the receiver uncorrupted, or the sending station will at least know when transmission has been unsuccessful. The convenience of automatic error detection and correction usually outweighs the added overhead, so most internet communications use the TCP protocol.

A third protocol, the internet control message protocol (ICMP) is not used to transmit application data, but

**UDP* is commonly used both as an acronym for the generic concept of an unreliable datagram protocol, and for the *User Datagram Protocol*, which is the specific implementation of an unreliable datagram protocol that is commonly used on the internet.

is nonetheless important for data communications.[3] It provides a means for the sending and receiving systems to coordinate their operation to optimize use of available bandwidth and network capacity.

PORTS

In the TCP and UDP protocols, an abstraction referred to as a *port* is used to allow the communicating systems to keep track of which data packets belong to which software on each system, allowing two computers to maintain multiple simultaneous data streams as different pairs of programs communicate with each other. The TCP and UDP protocols each provide 65,535 logical ports to be used by programs to communicate through the network. When a program wants to use the network, it asks the operating system to assign it a port number to use. Programs operating as servers usually request a specific port number that is standard for the service they provide, while client programs usually use any arbitrary port number provided by the operating system. Web servers, for example, normally use port number 80. A web browser communicates with a web server by asking the local operating system to connect it to port 80 on the web server. On the server, any data received at port 80 is handed to the web server software for processing. By convention, the first 1024 port numbers are reserved for specific services and on some operating systems can only be accessed by programs running with system administrator privileges. For this reason, the first 1024 port numbers are known as *reserved ports*, although many services now operate on ports above 1024, and the security once provided by limiting access to system administrators is no longer meaningful in a time when many desktop computer users are their own system administrators.

Note that there is a difference between how the data are handled at the server end and at the client end. Servers generally claim a port number and accept any data directed at that port, regardless of the source.* Clients request a port number for the specific purpose of using it to open a connection to a server, and do not accept data from other sources. Because the operating system sorts traffic based on the combination of addresses and port numbers used at both ends of the connection, only data from the server's port to the client's port is sent to the client software: data from other addresses or other port numbers sent to the client's port number will be ignored. This distinction, that server software typically accepts data from any source while client software receives data only from the server to which it is connected, is why servers are usually considered a greater security risk than are clients.

*Although as a general concept it is accurate to say that servers accept connections from any source, exceptions are common. Servers are often configured to only accept connections from specific internet addresses or networks, and in fact that is a recommended security precaution when it is appropriate.

TROJAN HORSES AND OTHER ATTACK VECTORS

In the nomenclature of computer security, the various forms of logical attack are frequently given names based on analogies to the physical world. *Virus*, for instance, is the name given to a program that can copy itself into a victim system, but only if a human takes some action to facilitate the infection. For example, many email viruses infect workstations when the human operator opens an infected email attachment. The virus then proceeds to email copies of itself to addresses found in the victim's email address book, thus perpetuating itself. A rung higher on the virtual evolutionary ladder is the *worm*, a program that can propagate itself without human intervention. Worms most commonly infect servers, and usually do so by exploiting bugs in the server code. In some cases, a worm may propagate itself by attempting to exploit multiple vulnerabilities, such as attacks on servers combined with virus-like email propagation. One example of this is the Nimda worm, which hit the internet in the fall of 2001 and, depending on the host on which it found itself, attempted to actively propagate itself by exploiting any of several bugs in Microsoft IIS web servers, emailing copies of itself to other victims, or inserting itself in web pages on an infected web server and taking advantage of a bug in some versions of Microsoft Internet Explorer to infect the computers of users who viewed the infected web pages.**

The motivation of the person who authors virus or worm software can vary considerably. Some of the most destructive were written by students who claimed they

**A good discussion of the Nimda worm is available at http://www.thesitewizard.com/news/nimbdaworm.shtml. Nimda and its less-capable predecessor "Code Red" are still propagating on the internet and are likely to infect a vulnerable version of Microsoft Internet Information Server (IIS) within hours of placing the server on the internet.

were merely trying to prove a concept and didn't intend for their creations to get loose on the general internet. The first worm to cause substantial disruption of the internet, the Morris worm, was just such a proof of concept demonstration that was not intended to infect systems outside its local network.* These unintentionally damaging worms and viruses usually do nothing more than propagate themselves, and are destructive because they propagate so quickly that they overwhelm by sheer volume the host systems or the networks they infect. More malicious authors may have a variety of less defensible motives. Worms and viruses written by these authors typically install some sort of server software on the victim computer that will allow a human to come back later and remotely access the system. These remote-access programs are in some instances used to steal or modify data that is on the victim system, but more commonly they are used to launch remotely-controlled network attacks from the victim system. By hiding behind several layers of victim systems, an attacker can usually hide their identity during attacks. Often, the attacks launched from remotely-controlled victim systems are coordinated so that many, sometimes hundreds or even thousands, of these slave systems simultaneously attack the target system. These coordinated attacks are typically intended to overwhelm the target system so that it is not available to normal users of the system. Such attacks, intended to deny use of a system to its normal users, are known as *denial of service* (DoS) attacks, and when the source of the attack is distributed among multiple slave systems under the control of a single master, they are known as *distributed denial of service* (DDoS) attacks.

Another common technique for installing remote-access software on a victim system so it can later be exploited is the *Trojan horse* program, so named because it is disguised as a program that a user would want to have on their system. For instance, there have been cases in which such programs were disguised as humorous animated jokes, but which installed remote access "back doors" on the victim's system when run. The unknowing victim is likely to email a copy of the Trojan horse program to a friend who also runs it, installing yet another copy of the remote-access program on another victim computer.

The challenge for the writers and users of Trojan remote access programs is that of locating their victims. How do you locate thousands of infected computers distributed among the many millions of systems on the internet? One approach is to scan large segments of the internet attempting to connect to the port number used by the remote-access program in the hope that you will run across a few by random luck, and another is to have the Trojan program "phone home" to tell its master where it is. For example, some Trojans simply email the identity of the infected host system to their master, while others post information about themselves to various public forums on the internet which their masters can monitor for messages from new slaves. Good network security practices attempt to thwart attackers at every step of this process: by preventing worms and viruses from infecting computers on the network; by properly maintaining servers so security bugs are patched before they can be exploited; and by preventing any "trojaned" systems on your network from being identified or controlled by their masters.

SYSTEM CONFIGURATION AS A SECURITY MEASURE

Because most operating systems, including *Windows*, open server ports even on workstations, an important step in securing a network or a specific system is to review the configuration of all systems on the network and turn off any unneeded servers. The procedures and knowledge required to accomplish this vary depending on the operating system in use, and cannot be covered in detail here, but there are a few concepts that apply in general.

System Updates

One measure that cannot be overemphasized is to keep every system on your network up-to-date on security patches and upgrades. A good rule of thumb is to check weekly for any security patches or upgrades that are recommended by the operating system vendor, but even better, when available, is a system that checks for upgrades automatically. An example of the latter is Microsoft's Windows Update (when the feature is enabled), but be cautioned: Windows Update is intended to maintain workstations, and cannot be counted on to automatically install patches required by server software that may be running on a system. It is also important to subscribe to the email alert list maintained by the vendor of each of the operating systems in use on the local network, as well as a good general-purpose security mail-

*An historical discussion of the Morris worm is found at http://www.sans.org/rr/malicious/morris.php. A technical description of the worm is at http://www.ee.ryerson.ca:8080/~elf/hack/iworm.html.

ing list such as the Security Alert Consensus available from SANS Institute at http://www.sans.org/newsletters/.

Server Software

It is important to understand that server software can run on any computer, whether that computer is called a server, or a personal workstation. Operating system vendors usually include some server software in the default configuration of their operating system. In *Windows*, for instance, it is common for Windows Networking services ("File and Printer Sharing") to be enabled. If "Internet Publishing" is enabled, several more services (a web server, an FTP server, and a mail server) may be running. In Linux, a different variety of services are often enabled by default. Other user-installed software packages may install server software; for example, anything that is "peer to peer" includes a server; and several Microsoft packages install either web or SQL (database) servers. The fact that a computer is used as a personal workstation does not mean it is not running server software.

Because the programmers that create them are human, and thus less than perfect, any server software can have design or programming errors that lead to security problems. When these problems are discovered by attackers, they can be used to break in to the system hosting the server software. As there is no way to predict which software will eventually be shown to have security holes, it is prudent to minimize exposure to attack by disabling any server software that is not absolutely necessary on each computer on your network.

Port Scanner

In addition to the configuration tools provided with the operating system, another important system configuration tool is a *port scanner*. A port scanner is a program run on another computer which attempts to connect to various TCP or UDP ports on the computer being tested. The result is a list of ports that are accepting connections. Once you have obtained such a list, you can use the system's own configuration tools to determine which server software is accepting connections to each of the listed port numbers, and disable those services which don't need to be running on the system.

Although many excellent port scanners are available, perhaps the most popular port scanner is *nmap*, originally developed for Linux and Unix-derivatives and now also available for Windows. *nmap* or an equivalent program belongs in any network administrator's toolkit.*

Virus Detection Software

In addition to port scanning to locate unexpected servers, both servers and workstations should have good virus detection software. There are many vendors of virus detection software, and every vendor has both fans and detractors. In practice, the choice of anti-virus software is largely personal, as every package will miss a few virus infections that another vendor's tool would have detected. In some networks with a large number of client workstations, the ability to remotely manage virus detection software on the workstations is important, while other networks may have only a few workstations and select anti-virus software for compatibility with the particular mail server used on that network. In either case, on any network with user workstations an important step in preventing email virus infection is to repeatedly stress to the users that they should never open unexpected email attachments, even if they come from someone known to the user. If there is any doubt about an attachment, the user should contact the sender of the message to make sure they meant to send it before opening the attachment.

FILTERS AND FIREWALLS

Another valuable tool for protecting networks from intrusion is the use of *filters* to prevent harmful data from reaching your network. Filtering capability can be provided by dedicated firewalls, by router filters, by host-based firewalls, or integrated into the networking software on the computer being protected.

Firewalls are dedicated systems that provide an interface between a Local Area Network (LAN) and the general internet. By inspecting the contents of each data packet traversing the interface, a firewall attempts to limit access to (and from) the LAN so that only data considered safe and desirable crosses the interface. Simpler firewalls inspect only the internet addresses and the port numbers associated with each data packet when deciding which data to pass, while more sophisticated firewalls inspect the data itself to provide more advanced capability such as the ability to block access to specific web sites or even specific web pages. The act of blocking or passing data packets based on addresses, port numbers, or other content is known as *filtering*,

*nmap is a free tool available for download from http://www.insecure.org/.

while the noun *filter* usually refers to the more abstract concept of the non-physical thing that does the filtering, i.e. the configuration commands that determine what data are filtered. These commands are also known as *filter rules*.

Host-based firewalls are simply firewalls that run on the computer (or *host*) they are protecting, rather than in a separate system dedicated to the firewall function.* Although host-based firewalls are often called "software" firewalls, this is somewhat of a misnomer. All firewalls are software based, and the distinction is whether the software is running on dedicated hardware (referred to as a *hardware firewall* when the hardware is designed specifically to host firewall software), or on hardware that is also used for some other purpose. Host-based firewalls have disadvantages relative to dedicated firewalls: perhaps most significant is that any attack that penetrates the firewall (e.g. an email virus) is in a position to not only infect the target system, but to disable the firewall that is running on the target system. Host-based firewalls also tend to generate warning messages that are confusing to users of the host system (when it is a workstation), although in some implementations that can be controlled by having the warning messages directed to a master system staffed by experienced IT staff. To offset these disadvantages, host-based firewalls do present some unique advantages, not the least of which is that a host-based firewall is able to determine which program is responsible for each data packet. For instance, some streaming video servers use the same port number (port 80) as web servers in order to bypass typical firewall rules. A good host-based firewall can be configured to permit traffic from port 80 to be delivered only to the workstation's web browser, and drop packets arriving from port 80 that are destined for any other software, providing a tool for IT staff to use to enforce policies regarding network use.

Routers and *switches* are network infrastructure devices that determine how to route data packets through the network so they reach their destination.** Inherent in this routing function is the ability to filter data packets and provide rudimentary firewall capability. The filtering capability varies from vendor to vendor and model to model, and the filtering on more advanced routers can be more sophisticated than some dedicated firewalls. Even the simple routers marketed for residential use with DSL or cable modem connections typically have effective filtering capability, and may in fact be marketed as firewalls.

As an alternative to firewalls and router filters, some operating systems provide limited filtering capability either as part of their core networking capability or as an option. A common example is the *TCP Filters* package on Linux and other Unix-like systems. Although generally limited in capability, these filters are nonetheless useful tools for protecting systems from attackers.

Port Filtering

Port filtering is a term for very simple filtering techniques that are based strictly on the TCP or UDP port numbers that the data packets are directed to (or from). Despite the simplicity, port filtering is quite effective and a very popular method for reducing exposure to network attacks. The basic concept of port filtering is that because the greatest risk to a network is usually the servers that operate there, and most servers operate on fixed, well-defined port numbers, filtering out data destined to specific ports will greatly reduce the risk of an attacker breaking in to a network. Because of the simplicity, basic port filters can be implemented on almost anything capable of doing any filtering at all. By setting up appropriate port filters, you can greatly reduce the risk posed by numerous worms, Trojans, and other attacks. If you have a router or firewall that supports more advanced techniques, you may want to start by implementing a set of basic port filters to provide protection while you learn to use the more advanced tools.

Filter Strategies

Regardless of whether your filtering will be done by a full-featured firewall or a set of simple router filters, there are two basic strategies for developing filter rules. The most secure approach is to set up filters that block all data except that directed to (or from) specific ports that you wish to allow. This strategy is often summarized as "prohibit everything, except that which is expressly permitted."

An alternative approach is to only block data that you specifically want to prohibit, and to permit everything else. Although this strategy is notably less secure, it is widely used because of its ease of implementation and the low demand it places on the filtering device.

*Examples of the many host-based firewalls available include *ZoneAlarm* for Windows, *iptables* on Linux systems, and *ipfw* on BSD systems. The latter two can also be used as dedicated firewalls to protect entire networks.

**A simple description of the difference between a router and a switch is that switches determine how to route traffic through the local area network, while routers determine how to route traffic to, from, or within the internet. Although it is now common for routers to have advanced filtering capabilities, it is less common to find such features in switches.

In the "prohibit everything" strategy, the data you normally permit is (a) outgoing data from clients on your network to servers outside your network; (b) replies from those servers to your local clients; (c) incoming data from clients outside your network to servers on your local network; and (d) replies from your local servers to outside clients. The primary disadvantage of this approach is that it may, depending on your needs, require some moderately sophisticated features on the part of the device doing the filtering. Because clients on your network trying to connect to servers outside your network can use almost any port number as the source of their connection request, the most basic requirement of such a filter strategy is that outgoing data must be allowed to originate from almost any port, while incoming data must be blocked on all ports that are not intended to be used by servers on your network, but must still allow replies to your clients' requests.*

Some network protocols are difficult to handle with simple filter rules. For instance, in active-mode FTP the client opens a command channel to the server; the server and client then negotiate a port number to use for the data channel; and finally the server opens a data channel to the client. If the client is behind a firewall, the attempt by the server to open a data channel to the client looks (to the firewall) like an attempt by an outside system to connect to an internal server that shouldn't exist.** This poses a challenge if you need to permit FTP connections to or from your network; the only way to properly handle such a situation is with a filter device that can recognize an FTP connection when the command channel is opened, monitor the conversation between the client and the server to determine what port numbers they negotiate for the data channel, and modify its own filter rules to allow that data through the network interface. Similarly, some streaming video protocols use a different port to receive data from the server than the port the client originally used to request the video, although they are less challenging than FTP in that the port numbers used for both connections are fixed by convention and need not be determined by monitoring the connection. In this case, it is merely necessary for the filter rules to open a specific incoming port in response to activity on an outgoing "trigger" port. Fortunately, it is now common for both of these capabilities to be supported by even simple router/firewall appliances intended for residential or small office use, allowing support of FTP and streaming video clients.

A good intermediate strategy that combines some of the benefits of the "prohibit everything that is not explicitly permitted" and the "permit everything not explicitly prohibited" strategies is to block access from the outside world to ports 1-1024, except to servers that specifically provide services on those ports, and to also block any ports above 1024 that are used by servers on your network. This strategy takes advantage of the fact that port numbers 1-1024 are reserved for specific services, and you can safely block access to them on systems that are not providing those services. You cannot block outgoing connections to those ports without affecting your users' ability to use the corresponding services on external servers.

When you configure a filter rule set, you will usually have a choice between *dropping* incoming packets, and *rejecting* them. When a firewall or filter drops a packet, the sending system receives no indication that the packet was lost, and must wait some time-out interval before deciding that the packet must be re-sent. The result, from the point of view of the sending system, is that packets are being lost just as if there were no system at all at the destination address. This sort of network invisibility often discourages follow-up probing and attacks. In some cases, though, you do not want to introduce the time-out delay associated with dropped packets. For example, *sendmail*, one of the most common mail servers on the internet, responds to incoming mail by sending an *ident* query back to the source of the mail. The normal response to the *ident* query would be to provide information about the user that sent the mail, but many network administrators do not want to reveal the sort of user information returned by *ident*. If your filter drops the incoming *ident* queries from an outside mail server, the mail system sending the queries will need to wait for the packet time-out period before it can continue to process your mail message, thus significantly delaying the processing of the message. If you instead reject, rather than drop, the incoming *ident* query, the mail server immediately receives a packet indicating that the query was rejected, and can continue to process your message without delay. Very few, if any, mail servers will reject mail if they do not receive a response to the *ident* query, so your mail will not be lost if you drop or reject ident queries.

Appendix A contains a list of ports that should be considered for filtering, or in some cases, ports that

*In practice, most legitimate servers use port numbers below 1024, and clients normally only use port numbers above 1024. Some legitimate servers do, however, use port numbers above 1024.

**A newer form of FTP known as passive mode was developed to simplify the use of FTP with firewalls. In passive mode FTP, the client opens both the command and the data connections so that a firewall protecting the client needs no special knowledge of the FTP protocol.

should not be filtered, when setting up a network. Although the list is not exhaustive, it includes most of the ports commonly involved in security issues, and explains why each is a candidate for filtering.

ICMP Filtering

Many devices that can provide port filtering for TCP and UDP packets can also filter ICMP packets. Recall that ICMP is the protocol used to convey out-of-band information about the status of data connections. As such, ICMP includes several packet types that are used to convey status information. For example, there are ICMP packet types that inform the sending system that the destination was unreachable, and give some indication as to why. Unfortunately, there are also ICMP packet types that can be used to reduce or breach the security of your network, so an important part of setting up your filter rules is to block the more dangerous ICMP packets from entering or leaving your network. Appendix B is a list of ICMP packets that are important to the normal operation of a network. A good strategy is to permit these types and block all other ICMP packet types.

CONCLUSION

Although network security can be an extremely challenging endeavor, a few basic practices will provide protection that is adequate for all but the most sensitive networks. By eliminating unnecessary server software on all computers in the network, promptly installing all security patches for the operating systems in use, providing users with training in basic security practices, providing virus detection capability on any system that is used to receive email, and operating an effective firewall or other packet filter system, the local network administrator can eliminate the threat from the great majority of routine attacks on the internet.

References

[1] Postel, J., "User Datagram Protocol," RFC 768, August 1980, http://www.ietf.org/rfc/rfc0768.txt.

[2] Postel, J., "Transmission Control Protocol," RFC 793, September 1981, http://www.ietf.org/rfc/rfc0793.txt.

[3] Postel, J., "Internet Control Message Protocol," RFC 792, September 1981, http://www.ietf.org/rfc/rfc0792.txt.

APPENDIX A: PORT LIST

The following is a list of ports that should be considered for filtering as a basic security precaution on any network connected to the internet. Not all of these ports should be filtered on every network, but you should evaluate each port and decide whether it is appropriate to leave it unfiltered on your network. You should not arbitrarily block outbound ports because your users' systems need to use those ports to access services outside your network. Most inbound ports, however, can be blocked without affecting services other than that specifically targeted by the filter.

Because port numbers below 1024 are reserved for various servers and not used by clients, a common practice when static filters are used is to block inbound packets to all port numbers below 1024 except those used by servers on the local network, plus any additional ports considered a threat to the network.

If the filtering system is capable of dynamic filtering, block all inbound ports except those used by local servers, but be sure that the filter system dynamically adjusts to allow users to receive replies from external servers to which they connect.

When consulting the following list, keep in mind that Trojan programs, viruses, and worms are constantly being developed and revised, and there are many that are either not listed here, or that will have different behavior in the future.

1/tcp

TCP port 1 is the *TCPMUX* service. It was intended to provide a functionality similar to RPC (see port 111): it allows a TCP service to be dynamically assigned a port when it is started, rather than having to be assigned a fixed port number. It is seldom used. There are some exploits that attempt to take advantage of this service, although few systems provide services through it. This port is also used by some remote access Trojans, so on high-security networks it should be blocked if not in use.

7/tcp

TCP port 7 is the *echo* service. It is normally used only for testing, and can also be used in certain attacks. It is often enabled by default, particularly on older version of Unix-like operating systems. This port should be blocked except when actually needed for testing, or the service should be disabled on all systems on the local network.

11/tcp

TCP port 11 is the *sysstat* service. It reveals a great deal of information about a system that it is running on. Although it is not commonly found running on contemporary systems, it might be prudent to block this port if you have Unix-like systems (including Linux) on your network and you aren't sure what services they provide.

17/tcp

TCP port 17 is the *quote of the day* service. This is seldom used, but has potential for abuse by attackers. It is also enabled by default on some operating systems. You should block inbound access to this port unless you have a real need for it.

19/

Port 19 is the *character generator* service. It can be used in denial of service attacks, so block it (both inbound and outbound) unless you have a real need for it.

20,21/tcp

TCP port 20 is the *FTP* service data transfer port, and **port 21** is used for FTP control commands. Because a misconfigured FTP server is a significant security hazard, all incoming packets to TCP port 21 should be blocked, except to authorized FTP servers. Similarly, outgoing TCP port 20 packets could be blocked, although that is both less critical, and less effective, because alternate data ports can be used. FTP server software is widely available for all operating systems, and users frequently install them for various purposes.

23/tcp

TCP port 23 is the *telnet* service. It allows remote access to computers, and is frequently enabled by default on Linux and other Unix-like systems, as well as being available for Windows. Standard telnet does not encrypt passwords or data, and is thus a security hazard if either the user's system or the host system is on a "sniffable" network. With the rapid proliferation of wireless networking, you should treat all incoming connections as "sniffable," even if you consider your local network to be unsniffable (it probably isn't, anyway). Block incoming access to TCP port 23 except on those systems that are supposed to be providing telnet service. This will help protect you from users who (often unknowingly) set up telnet for remote access to their desktop systems.

25/tcp

Port 25 is the *SMTP* (Simple Mail Transfer Protocol) service, which is the most common inter-network mail service used on the internet. You should block incoming access to this port except to local systems you specifically intend to use as incoming mail exchangers. You can force users to relay mail through your local outbound mail exchanger if you block outbound access to this port except for your local mail exchanger. This can be useful if, for example, you want to archive copies of all outbound mail as an approach to complying with public record retention laws or corporate policy, and can also prevent propagation of email viruses that may infect systems on your network.

37/

Port 37 is the *time* service (an early time-of-day service not associated with the Network Time Protocol used by most modern operating systems, see port 123). Block inbound access to this service (and possibly outbound access) unless you are sure your users need it.

53/udp

UDP port 53 is used by the domain name system, which translates host names to IP numbers and is essential for proper network operation. You should allow incoming UDP packets to port 53 only to your name servers, or in response to outbound DNS queries from your users if you have no local name servers. This is particularly important if you have users who set up their own systems, as several operating systems will offer to set up a name server as part of the installation. A name server that is not properly managed is likely to be vulnerable to various remote exploit attacks.

53/tcp

TCP port 53 is primarily used for DNS zone transfers. Since you generally don't want to provide zone transfers except to trusted DNS servers, you should block external access to TCP port 53 on your entire network, unless you operate a DNS server. The DNS server will need TCP port 53 access to trusted secondaries, and primaries for which it is a secondary. There is, however, an exception: DNS answers that are too long to be contained in a UDP packet (512 bytes) will be handled by TCP instead. For most networks, this won't be a problem, but you should be aware of it in case you are setting up an unusual DNS configuration or need to diagnose a DNS problem caused by the filter.

57/tcp, udp

Port 57 is reserved for private terminal servers. TCP port 57 is often the target of probes that are presumably searching for systems infected with some worm or virus, so it would be wise to block this port as a precaution.

67,68/

Port 67 is the *DHCP Server* service, while **port 68** is the corresponding *DHCP Client* port. Block inbound and outbound access to these ports. If you have a DHCP server that serves more than one subnet, you will need to adjust your filters accordingly.

69/udp

UDP port 69 is the "Trivial FTP" (TFTP) service. It rarely has legitimate uses in modern networks, being primarily intended to allow loading of boot images on diskless workstations, and in that application it is unlikely to need to traverse a router. It is, however, used by some malicious worms to upload copies of themselves to systems they are infecting. You should block access to UDP port 69 for both incoming and outgoing traffic.

79/

Port 79 is the *finger* service. On Linux and Unix-like systems, it can reveal details about user accounts. It is generally a good idea to block access to this port from outside your network.

80/tcp

TCP port 80 is HTTP (HyperText Transfer Protocol), i.e., the www service. You generally don't want your average user to run a web server because of the high risk that they will misconfigure the server and be infected by a worm such as Code Red, or worse. You should block inbound access to TCP port 80 except to your authorized web servers. When you block port 80, you might want to also block common alternate ports such as 1080, 8000, 8080, and 8888; as well as the secure HTTP protocol port, 443/tcp. If you block outbound access to port 80, your users will not be able to visit web sites unless you provide a local proxy server that relays web access for them.

81/tcp

Port 81/tcp (and port 444/tcp) is used by Sun's Cobalt Administrator package. You may wish to block these ports if you use Cobalt Administrator.

87/tcp

TCP port 87 is the *ttylink* (a.k.a. "network chat") service. Although it is not used often on contemporary networks, it may be prudent to block this port unless you are sure you need it.

98/tcp

TCP port 98 is used by *linuxconf* to allow configuration of Linux systems via a web-based interface. It has a history of security problems, and external access to it should be blocked for networks on which Linux systems might be found.

109,110/tcp

TCP port 110 is the *POP3* (Post Office Protocol, version 3) service, a popular email protocol. Many POP3 servers have a history of security problems, and they often end up being inadvertently installed on various operating systems. It's a good idea to block incoming traffic to port 110 except for your authorized mail servers. **Port 109/tcp** was used by the older POP2 protocol, and some POP3 software automatically listens on this port also.

111/tcp & udp

Port 111 is the *RPC portmapper* and the *rpcbind* service. These are used by Unix-like systems (including Linux) to direct incoming connections to specific RPC services, which don't operate on fixed port numbers. The most common use of RPC is probably to provide NFS networking, which is not secure across the general internet without special measures (see port 2049 below). If anyone on your network is using Unix-like systems, or might set one up without your knowledge, you should block port 111, both TCP and UDP.

113/tcp

TCP port 113 is the *auth* (or *ident*) service. The standard implementation of this reveals a lot of information about your system. Unfortunately, many mail exchangers expect to be able to make auth queries into your network when your mail server (or your users) connect to them, so if you simply drop packets to port 113, they will have to time out before mail transfer can continue, thus causing delays when transmitting mail. The solution is to set your filter set to send an RST (reset packet) as a response to any incoming connection to port 113 instead of simply dropping the incoming packets.

119/

Port 119 is *NNTP* (Network News Transfer Protocol, i.e. the Usenet News service). It is not generally a great security risk, although there are some remote access exploits that can be used against some news servers. It is generally a good precaution to block incoming connections to port 119 except for those to authorized news servers on your network. News servers also have a tendency to be huge bandwidth hogs unless they are carefully managed.

123/tcp,udp

Port 123 is the *NTP* (Network Time Protocol) service. Some implementations of NTP have significant security bugs, so it is wise to block access to and from this port except for any NTP server you run that needs to be able to synchronize to an external NTP server, or that provides time service to users external to your network.

135,137,138,139/tcp & udp

These ports are used for various purposes by Windows Networking (a.k.a. CIFS). Because so many Windows system are configured either by default or by user action to share directories or whole drives with the entire world, it is very important that you block these ports (both inbound and outbound) to prevent abuse by attackers, and infection by worms such as Qaz or Nimda (some attacks don't even require open shares). Allow these ports only to and from subnets that you intentionally share Windows networking with.

These ports are used as follows (you may find them using either TCP or UDP depending on circumstances): **Port 135** = Windows RPC Endpoint Mapper: used to locate DCOM services that are not on fixed port numbers. Vulnerable to denial of service attacks, and perhaps worse. The Windows Messaging Service uses port 135/udp, and can be used to send unwanted advertising or other pop-up messages to your users unless you block this port. **Port 137** = NetBIOS Name Service: used for name lookups in Windows networking. Since many Windows systems try to resolve the name of any system they talk to, you will see a lot of these (most UDP connections to port 137 are annoying but harmless byproducts of the Windows operating system). Access to port 137 can be used to collect a lot of valuable information about your system, e.g. the name of the administrator account, so you should block this port. **Port 138** = NetBIOS Datagram Service: used to transfer data in Windows networking. **Port 139** = NetBIOS Session Service: used to control Windows network shares.

143/

Port 143 is the *imap* mail service. Many systems inadvertently end up with IMAP servers as part of the normal installation process. Because IMAP has a history of remotely exploitable bugs, it is wise to block access to this port on all local systems except those that act as your official mail servers.

161,162/

Port 161 is the *SNMP* (Simple Network Management Protocol) service. **Port 162** is the *SNMP trap* service. SNMP is installed by default on many operating systems, and the default passwords (known as "community strings" are widely known. You should block access to these ports except from subnets that you intend to use for remote management of your systems, or you should be very careful to disable or properly configure SNMP on all devices in your network.

177/

Port 177 is the *xdmcp* service. It provides log-in services for the X window system. There are numerous attacks that can be used to exploit this service, but they are not likely to do much damage if the X ports (see port 6000) are not also open.

194/

Port 194 is the *IRC* service (internet relay chat). This service is often abused, and it is wise to block incoming access to it except to authorized IRC servers on your network. It is also common to block outgoing access to this port to prevent employees from using IRC, and the associated security risks. Some network worms and Trojan horse programs use IRC to communicate with their master controllers, so blocking outbound packets to port 194 will minimize damage if one penetrates your network. Some IRC servers use port 6666 instead: they tend to be unauthorized or "hacker" servers.

443/tcp

Port 443 is the *https* (secure HTTP) service, i.e., it is HTTP tunneled through SSL. See the discussion of port 80.

444/tcp

Sun Cobalt Administrator. See port 81/tcp.

445

Port 445 is the *Microsoft Datagram* service, used by Windows 2000 to replace the older NetBIOS services (see ports 135-139). You should block both inbound and outbound access to these ports except for subnets that you want to provide Windows networking shares to (or from).

512,513,514/tcp

TCP ports 512 through 514 are the old BSD "r" services, *rexec*, *rlogin*, and *rsh*. They all have serious security problems, and if you use them at all, you should be extremely careful. Because these services are often inadvertently enabled, you should block inbound access to them.

515/tcp

TCP port 515 is the *lpd* (line printer) service. It is widely used by Unix-like systems, and some versions have bugs that allow remote exploitation of your system. You should block access to this port except from subnets that you intend to provide printing services to, and you might also consider blocking access *to* this port on all systems that are not intended to be print servers.

535/udp

UDP port 535 is used by the *CORBA* remote procedure call system. CORBA servers are not likely to be running by accident, but if you have any on your network, you should probably block external access to this port. CORBA has a history of security problems.

540/tcp

TCP port 540 is the *UUCP* service. Although largely obsolete, it may still be found on some Unix-derived systems, and in those cases has significant potential for abuse. You may wish to block TCP port 540 if you have any older Unix systems (or early Linux releases) on your network, or are not sure.

631

Port 631 is used by the internet printing protocol. It is supported by most contemporary operating systems, and some implementations have substantial security problems. It would be prudent to block access to this port from outside your network, except from those subnets to which you intentionally provide printing service.

777

Port 777 is used by the Aim Spy Trojan. This is a remote-access Trojan that allows an attacker to gain control of infected computers on your network. You should consider blocking this port to incoming traffic.

1080/tcp

Port 1080 is used by the SOCKS web proxy. See port 3128 for a discussion of a similar proxy server. Port 1080 is also used by some versions of the Sub Seven Trojan, so it is prudent to block this port.

1214/tcp,udp

Port 1214 is the default port for the *KaZaA* file sharing protocol. It is used for peer-to-peer sharing of files across the internet, often MP3 or video files. The large size of the files involved, and the design of the protocol, can combine to use a huge amount of your network bandwidth. You may wish to reduce such traffic, or reduce potential liability for copyright violation, by filtering this port.

1433/tcp,udp

Port 1433 is used by the *Microsoft SQL Server*, which is installed as a component of several Microsoft packages, and is thus often installed inadvertently. This port is a frequent (and often successful) target of attacks, and should always be blocked from incoming traffic unless you intend to allow external systems to access MS-SQL servers on your network. Even in this case, you should limit access as much as possible. Even better would be to prohibit all external access to this port, and require that external access be performed through a www interface that generates queries indirectly instead of passing user queries directly.

Some versions of MS SQL Server have blank administrator passwords by default: be sure to check for this if you operate an MS SQL server!

1645, 1646/udp

RADIUS authentication. See port 1812/udp, below.

1763

Port 1763 is used by Microsoft Systems Management Server clients. As this service is intended to allow remote management of various Windows systems, you should block access to this port from outside your network if you have systems that have the client software installed or enabled.

1812, 1813/udp

These ports are the default ports used by RADIUS authentication servers, and whether, and to whom, the ports should be open or blocked depends on whether you use RADIUS and for what purpose (some RADIUS servers use ports 1645 and 1646 instead). Some versions of the "Slapper" worm also use port 1812/tcp to receive commands from their master (Slapper is a worm that infects Linux systems running vulnerable versions of the Apache web server).

1900/udp

Microsoft UPnP Discovery Service. Unpatched versions of recent Windows installations have a major security vulnerability on this port. All incoming traffic to UDP port 1900 should be blocked.

2000/tcp,udp

Port 2000 is used by *OpenWindows*.

2049/tcp,udp

Port 2049 is the NFS (Network File System) service. If you have any Unix-derived operating systems on your network (e.g. Linux), you should block UDP and TCP access to this port except from subnets to which you intend to provide NFS file sharing.

3017

Port 3017 is used by Novell Netware printing clients and probably does not need to be reachable from outside your local network.

3128/tcp

Port 3128 is the default port used by the Squid proxy server, which provides local caching for several protocols so that subsequent accesses to the same item will be served from the cache, thus reducing internet traffic. Squid is often used as a web server cache. Because all local network connections to web servers are routed through Squid, you can also use it to track web server usage by employees, and to block access to specific web sites. If you use Squid, you will need to block access from your network to port 80 outside your network to force users to use the proxy server rather than accessing the remote server directly. You should also configure your filters so that port 3128 is reachable from inside your network, but not from outside.

Port 3128 is also used by the "Ring Zero" Trojan, which can provide an attacker with remote access to a computer.

4000/tcp

Port 4000 is used by several remote-access Trojans. If you feel there is a significant risk that one of your users might become infected by a Trojan horse program, you might want to block this port.

4899/tcp

Port 4899 is used by *radmin*, a popular remote administration tool that allows users to remotely access and control Windows systems. If configured incorrectly, or installed by an attacker through a Trojan horse attack, it can give an attacker complete control of the system. You may wish to block inbound access to this port to reduce the risk of abuse.

5000

Port 5000 is used by Microsoft's "Universal Plug and Play" protocol (UPnP). This protocol has significant security problems, so you should either block port 5000 from incoming traffic (and perhaps outgoing as well), or disable UPnP on all of your systems.

5500

Port 5500 is the default port for VNC, a cross-platform remote access protocol. Although not generally considered a major security risk, it can be if not installed with a strong password. There is also some risk if an outsider can monitor VNC packets, e.g. when the remote user is on a wireless network. It might be wise to block the port and require legitimate users to tunnel VNC connections through SSH.

5632

Port 5632 is the default *PC-Anywhere* port. PC-Anywhere can be used for legitimate remote-access purposes, but users frequently install it without a password. This practice is extremely dangerous. You may wish to block inbound access to port 5632 (tcp and udp) except for specific systems that you authorize to use PC-Anywhere, and have confirmed that they are properly configured.

6000,6001,.../tcp

TCP ports starting at 6000 are used by the "X" window system for remote access. The X security model is rather weak, so if you have any Linux or Unix-like systems on your network, it would be wise to block external access to these ports except subnets from which you intend to allow remote access. The possible range of port numbers used is 6000-6255, but the higher numbers are not normally encountered. Blocking only the first ten or twenty (or forty) ports in the range is sufficient to block normal use, and reduces the chance of interfering with some other legitimate program that has claimed a port in that range. If you want to be thorough, block the entire range. If you need to provide X access to users outside your network, they should tunnel it through SSH rather than connecting directly to the X server port.

6101, 6103

Ports 6101 and 6103 are used by the BackupExec backup utility, and should not be accessible from outside your network if you use BackupExec.

6111/tcp,udp

Port 6111 is used by the HP Softbench Subprocess Control system.

6346, 6347/tcp

Port 6346 is the default port for the *Gnutella* file sharing protocol. It is used for peer-to-peer sharing of

MP3, video, and other files, often in violation of copyrights. The inefficiency of the protocol combined with the size of many of the files transferred can combine to use huge amounts of your network bandwidth. You may wish to filter this port to reduce the amount of such traffic, although users can potentially get around the filters by using different ports. As with all peer-to-peer protocols, there is significant security risk if a previously unknown flaw is discovered and exploited by attackers. Gnutella also uses port 6347.

6588/tcp

Port 6588 is used by some web server proxies. If you block access to web servers (port 80/tcp), you should also block this port.

6666, 6667/tcp

Unofficial or "hacker" internet relay chat (IRC) servers are often found on these ports (the normal IRC port is 194). Often the target of attacks, and often installed by attackers after they have successfully compromised a system. Also used by numerous illicit remote-access programs.

6970/

Port 6970 is the default port for the *RealPlayer* protocols. It is generally not considered a security problem except for the bandwidth it uses, so there is usually no need to specifically filter this port. Note that these protocols will automatically try other backup ports (including port 80) if this port is blocked. Port 6970/tcp is also used by the Gate Crasher Trojan, and for that reason you may wish to block incoming packets to port 6970.

8080/tcp

Port 8080 is often used for SOCKS proxy servers. See port 3128 for a discussion of proxy servers. It is also used by several dangerous Trojans, and should probably be routinely blocked.

17300/tcp

Port 17300 is used as a remote access port by the Kuang2 virus. More recently, a worm that infects systems previously infected with Kuang2 has appeared and is frequently encountered attacking this port. Incoming traffic on port 17300/tcp should be blocked.

27374/tcp

TCP port 27374 is a default port for the *SubSeven* remote access Trojan. SubSeven is an extremely dangerous program which allows its remote master almost complete control of the infected computer, including being able to watch what is happening on the screen, monitor what is typed at the keyboard, and even speak through the sound system and monitor the local microphone if one is connected. It is wise to routinely block access to or from this port.

31337/

Port 31337 is a default port for *Back Orifice* and several other remote access and backdoor Trojans. It is wise to routinely block access to or from this port.

43981/udp

Port 43981 is used by Novell Netware IP. If you use Netware, you should consider blocking this port from external access.

APPENDIX B: ICMP PACKET TYPES

ICMP does not have the concept of "port numbers" found in UDP and TCP. ICMP packets do, however, have "types" that are usually treated similarly when setting up filters. In addition, some types have additional "codes" that further specify the meaning of the packet. Some ICMP packet types are dangerous to your network, and many others are either not necessary for proper operation of your network, or are only needed internally within your LAN. You should consider blocking all ICMP packet types, both inbound and outbound, except the following:

Type 0

Echo Reply

These are the response to *Echo Request* (type 8) packets. If you block Echo Reply packets, you should also block Echo Request packets (Type 8) in the opposite direction. See the discussion of Type 8 packets, below, for more information.

Type 3, Code 4

Host Unreachable, Fragmentation Needed

These packets are necessary for the proper operation of MTU discovery protocols, which determine the largest packet size that can traverse a given route without being broken into smaller packets. If you block these packets, you may have strange problems in which some data from a particular host or network will reach you, and some won't. This is because you will be preventing packets above some size from reaching you. Do not block these packets!

Other type 3 packets indicate that the host was unreachable for other reasons. Although less likely to cause significant problems if you block them, it may cause some time-out delays and other possible minor disruption of your normal operation. Unless you are particularly concerned that they will be used in a denial of service attack, you do not need to block any Type 3 packets.

Type 4

Source Quench

These packets are used as part of the IP flow and congestion control mechanism. If you block these packets, your systems will blindly send packets into a congested route, thus helping to perpetuate the congestion. You should not normally block these packets for that reason. There is, however, the possibility that an attacker can implement a denial of service attack against a host on your network by sending spoofed Source Quench packets to the host. If you feel that the risk of this is greater than the damage caused by congested routes, go ahead and block Type 4 packets. For most networks, though, it is best to normally allow Type 4 packets into your network, and block them only when you believe you are actively being subjected to a denial of service attack. Source quench packets are also used when hijacking an existing TCP connection. This is a fairly difficult attack to mount, but if you are trying to maximize your security, you should make sure source quench packets cannot enter your network.

Type 8

Echo Request

These are used by some implementations of the "ping" command, which is used to determine whether there is a system using a specified IP number. Many administrators block these packets, reasoning that *ping* can be used to map their network and provide attackers with information. While this is true, there are other ways to gather the same information, so blocking ICMP Echo packets provides marginal benefit. The ability to use ping to diagnose network problems can offset the security penalty, so a good compromise is to allow *outbound* Echo Request packets and *inbound* Echo Reply packets.

Type 11

Time Exceeded

These packets indicate that some of your outgoing packets cannot reach their destination because the route is too long, or is a loop that goes nowhere. If the originating host never receives this message, it will continue to blindly send packets that can never reach their destination. Many administrators block *outgoing* Type 11 packets because they are used by the trace-route utility. They reason that *trace-route* is used by attackers to map a network. You may elect to do so, but you should not block *incoming* Type 11 packets.

Chapter 20

Network Security for EIS and ECS Systems

Joel Weber
Weber and Associates, Inc.

AS ENERGY and facility managers increasingly utilize web-based information systems, they must also develop security for their information and their systems. This chapter describes some of the security nightmares that lurk in cyberspace, waiting like evil villains, to attack unsuspecting system operators and users. Methods for avoiding attacks, recovering from them, and minimizing their impact are covered.

INTRODUCTION

During the past two decades the energy sector and operations managers in charge of energy management in industrial sectors have become increasingly dependent on information technology (IT). At the operational level, IT has facilitated higher yields, more efficient quality control, and more effective inventory management. At the managerial level, IT has proven critical in customer relationship management, marketing, finance, accounting, and strategic planning. In short, IT systems are an essential element in maintaining and improving the integrity of any energy-focused or energy-dependent organization. Consequently, the security, reliability and availability of these systems are crucial to their successful implementation.

Many diverse IT systems are found within the energy sector, including energy marketing and trading, exploration and production, refining operations, power generation, and power and gas distribution. Such systems are also used by industrial, commercial and residential energy management applications. Energy IT systems can operate on desktop workstations, servers, laptops, portable devices, and embedded systems; they may be run on platforms such as Microsoft Windows 3.x/95/98/NT/2000/XP/CE/.NET, Linux, Unix, BSD, and Solaris.

Energy IT systems are subject to the same internal and external threats and vulnerabilities that threaten the integrity of other computer systems. Many high profile incidents in recent years such as the Melissa, Code Red and Nimda viruses, have underscored the need for increasing energy IT security awareness and making security a major feature of an energy manager's IT policy.

Types of Security Threats and Vulnerabilities

The following thirteen categories have been developed for classifying IT security violations:[1]

- Theft of proprietary information
- Sabotage of data or networks
- Telecom eavesdropping
- System penetration by outsider
- Insider abuse of network access
- Financial fraud
- Denial of service
- Spoofing
- Virus
- Unauthorized insider access
- Telecom fraud
- Active wiretapping
- Laptop theft

Each type of violation requires an intrusion or attack in order to succeed and the nature of such intrusions and attacks must be understood before energy IT assets can be secured and security policies can be implemented.

The Nature of Security Threats

To manage security threats we must understand the nature of threats, how intrusions and attacks are mounted and how they proceed. By definition a security intrusion requires some type of unauthorized access. An attack does not require entry into a system or network, but can cripple a system as fatally as an intrusion can. Intruders are either external or internal with respect to the organization. External intruders are those who are attempting to gain access to a system and do not have access authorization. Internal intruders are authorized users of the system who access data, resources or programs to which they are not entitled.[1] Whether or not an attacker is intruding upon a system, there are certain techniques he implements and processes he utilizes.

Security attacks are launched in either a direct or distributed manner. A *direct attack* is launched from a computer used by the attacker. A *distributed attack* makes use of other computer systems to cloak the origin of the attack and to potentially amplify the affect of the attack. Multiple systems can be enlisted to execute a distributed attack in a coordinated fashion. Consequently, the number of victims in this scenario can become quite large.

The most common form that distributed attacks take are denial of service (DoS) attacks. The purpose of a DoS attack is to render a network inaccessible by generating such a large amount of network traffic that the servers crash, the routers are overwhelmed, and the network's devices cannot function properly.[2] Many distributed security attacks are automated through the use of software attack tools. Such tools make launching attacks considerably easier by lowering the technical difficulty level and reducing the number of steps required to initiate an attack.

Until 2000, most automated distributed attacks would only execute one attack sequence. That year, the Nimda and Code Red attacks presented a new breed of automated distributed attack in which user interaction on a compromised system would initiate subsequent attacks and compromise more systems. The advent of such attacks underscores the difference between deliberate and accidental attacks. The parties that originally initiate such attacks clearly intend to harm other systems or users. However, the people who might accidentally perpetuate such attacks through otherwise normal computer interaction represent a source of "accidental" attacks. The success of automated distributed attacks is increasingly dependent upon accidental attacks made by unwitting users. IT security staff must remain aware of these types of attacks and must educate energy IT staff in order to mitigate the effect of such attacks on the organization's computer systems.

Phases of Attack

Each security attack has certain phases in common: Preparation, Initial Access, Full System Access, Establishing Future Access Opportunities, and Obfuscation. Understanding how security can be compromised and what steps should be taken to prevent security breaches is important for energy managers who are deploying and utilizing energy IT systems.

Preparation

The first step in mounting an attack on a system (the preparation stage) is to acquire information and develop a plan. This phase usually involves activities such as port scans and spoofing. A port is a service that runs on a computer and is a virtual door through which information enters and exits the system. There are 65,535 TCP (transmission control protocol) ports. Some ports are reserved for use by certain protocols and others are available for custom applications. See Appendix 1 for an abridged listing of these ports and their corresponding protocols and applications. A thorough listing of well-known port/protocol assignments can be found at *www.freesoft.org/CIE/RFC/1700/4.htm.*

Port scanning software tools are widely available on the internet as freeware. Such tools are not illegal in the United States because IT security personnel and law enforcement frequently use them to insure that only ports that are necessary for operations are open. However, these tools are frequently used and detected in the preliminary phases of attacks on computer systems. Much like weaponry in the physical world, port scanners can serve law-abiding citizens and heroes as much as they can serve criminals and villains. When a port scanner is used, it is roughly analogous to walking around the perimeter of a building and testing for whether the outer doors and windows are locked and/or open. For those preparing an attack, the scanner is used to locate and identify which ports (or doors and windows) are open and which protocol(s) is used by that port.

"Spoofing" is defined as the act of getting one computer on a network to pretend to have the identity of another computer, usually one with special access privileges, so as to obtain access to the other computers on the network.[3] There are three general subclasses of spoofing activity: ARP spoofing, IP spoofing, and DNS spoofing.

Address resolution protocol (ARP) is the communications protocol through which the network determines what transmissions should go to which computers. When a computer is logged onto a network, it receives an internet protocol (IP) address, which will serve as its address on the network, and all transmissions to or from that computer will be routed using that address. Each computer's network card has its own unique media access control (MAC) address, which serves as a physical address of the computer. The address resolution protocol maintains a table of IP-MAC address relationships called the ARP cache. ARP spoofing involves altering the ARP cache so that communications on the network are routed to a machine that the attacker chooses.

Internet protocol (IP) facilitates speedy, reliable and asynchronous communications between machines by dividing information into small packets that are easier to transmit across a network or between networks than if a dedicated communications channel were used. Each of these packets contains the IP address of the source of the transmission in a "header." Since the IP header can act as type of return address, many attackers will engage in IP spoofing in order to falsify the identity of the network and computer from which they are attacking a targeted system. IP spoofing involves changing the IP address in packet headers in order to impersonate a different machine. When an attacker knows the address of other trusted computers with which the targeted machine communicates, he can route his impending attack through one of those machines and conceal his identity and location.

The ***domain name system*** (DNS) is used to make the internet more efficient and user-friendly. This is the database system that keeps track of all of the domain names such as intel.com, and their associated static IP addresses. The DNS is by far the most critical single system for the operational integrity of the internet. Businesses, organizations, and individuals with an internet presence therefore depend on the DNS to insure that internet traffic intended for them is directed to their network. DNS spoofing is similar to ARP spoofing in that the address associations are altered. With DNS spoofing, the impact can be quite severe as web site traffic, file transfers, emails, etc. can be re-routed and web sites can be replaced with impostors. This method can even be used to con an unsuspecting individual into providing personal information through web forms.[4]

Initial Access

In order to gain initial access to a system that has been marked in the preparation phase, an attacker must either gain a password that yields access to a computer's operating system or application running on that platform, or exploit some known or newly discovered weakness in an operating system, application, or protocol.

Passwords can be compromised in many ways. The simplest method is that of gross enumeration or "brute force" in which the password is guessed until access to the target system is granted. The process of guessing words is tedious at best and requires a large helping of luck. To speed this process an attacker might perform a dictionary attack in which a software program automates the process of trying each and every word in a dictionary as a candidate password. This attack, however, is only successful against the least sophisticated computer users and those organizations without password policies. Password cracking has legitimate applications in the case of employees who forget their passwords or are terminated or deceased. In this context, however, password cracking is called "password recovery."

Operating system and application passwords are frequently stored on a machine. Some older operating systems store encrypted passwords in files but the encryption algorithms have proven to be quite weak. Newer operating systems such as Windows 2000/XP implement Kerberos authentication, which is currently one of the most robust encryption algorithms in existence, and store the encrypted passwords in a database that cannot be accessed through the file system. In the face of such safe password storage, an attacker must resort to intercepting a password as it is sent across a network in clear text format. This will require a network intrusion which, if appropriate network hardware and software solutions have been installed as part of the IT security strategy, will require significant resources, time and effort.

If a password cannot be decoded, an attacker may attempt to gain initial access to a system by exploiting a software weakness. Some operating systems allow for remote access, thus presenting an opportunity for unauthorized entry if these access points are not properly secured. One of the most common software vulnerabilities is called a buffer overflow. A buffer overflow occurs when input data exceed the size of their allocated program memory buffer because checks are lacking to ensure that the input data are not written beyond the buffer boundary.[5] Unfortunately, due to poor coding and quality assurance practices many operating systems and applications contain this vulnerability. An attacker exploiting a buffer overflow can remotely escalate his privileges on the target system. Through a complex process of analyzing memory addresses of applications with known vulnerabilities, the attacker can gain administrator privileges on the target system by either launching a command shell application or crashing an application or operating system such that the system restarts in 'Administrator' mode. The cost of a single compromise can be astronomical if the attacker is able to further infiltrate a system and access valuable information.[6] Exploitation of buffer overflows can generally be prevented by active maintenance of software patches as they become available and by improved software development processes within the organization.

Applications can also present security vulnerabilities if unexpected input from users leads to application failures. Protocols are often exploited via a denial of service (DoS) attack in which the protocol is overwhelmed with a high volume of requests or large single requests.

Full System Access

Once an attacker gains access to a targeted system he may engage in a variety of activities ranging from simply browsing the file system to introducing software or spyware to stealing data. Malicious software is commonly described as a Trojan (or Trojan horse), worm, or virus.

A Trojan is a software program that is introduced into a system under the pretense of serving a legitimate purpose when it actually aims to compromise system security and gain unauthorized access. It may be hidden in an application file, or disguised as a legitimate text file or email attachment. The 2000 "I Love You" Trojan appeared to be a text file attachment to an email when it was actually a visual basic script that forwarded the original email with the Trojan attachment in tow to every address in the user's Microsoft Outlook address book. The effects of this Trojan were felt worldwide as mail servers crashed from the overload. The end result of this attack was denial of service for email users worldwide.

Worms were the first incarnation of malicious software. A worm is a program that is capable of traveling across a network. Their original purpose was to distribute legitimate software across networks. In their malicious form, worms can make multiple copies of themselves and spread through a network. Their primary purpose is to replicate.[7]

A virus is an intrusive program that infects computer files by inserting copies of itself into those files. The virus program is usually executed when the infected file is loaded into memory. This allows it to infect other files, and so on.[8]

Viruses have several different morphologies. They can be logic bombs that "detonate" and execute their code when a series of conditions is met or a specific point in time occurs. The Michelangelo virus was the first incarnation of this virus class; it was programmed to erase the hard drives of the infected computers on the date of the artist's birthday. *Macro viruses* are another type that are embedded in documents and have been seen most commonly in Microsoft Word documents. *Application viruses* sometimes infect programs and execute their malicious code when the programs are run. *Boot sector viruses* are transmitted by portable media such as floppy diskettes. This type of virus inserts itself into the master boot record on permanent storage media like hard drives and is loaded into and executed in memory whenever a computer is booted. Fortunately Trojans, worms and viruses have caused so much upheaval in society that people are increasingly aware of the risks of opening unknown attachments and the means by which viruses are transmitted; therefore, they are implementing effective risk management practices including use of virus recognition software.

Upon gaining full access to a computer system, some attackers install *spyware* that allows the attacker to continually acquire data from the compromised system. In recent years, "key logger" software has become a tool of choice for attackers. Such software will actually log all of the keyboard keystrokes made by an unsuspecting user, thus giving the attacker complete information on the user's data, passwords, email, and other procedures.

An attacker with full access privileges can engage in many different activities. He may install or execute malicious software or steal privileged data or engage in other unanticipated misdeeds. Awareness of these major activity classes is a significant contribution to maintaining a credible IT security strategy.

Establishing Future Access Opportunities and Obfuscation

After an attacker has gained full access to a system, he is very likely to create difficult-to-detect doorways for his own use. This will make it much easier for him to use that same computer system again. He is likely to open additional ports or create separate user accounts for his use. Once his attack is complete, he will attempt to "cover his tracks" and obfuscate (conceal) his activities. To do this, he will alter and delete log files and eliminate all evidence of his unauthorized presence.

Information Security and Risk Management

The process of information security and risk management can be characterized by a modified version of the Deming cycle model of *Plan-Do-Check-Act* (Figure 20-1).

In this chapter, the hardening/securing and preparation phases of the information security assurance process will be our primary concern; the detection, response and improvement

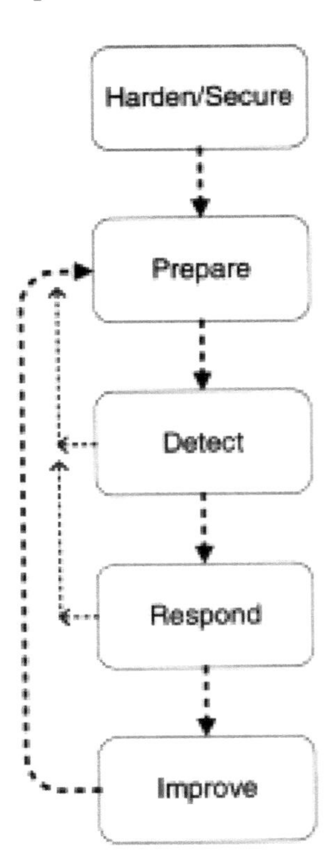

Figure 20-1. Deming Cycle for Securing Information Assets *CERT Guide to System and Network Security Practices.*

activities are largely a function of a specific organization's resources, needs, and strategic objectives. There are many reputable resources for managing these phases of information security management. One of the best guides for information security management is the Operationally Critical Threat, Asset, and Vulnerability Evaluation (OCTAVE^{SM}) framework for Information Security Assurance and Risk Management developed by the Software Engineering Institute at Carnegie Mellon University.

To be successful, an IT security strategy must have support at all levels of the organization and must be viewed as a process of continuous assessment and improvement. An organization that wishes to improve its security posture must be prepared to take the following steps:[9]

1. Change from a reactive, problem-based approach to proactive prevention of problems.
2. Consider security from multiple perspectives
3. Establish a flexible infrastructure at all levels of the organization capable of responding rapidly to changing technology and security needs.
4. Initiate an ongoing, continual effort to maintain and improve its security posture.

Effective definition and implementation of an organization's IT security strategy will require a forward-looking view, teamwork, senior management participation, vulnerability assessment, threat classification and prioritization, information systems audits, and information security risk analysis. Most importantly, ownership of the IT security strategy must be established at all levels as all personnel must bear a degree of responsibility for maintaining the integrity of the organization's information systems that is appropriate to their position and job within the organization.

In establishing an enterprise-wide IT security policy, the following elements must be addressed:

- Maintenance of the confidentiality of privileged digital records.
- Address risks associated with recently terminated and former employees.
- Email and internet usage policy.
- Document management and file storage and encryption policy.
- Password Maintenance and Change policy.
- Incident response methodology, policy and procedures.
- Information security audit and assessment methodology, guidelines, and procedures.
- What security functions are internally addressed versus outsourced.
- Specific and general duties of employees within distinct functional areas.
- Workstation, server, and other digital device usage policies.
- Application use and maintenance policy.
- Backup and recovery policy.

To provide credibility to the IT security policy, the human resources department must be involved. This department should oversee the assurance that each employee, as a condition of employment, has signed an agreement to abide by the IT security policy as it pertains directly or indirectly to his or her job.

Securing and Hardening IT Assets and Security Preparation

Effective information security management can be modeled by the three legs of an equilateral triangle with the legs corresponding to the principles of confidentiality, integrity, and availability. The following recommendations will prove useful in meeting these persistent objectives. However, please note that the following discussion is not meant to be an exhaustive treatment of ways to secure and harden information system assets from intrusion and attack, and even if an organization follows all of these recommendations, they may not be sufficient to make it impervious to intrusion or attack.

Hardware and Software Management

Managing information security requires the proper installation, maintenance, and protection of numerous hardware and software assets. This section will discuss the securing and hardening of such assets that are common to nearly every business operation.

Updating Operating Systems and Applications

It is absolutely essential that operating systems and applications be kept up-to-date by technical administrators. Because the essence of software engineering is complexity, there will always be defects and weaknesses that a skilled attacker may exploit. Furthermore, with widespread use of certain operating systems and applications, the likelihood of a security flaw being discovered in such software is remarkably high.

Administrators must keep themselves apprised of current operating system and application upgrades and patches. Whenever an update is released, technical staff must evaluate it, determine if it is applicable to their organization's computer systems, and if so install it.[10] Failing to install relevant upgrades or patches can have disastrous consequences. For example, in January 2003, the SQL Slammer Worm caused worldwide chaos across corporate networks and the internet. This small but malicious program rapidly exploited a flaw in Microsoft SQL Server even though a patch had been available for six months, underscoring a dirty secret in the information technology industry: Software bugs are common, and administrators are slow to fix even widely publicized problems, said Johannes Ullrich, director of the security information site, Incidents.org.[11] Administrators must also weigh costs and benefits of installing upgrades, patches, and service packs as well as when and how to install them. Install an update can itself cause security problems, such as the following:[12]

- During the update process, the computer may temporarily become more vulnerable.
- If the update is scheduled inappropriately a computer or information asset may not be available when needed.
- If an update must be performed on a large number of computers, there may be a period when some computers on the network are using different, potentially incompatible versions of software, which might cause information to be lost or corrupted.
- The update may introduce new vulnerabilities.

Before installing software updates, system and user data must be backed up, as updates sometimes do not go as well as planned. Also, because viruses and other malicious software can sometimes be unintentionally introduced when deploying upgrades, it might be worthwhile to invest in integrity checking software such as Tripwire (http://www.tripwire.com). Integrity checking tools can identify changes made to files and directories when updates are installed.[13] When using such tools, a system baseline assay is established which can be used as a point of comparison in monitoring system changes and identifying those that might warrant investigation. For system maintenance to be performed in a precise and meaningful way, the maintenance methodology must be actively managed and thoroughly documented.

Configuring Operating Systems and Applications

The primary principle that should guide system administrators in configuring any operating system or application is: "deny first, then allow." That is, turn off as many services and applications as possible and then selectively turn on only those that are absolutely essential.[14] In the effort to monitor unexpected machine, network, or user behavior, an administrator must consider enabling the logging capability that is available with most modern operating systems. Logging will keep track of system, application, security, and internet browser events, errors, warnings and information that may prove quite valuable in enhancing and maintaining system security. On certain high traffic systems, this might require additional hardware. But should this need arise the cost of introducing more hardware to distribute computing will pale in comparison to the cost of a compromised system with no trail for conducting an analysis.

Workstations

The principle of "deny first, then allow" should govern workstation configuration activities. It is recommended that the most minimal operating system and application image that meets business requirements be installed on workstations. All unnecessary software should be removed or disabled on workstations also. Furthermore, users should be allowed only the minimal privileges that allow them to do their jobs on these workstations. Network services for each workstation should also be enabled frugally. In cases where extremely sensitive data resides on a workstation and that data should not leave the machine under any circumstances, portable media devices (floppy drives, CD-ROM writers, etc.) should be removed if system administrators and management consider such action appropriate. All organizations should have a workstation use policy that addresses such issues as inappropriate or private use of IT assets, locking operating systems and applications when machines are unattended, data backup, and user maintenance.

Servers

All of the recommendations for securing workstations apply to servers as well. However, special consideration must be given to the fact that servers usually serve multiple users and offer multiple services and applications, thus presenting a significantly larger security risk if servers are deployed or maintained haphazardly. For each network server, file systems, server maintenance methodologies, and protocols (and ports) offered must be determined. Adherence to the security principle of "deny first, then allow" is extremely important in the administration of servers. Nowhere is adhering to this principle more critical than in the configuration of services (ports). Depending upon the service (port), several configuration options can be considered[15]:

- Limit the network hosts that can access the service.
- Limit the users who can access the service.
- Configure the service to allow only authenticated connections. The authentication should not rely solely on network data such as IP addresses and DNS names, which can be spoofed, regardless of whether the host is trusted.
- Limit the degree of access (especially in cases where that would permit a user to change the configuration of network services).
- If applicable, limit the facilities and functions offered by the service only to necessary ones (e.g., if files will be shared via FTP, permit only file download and restrict file uploads).
- Isolate the service's files (configuration, data files, executable images, etc.) from those of other services and the rest of the system.

Backup and Recovery

File backups allow users and administrators to restore the availability and integrity of information assets following security breaches and accidents. Without a backup, organizations may be unable to restore a computer's data after system failures and security breaches.[16] A backup and recovery plan is a critical element of any successful information security policy. By implementing the principle of "deny first, then allow," which is also known as the principle of minimal privilege on workstations and servers and adequately documenting what software resides on each machine (the software "image"), the costs of the backup and recovery process will be better controlled. The backup and recovery procedures should be equally well documented, and users and administrators should be well aware of their responsibilities in the context of regular backup activities. When performing recovery operations users should watch for unexpected changes to their files or data.

For workstations there are two common approaches to backup processes:[17]

- Files are backed up locally at each workstation, often by the user(s) of that workstation. The advantage of this approach is that protected data does not have to traverse the network, which reduces the chances of its being monitored, intercepted, or corrupted. The disadvantage is that each workstation must have additional storage devices, which must be kept secure, and users must be trained to perform the backups.

- Backups are centrally administered, with data copied from workstations by a network-based backup program. Encryption tools can be used to protect data passing from a user workstation to a central backup host server.

The server itself must have a backup system to protect its data and applications. Backup processes for servers generally involve establishing and maintaining a copy of the information content of the server on a separate secured server. Although implementation of encryption technology in backup processes is not commonplace, it would provide additional protection against intrusion, physical security failures, etc.

Malicious Software Protection

As mentioned earlier, malicious software such as Trojans, worms, and viruses represents a significant threat to system integrity. The best method for managing this risk involves both technology and human stewardship. Many anti-virus tools, such as McAfee VirusScan and Norton Anti-Virus, are available for both workstations and servers. These programs use sophisticated algorithms to recognize code patterns that are characteristic of malicious software. Anti-virus programs use a database of known malicious software and code patterns and the database must be updated frequently as new malicious software threats emerge literally every day. All levels of the organization should be involved in protecting the IT system from viruses and malicious software scanning on a regular basis. Users must therefore be trained to understand how malicious software works and how the available tools can be used to prevent damaging attacks. Administrators should actively discourage users from receiving, much less running, executable files received through email and should disallow such privileges if at all possible.

Secure Remote Administration

Energy management control systems (EMCSs) can be operated by a single computer at a central location within a facility or at another site belonging to the company. This provides significant cost benefits because many tasks can be automated, and the administrator does not physically have to visit each computer.[18] This convenience for the administrator can also create a significant security risk when the computers are network-based or linked to the internet. Authentication of user identity is the method for addressing this risk. User identity can be determined through public key authentication and use of secure protocols such as SSL and HTTPS. In addition, all remote data transmissions between administrator and computer must be made over an encrypted connection so that eavesdroppers cannot intercept confidential data.

Network Management

Routers and Firewalls

Networks by their very nature have several nodes, each of which may need to communicate with other nodes on the same network or with nodes on separate, outside networks. In networks with multiple segments using different protocols, the best device to use for efficient communications is a router.[19] A router is an intermediary device on a communications network that expedites message delivery. On a single network linking many computers through a mesh of possible connections, a router receives transmitted messages and forwards them to their correct destinations over the most efficient available route. On an interconnected set of local area networks using the same communications protocols, a router serves the somewhat different function of acting as a link between these local area networks, enabling messages to be sent from one network to another.[20] Because a router essentially handles all traffic across a network, it is a logical target for an attacker. Therefore, all routers used within an organization's computer network must have firewall capability.

Like the information security management cycle, implementation of firewalls is a Deming Plan-Do-Check-Act cycle; except here the process phases are Prepare, Configure, Test and Implement. See Table 20-1. There are numerous firewall topologies available that will significantly enhance network security. It is one of the network administrator's most critical responsibilities to be familiar with network and firewall topology and to implement the one that best fits the organization's security needs.

As with all other hardware and software assets, routers and firewalls must be configured with minimal privileges and functions that meet the organization's needs. It is absolutely necessary that the firewall logging and alert mechanisms are enabled during the configuration phase so that the firewall will monitor and report attempted attacks in addition to enhancing network security.

Before the firewall system is implemented, it should be thoroughly tested. This will ensure that the design specifications and installation settings operate as intended and reveal any failures or weaknesses in the system. The features that must be tested include the following:[22]

- Hardware (processor, disk, memory, network interfaces, etc.)
- Operating system software (booting, console access, etc.)
- Firewall Software
- Network interconnection equipment (cables, switches, hubs, etc.)
- Firewall configuration software—including routing rules, packet filtering rules, and associated logging and alert options.

Testing the firewall system should also provide an opportunity for testing the network data backup and recovery processes and identifying strengths and weaknesses.

When installing a newly designed, configured, and

Table 20-1. Firewall Implementation Process Summary[21]

Process Phase	*Practice*
Prepare	Design the Firewall System
Configure	Acquire Firewall Hardware and Software
	Acquire Firewall Training, Documentation, and Support
	Install Firewall Hardware and Software
	Configure IP Routing
	Configure Firewall Packet Filtering
	Configure Firewall Logging and Alert Mechanisms
Test	Test the Firewall System
Implement	Install the Firewall System
	Phase the Firewall System into Operation

tested firewall system, administrators must consider the effects that such a system will have on the networks with which the firewall system will connect and how newly available connectivity will affect normal operations. Consequently, a firewall system should be phased in across the other network hosts that are connected to the new firewall system. Each host that is intended to send traffic through the firewall must be made aware of the new firewall's existence.[23] If they are not, the router will still function properly but no security benefits from the firewall will be realized. As for the hosts controlled by the new firewall system, those hosts must also be configured for sending data to and receiving data from the firewall system. Finally, all users should be notified of the firewall system's installation. The new network configuration will generally be transparent to users, but making users aware of the installation will expedite and ameliorate troubleshooting efforts.

Virtual Private Networks

Any organization with a network, and especially an organization that has significant data transmissions between internal and external networks, should use routers with Virtual Private Network (VPN) capability. A VPN is a set of nodes on a public network such as the internet that communicate among themselves using encryption technology. Their messages are as safe from being intercepted and understood by unauthorized users as if the nodes were connected by private lines. By installing a router with both firewall and VPN capability, the risk that data transmissions will be intercepted is markedly reduced. The VPN capability takes care of securing and encrypting transmissions and communications. This is especially necessary now that many computer users are highly mobile, and may travel to distant offices. A VPN can allow an organization with different regional offices to avoid the costly startup and maintenance of a Wide Area Network. As a general policy, employees should be required to use the company's VPN when remotely transmitting or receiving data over the organization's network.

Wireless Access

Since 2000 there has been a rush to implement wireless network access in the workplace. As a result, convenience has taken priority over security; in fact, as of September 2002, security was lacking in 80% of wireless networks.[24] Anyone with a wireless network card and a laptop can access a wireless network through an "access point" since the signal from nearly all "access points" spills onto the street. In fact, this spillover is so widespread that buildings are routinely "chalked"*[25] with information indicating known unsecured access points into corporate and private networks.[26] This behavior is known as "war chalking." There are many resources on the internet illustrating equipment that is helpful in addressing this security issue.[27]

A typical scenario might be the following. Unauthorized access starts when an intruder pulls his car into a public parking lot equipped with a laptop, wireless network card and free "sniffer" software. If the wireless network is not secured, the intruder changes the wireless card Service Set Identifier (SSID) on his laptop to match that of the wireless "access point." The intruder then requests an IP address from the network and opens Microsoft's Network Neighborhood to see what computers are unprotected. Any PC that has file and print sharing enabled could be accessed. From this point, an intruder can use advanced tools to probe into the network without the trouble of a firewall to prevent access. These access tools are available on the internet at no cost. Even if no other computers are exposed, the intruder has unfettered internet access and can impersonate a legitimate corporate computer or user. Any illegal computer activity will be traced back to the network owner who may not be able to determine the intruder's identity or location.

Currently, transmissions sent over a wireless IEEE 802.11b connection are not encrypted. Transmissions to pagers and between cellular telephones are also not encrypted. Consequently, the risk of eavesdroppers intercepting such transmissions is considerable. Moreover, the increasing popularity of "war-chalking" demonstrates the magnitude of the risk. Since many data transmissions are unencrypted, an intruder who acquires access through the wireless system could gain undetected and unfettered access to critical data transmissions and network resources. If the organization can afford to do so, establishing a sufficient number of static IP addresses to accommodate the needs of all devices will provide a significant barrier to intruders. If this is feasible for the organization, a network administrator would establish fixed relationships in the ARP (see Appendix B—Key Terms) cache between static IP addresses and individual device's network card MAC addresses. For this wireless security strategy to be complete, DHCP (see Appendix B—Key Terms) must be disabled at the host server level.

*Using chalk to place a special symbol on a sidewalk or other surface that indicates a nearby wireless network, especially one that offers internet access.

Wireless security can also be improved by making use of the wired equivalent privacy (WEP) protocol. Nearly all IEEE 802.11b-enabled cards and access points on the market implement the WEP standard, which makes it very difficult to use the wireless network without authorization. Since most access points are not protected by WEP, any organization that implements it will discourage all but the most dedicated intruders from gaining access to the network. After enabling WEP, businesses need to change the wireless network's default SSID to another character string that cannot be readily guessed. SSID broadcasting should also be disabled so that the SSID is not easily intercepted. At the hardware level, any wireless network should include or at least be contained within a network that has firewall and VPN capability.

Other Issues

Encryption of Critical Storage Media and E-mail

In order to mitigate the risk of eavesdroppers and interlopers gaining access to the exchange of operationally valuable data via email, the email should be encrypted using PGP, Kerberos, or some other public key encryption protocol. These methods implement significant barriers to decrypting ciphertext to plaintext. In addition to this technological solution for securing email, such communications should be explicitly marked as private, confidential, etc. and should contain an appropriate statement to that end in the footer of each and every email message. When documents are transmitted by email as attachments, they should be converted to portable document file (PDF) format or rich text format (RTF) before sending them. Microsoft Word and Corel WordPerfect file documents contain metadata (information that is hidden within the document). Such metadata may contain every modification or the change history of the document. PDF and RTF document formats contain minimal metadata about the document itself and therefore should be the only acceptable formats for transmission of formatted attachments. Finally, what little metadata can be extracted from the PDF and RTF file formats cannot be easily obtained by someone who has not had significant technical training.

Computer use is no longer limited to desktops located securely inside the walls of the buildings. Many computer users have laptop computers that are used for communications (email and peer-to-peer), word processing, document management, file storage, etc. Because these devices are highly mobile and are sometimes left unattended or are stolen, there is a significant risk of outside parties gaining access to privileged communications and data. Laptop users can easily protect their data by enabling the encryption capability on their permanent (hard drive) and local storage media (floppy, CD, etc.). While this solution is certainly no substitute for a user being cautious and protective of a laptop that contains privileged data, it will help prevent the data being compromised. As mentioned earlier, encryption protocols should be implemented wherever feasible, as doing so only enhances security by insuring greater data confidentiality and integrity.

In addition to data encryption, a user may incorporate the use of the new biometric devices that use fingerprints or other anthropocentric data for authentication purposes. Early devices allowed for only one possible user. This posed a risk that the data might be lost to other parties within the firm if the original user was not available to authenticate the data. Newer biometric device offerings allow multiple users to be able to authenticate a storage device. If an organization were to use such devices, the fingerprints of the primary user and a system administrator would be sufficient to provide a high level of security without excessively restricting access.

User Authentication Policy

One of the simplest methods of thwarting intruders and attackers is by having an effective user authentication policy. Such a policy should use hardware authentication features that are available on any computer's BIOS. User accounts should be actively managed at the administrator level with the principle of minimum privilege in mind. Unnecessary accounts and those belonging to recently terminated or deceased employees should be eliminated. Administrators should require users to re-authenticate themselves on machines that have been idle a significant period of time and should deny log-in to any device after a small number of failed attempts (three to five is common). Development of a robust password policy is probably the most significant element of user authentication policy administration; humans are notoriously weak protectors of information, and intruders and attackers frequently try to acquire a password in order to gain initial system access. A password policy should cover five elements:[28]

1. Length—passwords should have a minimum length of eight characters.
2. Complexity—passwords should contain a mix of characters, that is, both uppercase and lowercase letters and at least one non-alphabetic character.
3. Aging—users should change their passwords peri-

odically (every 30-120 days). The policy should permit users to do so only through approved authentication mechanisms.

4. Reuse—administrators should decide whether a password may be reused. Some users try to defeat a password-aging requirement by changing the password to one they have used before.
5. Authority—administrators should decide who is allowed to change a user's passwords.

The password policy should be documented and communicated to users, and users should be trained and expected to follow the policy.

Redundancy

If a system is critical to a company's operations, there should be systems and storage redundancy. By having a redundancy strategy in place, low probability/high cost events are addressed. Major systems should have identical systems on-line that will seamlessly continue operations in the event of interruption for the primary system. The most critical area for implementing redundancy is in storage media. If a primary storage media systems failure occurs without having a redundant device immediately available, it can be disastrous. A daily backup policy alone is insufficient for addressing random failures between backups. This is true for both file systems and databases. Storage media are continually decreasing in cost per unit of storage and a redundant array of inexpensive disks, or RAID, technology is practical for addressing this need. RAID is a disk system that is comprised of an array of disk drives to provide greater reliability and storage capacity and better performance at a lower cost.

On a related note, redundancy is not only applicable to systems. It should also be considered when staffing system administrator positions. While one system admin may be cost effective in the short run, if that person leaves the firm, is incapacitated, or simply on vacation or sick, the firm could experience catastrophic results in the event of system failure. It is critical for any organization to have a redundancy strategy that addresses these issues.

Physical Access

Serious restrictions should be placed on physical access to the office facility, operations and computer facilities of any organization. Companies should use biometric or keycard systems as well as human security guards for maximum security. They should also employ or outsource an information security staff that is complementary to their regular IT staff. These individuals would conduct regular training as well as manage the daily physical security of all information technology employed by the organization. Security personnel should undergo strict background checks and regular controlled substance abuse testing. Active management of physical security, digital security and employee security will significantly reduce an organization's overall information security risk profile.

In addition to hiring security staff, IT assets should be used in secured facilities. Furthermore, the level of security associated with each IT asset should be commensurate with its importance to the continued operations of the business. System administrators should also limit the installation of unauthorized hardware as this is a frequently overlooked element of physical security. Such unauthorized hardware might include removable media storage devices, modems, or devices that might intentionally or unintentionally be used to bypass any security measures.

DETECTION, RESPONSE, AND IMPROVEMENT

The activities of detection, response, and improvement are important phases of the information security management process. Throughout the process, constant communication within the organization is absolutely necessary. Otherwise, any information security policy or strategy will fail to produce favorable results.

Detection

Intruders are always looking for new ways to break into networked computer systems. Even if an organization has implemented a number of information security protection measures, such as firewalls and intrusion detection systems, employees must closely monitor the organization's information assets and transactions involving these assets to check for signs of intrusion.[29] System administrators should regularly review and monitor the following:

- Network Alerts
- Network Error Reports
- Network Traffic
- Software image checksums where current software images on computers are compared to authoritative software images via complex mathematical algorithms.
- System performance statistics on all computers

    - CPU, memory, storage media
    - Message and print queues
    - Changes in file system status or warnings
- Any unusual behavior by a system or by individual personnel

If a system administrator notes any anomalies during his reviewing and monitoring activities, he should ask himself the following questions:[30]

- Is the apparent anomaly the result of a legitimate new or updated characteristic of the system? (e.g., the unexpected process is executing a recently added administrative tool.)
- Can the anomaly be explained by the activities of an authorized user? (e.g., the user really was in Cairo last week and connected to the network; a legitimate user made a mistake.)
- Can the anomaly be explained by known system activity? (e.g., there was a power outage that caused the system to reboot)
- Can the anomaly be explained by authorized changes to programs? (e.g., the mail log showed abnormal behavior because the system programmer made a mistake when the software was modified)
- Did someone attempt to break into the system and fail?
- Did someone break in successfully? Does the administrator have the data that will tell him what the intruder did?

Response and Containment

In responding to any security incident, the administrator and management must remain disciplined in their procedures and use of documented processes and labor to generate sufficient documentation. This is critical to driving the continuous review and learning element of the information security process as well as collecting digital evidence of the anomalous activities.

At the outset of any incident response and containment process, administrators must assess the severity of the incident in terms of scope, impact and damage. From this analysis, it can be determined what subsequent actions must be taken. A worst-case scenario should be assumed in order to enumerate the major actions that must be taken when faced with such an incident.

Administrators and management should inform other sites, networks and organizations that might have been affected by the incident. Continuous communications with potentially concerned parties should be maintained until the incident is contained and normal system operations have resumed. Throughout the entire incident response process, information must be collected at all affected levels. The following data should be collected from all relevant system and network logs:[31]

- The name of the system
- The date and time of each incident affecting that system
- What actions were taken
- What was said during the individual system investigation
- Who was notified
- Who had access
- What data were collected
- What information was disseminated—to whom, by whom, when, and for what purpose
- What was submitted to legal counsel—to whom, by whom, and how it was verified (e.g., notarized)

Upon this investigation administrators and management might find it necessary to take further actions such as temporarily shutting down affected systems, disabling access, services and accounts, and contacting law enforcement agencies.

Once the root cause of the incident has been established, the damage has been contained, the system vulnerability addressed, and the appropriate protective and preventive measures taken, administrators and users must take actions to assure data and system integrity, particularly of backup and redundant systems. Also, every user on every affected system should be required to change his password(s). This should be done for all software at all affected levels too. Finally, because information security assurance is an ongoing process, system-wide monitoring should continue and lessons learned should be articulated and applied to the continuous improvement of the organization's information security stance.

Improvement

It is important to learn from the successful and unsuccessful actions taken in response to an intrusion. Capturing and disseminating what worked well and what did not will help reduce the likelihood of similar intrusions and will improve the security[32] of the organization. Organizations should also try to learn from actions taken during normal system activities and security assessments. If an organization fails to learn from its actions during both normal and abnormal operations and does not document and disseminate its discoveries, it will continue to maintain a reduced system security stance. One mechanism by which the learning process can be formalized is the postmortem review meeting. At such meetings, policies, procedures, system and administrative successes and failures should be evaluated, revised, and improved in order to strengthen the organization's security posture. By having event postmortem discussions the learning process will be served and vivid recent experiences from those involved will be captured and documented.

CONCLUSION

The importance of information of information security management for every energy professional cannot be underestimated. As the field of industrial energy management becomes increasingly automated and information technology becomes a more integral part of operational efficiency and strategic advantage, the breadth and depth of the risks that industrial organizations face will continue to expand. The diversity of threats and risks that the industrial energy manager faces must be managed at all levels of the organization. While it is impossible to manage all possible information security contingencies, continuous education about information security throughout the organization will facilitate the effective management of known and unknown security risks. In addition to education about information risk management and best practices, information security policies must be developed and continually reviewed. These policies must address the most probable security issues without constraining the organization to the extent that it fails to adapt to future challenges presented by the continually changing technological landscape. As any operations manager or energy manager remains ever vigilant about his physical inventories and resources, so he must also be about protecting his information systems assets. As the old Boy Scout motto advises: "Be Prepared."

Endnotes/References

[1] "2001 CSI/FBI Computer Crime and Security Survey," Computer Security Issues and Trends, vol. VII, no. 1. Computer Security Institute, Spring 2001.
[2] Cybercrime: Vandalizing the Information Society. Furnell, Steven. Pg. 25. 2002.
[3] Scene of the Cybercrime: Computer Forensics Handbook. Shinder, Debra Littlejohn; Tittel, Ed. Pg. 317. 2002.
[4] Information Assurance: Managing Organizational IT Security Risks. Boyce, Joseph G.; Jennings, Dan W. Pg. 207. 2002.
[5] Scene of the Cybercrime: Computer Forensics Handbook. Shinder, Debra Littlejohn; Tittel, Ed. Pg. 300. 2002.
[6] The CERT® Guide to System and Network Security Practices. Allen, Julia H. Pg. 101. 2001.
[7] Writing Secure Code. Howard, Michael; LeBlanc, David. Pg. 63. 2002.
[8] Scene of the Cybercrime: Computer Forensics Handbook. Shinder, Debra Littlejohn; Tittel, Ed. Pg. 338. 2002.
[9] Microsoft Press Computer Dictionary, Third Edition. 1997.
[10] Managing Information Security Risks: The OCTAVESM Approach. Alberts, Christopher; Dorofee, Audrey. Pg. 8-9. 2003.
[11] The CERT® Guide to System and Network Security Practices. Allen, Julia H. Pg. 39. 2001.
[12] "Work exposes apathy, Microsoft Flaws." Lemos, Robert, CNET News.com, January 26, 2003.
[13] The CERT® Guide to System and Network Security Practices. Allen, Julia H. Pg. 40. 2001.
[14] *Ibid*. Pg. 42.
[15] *Ibid*. Pg. 43.
[16] *Ibid*. Pg. 44.
[17] *Ibid*. Pg. 59.
[18] *Ibid*. Pg. 60.
[19] *Ibid*. Pg. 67.
[20] MCSE Networking Essentials. Sportack, Mark and Glenn, Walter J. SAMS Publishing. Pg. 194. 1998.
[21] Microsoft Developers Network, July 2000.
[22] The CERT® Guide to System and Network Security Practices. Allen, Julia H. Pg. 123. 2001.
[23] *Ibid*. Pg. 161-162.
[24] *Ibid*. Pg. 173.
[25] www.landfield.com/isn/mail-archive/2002/Sep/0046.html.
[26] www.warchalking.org.
[27] Equipment to "war-drive": www.bitshift.org/wardriving.shtml.
[28] The CERT® Guide to System and Network Security Practices. Allen, Julia H. Pg. 52. 2001.
[29] *Ibid*. Pg. 231.
[30] *Ibid*. Pg. 262.
[31] *Ibid*. Pg. 283.
[32] *Ibid*. Pg. 296.

APPENDIX A. COMMON PORT/PROTOCOL ASSIGNMENTS

Port	Protocol
21	FTP
23	Telnet
25	SMTP
53	DNS
80	HTTP
88	Kerberos
110	POP3
119	NNTP
135	RPC
139	NetBIOS session service
194	IRC
389	LDAP
443	HTTPS
1,024—65,535	Open ports to be utilized by user processes or applications.

APPENDIX B. KEY TERMS

802.11—A family of specifications developed by the IEEE for wireless LAN technology. 802.11 specifies an over-the-air interface between a wireless client and a base station or between two wireless clients.**

ARP—Address Resolution Protocol. A TCP/IP protocol used to convert an IP address into a physical address, such as an ethernet address. A host wishing to obtain a physical address broadcasts an ARP request onto the TCP/IP network. The host on the network that has the IP address in the request then replies with its physical hardware address.**

BIOS—Basic Input/Output System. On PC-compatible computers, the set of essential software routines that test hardware at startup, start the operating system, and support the transfer of data among hardware devices. The BIOS is stored in read-only memory (ROM) so that it can be executed when the computer is turned on. Although critical to performance the BIOS is usually invisible to computer users.*

Ciphertext—Data that have been encrypted.***

DHCP—Dynamic Host Configuration Protocol. A TCP/IP protocol that enables a network connected to the internet to assign a temporary IP address to a host automatically when the host connects to the network.*

DNS—1. Acronym for Domain Name System. The system by which hosts on the internet have both domain name addresses and IP addresses. The domain name address is used by human users and is automatically translated into the numerical IP address, which is used by the packet-routing software. 2. Acronym for Domain Name Service. The internet utility that implements the Domain Name System (see definition 1). DNS servers, also called name servers, maintain databases containing the addresses and are accessed transparently to the user.*

Firewall—A security system intended to protect an organization's network against external threats coming from another network. A firewall prevents computers in the organization's network from communicating directly with computer external to the network and vice versa. Instead, all communication is routed through a proxy server outside of the organization's network, and the proxy server decides whether it is safe to let a particular message or file pass through the organization's network.*

FTP—File Transfer Protocol, the protocol used for copying files to and from remote computer systems on a network using TCP/IP, such as the internet. This protocol also allows users to use FTP commands to work with files, such as listing files and directories on the remote system.*

HTTP—HyperText Transfer Protocol. The client/server protocol used to access information on the World Wide Web.*

HTTPS—An extension to HTTP to support secure data transmission over the World Wide Web.**

IP—Internet Protocol. The protocol within TCP/IP that governs the breakup of data messages into packets, the routing of the packets from sender to destination network and station, and the reassembly of the packets into the original data messages at the destination.*

IRC—Internet Relay Chat. A service that enables an internet user to participate in a conversation on-line in real time with other users. An IRC channel, maintained by an IRC server, transmits the text typed by each user who has joined the channel to all other users who have joined the channel.*

Kerberos—A network authentication protocol developed by MIT. Kerberos authenticates the identity of users attempting to log on to a network and encrypts their communications through secret-key cryptography.*

LDAP—Lightweight Directory Access Protocol. A set of protocols for accessing information directories that supports TCP/IP, which is necessary for any type of internet access. LDAP makes it possible for almost any application running on virtually any computer platform to obtain directory information, such as email addresses and public keys. Because LDAP is an open protocol, applications need not worry about the type of server hosting the directory.**

MAC address—Media Access Control address. A hardware address that uniquely identifies each node of a network.**

NetBEUI—NetBIOS Enhanced User Interface. An enhanced NetBIOS protocol for network operating systems, originated by IBM for the LAN Manager server and now used with many other networks.*

NetBIOS—An application programming interface that can be used by application programs on a local area network consisting of IBM and compatible microcomputers running MS-DOS, OS/2, or some version of UNIX. Primarily of interest to programmers, NetBIOS provides application programs with a uniform set of commands for requesting the lower-level network services required to conduct sessions between nodes on a network and to transmit information back and forth.*

NNTP—Network News Transfer Protocol. The internet protocol that governs the transmission of newsgroups.*

NTP—Network Time Protocol. A protocol used for synchronizing the system time on a computer to that of a server or other reference source such as a radio, satellite receiver, or modem. NTP provides time accuracy within a millisecond on local area networks and a few tens of milliseconds on wide area networks.*

PGP—Pretty Good Privacy. A technique for encrypting messages that is one of the most common ways to protect messages on the internet because it is effective, easy to use, and free. PGP is based on the public-key method, which uses two keys—one is a public key that you disseminate to anyone from whom you want to receive a message. The other is a private key that you use to decrypt messages that you receive.**

Plaintext—Data that has not been encrypted.***

POP3—Post Office Protocol 3. A protocol for servers on the internet that receive, store, and transmit email and for clients on computers that connect to the servers to download and upload email.*

PPP—Point-to-Point Protocol. A data link protocol for dial-up telephone connections, such as between a computer and the internet.*

Public Key Encryption—An asymmetric scheme that uses a pair of keys for encryption: the public key encrypts data, and a corresponding secret key decrypts it. For digital signatures, the process is reversed: the sender uses the secret key to create a unique electronic number that can be read by anyone possessing the corresponding public key, which verifies that the message is truly from the sender.*

RPC—Remote Procedure Call. A type of protocol that allows a program on one computer to execute a program on a server computer. Using RPC, a client program sends a message to the server with appropriate arguments and the server returns a message containing the results of the program executed.*

SLIP—Serial Line Internet Protocol. A data link protocol that allows transmission of IP data packets over dial-up telephone connections, thus enabling a computer or a local area network to be connected to the internet or some other network.*

SMTP—Simple Mail Transfer Protocol. A TCP/IP protocol for sending messages from one computer to another on a network. This protocol is used on the internet to route email.*

SNMP—Simple Network Management Protocol. The network management protocol of TCP/IP. In SNMP, agents, which can be hardware as well as software, monitor the activity in the various devices on the network and report to the network console workstation.*

SSID—Service Set Identifier. A 32-character unique identifier attached to the header of packets sent over a wireless LAN that acts as a password when a mobile device tries to connect to the network. The SSID differentiates one wireless LAN from another, so all access points and all devices attempting to connect to a specific wireless LAN must use the same SSID. A device will not be permitted to join the network unless it can provide the unique SSID.**

SSL—Secure Sockets Layer, a protocol developed by Netscape for transmitting private documents via the internet. SSL works by using a public key to encrypt data that's transferred over the SSL connection. Both Netscape Navigator and Internet Explorer support SSL, and many web sites use the protocol to obtain confidential user information, such as credit card numbers. By convention, URLs that require an SSL connection start with https: instead of http:.**

TCP—Transmission Control Protocol. The protocol within TCP/IP that governs the breakup of data messages into packets to be sent via IP, and the reassembly and verification of the complete messages from packets received by IP.*

TCP/IP—Transmission Control Protocol/Internet Protocol—A protocol developed by the U.S. Department of Defense for communications between computers. It is built into the UNIX operating system and has become the de facto standard for data transmission over networks, including the internet.*

Telnet—A protocol that enables an internet user to log on to and enter commands on a remote computer linked to the internet, as is the user were using a text-based terminal directly attached to that computer. Telnet is part of the TCP/IP suite of protocols.*

UDP—User Datagram Protocol. A connectionless protocol that converts data messages generated by an application into packets to be sent via IP but does not verify that message have been delivered correctly.*

VPN—Virtual Private Network. A set of nodes on a public network such as the internet that communicate among themselves using encryption technology so that their messages are as safe from being intercepted and understood by unauthorized users as if the nodes were connected by private lines.**

WEP—Wired Equivalent Privacy. A security protocol for wireless local area networks defined in the IEEE 802.11b standard. WEP is designed to provide the same level of security as that of a wired local area network.**

---

* Microsoft Press Computer Dictionary, Third Edition. 1997.
** Webopedia: On-line Dictionary for Computer and Internet Terms. www.webopedia.com
*** Microsoft Developer Network, July 2000.

Section Seven

Relational Database Choices and Design

Chapter 21

Fundamentals of Database Technology and Database-Driven Web Applications

Fangxing Li
ABB Inc

DATABASE TECHNOLOGY involves the access and manipulation of information. It is critical to the development of highly efficient information systems. It also plays an important role in the development of web-based applications that require information processing and distribution between web-browsers and web-servers. This chapter provides a quick guide to database technology and illustrates common structures of database-driven web applications.

INTRODUCTION

Like many other information systems, utility information systems contain and process a large amount of data. Without the ability to manage that data efficiently, it is difficult for utilities to provide satisfactory services to customers. The development of information technology has answered this challenge. Databases and database management systems (DBMS) have been broadly deployed to manage bulk data in enterprise information systems.

Database

A database is a self-describing collection of data. The term "self-describing" implies that the database contains not only the actual data, but also the structure of the data (or the meta-data).[1-3]. A database may achieve high integrity because the meta-data typically describes the relationships among different tables. This feature of "self-describing" is the main difference between a database and a flat file that was used in the early age of computing. A flat file does not contain any information about the structure and relationships among different pieces of data, and therefore is less integrated than a database.

DBMS

A DBMS is a software tool that helps users define, access, and maintain the underlying data contained within a database. As the definition shows, a database is the collection of structured data, and a DBMS is a software program that helps users manage the database efficiently. Figure 21-1 describes the logic interaction between a user, a DBMS and a physical database.

There are many commercial DBMS products available from different vendors. Microsoft's Access is a popular example of a desktop DBMS. Microsoft's SQL Server is an example of an enterprise DBMS that works across a network for multiple users. Other popular DBMSs are IBM's DB2, Oracle's series of database management products, and Sybase's products.

DATABASE TABLES

Typically, a database is organized as a collection of multiple two-dimensional tables. Each entry in a table is single-valued. Each row in a table represents an instance of a real world object. Each column in a table represents a single piece of data of each row.

Figure 21-2 shows two tables in a database of a utility information system. Each row in the first table represents the information about a different metering device. Each row in the second table represents the amount measured by a specific water meter at a given time.

In the first table, there is a specific column ID, which uniquely identifies a metering device in the real world. This column is the primary key of this table. The primary key of a table may be a combination of several columns. For example, the primary key of the second

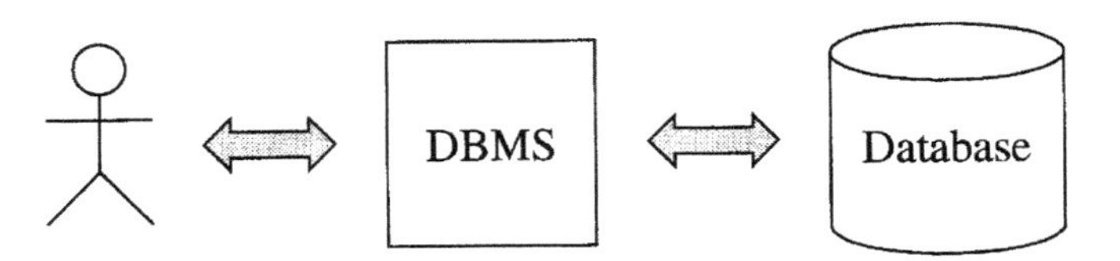

Figure 21-1: Information flows among a user, a DBMS and a physical database

Metering_ Equipment Table

ID	*Name*	*Utility*	*Facility*	*Manufacturer*
X1001	GasA1	GAS	BLDG1	M101
X1002	ElecA1	ELECTRIC	BLDG1	M201
X1003	WaterA1	WATER	BLDG1	M301
X1011	GasB1	GAS	BLDG2	M101
X1012	ElecB1	ELECTRIC	BLDG2	M201
X1013	WaterB1	WATER	BLDG2	M301

Water_Consumption Table

ID	*Measurement_Date*	*Measurement_Time*	*Gallons*
X1003	2/1/2003	12:00:00	341.23
X1003	2/1/2003	12:15:00	355.68
X1003	2/1/2003	12:30:00	362.42
X1003	2/1/2003	12:45:00	377.81

Figure 21-2. Two tables in a utility information system

table is the combination of ID, Measurement_Date and Measurement_Time. That is, the remaining columns (in this case only one column—'Gallons') are uniquely determined by the combination of the first three columns.

RELATIONAL DATABASES

As the previous definition shows, a database contains structural information about the data as well as the actual data. How is the structure defined? Or, what is the structural model? The most popular model over the past 20 years is the relational model. Other models include the hierarchy model and the network model that existed for some time but did not gain considerable market share. There are also some emerging models like the object-oriented model and the object-relational model, both of which are gaining some market share. However, in general the relational model is still the dominant force. Hence, this chapter focuses on relational databases and related technologies.

What is a relational database? Theoretically, a relational database is a database comprised of a set of tables that follows the rules of normalization. The definition of normalization in database theory is complicated if expressed in a mathematical way. For simplicity, the primary principles of normalization can be roughly interpreted as the following guidelines:

1. When you design a table in a database, your table should avoid repeating groups.
2. The columns in a table depend on the primary key only.
3. There is no column depending on another column that is not part of the primary key.

The "repeating groups" problem is illustrated by the following example. A table is created to store all purchase orders, while each order may have different items. The first table in Figure 21-3 shows an un-normalized design, which can handle orders with no more than two items. If the maximum number of items is 30, then the columns must be expanded to contain 30 groups of item and quantity, i.e., from {Item1, Qty1} to {Item30, Qty30}. This could cause a serious waste of space if most of the other orders have less than 5 items.

The second table in Figure 21-3 shows a normalized design, which involves only four columns. The column Sub_Order_ID is used together with Order_ID to avoid the repeated grouping problem in the first table. In other words, if two rows have the same Order_ID, then the items in these two rows are associated with the same order. The column Sub_Order_ID can be used to identify different items within the order. With this design, there is no limit on the number of items within one order.

Purchase_Order Table: Un-normalized Design

Order_ID	*Item1*	*Qty1*	*Item2*	*Qty2*
1	Circuit Breaker	4	Transformer	1
2	Sectionalizer	2	Distributed Generator	1

Purchase_Order Table: Normalized Design

Order_ID	*Sub_Order_ID*	*Item*	*Qty*
1	1	Circuit Breaker	4
1	2	Transformer	1
2	1	Sectionalizer	2
2	2	Distributed Generator	1

Figure 21-3. Two designs for a purchase order table

Further, there is no redundant information stored and efficiency is achieved.

The term "depending on" mentioned in the above guidelines can be interpreted as "being uniquely determined by." That is, if the column A depends on the column B, then the value of A is uniquely determined by the value of B, but not vice versa. For example, if we know the social security number (SSN) of a person, then we know his or her name, title, address, etc. Thus, the column of name, title, or address depends on the column SSN. However, if we know the name of a person, it is possible that we cannot identify his or her SSN since people may have the same names. In the second table in Figure 21-3, the columns Item and Qty each depend on the combination of the columns Order_ID and Sub_Order_ID.

In order to achieve efficient database systems, the normalization rules or the above rough guidelines should be followed. Practical tests show that an un-normalized database may result in much poorer performance (5+ times slower) and need much more programming involvement. Although database designers may consider normalization by intuition without knowing it, they should be required to explicitly follow the rules to ensure high performance and efficiency. The performance and efficiency issue is particularly important for web-based database-driven applications, since users of web-based applications may experience longer delays than users of stand-alone applications. The longer delays may be attributed to the following features of web applications:

- There may be many clients (users) concurrently accessing the server.
- The users may be geographically distributed across the country.

SQL

SQL is a standard to create, retrieve and manipulate a relational database. Originally, SQL stood for the acronym of "Structured Query Language," but now it has become generally accepted as a non-acronym standard to access the internal data of a database, usually a table-based relational database.

Unlike full-featured programming languages such as C/C++, VB and Java, SQL is not a full-fledged programming language. It may be considered as a sub-language to create, retrieve, and manipulate information contained in databases. It can be dynamically coded into high-level languages like C/C++, Java or VB to facilitate the control of the underlying databases.

SQL consists of a set of text-based commands or queries to control data. The SQL commands can be classified into two major categories, Data Definition Language (DDL) and Data Manipulation Language (DML). DDL is used to create tables or change table structures, while DML is used to insert, delete or modify rows of tables. The DDL commands include statements of CREATE, ADD, DROP, ALTER and others. The DML commands include statements of SELECT, INSERT, UPDATE, DELETE, JOIN, UNION and others.

The following brief examples are given as a quick

guide to explain how the SQL commands work. The examples are based on the needs of a utility information system administrator who wants to create a database table to host information, populate the table, manipulate the table, etc.

1. Create a Table

The following command creates a blank table with the similar schema as the second table in Figure 21-2.

```
CREATE TABLE Water_Consumption(ID TEXT(20), Measurement_Date DATE, Measurement_Time TIME, Gallons DOUBLE)
```

The above CREATE command creates a table named Water_Consumption. The table has four columns called ID, Measurement_Date, Measurement_Time and Gallons. The data types of these four columns are a text string of 20 characters, Date, Time and Double, respectively. The Date data type is usually input in the format of "MM/DD/YEAR" or "YEAR/MM/DD." The Time data type is usually input in the format of HOUR:MINUTE:SECOND with the HOUR filed using a 24-hour clock. For example, "18:45:00" is input for 6pm, 45 minutes and 0 seconds.

2. Populate a Table

The following command adds a row into the table Water_Consumption. It should be noted that a text column is enclosed by opening and closing quotes (single or double). Columns in Date or Time data type should be enclosed in quotes as well.

```
INSERT INTO Water_Consumption VALUES ('X1003', '1/1/2003', '18:45:00', 165.82)
```

To add many rows into the table, users may use the command in the format like "INSERT INTO target SELECT ...†FROM source," in which the "SELECT...FROM" statement will be mentioned next.

3. Select Data

The most popular command used in SQL probably is the SELECT statement. The following command selects all rows and all columns from a table.

```
SELECT * FROM Water_Consumption
```

The * represents all columns in the selected table. Users may select partial columns by specifying the actual column names. For example, the following SQL command selects all rows but only 'ID' and 'Gallons' columns.

```
SELECT ID, Gallons FROM Water_Consumption
```

There are also various clauses that can be appended after the above SELECT statements to filter some rows. For example, the following command selects the information only related to Meter X1003 using WHERE clause.

```
SELECT ID, Gallons FROM Water_Consumption WHERE ID='X1003'
```

Delete Data

The following command deletes all rows from the table Water_Consumption.

```
DELETE * FROM Water_Consumption
```

The WHERE clause can be used as a filter for DELETE statement. The following command deletes the rows from meter X1003 and with a date no later than 12/31/2001.

```
DELETE * FROM Water_Consumption WHERE ID='X1003' AND Measurement_Date<'01/01/2002'
```

Since this chapter is not a detailed SQL guide, the above examples do not cover all aspects of SQL commands. For details about SQL, users may check references 1 and 2. Despite its simplicity, this section is expected to serve as a quick start for further SQL studies.

SQL CODED IN OTHER PROGRAMMING LANGUAGES

Although SQL is a powerful tool specifically designed for database access, it is not a full-featured programming language. Hence, to maximize the benefit of SQL in applications, embedded SQL is used. That is, SQL is coded into a programming language like C/C++, VB or Java in a database-driven application. The programming language is employed to perform the common "programming" tasks while the embedded SQL queries are utilized to access the database. This interactive process can be described as follows:

1. The application creates a connection to a database so that the host application can "talk" with the database and its tables.
2. The application generates a SQL query to obtain a table in the database.

3. The content in the table is retrieved and then mapped to the internal data structure of the application.
4. Operations on the data are carried out. This could be very simple or complicated depending on the application's requirement.
5. The updated data may be saved back into the database and output may be generated.

The VB code shown in Figure 21-4 illustrates a sample process of the five steps above, including how to set up a database connection, retrieve data from a database, identify rows with gallon amount over a predefined threshold, and generate a warning report. The report file contains all metering records with a gallon amount over the threshold.

In short, SQL queries could not do much but access the database. Programming languages are more flexible and powerful in many other functions, but not in direct database access. Hence, the combination of these two is a good choice to create fast, efficient, and easy-to-program applications involving underlying databases.

DATA ACCESS INTERFACES

After reading the previous example, readers may have this question: "How does the database receive the SQL query, interpret it and then send the response back to the VB application?" To answer this question, the mechanism of database access interfaces is explained.

Database access interfaces are software modules

```
Dim dbs As Database, rst As Recordset
Dim OutputFileName As String
Dim k As Long
Dim TheID As Long, TheDate As String, TheTime As String, TheGallons As Double

'Connect to a database
Set dbs = OpenDatabase("MyTestEIS.mdb")
'Create a SQL query to obtain the database table Water_Consumption
Set rst = dbs.OpenRecordset("SELECT * FROM Water_Consumption")
If rst.RecordCount = 0 Then
      MsgBox "No record in the table."
      Exit Sub
End If

OutputFileName = "OutputTest.csv" 'Output to a CSV spreadsheet file
Write #1, "ID", "DATE", "TIME", "GALLONS" 'Output a header

Open OutputFileName For Output As #1 'Open an output report file
For k = 1 To rst.RecordCount
      'Retrieve the field and store it in VB's internal data structures
      TheID = rst("ID")
      TheDate = CStr(rst("Measurement_Date")) 'Get date in string format
      TheTime = CStr(rst("Measurement_Time")) 'Get time in string format
      TheGallons = rst("Gallons")

      'Perform operations on the extracted data. Here, GetThresholdFor()
      'is a function to obtain the warning threshold of a metering device
      Threshold = GetThresholdFor( TheID )

      Write #1, TheID, TheDate, TheTime, TheGallons
      rst.MoveNext
Next k
Close #1

'Close the database connection
rst.Close
dbs.Close
```

Figure 21-4. An Example of VB and SQL Queries

that provide connections between an application and a specific database. They play a key role in implementing database-driven applications. Different vendors may have different database drivers. At times, this could cause portability and extensibility problems since users may have to deal with different vendors and even different platforms. An early solution to this problem was the ODBC (Open Database Connectivity)[4] technique provided by Microsoft. The ODBC module sits between applications and vendor-specific databases to provide the necessary connectivities.

Microsoft has recently replaced ODBC with Universal Data Access (UDA), which provides access to all kinds of data sources like ODBC databases, traditional SQL data, non-SQL data like spreadsheets, etc. UDA is the database access part of Microsoft's Component Object Model (COM), which is an overall framework for creating and distributing object-oriented programs in a network. UDA consists mainly of the high-level application program interface (API) called ActiveX Data Objects (ADO) and the lower-level services called OLE DB. SQL queries are sent to ADO interfaces from applications and then forwarded to OLE-DB interfaces. OLE-DB communicates with vendor-specific data providers to retrieve information from physical databases. The retrieved information is then sent back to the applications. This database access strategy is shown in Figure 21-5.

Sun Microsystems presents another data access technology, Java Database Connectivity (JDBC)[5], which is an API that can be used to access almost any tabular data source from the Java programming language. The JDBC interface provides cross-DBMS connectivity to a wide range of SQL databases. The latest JDBC API also provides access to other tabular data sources, such as spreadsheets.

JDBC architecture contains a driver manager and database-specific drivers to provide transparent connectivity to databases from different vendors. This is shown in Figure 21-6. A JDBC driver translates standard JDBC calls into a protocol that the underlying database can understand. This translation function makes JDBC applications able to access many different databases. There are four distinct types of JDBC drivers. Details about the four drivers and their mechanisms are beyond this chapter but can be found in reference 5.

An interesting and noteworthy point is that Sun's JDBC technology supports multiple operating systems but is restricted to the Java programming language. As opposed to JDBC, Microsoft's UDA technology is restricted to the Windows platforms but supports multiple languages like VB, C/C++, J++ and the latest .NET technology.

Although a thorough discussion of the technologies on database access interfaces is beyond the scope of this chapter, it is helpful to readers to understand the basic concepts. With the assistance of these interfaces,

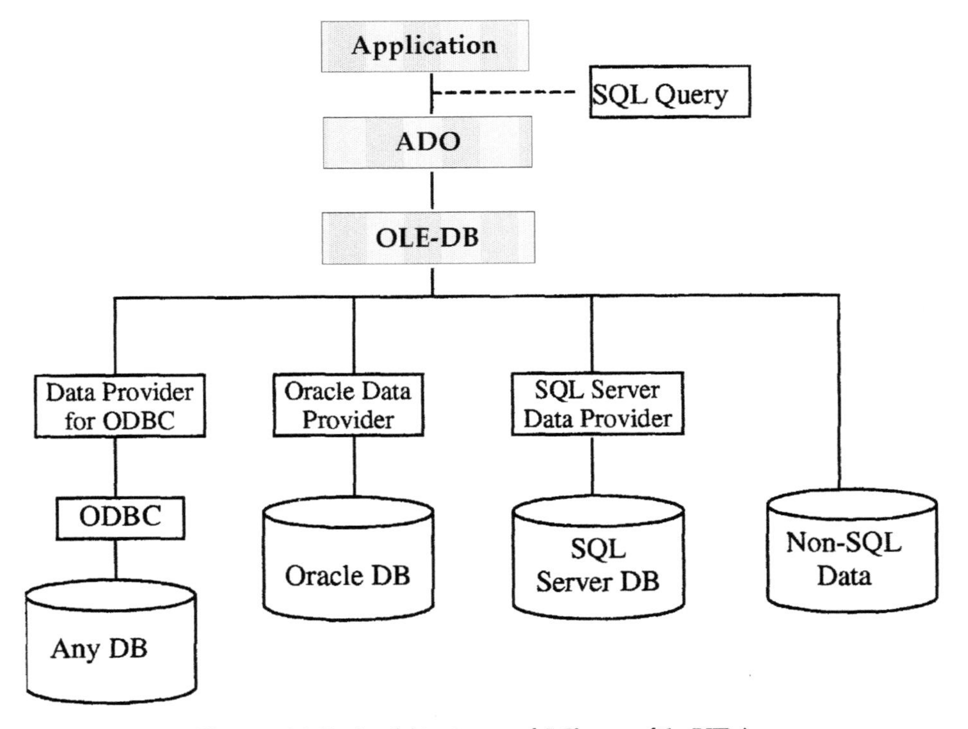

Figure 21-5. Architecture of Microsoft's UDA

applications can essentially "talk" with physical databases and can efficiently retrieve the information contained in various databases. Also, database access interfaces like ADO and JDBC can minimize application developers' efforts, because developers only need to deal with the ADO or JDBC rather than do the detailed work that the interfaces perform.

DATABASES AND WEB APPLICATIONS

Like stand-alone applications, web applications[6-8] may rely on database technologies for efficiency. The architecture of a database-driven web application is illustrated in Figure 21-7. The information flows and activities are also illustrated with the arrow-lines. Here is the description of the activities in the figure.

1. A client browser sends an HTTP request to a web server for a specific web page that may contains regular HTML code as well as code written in server-side script (SSS).

2. The web-server receives the request. If the requested web page involves SSS code, the web server invokes a SSS engine to process the SSS code.

3. If the SSS code involves database operations, the SSS engine queries the database through database interfaces to obtain the needed results, and may generate part of the returning HTML code based on query results.

4. The web server creates a response HTML page that is a combination of the returning HTML code from the SSS engine and the regular HTML code in the original web page.

5. The response HTML page is sent back to the client browser and the browser displays it for users.

The server-side script (SSS) is employed to generate dynamic web pages, which may require database manipulations. Since HTML is a markup language designed mainly for information displaying, it cannot handle complicated computation and database access. To make a web-server more powerful, some scripts are usually embedded into an HTML page to direct the web-server to perform some specific tasks. The web-server invokes the SSS engine to handle complicated tasks like database access. The SSS engine passes the database queries to the database interface/driver that communicates with the physical databases.

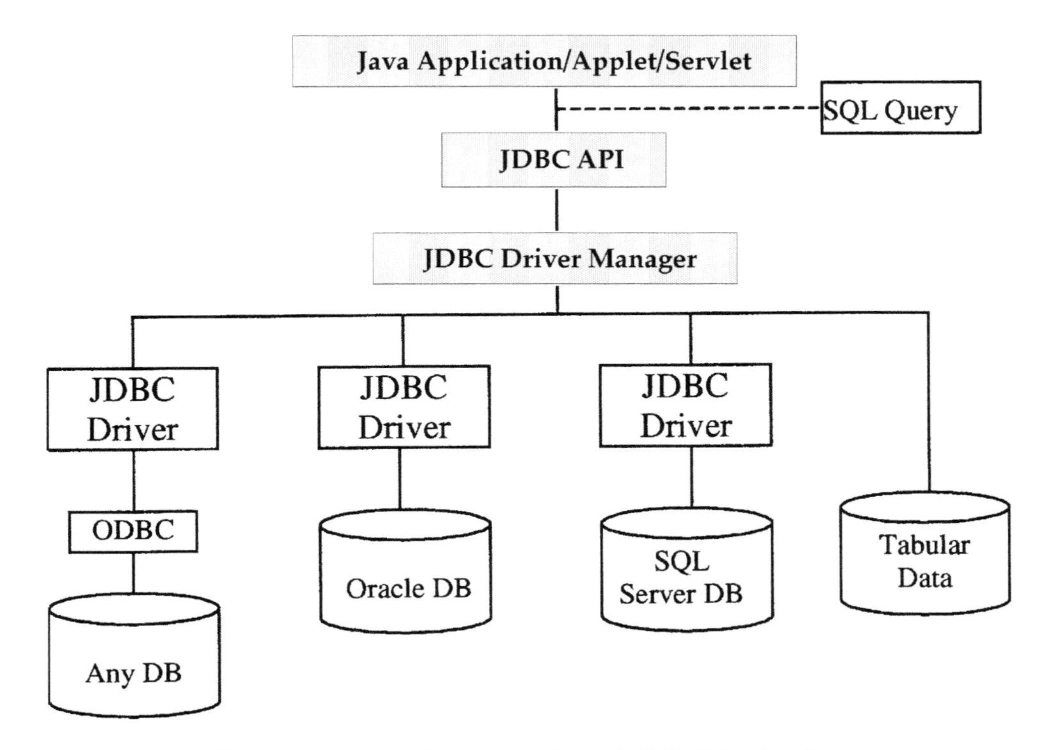

Figure 21-6. Architecture of Sun's JDBC Technology

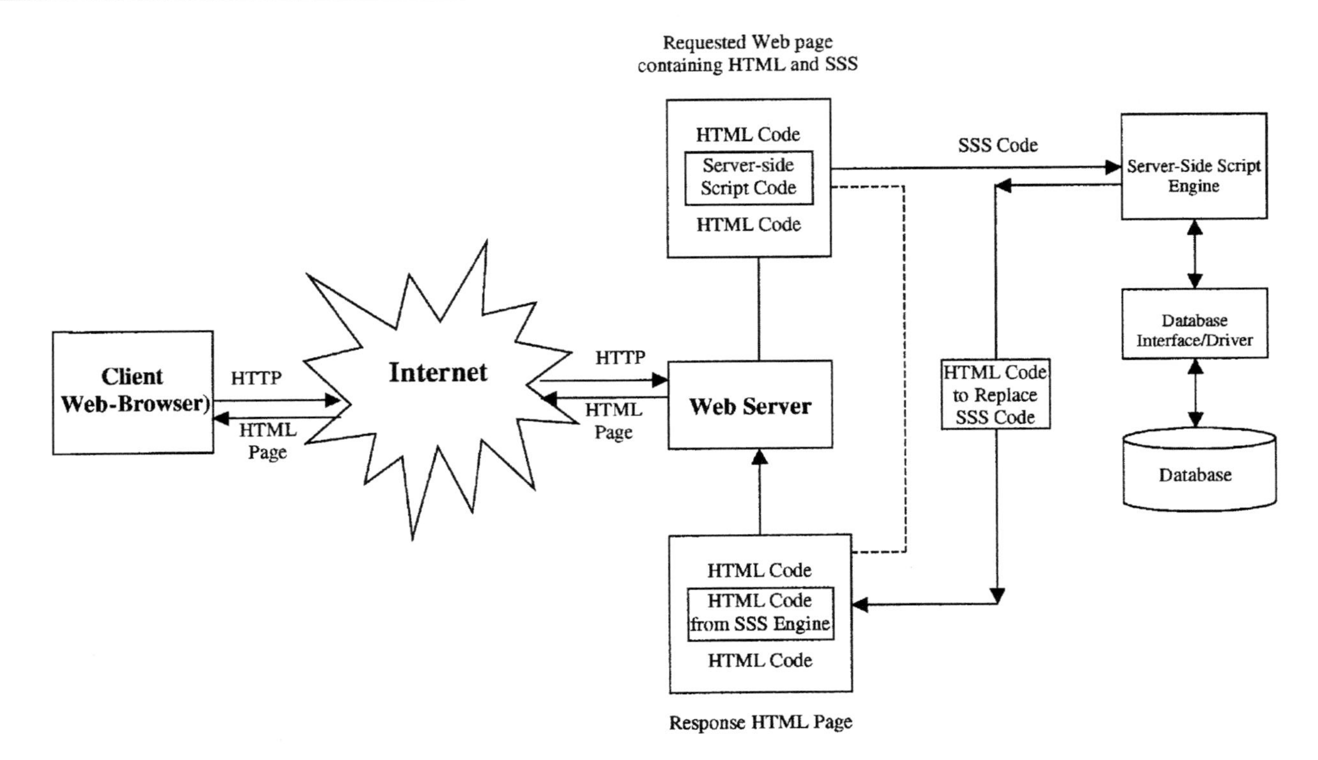

Figure 21-7. Generic Architecture of a Database-driven Web-based Application

Typically, the associated database interface/driver is located at the server side. This makes many database-related operations at the server-side transparent to the client side. This is an advantage of web-based applications, because users do not need to worry about the complicated database access and set-up processes. Once everything at the server side is set up, users from any place with internet access can benefit from the web application.

SERVER-SIDE SCRIPT TECHNOLOGIES FOR DATABASE-DRIVEN WEB APPLICATIONS

There are several server-side script technologies that can assist developers to implement database-driven web applications. Two of them, Active Server Pages (ASP) and FoxWeb, are briefly reviewed in this section.

ASP

ASP is the Microsoft's technology for building dynamic and interactive web pages. The overall architecture of ASP-centered web applications is similar to the generic architecture depicted in Figure 21-7. The main difference is that the so-called ASP script engine replaces the SSS engine in Figure 21-7. The ASP script engine can handle various requests including intensive database manipulations. The requests are coded in ASP scripts that are usually written in the VB Script language. An ASP web page is a text file with the extension of *.asp that contains HTML code and ASP scripts.

FoxWeb

FoxWeb is another technology that enables developers to create dynamic web pages, especially if the pages involve underlying FoxPro databases. The overall architecture of FoxWeb-centered web applications is also similar to that depicted in Figure 21-7. The FoxWeb script engine is a specific SSS engine. Like the ASP script engine, the FoxWeb script engine can handle complicated database manipulations, especially for FoxPro databases. This makes FoxWeb very attractive to developers who need to convert legacy desktop applications powered by FoxPro databases to web-based, FoxPro-driven applications. Similar to an ASP web page, a FoxWeb web page is essentially a text file with extension of *.fwx that contains HTML code as well as FoxWeb scripts. Also, the latest FoxWeb scripting object is compatible with Microsoft's Active Server Pages (ASP) objects. This makes it easier for developers who are already familiar with ASP to become familiar with FoxWeb.

There are also other similar server-side technologies such as JavaServlet, JSP, PHP, WestWind, etc., which use mechanisms similar to (but implementations different from) ASP or FoxWeb technologies to process server-side tasks. Since all of the above technologies are developed to carry out server-side tasks including database access, those technologies, together with database technologies, are the driving force of the evolution of web applications. For details of the server-side technologies, readers may refer to other chapters in this book.

CONCLUSION

Like many other information systems, utility information systems usually employ database technology to store and retrieve data to achieve high performance and efficiency. Database technology is especially important to web-based information applications since a large amount of data needs to be processed at the server-side and distributed to geographically remote clients. As a quick tutorial and guide, this chapter reviews the basics of database technology such as relational databases, SQL, and database access interfaces. The chapter also provides an illustration about common architecture of database-driven web applications.

References

[1] Raghu Ramakrishnan, *Database Management Systems*, McGraw-Hill, 1997.
[2] Jesse Feiler, *Database-Driven Web Sites*, Morgan Kaufmann Publishers, Inc., 1999.
[3] Paul Dorsey and Joseph. R. Hudicka, *Oracle 8—Design Using UML Object*, Oracle Press, 1999.
[4] Kyle Geiger, *Inside ODBC*, Microsoft Press, 1995.
[5] Cay Horstmann and Gary Cornell, *Core Java*, vol. 2, The Sun Microsystems Press, 2001.
[6] Chris Ullman, et al, *Beginning ASP 3.0*, Wrox Press Ltd., January 2000.
[7] Fangxing Li, Lavelle A.A. Freeman, Richard E. Brown, "Web-Enable Applications for Outsourced Computing," *IEEE Power and Energy Magazine*, vol. 1, no. 1, (Premier issue) January-February 2003.
[8] Dustin R. Callaway, *Inside Servlets: Server-Side Programming for the Java® Platform*, Second Edition, Addison Wesley Professional, May 2001.

Acknowledgment

The author would like to thank Mr. David Green and Dr. Barney L. Capehart for their valuable comments and suggestions. The author would also like to thank Ms. Lynne Capehart for her careful editing and formatting of the final document.

Chapter 22

Relational Database Choices and Design

Dale Fong, Itron, Inc.
Allan Schurr, Itron, Inc.

ENERGY MANAGEMENT systems have evolved from operational control systems to management and decision support tools. As a result of information technology advancements and energy market deregulation, a specialized set of software applications designed to solve real-world business problems has emerged. These unique applications are supported by sophisticated databases and data analytics that address the challenges of the modern energy managers wishing to achieve best practices in their field. This chapter addresses the issues driving database design and selection including user requirements, business processes, data types, database performance, scaling, data processing, and security.

INTRODUCTION

The field of corporate or enterprise energy management has undergone significant advancement in the last five years. This is largely due to applying information technology to the immense challenge of understanding and managing consumption, pricing, and risk to achieve overall objectives of an affordable, reliable energy supply. Designing systems to manage energy is far more complex than storing meter data in a relational database. Indeed, best practices for energy management require sophisticated data collection and management, purpose-built analytics, and decision support tools to affect the necessary management system and business processes. Just as form follows function, database design follows application requirements.

The Enterprise Energy Management Challenge

Energy use has grown dramatically over the past 20 years. In fact, energy management is an increasingly important function for most large public and private sector organizations. Besides increasing costs, this growing energy consumption reduces profits, operating budgets, and cash flow available for capital investment in manufacturing products or delivering services in a quality and cost effective manner. Ultimately, this impacts economic growth and efficiency. Yet energy costs in an organization are frequently regarded as uncontrollable until problems arise, at which point it is usually too late to implement effective change.

It's no wonder then that many organizations have added energy managers or energy management groups to focus on key energy management issues: financial management, operations or procurement. The challenge these managers face is often one of scarce information and ineffective management tools to manage the purchase, use and settlement of energy across sometimes hundreds of facilities, from dozens of suppliers in highly variable and volatile conditions. Common problems energy managers face include:

Finance

1. How can my organization accurately develop and monitor budgets for its various department and costs centers?
2. How can the organization manage the plethora of utility invoices to ensure accuracy, comply with dispute resolution guidelines, and pay on time to avoid penalties or service shut-offs?
3. How can the organization accurately allocate or assign costs to departments, cost centers, or other responsible parties? How can I present these costs in an understandable way so that appropriate action can be taken?

Operations

1. How do I determine how my plants or facilities compare to one another for energy performance? How can I factor out differences in weather, occupancy, sales, or production to make an accurate assessment?

2. How do I know if my energy management efforts have been successful? Is there an easy way to compare energy costs across time when production levels or other energy cost drivers are in constant flux?
3. How is my operation performing at the product line, major system, or equipment level? Are there operating parameters that I can track that indicate relative efficiency?
4. What is my true marginal cost of energy, when factoring in all the complex charge structures I have from each different utility? How would my operation change if I knew the true cost of the last unit of energy?
5. How can I optimize my operating schedule to maximize profits?

Procurement

1. In regulated and unregulated markets, how can I be confident the organization is receiving the best risk-adjusted price for the commodity, delivery, and accounts services?
2. Once markets are open to competitive suppliers, how can that opportunity be most expeditiously and accurately exploited?
3. For energy service, either regulated or unregulated, how can the organization monitor and otherwise manage its contract or tariff requirements, such as nominations and scheduling, or keep under volume limits and avoid penalties?

DATABASE DESIGN

Requirements

The design of the database reflects the value of the information contained in it and the applications that access the data. User characteristics, data volume and frequency, and the business processes supported are all factors in the design of the database.

User Characteristics

The variety of energy management information system users dictates certain design characteristics of the database. It will also drive security and access considerations, and address diverse user needs.

For example, users can be classified into three types:

- **Executive/Management Users**. These users expect enterprise-wide information and analysis that focuses on financial metrics such as unit cost, budget performance, benchmarking, and financial reporting period.
- **Site-Specific Users**. These users expect site-specific information that alerts them when a predefined operational issue is identified so routine operations are not disrupted by adding a new responsibility to the staff at each site.
- **Energy Manager/Analyst Users**. These users expect easy to use, yet powerful analytical and reporting tools that cover the entire enterprise, yet also allow for "drill down" into specific site related information.

Data Volume, Frequency

In an energy information system, the bulk of the data managed within the database consists of time-series energy data values. Typically, this data dwarfs all other data in the system, especially in interval metering, i.e., meters which sample more frequently than once per day, typically every hour or every 15 minutes. Although systems containing many interval meters are not yet common, they are becoming popular as energy prices continue to rise and real-time energy information relating to usage becomes a viable option in controlling energy consumption and costs during peak usage periods.

Naturally, interval meter implementations will find their first application with large energy consumers in the commercial and industrial sector, leaving most residential meters for monthly metering or at best, daily time-of-use metering. Deployments are limited mostly by economics as interval meters and the communication networks required to transmit their collected data continue to be the greatest barriers to entry. However, even the data volume associated with as few as 100,000 interval meters can be a challenge for most computing platforms to date. For example, the uncompressed storage required for a single data sample comprised of only a time stamp, a value and a status indicator is 20 bytes. Sampling every 15 minutes produces roughly 100 individual samples each day. If the system were required to store up to two years of data for seasonal comparison purposes, then for 100,000 such meters containing four channels of data each, the required data storage would be approximately 600 gigabytes. Adding in the required disk storage for indexes, backups, and log files, etc., the system then approaches physical storage volumes relegated to only a few commercially available database engines.

Lack of interval data doesn't preclude the use of available non-interval metering data for an energy information system, however. Even with simple residential monthly billing data, load profiles can be algorithmically generated which approximate such profiles in aggregate with enough accuracy for them to be used in load and distribution planning studies.

Business Processes Supported

Best practices energy management is a collection of inter-related business processes that can be categorized into three general business areas: Finance, Operations and Procurement. These business process areas are inter-related because they share the same information and because the impact of a decision made in one area affects decisions made in another. For example, the procurement function is influenced by operations decisions regarding equipment selection, production or store openings. The financial decision concerning cost allocation methodology is affected by the contract structure driven from procurement.

This interdependency calls for data models that are comprehensive and structured to deliver value on the various business processes involved.

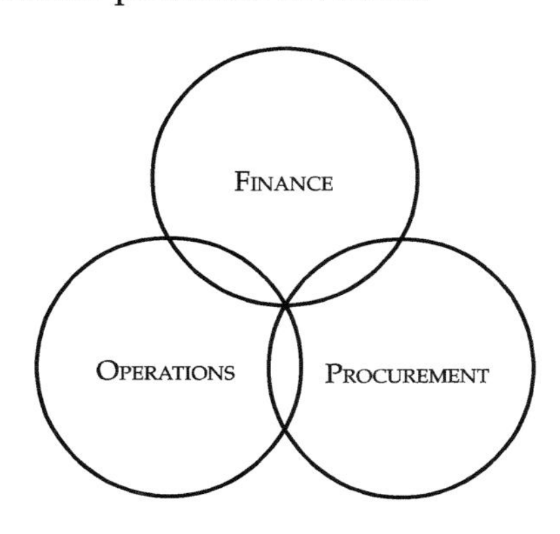

Within each of the broad functional areas, sub-processes can be defined as follows:

Finance

- *Bill Processing*—A/P related function, ensuring on-time payment, creating accruals and estimates
- *Bill Validation*—ensuring the bill is calculated correctly
- *Cost Allocation*—activity-based costing or rebilling for tenants
- *Budget Management*—setting and tracking actuals to budget

Operations

- *Baseline Measurement*—comparing actuals to targets or historic levels: can be at the enterprise, site, system, or equipment level. Often used for validating energy projects savings claims.
- *Benchmarking*—comparison of like facilities or systems, often normalizing for weather, occupancy, production variances.
- *Performance Monitoring*—monitoring of operating conditions compared to defined parameters to identify out-of-spec operations; can be used to set RCM maintenance schedules.
- *Demand Management*—used to avoid unnecessary demand charges, often the most costly component, through better load balancing and higher load factors. May involve reducing load in response to real-time process.
- *Cost/Use Optimization*—involves integrating accurate marginal energy costs into product and operation planning decisions. Can be hourly, daily, weekly or unit cost data.

Procurement

- *Load Projection and Aggregation*—creates the forward view of site specific and market aggregate load requirements, with standard deviations. It's very useful to get the best price in competitive markets.
- *Bid Analysis and Risk Assessment*—allows for price comparison of various price structure alternatives and assessment of volume and index price risk.
- *Best Rate Selection*—the process of comparing all applicable alternatives for regulated tariff service.
- *Contracting and Compliance*—the process of selecting the preferred supply option, and monitoring the contract during the term to avoid penalties.

Each of these sub-processes can be fully or partially automated through the use of database applications. Each drives a unique design element of the database schema.

Data Types for Energy Management

Time-Series Data

The most obvious data types for energy management systems are interval meter data. Other time series data are also useful.

1. **Energy consumption data** from primary utility gas, electric, and steam meters, at intervals ranging from five minutes to daily, are common. Also, reactive power can be collected to assist in validating billing or determining power factor. In addition, sub-metering data that isolates consumption at load centers, production line, tenants, or other useful subdivisions are particularly valuable in determining cost allocations and identifying operational efficiencies. Often, gathering this data once per day is acceptable.

2. **Billing data** are critical to an energy management system because it includes total and line item charge amounts. There are also rate schedules, payment history, due dates, eligibility for competitive procurement, meter number, and other relevant data.
3. **Operational data** such as temperature, flows, pressures, and system status are valuable for diagnosing performance improvement areas. In particular, the ability to use this operational data as a proxy for sub-meter information to help determine cause and effect, or to assist in the investigation of peak demand drivers is especially critical. This information can be collected at much faster intervals than five minutes, though techniques such as storing min/max/average values can reduce the volume of data substantially.

Pricing

Inasmuch as the goal of energy management is to reduce energy costs, pricing data is critical to a complete energy management system database design.

1. **Rates data** are the complex pricing structure for regulated utility energy supplies or unregulated contract structures. For example, in addition to energy, meter, and demand charges, the largest customers now have time-of-day or hourly prices, fuel cost adjustments, myriad taxes, competitive transition charges, power factor adjustments, and other charge elements. In some cases, more than 20 different prices apply to create a single invoice.
2. **Real-time prices (RTP)** are streamed once per day for each hour and sometimes previewed the day prior, as well as adjusted in the hourly market during the day. The values have locational attributes too, since transmission congestion can provide an overlay cost to the otherwise wholesale market-clearing price.

Other Important Data

1. **Weather** is a key driver of energy consumption, therefore integrating weather data into the database design is critical to forming a relevant understanding of energy consumption and for forecasting future consumption. Wet bulb and dry bulb temperatures are most common, but wind speed and direction are sometimes useful data elements.
2. **Indexing by other consumption drivers** is necessary to provide insight into period-to-period or site-to-site comparisons, as well as for forecasting consumption into future periods. **Production data** are the most common by product line. In addition, **occupancy** is relevant for schools, hospitals or offices, and hours open for business is useful for retail, especially to compare periods where extended-hour sales would otherwise skew the analysis.

Database Selection

The database engine behind any energy information system is typically that of a relational design. There are a variety of ad hoc queries and complex data associations and manipulations that various data users and applications must perform. Choosing an appropriate database engine includes parameters such as:

Storage volumes

As mentioned previously, the massive storage volumes required for a large system will limit the choices to only a few commercially available engines. Large deployments can involve terabytes of data storage encompassing thousands of interval meters or millions of time-of-use or monthly meters. Challenges include not only physical hardware limitations but also the logistics associated with maintenance and backups of such large volumes.

Performance

Performance considerations arise in the design of the database in two basic areas: data loading and data access. Energy information systems need current data to operate effectively. When current data are analyzed in conjunction with historical profiles stored in the data base, energy managers can make informed decisions to reduce energy consumption and costs in anticipation of near term events such as rising temperatures, anticipated high spot-market energy costs or shortages in energy delivery capacity. Multiple streams of current data ranging from weather to real-time prices to energy consumption will need to be imported into the system, and the database implementation must be able to handle such volumes of data loading, often in real-time or in large batch processes. High performance is hindered by large batch processes, as there is an even more compressed processing window by which the data must be loaded into the system. Activating the appropriate processes produces proactive decisions and timely recommendations based upon real-time information.

After data are loaded into the system, automated processes and user requests on this data will compete for database resources in which to access the data. It is not uncommon for a web-based system to have dozens or even hundreds of simultaneous users actively accessing data from a system hosted by a large utility. Therefore, the database engine and the implementation of the applications surrounding it must be designed to handle large volumes of simultaneous usage, both loading data into the system as well as providing access to this data.

Scale Up/Down/Out

There are several architectures that may be considered when deploying an energy information system database. *Scaling Up* (Figure 22-1) is the practice of accommodating more capacity by increasing physical hardware resources on a single server. Such resources may include the addition of CPUs, memory or disk storage. Lower administrative costs are an advantage of scaling up, as the entire data subsystem is still contained in a single physical server. The disadvantage is that there is always a limit to the maximum number of resources that can be added to a single physical server. At the upper end, the vendor selection becomes extremely limited and hardware costs can become exorbitant, often outweighing the savings associated with reduced administrative costs.

Scaling down (Figure 22-2) is the practice of accommodating small systems requiring little storage capacity as well as being able to perform adequately on less costly hardware platforms. In such scenarios, the licensing costs of the database engine should also scale down appropriately. Vendors will support various database engines to meet such criteria only if their target markets encompass lower-end systems.

In *scaling out* (Figure 22-3), capacity is expanded by adding additional servers to the system. The vendor software must be developed specifically to support such distributed configurations, which may become quite complex. This architecture preserves initial capital expenditures for hardware and provides virtually unlimited capacity expansion capabilities, using lower cost hardware in aggregate than a scale-up server scenario to achieve the same level of capacity and performance. Disadvantages typically include the higher administrative and setup costs associated with configuring numerous physical machines.

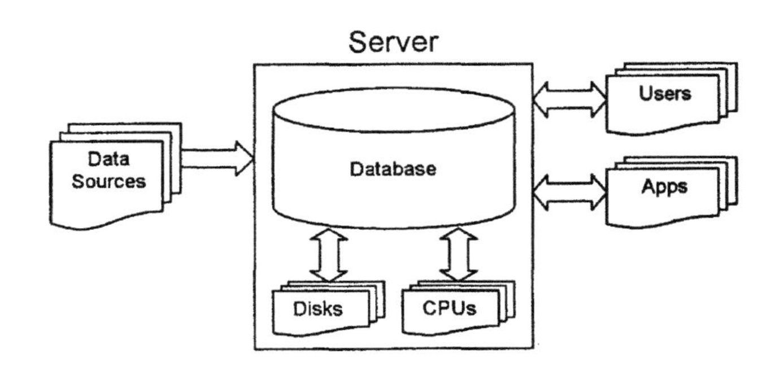

Figure 22-1. Scale-up

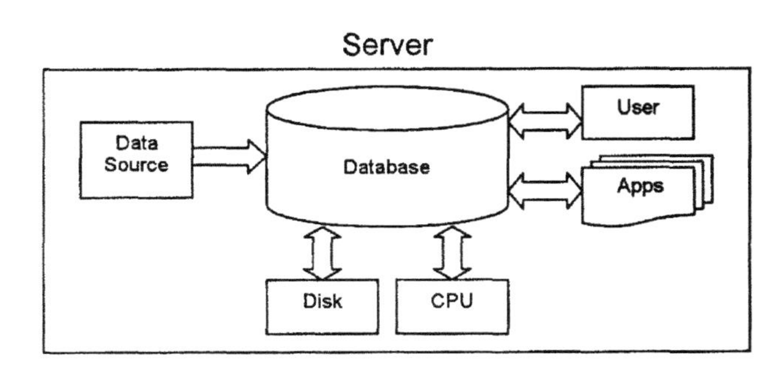

Figure 22-2. Scale-down

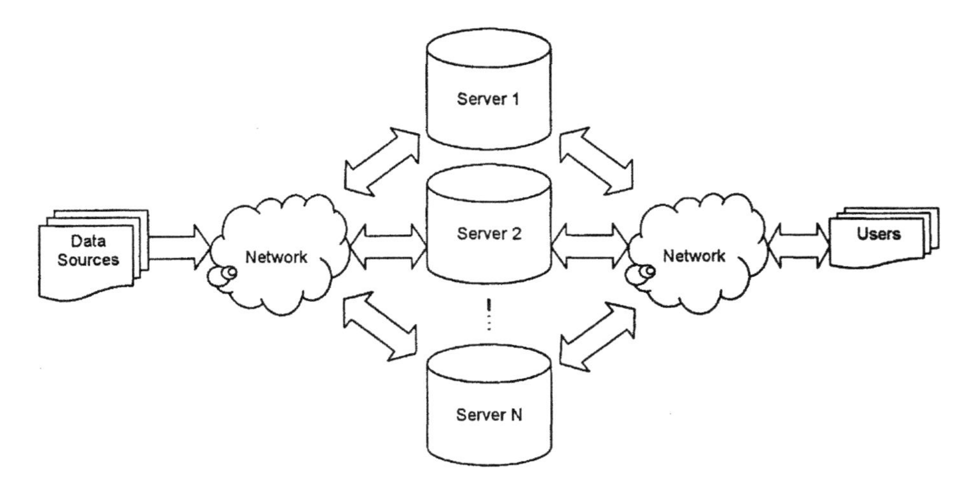

Figure 22-3. Scale-out

DATA PROCESSING

Energy information systems should not be limited to only the primary time-series energy consumption data they collect. There are several additional calculations that should be performed after the data collection process to extract the greatest benefit. The information needed for these calculations must be factored into the design of the database.

Alarms

As data samples arrive in the system, basic alarm threshold can be checked in order to alert end users to various important conditions. For example, the system may notify a user that an impending new demand peak is about to be reached; avoiding this peak can prevent costly energy rate ratchets. Different thresholds can be set for different periods of the day or for different seasons. If real-time energy prices are imported into the system, these values could trigger alerts to end users or to automated processes, which are designed to respond appropriately.

If alarm events are stored in the database, their history can be analyzed to produce additional important information. For example, by tracking energy consumption events for an on/off continuously cycling motor, a user can track the actual hours of motor operation. This alarm data (i.e. on/off alarms for the motor) can be extrapolated for use in preventive maintenance. As a result, users gain a more accurate metric for scheduling maintenance on the motor.

Synthetic Data: Calculations/Summations/Aggregations

Many times additional data can be derived from actual data collected in an energy information system. For example, the power factor can be calculated from individual measurements of wattage, voltage and current without the expense of a power factor sensoring device. Summations are calculations which sum data from individual samples together over a defined time window, e.g., calculating the total consumption between the hours of 8:00 a.m. to 5:00 p.m. Aggregation involves adding together data from different sensors for the same time interval, e.g., calculating the total load on a building at 12:00 p.m. by adding the load measured by individual sub meters at the building.

By storing data in the database for "synthetic" or derived data, alarm thresholds can be applied as well as historical analysis performed, just like any other actual data in the system that was generated by a physical sensor.

OLAP/Warehousing/Aggregation/ETL

Although relational database engines are necessary for supporting ad hoc queries of data involving complex relationships, often there are known relationships that need to be reported involving large volumes of data in which other technologies are better suited. One such technology is on-line analytical processing (OLAP) which creates and populates "cubes" of data that are optimized specifically for known queries involving fixed and predetermined relationships. OLAP technology is common in large data warehousing applications where large volumes of data can be summarized or "aggregated" for specified time intervals. Time-series energy information is a natural fit for this technology as data summaries of this information are desired for daily, weekly and monthly time boundaries for identifying consumption totals or peak demands.

In addition to aggregation, OLAP technology is also optimized for drill-down capabilities so that a user can extract increasing orders of detail for a single meter channel, e.g., drilling down from a monthly summary to a weekly, daily or even to an hourly profile. None of this is automatic however. The energy information system must specifically have processes designed to move and process the data from the relational storage into the appropriate OLAP cubes. Such a process is referred to as the extraction, transformation and load (ETL) process (Figure 22-4.).

Data residing in either the relational database engine or the OLAP engine should be accessible to the end-user reporting applications via a common medium that hides the nuances of their respective data access requirements. Such a medium is commonly referred to as a data access layer (DAL). Proper design of the DAL ensures that the storage mechanics of the data are properly isolated away from the consuming applications of the data.

DATA ACCESS AND REPORTING

Data navigation/hierarchy

Users may have difficulty specifying the data to report on and navigating an energy information application in a large system unless some form of hierarchical data navigation or filtering scheme is deployed. The best approach for specific end-users is to filter the data that is accessible to them, but for a system operator who must access all the data, the task can be quite daunting. Using search criteria or a hierarchical tree navigation scheme are typical paradigms employed to assist in such tasks.

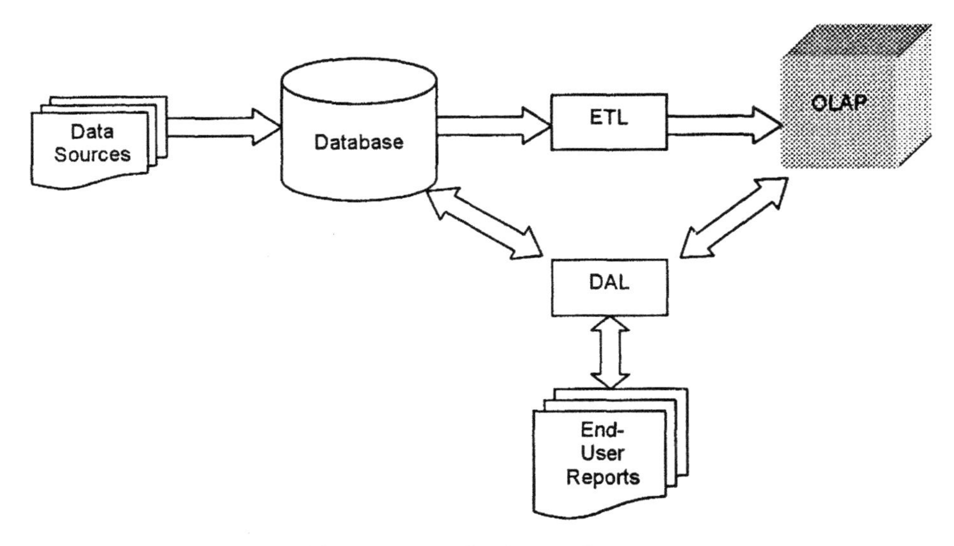

Figure 22-4. OLAP and ETL

The database implementation of either paradigm becomes challenging when the amount of discrete data accessible to a user is quite large, for example, discrete data points can number in the millions for a system operator for a large utility deployment. In addition, a user security mechanism must be fully integrated with the navigation scheme. This is probably a greater challenge than management of the time-series data itself.

From an end-user perspective, the primary measure of the application's quality is speed and ease of accessing information, secondary only to the quality of the information itself.

Data Access Layer—DB Engine Independence

As mentioned earlier, the data access layer (DAL) provides a layer of separation between the consuming applications of the data and their storage engine specifics (Figure 22-5.). This means that those who are developing the applications do not need specific knowledge of the underlying engine technologies, and can use engines from more than one vendor. Database engine selection can also influence a customer's choice of application vendors. Proper design of a DAL can assist in providing support for a customer's chosen database engine, thus eliminating the problem.

Software Layer	Examples
Presentation Layer	Web Browser, Thick-Client etc.
Business Application Layer	Reports, Alarming, VEE, etc.
Data Access Layer	
Data Services Layer	SQL Server, Oracle, OLAP etc.

Figure 5. Typical Layered Application

DATA SOURCES, QUALITY AND INTEGRATION

Third-party Integration

With any vendor application, support for integration with other vendor applications is imperative; no single vendor can reasonably be expected to accommodate every user-desired function pertaining to the data or information contained within a particular application. It is also virtually impossible to modify a vendor application to specifically integrate with another application for which there is no explicit control over the integration effort or support for ongoing maintenance of the specific modification by both vendors. To enable such integrations, vendors should support a standard data interchange format or a custom application programming interface (API). Each of the respective vendors should ensure that such interfaces continue to work and are supported with ongoing releases.

The concept is simple: by adhering to a supported protocol for data interchange between two parties, each party has the ability to change implementations behind the scenes without the other's knowledge or necessary cooperation, as long as both parties continue to agree to communicate with each other only through the supported data interchange protocol. Another important aspect of the data interchange protocol is an agreement of how to identify different pieces of data, whether by a standard naming convention, a unique number or symbol, etc.

Various technologies can be deployed in this regard including standard electronic data interchange (EDI) formats, public domain communication protocols, simple object access protocol (SOAP), web services, etc. In addition, third-party data hubs supporting these technologies may be appropriate to broker such data exchanges between disparate vendor applications, e.g., BizTalk, RosettaNet, Tibco Rendezvous, etc. Much like performance considerations, data integration between disparate systems manifests itself in the same two basic areas as well: data loading and data access, or in other words, data import and export respectively.

Although it is simple and straightforward to use a variety of third-party tools and technologies to access data directly from tables stored by the database engine within the energy information system, this practice violates the basic premise behind the data interchange concept. The integration is tied to a specific database schema which could be changed by the vendor, or the owner. This could mean that the access programs no longer work as expected.

Change Tracking

The information related to the configuration and environment of an energy information system is important. For example, a user may wish to track meter change-outs at a particular site or the addition of new customers to a particular distribution feeder, which is especially important when using the system for load planning. Errors in specifying multiplier constants, vendor type, etc. may occur when configuring a meter. Therefore, the ability to determine when a correction was made is important so that prior recorded data can be appropriately adjusted.

When used in distribution planning, an energy information system must be able to provide "snapshots" of the system state for particular periods of time. To ensure that the snapshots are from a "normalized" system, a change-tracking mechanism must be deployed across various data tables in the database or to various elements within a table. Various methods can be used from simple logging of changes to more elaborate designs that enable the automated tracking of specific changes.

VEE

Validation, editing and estimation (VEE) consists of various functions that are applied to time-series energy data samples after data collection but before delivery of that data to an end-user or external system, e.g., a utility billing system. Validation consists of performing various reasonability checks against the data that typically include high/low threshold checks and spike checks. Editing is the system's ability to support manual or automated data edits for data corrections. Estimation is the practice of estimating the values of missing data samples. Here, the algorithms typically vary by utility or associated Public Utility Commission and commonly involve averaging against historical data for the same meter and channel by day of week, month or season. With such variety, a VEE engine must be able to accommodate the addition of new algorithms as they are created and mandated by the utility company.

With any change to data samples as a result of any or all of the functions available within the VEE engine, tracking of such changes for auditing, regulatory or legal purposes becomes a design issue in database implementation. For example, typical audits may involve tracking changes to data for a particular meter, a particular billing cycle, a particular VEE function or even an individual customer or site.

Maintenance

With any database-centric system that involves an ever-increasing amount of data, or is subject to substantial additions or changes to data over time, regular database maintenance is a necessary function. Maintenance may range from simple backups to selective truncations of data tables to maintenance procedures on the data tables themselves, such as re-indexing, optimizations, etc. Maintenance procedures are best handled automatically on a predefined schedule, but typically involve at least some manual intervention for backup media change outs. The application may contain its own built-in maintenance procedures especially when application-specific functionality is concerned (e.g., truncation of a data history table). However, most people use off-the-shelf generic database maintenance and backup programs, either contained within the database engine itself or available from a third-party vendor.

Security

Protecting energy information has become more sensitive over time as such information can reveal much about the inner workings of a facility, its business practices, or the habits of the occupants, particularly residential premises. For example, the electrical load profile of a residence can indicate when the home is least likely to be occupied during the day—information that could be useful to those planning criminal activity. For commercial facilities, energy information can give competitors insight about operational costs. Keeping such data secure is fundamental to any energy information system.

Authenticated access to the database and its data should be required at both the system level and the application level. System level security relates to securing the access to the database to authorized applications only. Most commercially available database engines on the market today implement system level security within the engine itself or integrate it into the same system level security that is used by the operating system. Some database engines can use either method concurrently. The advantage of using the same security method for access to both the database and the operating system is that the administrator will only need one set of authentication credentials to maintain and administer the database.

Application level security refers to limiting access to only certain elements of the database or to certain pieces of data that have been authorized for a particular end-user, typically via the credentials that have been manually supplied by the end-user upon a log-in to the application. Application level security can become quite complex to implement and administer as the business rules for limiting end-user access may be hierarchical, linked to various user characteristics, grouped with other classes of end-users, or limited by various time periods of the day, etc.

With application level security, it is also possible to integrate the authentication mechanism with that used by the operating system. Integration of application security with the operating system is typical for intranet applications where there are benefits for having users only needing to log on once in order to access multiple applications. However, for an internet application, it is advantageous to have separate credentials for the internet application so that those credentials cannot be used to access other applications and resources maliciously or inadvertently. In such circumstances, the application is typically responsible for validating and storing the authentication credentials. Great care needs to be taken during implementation to ensure the credentials themselves have been secured. For example, it is not uncommon for end-users to use the same passwords for accessing disparate systems. A common technique is to use a one-way hash algorithm (e.g., MD5) to encode the end user passwords. In this fashion, even the system administrator will not be able to gain access to the users' original passwords. The log-in portion of the application is responsible for using the chosen algorithm to convert the password before comparing it to the hash value already stored for the user in the system, then approving the user for access to data in the system.

The log-in application itself should also be secured against various forms of "hacking" by using a variety of techniques, including but not limited to encrypting the authentication handshake itself, using public-key/private-key technologies, and lockouts for repeated unsuccessful attempts, etc.

SUMMARY

Databases are a necessary component of today's energy management systems. The amount and variety of data that must be stored, and the energy management process that are supported by the database and supported applications, means that careful planning must go into designing the database. Accessibility of the data is an important consideration. Questions of who can access it, how it can be accessed and how much of it can be accessed must all be addressed. Finally, security of the data and the database is a paramount concern. A properly designed database can be a valuable tool for energy and facility managers.

Section Eight

Techniques for Utility Data Web Page Design

Chapter 23

Utility Data Web Page Design—An Introduction

David C. Green, Green Management Services, Inc.
Paul J. Allen, Walt Disney World

IN ORDER TO MAKE a conscious effort to conserve energy at large scale facilities it is necessary to track utility use on a daily basis. An added benefit of utility tracking is that discovering and troubleshooting equipment malfunctions becomes much easier. Utility data has unique and complex characteristics. Collecting, analyzing and publishing utility data to a large number of people puts more eyes on the task of managing utilities. Different people use the data in different ways. Internet technologies allow us to publish utility data, in a meaningful way, to all of those people simultaneously. This chapter describes the characteristics of utility data and utility data users. It also introduces some web programming languages to show how web pages can be designed effectively.

INTRODUCTION

Utilities seem to be the substance of our existence. Almost every minute of our busy lives we consume some sort of utility. Electricity, water, natural gas, fuels and others all play an important part in our lives. To track the use of these commodities minute by minute is a task vital to the goal that we conserve and use these resources as efficiently as possible. The amount of utility data collected everyday is vast. The data itself is somewhat complex in nature due to the multitude of utility types and means of measurement.

The philosophy, "If you can measure it, you can manage it," is critical to a sustainable energy management program. Continuous feed back on utility performance is the backbone of an energy information system (EIS). At the center of today's energy information systems is the web technology that publishes data to a wide audience on a daily basis. Energy managers, technicians, department managers and directors all become involved in utility monitoring by using these systems.[1]

The capability and use of Information Technologies and the internet in the form of web based energy information and control systems continues to grow at a very rapid rate. Utility data can be published on an intranet or the internet using client/server programming. The data are stored on a central computer, the server, and waits passively until a user makes a request for information using a web browser, the client. A web publishing program retrieves the information from a relational database, and sends it to the web server, which then sends it to the client that requested the information.[1]

The purpose of this chapter is to introduce how to publish utility data using a web page. We explain the characteristics of utility data, who needs to see the data, and why. Then we explain the web page design requirements and describe some of the internet technologies used to present utility data.

CHARACTERISTICS OF UTILITY DATA

Utility data have unique characteristics that make web applications especially suited to reporting it. First, utility data are usually collected on an hourly basis, if not even more often. This creates a huge amount of data. Only a powerful database application can store and manipulate that much data. EISs collect data from many different locations and report it on different levels. The data are usually date specific and requires trending over time to be useful. The data comes from many types of utilities and is collected in a variety of units of measure.

Systems collect utility data quite often from many different locations. It may need to be broken down into groups for reporting purposes. Looking at the utility usage for different "levels" is a reasonable expectation. The electric consumption of a single building can be just as important as for a whole complex of buildings. It might also be necessary to combine data from several meters and report it as one data point or fractionalize data from one meter and report it as more than one data point. EISs record all data with a time stamp. Figure 23-1 is a diagram of a typical complex of buildings and the some of their data.

The data are only meaningful when presented in the context of time. Hourly data for specific days and daily data spread over the period of a month provide the most meaningful reports and graphs. Comparisons to other data, such as outside temperature, are also helpful. Sometimes averaging data or summing for each utility is required. EISs can average or sum data over periods of time or by combining meters at various locations.

Figure 23-2 is a graph of average hourly electric consumption for all of Building-1 over a 24-hour period along with average outside temperature.

Figure 23-3 is a graph of total daily electric consumption for one meter in Building-1 over a 30-day period.

A typical facility can produce data for every electric, water, and gas meter as well as others. Temperature sensors collect temperature data. Different utilities produce data in different units of measure. Tracking the units at the lowest level is important so that applications sum and average data over time correctly. It is important to keep in mind the units of measure may change as metering requirements or reporting requirements change. Table 23-1 shows some examples of utility data units of measure.

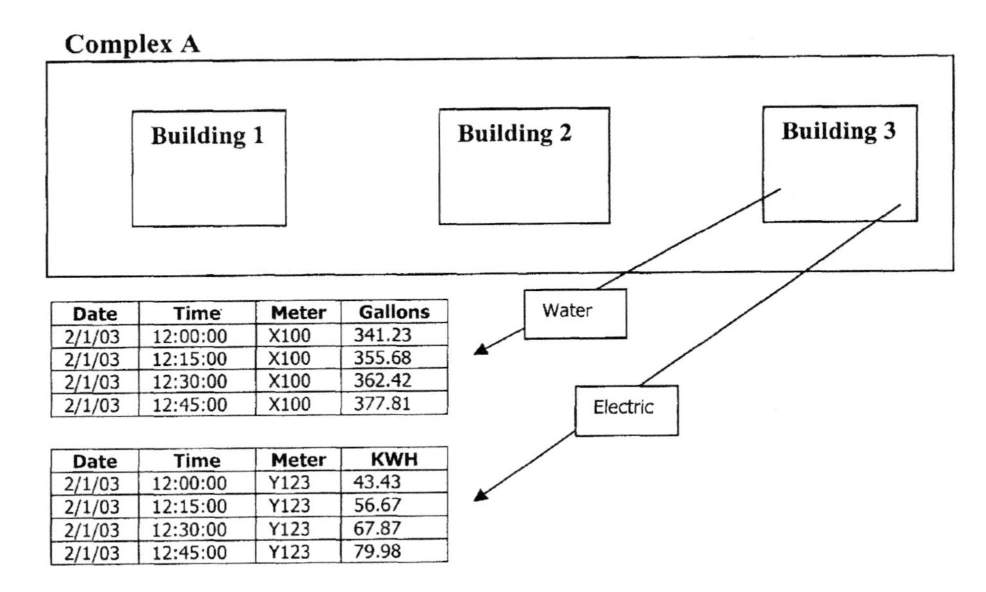

Date	Time	Meter	Gallons
2/1/03	12:00:00	X100	341.23
2/1/03	12:15:00	X100	355.68
2/1/03	12:30:00	X100	362.42
2/1/03	12:45:00	X100	377.81

Date	Time	Meter	KWH
2/1/03	12:00:00	Y123	43.43
2/1/03	12:15:00	Y123	56.67
2/1/03	12:30:00	Y123	67.87
2/1/03	12:45:00	Y123	79.98

Figure 23-1. Data Collection

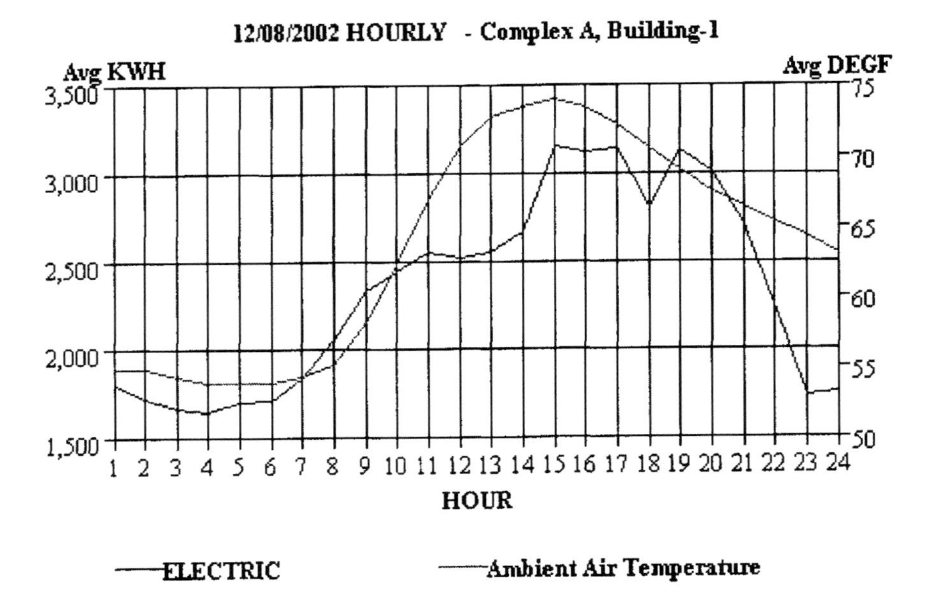

Figure 23-2. Average Hourly Electric Consumption vs. Average Outside Temperature

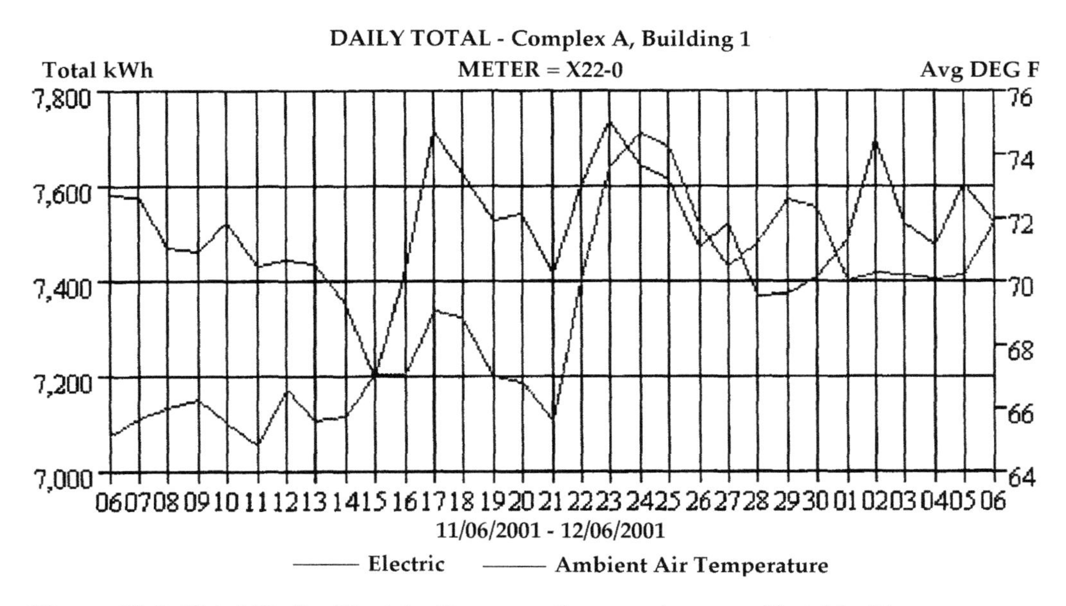

Figure 23-3. Total Daily Electric Consumption vs. Average Outside Temperature

Table 23-1. Utility Data Units of Measure

UTILITY	*UNITS OF MEASURE*
Hot Water	MMBtu, kBtu
Cold Water	gal, kgal
Natural Gas	kscf
Electric	kWh, MWh
Compressed Air	kscf, scfm
Flow	gpm
Pressure	psi
Fuel oil	gal
Chilled Water	tonhrs
Temperature	Degf degc
Efficiency	kW/ton

USING UTILITY DATA

Uses for utility data are as diverse as the users. Energy managers probably look at this data more than any other users do. An energy manager might look for spikes in the data signifying a mechanical failure. Department managers will look for conservation opportunities in hourly data and track their performance month to month. Facility directors may be interested in some critical data effected by temperature extremes or economic circumstances. Technicians typically respond to evaluations of data results but may find it helpful to "zoom in" on data to get an idea about what a particular piece of equipment is doing.

Energy managers must communicate with maintenance and operations personnel on a daily basis. It is helpful in communicating to know the performance data of the equipment and systems involved. Energy managers use the data to evaluate the performance of systems such as HVAC. They compile, maintain and distribute reports from the data. These reports help to identify conservation opportunities and the effectiveness of conservation measures already taken. They also may use the data to identify equipment malfunctions.

Department managers set procedures in place to conserve energy and then use the utility data to monitor the progress of those procedures. They can identify places and times to improve conservation efforts. Monitoring utility data also provides them an opportunity to watch for abnormalities in equipment performance.

Directors use utility data to track critical areas of concern that may effect budget requirements. They try to use the data to predict the future impact of certain conditions. They plan according to trends in data. Directors might use the data to report on reasonable expectations of costs for future development.

Technicians typically respond to action requests brought on by evaluating the utility data. They then use the data to monitor detailed system functions and make adjustments accordingly. They may use the data to set procedures for operating and maintaining the equipment properly.

DESIGNING WEB PAGES FOR UTILITY DATA REPORTS

Web pages that display utility data must be informative, intuitive, reliable and robust. The display should have as much data as can fit on the page and still be readable and printable. Pages that provide as many ways as possible to change the presentation of the data quickly and easily are robust. A good intuitive design provides results with the user having to make only a few choices. Developers should create pages in such a way as to be compatible with many types of browsers.

Because utility reports involve so much data, it is impossible to display all of it at one time. The source of the data as well as its utility type and units must be displayed. Other values to show costs, averages, efficiencies, etc. are all more meaningful when set alongside the data values themselves in one report. Figure 23-4 shows how this report might look.

Sorting and filtering from links in the display lets one easily change the presentation with few selections. *Sorting links* associated with the headings change the order in which the rows of the report are displayed. Selecting a heading quickly sorts the report by that column. *Filtering links* can be associated with any value in the report. However, they are most useful as a way to focus in on a single source. Selecting a source from the list, such as METER X122-0, would eliminate all rows except those with a source equal to that selected. *Option links* provide the ability to change values instantly. These methods provide an intuitive utility data web page design.

Hypertext Markup Language (HTML)

Using well-proven and accepted web page creation tools is the way to insure reliability in a web page. There are many ways to present data using a web page but not all browsers interpret those methods the same. Using basic hypertext markup language (HTML) and following with simple enhancements will insure reliable results. You create a file and put special character sequences called *HTML elements* into your file. These elements identify the structural parts of your document. When a web browser displays the file, it will display the content, but not the characters that make up the structure.[2] Below is a very simple HTML example with HTML element tags.

<html>
<head>
<title>
A Small Hello
</title>
</head>
<body>
<h1>Hi</h1>
<p>This is very minimal "hello world" HTML document.</p>
</body>
</html>

This is what the HTML page above looks like when displayed in a browser:

Hi

This is a very minimal "hello world" HTML document.

One way to learn HTML is to go through the tutorial on the W3Schools web site at http://www.w3schools.com/html. Another way is to enroll in a web-page design class at a local community college.

Utility Report 12/08/2002

COMPLEX A

BUILDING-1 - ELECTRIC	UNITS	DAILY Total	7-DAY AVG	% DIFF	COST
METER X122-0	KWH	948	927	2.2 %	$62.83
METER X122-1	KWH	828	900	8.0 %	$54.89
WEATHER - TEMP	**UNITS**	**DAILY Avg**	**7-DAY AVG**	**% DIFF**	**COST**
AMBIENT AIR TEMPERATURE	DEGF	63.64	58.02	9.7 %	

Figure 23-4. Utility Report

Extensible Hypertext Markup Language (XHTML)

Extensible hypertext markup language (XHTML), the latest version of HTML, provides even more reliability with its strict rules. XHTML gives you the opportunity to write "well-formed" documents now, that work in all browsers and that are backward browser compatible. Developers can validate XHTML pages for correctness using tools available on-line. Below is an example of a very simple XHTML file.

```
<?xml version="1.0" encoding="iso-8859-1"?>
<!DOCTYPE html PUBLIC "-//W3C//DTD XHTML
      1.0 Transitional//EN" "DTD/xhtml1-
      transitional.dtd">
<html xml:lang="en" lang="en" xmlns="http://
      www.w3.org/1999/xhtml"> <head>
            <title>Hello World</title>
      </head>
      <body>
            <p>My first Web page</p>
      </body>
</html>
```

The XHTML page above looks like this when displayed in a browser:[3]

My first Web page.

You might also consider completing the XHTML tutorial on the W3Schools web site at *http:/ www.w3schools.com/xhtml*.

Cascading Style Sheets (CSS)

The World Wide Web Consortium (W3C), the non profit standard setting consortium responsible for standardizing HTML and XHTML, created STYLES in addition to HTML 4.0. Styles define how to display HTML elements. Using cascading style sheets (CSS) can minimize the amount of HTML code needed to display data and therefore make the pages less complex, load faster and easier to maintain. Below is a very simple CSS example.[4]

```
<head>
      <style>
            b {color: red}
            i {color: blue}
      </style>
</head>
<body>
This is how to use <b>CSS</b> to brighten up your
      <i>World</i>!
</body>
```

The page looks like this when displayed in a browser:[5]

This is how to use CSS to brighten up your *World*!

You can find a CSS tutorial on the W3Schools web site at *http://www.w3schools.com/css*.

Dynamic HTML (DHTML)

HTML 4.0 also introduced the document object model (DOM). The DOM gives us access to every element in the document. By using HTML, CSS and JavaScript web pages appear to be "dynamic." Dynamic HTML (DHTML) is not a standard defined by the World Wide Web Consortium (W3C). DHTML is a combination of technologies web developers use to control the display and position of HTML elements in a browser window. Below is a simple example of DHTML.[6]

```
<html>
<body>
<h1 id="header">My header</h1>
<script type="text/javascript">
header.style.color="red"
</script>
</body>
</html>
```

The page looks like this in the browser:

My header

The DHTML tutorial on the W3Schools web site is at *http://www.w3schools.com/dhtml*.

Sorting, Filtering and Graphing

The headings for the data can be designed with links to sorting, filtering and graphing routines. Links in the display provide plenty of robust functionality to the user.

The utility heading in the display at the top of page 272 has a link to a sorting routine.

The next display shows that the rows have been sorted alphabetically by utility. You can then select the facility heading which links to a filter routine that will filter the results into only those for Building 1.

Select this link to sort by utility

Facility	Utility	Consumption	Units	Cost
BUILDING-1	ELECTRIC	6480.33	KWH/DAY	$ 18144.92
BUILDING-1	WATER	6291.14	GALLONS/DAY	$ 4592.53
BUILDING-2	ELECTRIC	6938.27	KWH/DAY	$ 19427.16
BUILDING-2	WATER	6653.16	GALLONS/DAY	$ 4856.81

Select this link to filter for Building 1

Facility	Utility	Consumption	Units	Cost
BUILDING-1	ELECTRIC	6480.33	KWH/DAY	$ 18144.92
BUILDING-2	ELECTRIC	6938.27	KWH/DAY	$ 19427.16
BUILDING-1	WATER	6291.14	GALLONS/DAY	$ 4592.53
BUILDING-2	WATER	6653.16	GALLONS/DAY	$ 4856.81

The display below shows values filtered for Building-1. The Consumption heading has a link to a graphing routine.

Select this link to graph

Facility	Utility	Consumption	Units	Cost
BUILDING-1	ELECTRIC	6480.33	KWH/DAY	$ 18144.92
BUILDING-1	WATER	6291.14	GALLONS/DAY	$ 4592.53

Figure 23-5 below shows the result of the graphing link: a graph of electric consumption per day for Building-1 over the last 30 days.

Extensible Markup Language (XML)

Data stored in the form of extensible markup language (XML) allow for quick access to data attributes and the ability to customize the display based on those attributes using CSS.

Appendix A illustrates an example of an XML file with its schema included at the top of the file.

Appendix B illustrates an example of a CSS file:

Below is what the two files from Appendices A and B above look like in a browser:[7]

AH 101 ELECTRIC KWH
455.34 2001-09-13T11:30:06
578.29 1002-09-13T11:45:04
479.30 2001-09-13T12:00:05
AH 102 ELECTRIC KWH
347.59 2001-09-13T11:30:06
498.60.2001-09-13T11:45:04
444.57 2001-09-13T12:00:05

You can see how these files work together by visiting the UtilityReporting.com web site.

- The XML file is at *http://utilityreporting.com/utilitydata.xml*
- The CSS file is at *http://utilityreporting.com/utilitydata.css*
- The combination of the two files is at *http://utilityreporting.com/utilitydata2.xml*

You might also want to visit the XML tutorial on the W3Schools web site at *http://www.w3schools.com/xml*.

Server-side Technologies

Common gateway interface (CGI) provides a convenient method of running the intensive server operations required for utility data analysis using a web browser. The first generation of web server-side technologies was CGI. Many developers use the PERL programming language to create CGI applications. They also use nearly any higher level language. This technology employs server-side applications to complete a predefined task when the client makes a request. The following line from the address field of a web browser runs a server application to query a database according to the parameters and return the results to the browser.

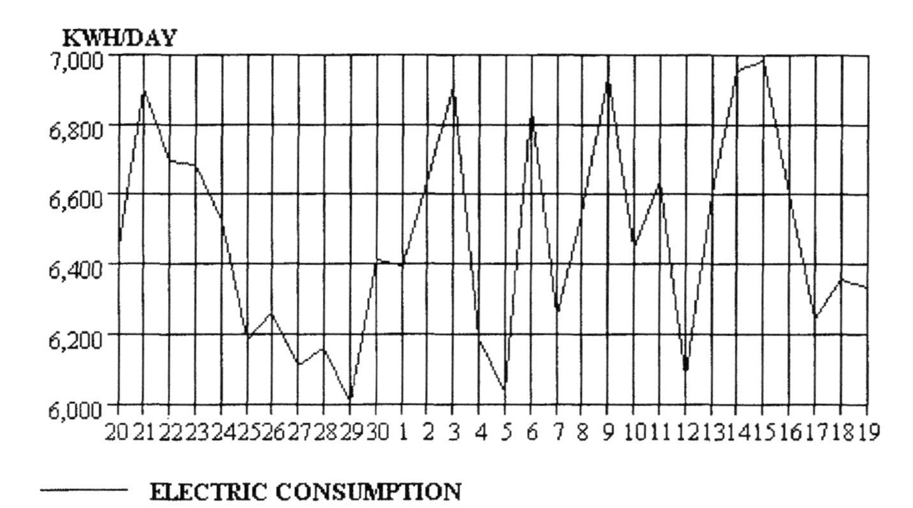

Figure 23-5. Electric Consumption over the Last 30 Days

http://localhost/scripts/foxweb.exe/
A1@view?sortby=cost~date=12/08/
2002~complex=A~building=1~level=building~

The next generation of server-side technologies is server-side scripts such as Active Server Pages, Java Servlet/Java Server Pages, PHP, and others. These technologies allow the server to generate a thread to handle a client's request. The thread is light-weighted and consumes much less resources than the application generated by CGI. On the other hand, server-side scripting languages often provide only a subset of the capabilities of full programming languages.[6]

Client-side Technologies

Unlike server-side technologies, client-side technologies execute tasks at the client-side. There are a number of popular client-side technologies available to implement extra functionality on top of HTML. Three of the major client-side technologies are client-side scripts, Java applets and ActiveX controls. Client-side scripts such as JavaScript, VBScript and Jscript are a collection of commands interpreted, rather than compiled, by the web browser. A Java applet is a Java program downloaded from a web server into client browsers and runs inside a Java Virtual Machine (JVM) that is integrated into the web browser. The term ActiveX generally refers to ActiveX controls that are software modules based on Microsoft's Component Object Model (COM) architecture. The power of ActiveX controls in web development is that any ActiveX-compliant web browser can download the controls by clicking on its link. This virtually turns web pages into software pages that can perform operations just like any desktop application.[8]

Below is an example of JavaScript:

```
<html>
<body>
<script type="text/javascript">
document.write("Hello World!")
</script>
</body>
</html>
```

This is what the above page looks like in a web browser:

Hello World!

Configuration Tables

Configuration tables allow for color definitions, text lookup and runtime adjustment of critical constants. See Table 23-2 for an example configuration table.

All of this leads to a more robust look and feel. Links allow the user to change the display with a single selection. HTML and XHTML provide reliable presentation of the data in the web browser window. XML combined with CSS provides a direct association between the raw data attributes and the presentation of the data. JavaScript manipulates elements of the document object model (DOM). Configuration tables allow developers to make major changes in a relatively short time with very little effort.

Table 23-2. Configuration Table

LANGUAGE	*NAME*	*VALUE*
English	Title	Energy Information System
English	Facility	Facility
English	Utility	Utility
English	Consumption	Consumption
English	Units	Units
English	Cost	Cost
English	Background color	#A5D1D1
English	Text color	#000000
English	Link color	#0000FF
French	Title	Système D'Information D'Énergie
French	Facility	Service
French	Utility	Utilité
French	Consumption	Consommation
French	Units	Unités
French	Cost	Coût
French	Background color	#A5D1D1
French	Text color	#000000
French	Link color	#0000FF

CONCLUSION

Facilities have an obvious need to track utility use in order to conserve and troubleshoot equipment problems. The more people that are involved in the process the easier the task. Good web page design and web publishing technologies make successful utility tracking a bright reality.

Making good use of basic HTML or XHTML along with XML combined with CSS can produce plenty of good reports. DHTML used sparingly can make the pages user-friendlier. CGI scripts are a simple and effective way to connect web pages with server applications. Organizing utility data on a web page in tabular form with links to sort, filter and graph provides an informative view toward utility monitoring. This allows people to make informed decisions in a timely manner concerning utility use and conservation.

An excellent source for learning these web publishing technologies is W3schools.com at http://www.w3schools.com. There you will find free web-building tutorials from basic HTML and XHTML tutorials to advanced XML. A number of books are available on the subject. A good place to start is the book *10 Minute Guide to HTML* by Tim Evans, published by QUE Corporation. Local community colleges might offer classes as well. Other chapters in this book discuss in detail web publishing technologies required for good utility data web page design and how to use those technologies to present utility data in way to make maximum use of its unique characteristics described above.

References

[1] Allen, Paul J and David C. Green, "Managing Energy Data Using an Intranet—Walt Disney World's Approach," Proceedings of the 2001 World Energy Engineering Congress Conference & Expo, Atlanta, Ga., October 2001.

[2] December, John, "Creating Web Documents"[article on-line] (January 20, 2003, accessed 12 February 2003); available from http://www.december.com/html/tutor/hello.html; Internet.

[3] UK Web Design, Internet Marketing, Usability Testing; "XHTML Web Design for Beginners: Hello World"[article on-line] (2003, accessed 12 February 2003); available from http://www.miswebdesign.com/resources/articles/web-design-xhtml-1-2.html; Internet.

[4] "Introduction to CSS"[article on-line] (2003, accessed 27 February 2003); available from http://www.w3schools.com/css/css_intro.asp; Internet.

[5] Wisman, Raymond F., "Cascading Style Sheets"[article on-line] (December 13, 2002, accessed 12 February 2003); available from http://homepages.ius.edu/RWISMAN/A346/html/CSS.htm; Internet.

[6] "Introduction to DHTML"[article on-line] (2003, accessed 27 February 2003); available from http://www.w3schools.com/dhtml/dhtml_intro.asp; Internet.

[7] Green, David C., "UtilityReporting.com"[article on-line] (January 12, 2003, accessed 12 February 2003); available from http://utilityreporting.com/utilitydata.xml; Internet.

[8] Fangxing Li, Lavelle A. A. Freeman, "Using Client-side Technologies to Develop Web-based Applications for Power Distribution Analysis," Distributech 2003 Conference, February 4-6, 2003, Las Vegas, Nevada.

APPENDIX A

```
<?xml version="1.0" ?>
<UtilityData>
<Schema name="Schema" xmlns="urn:schemas-microsoft-com:xml-data" xmlns:dt="urn:schemas-microsoft-com:datatypes">
<ElementType name="AirHandlers" >
<element type="AirHandler"/>
</ElementType>
<ElementType name="AirHandler" >
<element type="AirHandlerNo"/>
<element type="Utility"/>
<element type="Units"/>
<element type="Readings"/>
</ElementType>
<ElementType name="AirHandlerNo" dt:type="string">
<AttributeType name="type"/>
<attribute type="type" default="string"/>
<AttributeType name="size"/>
<attribute type="size" default="8"/>
</ElementType>
<ElementType name="Utility" dt:type="string">
<AttributeType name="type"/>
<attribute type="type" default="string"/>
<AttributeType name="size"/>
<attribute type="size" default="30"/>
</ElementType>
<ElementType name="Units" dt:type="string">
<AttributeType name="type"/>
<attribute type="type" default="string"/>
<AttributeType name="size"/>
<attribute type="size" default="10"/>
</ElementType>
<ElementType name="Readings" >
<element type="Reading"/>
</ElementType>
<ElementType name="Reading" >
<element type="Consumption"/>
<element type="DateRecorded"/>
</ElementType>
<ElementType name="Consumption" dt:type="float">
<AttributeType name="type"/>
<attribute type="type" default="float"/>
<AttributeType name="size"/>
<attribute type="size" default="15"/>
<AttributeType name="precision"/>
<attribute type="precision" default="2"/>
</ElementType>
<ElementType name="DateRecorded" dt:type="datetime">
<AttributeType name="type"/>
<attribute type="type" default="datetime"/>
<AttributeType name="size"/>
<attribute type="size" default="8"/>
</ElementType>
</Schema>
<AirHandlers xmlns="x-schema:#Schema">
<AirHandler>
<AirHandlerNo>AH 101</AirHandlerNo>
<Utility>ELECTRIC</Utility >
<Units>KWH</Units >
<Readings>
<Reading>
<Consumption>455.34</Consumption>
<DateRecorded>2001-09-13T11:30:06</DateRecorded>
</Reading>
<Reading>
<Consumption>578.29</Consumption>
<DateRecorded>2001-09-13T11:45:04</DateRecorded>
</Reading>
<Reading>
<Consumption>479.30</Consumption>
<DateRecorded>2001-09-13T12:00:05</DateRecorded>
</Reading>
</Readings>
</AirHandler>
<AirHandler>
<AirHandlerNo>AH 102</AirHandlerNo>
<Utility>ELECTRIC</Utility>
<Units >KWH</Units>
<Readings>
<Reading>
<Consumption >347.59</Consumption>
<DateRecorded >2001-09-13T11:30:06</DateRecorded>
</Reading>
<Reading>
<Consumption >498.60</Consumption>
<DateRecorded >2001-09-13T11:45:04</DateRecorded>
```

```
</Reading>
<Reading>
<Consumption >444.57</Consumption>
<DateRecorded >2001-09-13T12:00:05</DateRecorded>
</Reading>
</Readings>
</AirHandler >
</AirHandlers>
</UtilityData>
```

APPENDIX B

```
UtilityData
{
background-color: #ffffff;
width: 100%;
}
AirHandlers
{
display: block;
margin-bottom: 12pt;
margin-left: 0;
}
AirHandler
{
color: #FF0000;
font-size: 10pt;
}
Reading
{
Display: block;
color: #000000;
font-size: 10pt;
margin-left: 10pt;
}
```

Chapter 24

Utility Data Web Page Design—Learning Technologies

David C. Green, Green Management Services, Inc.
Fangxing Li, ABB Inc.

CREATING WEB PAGES requires the use of at least a few of the several technologies available to developers today. Reliable, robust, informative and intuitive web pages displaying utility data can be created using hypertext markup language (HTML) and common gateway interface (CGI). Other technologies enhance the presentation and functionality of the web pages. We discuss these technologies from the viewpoint that a great deal of application development today is undertaken through successive revisions. We present the technology descriptions along with a simple example of each to get a developer started or simply for better understanding. This is not an exhaustive guide to using each of the technologies but rather an "opening the door" approach. We then provide references and links to resources for further study.

INTRODUCTION

In order to web-enable utility data, the user interface (UI) must be embedded in a web browser like Microsoft Internet Explorer (IE) or Netscape Navigator. Existing and recently released technologies to implement this are:

- Plain HTML
- Server Side Scripts such as Common Gateway Interface, Active Server Pages, Java Server Pages, Java Servlets, and PHP
- Client Side Scripts using JavaScript and VBScript
- Java applets
- ActiveX controls

HTML and dynamically generated web pages are usually employed to create a UI embedded in a web-browser. In general, HTML is only able to display data and submit simple requests. Server or client side scripts can be a good choice for non-static web pages requiring more user interaction, database support, and/or constant updates.

Current scripting technologies can be used to develop and process UI components like forms, selection boxes, buttons, and other simple human-computer interaction (HCI) components. A Java applet is a program that is downloaded from a web server into client browsers and runs inside a Java Virtual Machine (JVM). Applets are almost as powerful as stand-alone applications and are therefore able to locally process intense user interactions such as drawings, which are often difficult or impossible to implement with HTML and scripts. Browsers like IE may provide their own JVMs, but Sun provides a standard JVM Java Plug-in, that has some significant advantages. The Plug-in absorbs the burden of supporting all browsers on all platforms and guarantees an execution environment. It can also cache applets to expedite the return visit to web page. In addition, the Java Plug-in supports Swing classes that are both powerful and convenient for GUI development. To take advantage of the Java Plug-in, the user only needs to initially download the Plug-in into the client machine. An alternative to Java applet is Java Web Start, which is a newly released web-deployment technology. As of this writing, Java Web Start has limited capability to interact with the parent web pages and requires more involvement from administrators and users alike. As this technology matures, it may be a good solution in the future.

ActiveX technology is an alternative to Java technology for complex web-based UI development. The term ActiveX generally refers to ActiveX controls that are software modules based on Microsoft's Component Object Model (COM) architecture. ActiveX enables a program to add functionality by calling ready-made components that blend in and appear as normal parts of the program, such as toolbars, calculators, and even spreadsheets. The power of ActiveX controls in web development is due to their ability to be linked to a web page and downloaded by any ActiveX-compliant web browser. This virtually turns web pages into software pages that can perform operations just like any desktop application. An ActiveX control can be written in a variety of programming languages from Microsoft's suite of Visual tools, including Visual Studio.Net.

Table 24-1 compares the various features of common server-side and client-side technologies. Each technology is useful under the right circumstances. Developers need to make a decision based on their needs and the features of those technologies. Often, the final application is a combination of technologies, taking advantage of the situational strengths of each one.

ACCESSING THE DATA SOURCE

Applications, whether running on a desktop or over the internet, often need to read and write information to and from data storage. For engineering applications, an electronic circuit model might be created by the GUI or a third party application, and later read by the analysis engines. Alternatively, the results of a computation such as load-flow or short circuit might be generated by the engine and stored into a data source to be retrieved by the user or displayed in the GUI. A desktop database like Microsoft Access or flat files may be suitable for a traditional single-user applications or even client-side web applications with relatively straightforward data requirements. A distributed database, such as Oracle or Microsoft SQL Server, is more suitable for enterprise applications requiring better performance, better scalability, and more features such as multiple controlled access, transaction roll back, and remote access and administration.

There are many different database drivers that provide access to databases. Microsoft has provided several technologies for MS Windows applications. Data Access Objects (DAO), Open Database Connectivity (ODBC) and Remote Data Objects (RDO) are earlier technologies, which are still utilized for many existing applications. Microsoft has recently replaced those technologies with a single model, Universal Data Access (UDA), which provides access to a variety of relational and non-relational information sources. UDA consists mainly of the high-level interface called ActiveX Data Objects (ADO) and the lower-level services called OLE DB.

Sun Microsystems offers Java Database Connectivity (JDBC), an application programming interface (API) that can be used to access virtually any tabular data source from the Java programming language. JDBC pro-

Table 24-1. Comparison of Common Web Technologies for UI Development

Features	HTML	Server Side Scripts	Client Side Scripts	Java Applet	ActiveX
UI Complexity	Simple	Simple to moderate	Simple to moderate	Advanced	Advanced
User Interactivity	Limited	Limited	Limited	Extensive	Extensive
Start-up Response	Seconds	Seconds	Seconds	Seconds to Minutes	Seconds to Minutes
Cached Locally	No	No	Yes	Yes	Yes
Security Risk	No	No	Medium	Low	High
UI Response Time	Dependent on network speed server load	Dependent on network speed and server load	Network independent after initial down	Same as local application after initial download	Same as local application after initial download
Browser Independent	Yes	Yes	Depends on language	Yes	No (may require plug-in)
Platform Independent	Yes	Yes	Depends on language	Yes	No (Win 32 only)

vides cross-DBMS connectivity to a wide range of SQL-based databases, and also provides access to other tabular data sources, such as spreadsheets.

XML, (extensible markup language), is a tag-driven technique to create structured information and share both the structure and the data on the internet, intranets, and elsewhere. Since its standard format is plain text, it may be inefficient when applied to engineering applications involving mass data processing.

A simple, but efficient way of storing and accessing data is the use of binary format data files. The main advantage is that they are generally small and efficient so that they can be transported very fast over the internet. Also, there are no additional requirements for database drivers. Object serialization techniques can be utilized to automatically create binary files. The main disadvantage in this context is that they are not human-readable. When the data are not explicitly exposed to and manipulated by users, this may be a viable solution.

PUTTING IT ALL TOGETHER

The developer may have many different options to implement web-enabled tool. Some of the factors that might influence a developer's design decision are:

- User requirements
- Platform constraints
- Migration from the legacy system
- Time to market
- Development cost
- Difficulty of integration.

For a tool to ultimately work, data must be passed efficiently and seamlessly between the various components. Communication between the components can range from basic scripting and simple text file transfers to more complex web-service techniques such as SOAP (Simple Object Access Protocol) and CORBA (Common Object Request Broker Architecture). However, generally speaking, the more complex the communication protocol, the more costly it may be to implement and maintain.

Figure 24-1 shows an example of a simple integration strategy that was implemented to web-enable a legacy power system application. The legacy application typically handles models with tens of thousands of components. The architecture of this legacy application consisted of a GUI developed in Microsoft Visual Basic (VB) which has CAD-like drawing capability, a set of analysis engines developed as C++ applications that are called by the VB GUI, and a data store comprised of a Microsoft Access Database and temporary text files.

For this application, the decision was made to rewrite the interface using HTML, JavaScript, Active Server Pages and a Java applet. The applet is the major

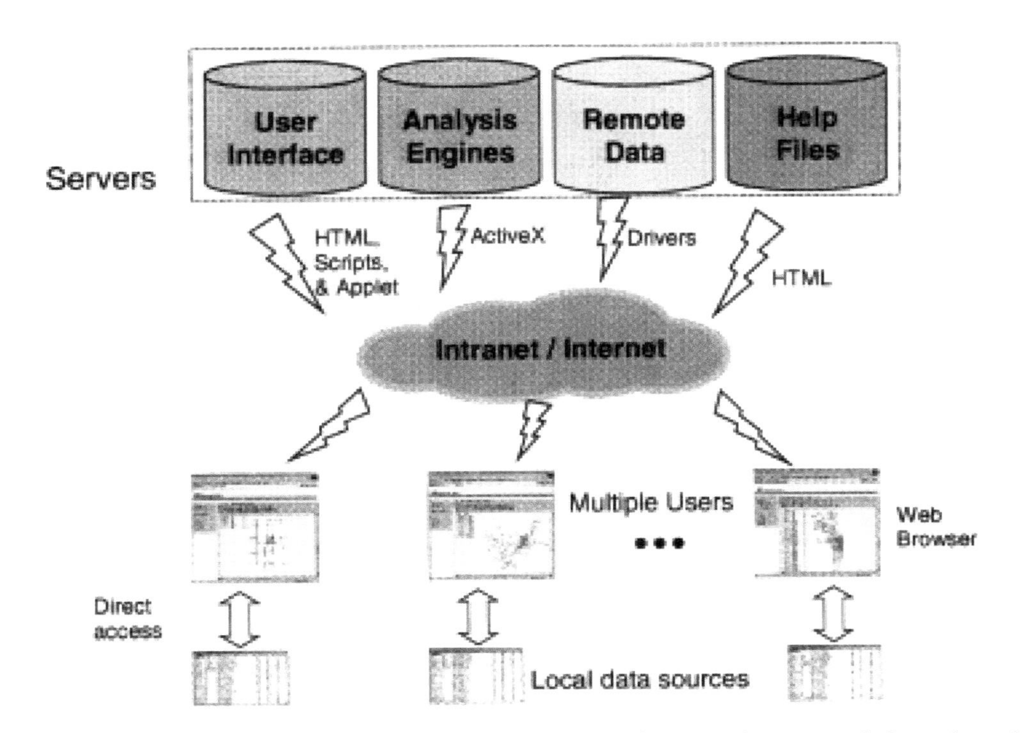

Figure 24-1. Integration of an Engineering Application into a Web-based Tool.

part of the GUI that implements the CAD-like drawing tool. This design decision was due to the extensive user interaction required, the size of the program, the requirement for enhancing the drawing tool, and the available resources. The engines were already tested and optimized so they were simply wrapped in a scriptable ActiveX control and embedded in a web page. To invoke the ActiveX engine, the GUI sends a request to the web server through Active Server Pages. Then the server updates the browser's page containing the ActiveX control to automatically invoke the engines. The MS Access database was replaced with a compressed, binary file that is much smaller. Subsequently, bulk data transfer between the new application and a remote data server is efficient, development cost is reduced and there is no need for a specific database driver at server and client sides.[1]

The developer can choose options from a wide range of technologies discussed in the remainder of this chapter. Some have advantages over others as to reliability, complexity and cost. Some technologies may be more on the "cutting edge" but less reliable and more costly. The difficult decision the developer has to make is whether to use the "latest and greatest" or stick with "well proven" technologies. We present some of these technologies to you in an objective form with little or no personal bias. It depends on the particular application as to which technologies will work best.

HTML

Hypertext markup language (HTML) is the authoring language used to create documents on the World Wide Web. HTML defines the structure and layout of a web document by using a variety of tags and attributes. Tags are also used to specify hypertext links. These allow web developers to direct users to other web pages with only a click of the mouse on either an image or word(s). HTML is a non-proprietary format, and can be created and processed by a wide range of tools, from simple plain text editors to sophisticated WYSIWYG authoring tools.

Tags

A tag is a series of characters, with no spaces, placed between the angle brackets < and >. It is usual to write the commands in upper case like <**TITLE**>, <**BODY**> in textbooks in order to show the tags more clearly, but in practice it is best to keep everything in HTML lower case. Most of the HTML editors produce lower case code.

Some tags are used singly, for example those that stop display on one line and have it continue on the next: <BR>. These are usually called *markup commands*.

Most tags are used in *pairs*. The end tag is identical to the start tag except for one difference: its name is preceded by a slash. These are called *container tags*. The command has an effect on the text placed *between* the *start* and *end* tags.

Example

Utility data for **<B>**4/13/2003**</B>**
will be displayed as follows:

Utility data for **4/13/2003**
The tag **<B>** indicates that the text be displayed in **bold**.

It is possible to add tags together to display more than one change.

Example

Utility data for **<B><I>**4/13/2003</I>**</B>**

will be displayed as bold and italic text:

Utility data for 4/13/2003

The tag **<I>** indicates ***italics***.

Basic Layout of Tags to Start Your Web Page

Your web page is divided into two distinct sections—the **HEAD** and the **BODY**. The entire file is enclosed in a **<HTML>** tag.

The Head

The head is placed in a **<HEAD>** container tag. This section often has only one other command—**<TITLE>**—which specifies a general title for the site. The title will be displayed in the title bar of the browser. No other part of the head is displayed on the screen.

The Body

The rest of the file is contained in a <BODY> container tag. It is this section that will contain the various elements that make up the web page (text, images, tables, line breaks and backgrounds etc.)

A skeleton page would be in the following order:

```
<HTML>
<HEAD>
<TITLE> Put your title here</TITLE>
```

</HEAD>
<BODY>
Everything else goes here. This is the visible part of the page.
</BODY>
</HTML>

Adding Interest to Your Pages with Images

Images can be used to make your web pages distinctive and greatly help to get your message across. The simple way to add an image is using the <img> tag. Let's assume you have an image file called "logo.jpg" in the same folder/directory as your HTML file. It is 200 pixels wide by 150 pixels high.

<img src="logo.jpg" width="200" height="150">

The src attribute names the image file. The width and height aren't strictly necessary but help to speed the display of your web page. People who can't see the image need a description they can read in its absence. You can add a short description as follows:

<img src="logo.jpg" width="200" height="150" alt="Our Logo">

The alt attribute is used to give the short description, in this case "Our Logo."

Adding Links to Other Pages

What makes the web so effective is the ability to define links from one page to another, and to follow links at the click of a button. Links are defined with the <a> tag. Let's define a link to the page "utilitydata.html." This is a link to <a href="utilitydata.html">Today's Utility Data</a>. The text between the <a> and the </a> is used as the caption for the link. It is common for the caption to be in blue underlined text. To link to a page on another web site you need to give the full web address (commonly called a URL), for instance to link to www.utilityreporting.com you need to write:

This is a link to <ahref="http://www.utilityreporting.com">UtilityReporting.Com</a>. You can turn an image into a hypertext link; for example, the following allows you to click on the company logo to get to the home page: <a href="/"><img src="logo.gif" alt="Home Page"></a>

Setting Colors

The color of a page background and default colors for text can be set in the **<BODY>** tag. The color of selected blocks of text can be set in the **<FONT>** tag. There are two ways to define colors. You can name them using a color name or give them a hexadecimal value. Unfortunately, only 16 color names can be recognized. However, hexadecimal can pick from 24 million colors. It is worth remembering, however, that many people have only a 256-color monitor.

Understanding Hexadecimal

We are used to a decimal system of numbers where the base unit is **10**. For example: 56 is seen as 5 × 10 plus 6. In hexadecimal, the **base number is 16**. Therefore, 56 now becomes 5 × 16 plus 6 which is 86. However, for this to work properly up to 255 we need 15 digits. These are comprised of the usual 0-9 and then A-F, where A = 10, B = 11, C = 12, D = 13, E = 14, F = 15.

Example 1: **FF** = 15 × 16 plus 15 which equals **255.**
Example 2: **00** = 0 × 16 plus 0, which equals **0**

RGB Colors

All colors on a TV screen or computer monitor are created by combining the three colors RED, GREEN and BLUE in varying intensities. For example, full red and full green display as yellow, add full blue and you get white. Half power red and blue give a deep purple. The intensity of each color can be set on a scale of 0 to 255—or 00 to FF in hexadecimal. The convention is to define the values in the order RED-GREEN-BLUE so for example FF0080 means RED at full power, no GREEN and BLUE at half power. The values are normally written with a # at the start to indicate that they are hexadecimal. They are sometimes enclosed in quotes—but this is optional.

Knowing all about HTML straight away is not essential—you will learn the basics as you build pages and with experience, you will be able to write extra commands into your code. You will eventually recognize the tags and be able to read HTML just as easily as reading a chapter in a book. Then you can begin to modify the code to suit your pages or make things happen that your software will not allow you to do. The best thing you can do is go to a discount book shop and buy the biggest book on HTML version 4 you can find at the lowest price. There are plenty of such books around and they are now out of print or are several years old and considered by the big book shops to be un-sellable. There are good resources on the web to explain the basics of HTML. Reference sites such as *The Web Designer's Virtual Library* [http://www.wdvl.com] allow you to look up the HTML tags.

Table 24-2. The Standard Color Set

Color Name	*Hexadecimal (RGB)*	*Color (RGB)*	*Hexadecimal Name*
Black	#000000	Green	#008000
Silver	#C0C0C0	Lime	#00FF00
Gray*	#808080	Olive	#808000
White	#FFFFFF	Yellow	#FFFF00
Maroon	#800000	Navy	#000080
Red	#FF0000	Blue	#0000FF
Purple	#800080	Teal	#008080
Fuchsia	#FF00FF	Aqua	#00FFFF

XHTML

XHTML is an acronym for "extensible hypertext markup language," a reformulation of HTML 4.0 as an XML 1.0 application. XHTML provides the framework for future extensions of HTML and aims to replace HTML in the future.[2]

To understand the future importance of XHTML, you first need to understand the current importance of HTML. HTML is one of the most widely used computer languages in the world. The popularity of HTML is because it is the coding technology used to publish content on the World Wide Web (also referred to as the internet or web). Programmers quickly discovered that HTML is a user-friendly language and is very easy to learn. This ease of coding significantly aided in the proliferation of web sites. The latest version of HTML is 4.01 which is defined by the standard published on 24 December 1999 by the World Wide Web Consortium (W3C).[3]

XHTML 1.0 is the first step toward a modular and extensible web based on XML (extensible markup language). It provides the bridge for web designers to enter the web of the future, while still being able to maintain compatibility with today's HTML 4 browsers. It looks very much like HTML 4, with a few notable exceptions, so if you're familiar with HTML 4, XHTML will be easy to learn and use.

XHTML in a nutshell

- XHTML tags are all lowercase.
- XHTML is a stricter, tidier version of HTML.
- Pages written in XHTML work well in most browsers.
- All tags, including empty elements, must be closed.
- XHTML is the reformulation of HTML 4.0 as an application of XML.
- The elements (tags) and attributes are almost identical to HTML.

Example:

```
<?xml version="1.0" encoding="iso-8859-1"?>
<!DOCTYPE html PUBLIC "-//W3C//DTD
XHTML 1.0 Transitional//EN"
"DTD/xhtml1-transitional.dtd"   >
<html        xmlns   =   "http://www.w3.org/
1999/xhtml">
<head>
<title>Quick Example</title>
</head>
<body>
<h1>        Quick Example
</h1>
<a          href     = "http://validator.w3.org/
check/referer">
<img        src      = "http://validator.w3.org/
images/vxhtml10"
    height    = "31"
    width     = "88"
    border    = "0"
    hspace    = "16"
    align     = "left"
    alt       = "Valid XHTML 1.0!"
    /></a>
<p>—Note that the layout (with tabs and align-
ment) is purely for readability—XHTML doesn't
```

```
require it.
</p>
</body>
</html>
```

Copy this example to your web server and use it as a template. Alternatively, simply view it at:

[http://utilityreporting.com/xhtml/example1.htm].[4]

XHTML 1.0 became an official W3C Recommendation January 26, 2000.

A W3C Recommendation means that the specification is stable, that it has been reviewed by the W3C membership, and that the specification is now a web standard.[5]

The W3C MarkUp Validation Service[http://validator.w3.org/] is a free service that checks documents like HTML and XHTML for conformance to W3C Recommendations and other standards.[6]

CASCADING STYLE SHEETS (CSS)

When cascading style sheets were introduced in late 1996, they represented an exciting new opportunity. They enabled much more sophisticated page design (typography and layout) than web developers had been used to, and they helped manage the complex tasks of developing and maintaining sites, and keeping them up to date. Looking at the code of an average web page in 1996 or today, you'd find the HTML itself remarkably similar. Above all, the appearance of the page, the fonts used, the color of text, effects like bold and Italics are all marked up with HTML elements. Back then, it was unavoidable. Now, there are many good reasons for avoiding that approach altogether.

Where Do Style Sheets Come From?

More than a decade before the world wide web, pioneers in the field of electronic documentation recognized the important distinction between what a document looks like (often referred to as its **appearance**) and the underlying **structure** of the document. Complex electronic publishing systems have since then been implemented in a way which separates information about how the document should appear from the document itself. When it was first imagined by Tim Berners-Lee, the web did provide a mechanism for this. However, in the heated rush of the web's early explosion, other, in some ways easier, but also problematic ways of coding page appearance took off. Among these were the <font> and <b> elements and other stylistic or **presentational HTML** elements. To this day, the majority of web developers implement web site appearance in this way, and most web tools, especially WYSIWYG web tools encourage the use of these devices. In short, CSS provides a means for web authors to separate the appearance of web pages from the content of web pages.

Cascading style sheets (CSS) is a recommendation of the World Wide Web Consortium at [http://www.w3.org] (the W3C). The W3C is a consortium of web stakeholders: universities, companies such as Netscape Communications and Microsoft, and experts in many web-related fields. Founded by Tim Berners-Lee (who can be said to have invented the web), it exists to enhance and promote the World Wide Web. One of its roles is to publish "recommendations." You can think of these W3C recommendations as a kind of standard.

Why Use CSS?

Style sheets exist to enable web pages to separate **content** from **appearance**. As a web developer, this means that the information in your web site should go into your HTML files, but HTML files should not contain information about how that information is displayed. In addition, you've probably guessed by now that information about how the pages should appear goes into CSS files.

The traditional HTML approach is to "hardcode" all of the appearance information about a page. So you want all your headings in courier, and at different point sizes to the sizes built into browsers. Then, for every heading, in every page, set the font size and face properties. You have more than 100 pages. That is a lot of editing, and a lot of re-editing if you decide to modify the appearance of your pages later. With all of that editing there is plenty of possibility for introducing errors.

With CSS, you can decide how headings should appear, and enter that information once. Every heading in every page that is linked to this style sheet now has that appearance. If you want to make every heading of level 3 more obviously different from those of level 2 just edit the style sheet and every such heading now has the altered appearance.

What Exactly is a Style Sheet?

A style sheet is simply a **text file** (which has a .css suffix); written according to the grammar defined in the CSS1 or CSS2 recommendations of the W3C.

Here is a simple example.

```
body
{font-family: Verdana, "Minion Web," Helvetica,
sans-serif;
font-size: 1em;
text-align: justify;
background:#00ff00;
color:#ffffff}
h1
{font-family: Verdana, sans-serif;
font-size: 1.3em}
```

Like an HTML document, a cascading style sheet is just a simple text file. However, unlike an HTML document, you don't need a special declaration at the top of the file to say that this is a style sheet. The name of the file should end with a .css suffix. In addition to being in .css files, style sheets can also be **embedded** into the head element of HTML files.

The Elements of a Style Sheet

Every cascading style sheet (whether it is contained in a .css file, or embedded in the head element of an HTML document) is a series of instructions called **statements**. A statement does two things:

- It identifies the **elements** in an HTML document that it affects
- It tells the browser how to draw these elements

In technical HTML terms, an element is anything marked up inside HTML tags.

The part of a statement, which identifies page elements, is called a **selector**. The part of a statement, which tells a browser how selected elements should be drawn, is called the **declaration**. A declaration can contain quite a number of **properties**, the individual pieces of style to be applied to the selected element.

Here is an example.

```
body
{
font-family: Verdana, "Minion Web," Helvetica,
sans-serif;
font-size: 1em;
text-align:justify;
}
```

This is a single **statement**, perhaps one of many in a style sheet. The **selector** is *body*. This means that the statement will affect the <body> element of any page linked to this style sheet. The statement has a **declaration** with three properties. That means that any element selected by this selector will have three of its properties affected. Each property has a name, for example "text-align" and a value, for example "justify." In this case, the font of the body text will be drawn in Verdana by the browser. If Verdana is not available, the browser will use Minion Web, or if this is not available, Helvetica. If none of the specified fonts are available, the browser will draw the body text in a default sans-serif font.

In addition to affecting the font of selected elements, this statement sets the font size to 1 em. If you are familiar with HTML, you'll know that you can't set point sizes with HTML, simply one of 6 relative font sizes. This is another distinct advantage of style sheets—much more sophisticated page layout and typographical control.

Lastly, the browser is instructed to draw body text as fully justified.

Embedding Style Sheets

Style sheets can be embedded into the head element of HTML documents. Quite simply, the style sheet itself is placed between a style tag like this:

```
<style type="text/css"> </style>
```

Embedding means, you lose one of the major advantages of CSS, which is the ability to modify the appearance of a whole site by making changes to a single file.

Linking to Style Sheets

Note that the web page links to the style sheet. The style sheet has no knowledge of the pages that are linked to it. To link a web page (HTML document) to a style sheet, you place a link to the style sheet in the **head** of the document, using the following syntax:

```
<link rel="stylesheet" type="text/css"
href="http://utilityreporting.com/styles/
style1.css" />
```

Let's take a look at each attribute briefly.

```
rel="stylesheet"
```

This says that it is a forward link, and tells the browser what to expect at the other end, namely a style sheet.

```
type="text/css"
```

Theoretically, style sheets might be written using any number of languages. The style sheets we've been talking about in here are cascading style sheets (CSS). Extensible Style Language (XSL) is another that may become important. This attribute tells the browser what format in which it is going to receive the style sheet. The content type is **necessary**.

```
href="http://utilityreporting.com/styles/style1.css"
```

This statement tells the browser where to locate the style sheet. This can be either a relative or an absolute URL. Remember that the relative URL is relative to the HTML document, unlike URLs within style sheets, which are relative to the style sheet.

Note that the "/>" closing of this element is the correct way of closing an empty element in XHTML 1.0. (7)

View the results of the example at:

[http://utilityreporting.com/css/example1.htm]

The W3C CSS Validation Service[http://jigsaw.w3.org/css-validator/check/referer] is a free service that checks documents like cascading style sheets for conformance to W3C Recommendations and other standards.

DYNAMIC HTML

DHTML is not a singular technology but a combination of three existing technologies glued together by the document object model (DOM):

- **HTML**—For creating text and image links and other page elements.
- **CSS**—Style sheets for further formatting of text and HTML plus other added features such as positioning and layering content.
- **JavaScript**—The programming language that allows you to access and dynamically control the individual properties of both HTML and style sheets.

The way JavaScript accesses the properties of an HTML document is through the document object model (DOM). The job of the DOM is to expose all the attributes of HTML and style sheets to JavaScript control. All you need to know about the DOM is what JavaScript commands it will accept. Not so easy, as different browsers have their slightly different versions of the DOM, so they access HTML properties differently as well as display them differently.

It's important to note that every piece of HTML has a location, much like a directory in a phone book. When finding that piece of HTML, go through the same hierarchy process as searching for a name in the phone book.

(state) Florida -> (City) Tampa -> (Listings) m -> (Name) Martin

In JavaScript, a reference to this would be equivalent to

florida.tampa.m.martin

Martin may have additional information such as his address and telephone number. Therefore, the JavaScript reference would be written this way:

florida.tampa.m.martin.address

or

florida.tampa.m.martin.phone

An excellent DHTML reference book is *Dynamic HTML—The Definitive Guide* by Danny Goodman (O'Riley Books). It lists all of the DHMTL properties and their cross browser compatibility's.[8]

THE DOCUMENT OBJECT MODEL (DOM)

Efforts have been under way, most noticeably at the W3C[http://www.w3.org/] to establish common standards for all browsers. The release of the CSS specification, and then of the DOM Level 0 and Level 1 specifications, has resulted in most of the major browsers falling in line with the proposed standards.

Below are some of the new DOM constructs, together with examples of how they can be used in "real" HTML documents.

We'll start with the basics—a very simple HTML page.

```
<html><head></head><body bgcolor="white"><div
id="a" style="font-family:
Arial; color: white; background: black">Wassup?</
div></body></html>
```

Under the new DOM, every element in an HTML document is part of a "tree." You can access each element by navigating through the tree "branches" until you reach the corresponding "node." Given that, here's the representation of the HTML document above, in "tree" form.

```
document
    |—<html>
        |—<head>
        |—<body>
            |—<div>
```

The DOM offers a fast and efficient method of accessing elements within the page using the getElementById( ) method.

Let's alter the font color of the text within the <div>.

```
<script language="JavaScript">
var obj = document.getElementById("a");
obj.style.color = "red";
</script>
```

In addition to the getElementById( ) method, which is typically used to obtain a reference to a specific element, the DOM also offers the getElementsByTagName( ) method, used to return a collection of a specific type of element. For example, the code

```
document.getElementsByTagName("form");
```

returns a collection, or array, containing references to all the <form> tags in the document. Each of these references is a node, and can then be manipulated using the standard DOM methods and properties.

Consider the following document, which contains three <div>s, each containing a line of text

```
<html>
<head>
</head>
<body bgcolor="white">

<div id="electric">
Electric Meter 112
</div>

<div id="water">
Water Meter 211
</div>

<div id="gas">
Gas Meter 312
</div>

</body>
</html>
```

and then study the code used to manipulate the text within the second <div>

```
<script language="JavaScript">
//get a list of all <div> tags
var divCollection =
document.getElementsByTagName("div");

//get a reference to the second <div> in the
collection
var waterObj = divCollection[1];

//verify that we are where we think we are
//alert(waterObj.getAttribute("id"));

//change the text string within the <div>
waterObj.childNodes[0].data = " Water Meter
212";
</script>
```

A collection of all the tags within a document (a lot like "document.all") can be obtained with

```
document.getElementsByTagName("*");
```

Now that you know how to find your way to specific HTML elements in the document, it's time to learn how to manipulate them. [9]

Consider the following example.

```
<html><body>
<div id="text" style="font-size:8pt">
Some text here!
</div>
<input type="button" value="A+" onClick="upSize(
)">
<input type="button" value="A-"
onClick="downSize( )">
<script type="text/javascript">
var max = 24
var min = 8
function upSize( ) {
size =
parseInt(document.getElementById("text").style.fontSize)
if (size != max) {
size = size + 2
}
```

```
document.getElementById("text").style.fontSize =
size + "pt"
}
function downSize( ) {
size =
parseInt(document.getElementById("text").style.fontSize)
if (size != min) {
size = size - 2
}
document.getElementById("text").style.fontSize =
size + "pt"
}
</script></body></html>
```

Try it out here:

[http://utilityreporting.com/dhtml/example1.htm]

XML

An XML document is a database only in the strictest sense of the term. That is, it is a collection of data. In many ways, this makes it no different from any other file; after all, all files contain data of some sort. As a "database" format, XML has some advantages. For example, it is self-describing (the markup describes the structure and type names of the data, although not the semantics), it is portable, and it can describe data in tree or graph structures. It also has some disadvantages. For example, it is verbose, and access to the data is slow due to parsing and text conversion.

A more useful question to ask is whether XML and its surrounding technologies constitute a "database" in the looser sense of the term, that is, a database management system (DBMS). The answer to this question is, "Sort of." On the plus side, XML provides many of the things found in databases. It has storage (XML documents), schemas (DTDs (Document Type Definitions), XML schema languages), query languages (XQuery, XPath, XQL (XML Query Language), XML-QL, QUILT, etc.), programming interfaces (SAX (Simple API for XML), DOM, JDOM), and so on. On the minus side, it lacks many of the things found in real databases: efficient storage, indexes, security, transactions and data integrity, multi-user access, triggers, queries across multiple documents, and so on.

It may be possible to use an XML document or documents as a database in environments with small amounts of data, few users, and modest performance requirements. However, it will fail in most production environments, which have many users, strict data integrity requirements, and the need for good performance.

Given the low price and ease of use of databases like dBASE and Access, there seems little reason to use an XML document as a database. The only real advantage of XML is that the data are portable, and this is less of an advantage than it seems due to the widespread availability of tools for serializing databases as XML. *Serialization* is the process of converting an object's state into a format that can be transported, persisted, and/or reconstructed.

Data Versus Documents

Perhaps the most important factor in choosing a database is whether you are using the database to store *data* or *documents*. For example, is XML used simply as a data transport between the database and a (possibly non-XML) application? Alternatively, is its use integral, as in the case of XHTML documents? This is usually a matter of intent, but it is important because all *data-centric documents* share a number of characteristics, as do all *document-centric documents*, and these influence how XML is stored in the database. The next two sections examine these characteristics.

Data-Centric Documents

Data-centric documents are documents that use XML as a data transport. They are designed for machine consumption and the fact that XML is used at all is usually superfluous. That is, it is not important to the application or the database that the data are, for some length of time, stored in an XML document. Data-centric documents are characterized by regular structure, fine-grained data, and little or no mixed content. The order in which sibling elements occurs is generally not significant, except when validating the document.

Data of the kind that are found in data-centric documents can originate both in the database (in which case you want to expose them as XML) and outside the database (in which case you want to store them in a database). An example of the former is the vast amount of legacy data stored in relational databases; an example of the latter is scientific data gathered by a measurement system and converted to XML.

In general, any web site that dynamically constructs HTML documents today by filling a template with database data can probably be replaced by a series of data-centric XML documents and one or more XSL stylesheets.

For example, consider the following document describing a flight:

```
<FlightInfo>
    <Airline>ABC Airways</Airline> provides
    <Count>three</Count>
    non-stop flights daily from <Origin>Dallas</
    Origin> to
    <Destination>Fort Worth</Destination>.
    Departure times are
    <Departure>09:15</Departure>, <Depar-
    ture>11:15</Departure>,
    and <Departure>13:15</Departure>. Arrival
    times are minutes later.
</FlightInfo>
```

This could be built from the following XML document and a simple stylesheet:

```
<Flights>
    <Airline>ABC Airways</Airline>
    <Origin>Dallas</Origin>
    <Destination>Fort Worth</Destination>
    <Flight>
        <Departure>09:15</Departure>
        <Arrival>09:16</Arrival>
    </Flight>
    <Flight>
        <Departure>11:15</Departure>
        <Arrival>11:16</Arrival>
    </Flight>
    <Flight>
        <Departure>13:15</Departure>
        <Arrival>13:16</Arrival>
    </Flight>
</Flights>
```

Document-Centric Documents

Document-centric documents are (usually) documents that are designed for human consumption. Examples are books, email, advertisements, and almost any hand-written XHTML document. They are characterized by less regular or irregular structure, larger grained data (that is, the smallest independent unit of data might be at the level of an element with mixed content or the entire document itself), and lots of mixed content. The order in which sibling elements occurs is usually significant.

Document-centric documents are usually written by hand in XML or some other format, such as RTF, PDF, or SGML, which is then converted to XML. Unlike data-centric documents, they usually do not originate in the database.

Data, Documents, and Databases

Characterizing your documents as data-centric or document-centric will help you decide what kind of database to use. As a rule, data are stored in a traditional database, such as a relational, object-oriented, or hierarchical database. This can be done by third-party *middleware* or by capabilities built in to the database itself. In the latter case, the database is *XML-enabled.* Documents are stored in a *native XML database* (a database designed especially for storing XML) or a *content management system* (an application designed to manage documents and built on top of a native XML database).

These rules are not absolute. Data, especially semi-structured data, can be stored in native XML databases and documents can be stored in traditional databases when few XML-specific features are needed. Furthermore, the boundaries between traditional databases and native XML databases are beginning to blur, as traditional databases add native XML capabilities and native XML databases support the storage of document fragments in external (usually relational) databases.

Storing and Retrieving Data

In order to transfer data between XML documents and a database, it is necessary to map the XML document schema (DTD, XML Schema) to the database schema. The data transfer software is then built on top of this mapping. The software may use an XML query language (such as XPath, XQuery, or a proprietary language) or simply transfer data according to the mapping (the XML equivalent of SELECT * FROM Table).

Mapping Document Schemas to Database Schemas

Mappings between document schemas and database schemas are performed on element types, attributes, and text. They usually omit physical structure (such as entities, CDATA sections, and encoding information) and some logical structure (such as processing instructions, comments, and the order in which elements and PCDATA appear in their parent). This is more reasonable than it may sound, as the database and application are concerned only with the data in the XML document. One consequence of this is that "round-tripping" a document, storing the data from a document in the database and then reconstructing the document from that data, often results in a different document, even in the canonical sense of the term.

Two mappings are commonly used to map an XML document schema to the database schema: the table-based mapping and the object-relational mapping.

Table-Based Mapping

The table-based mapping is used by many of the middleware products that transfer data between an XML document and a relational database. It models XML documents as a single table or set of tables. The structure of an XML document must be as follows, the <database> element and additional <table> elements do not exist in the single-table case:

```
<database>
    <table>
        <row>
            <column1>...</column1>
            <column2>...</column2>
            ...
        </row>
        <row>
            ...
        </row>
        ...
    </table>
    <table>
        ...
    </table>
    ...
</database>
```

Depending on the software, it may be possible to specify whether column data are stored as child elements or attributes, as well as what names to use for each element or attribute. In addition, products that use table-based mappings often optionally include table and column metadata either at the start of the document or as attributes of each table or column element. Note that the term "table" is usually interpreted loosely. That is, when transferring data from the database to XML, a "table" can be any result set. When transferring data from XML to the database, a "table" can be a table or an updateable view.

The table-based mapping is useful for serializing relational data, such as when transferring data between two relational databases. Its obvious drawback is that it cannot be used for any XML documents that do not match the above format.

Object-Relational Mapping

The object-relational mapping is used by all XML-enabled relational databases and some middleware products. It models the data in the XML document as a tree of objects that are specific to the data in the document. In this model, element types with attributes, element content, or mixed content (*complex element types*) are generally modeled as classes. Element types with PCDATA-only content (*simple element types*), attributes, and PCDATA are modeled as scalar properties. The model is then mapped to relational databases using traditional object-relational mapping techniques or SQL 3 object views. That is, classes are mapped to tables, scalar properties are mapped to columns, and object-valued properties are mapped to primary key/foreign key pairs.

(The name "object-relational mapping" is actually a misnomer, as the object tree can be mapped directly to object-oriented and hierarchical databases. However, it is used because the overwhelming majority of products that use this mapping use relational databases and the term "object-relational mapping" is well known.)[10]

For a complete description of the object-relational mapping, see section 3 of "Mapping DTDs to Databases[http://www.xml.com/pub/a/2001/05/09/dtdtodbs.html]."

SERVER-SIDE APPLICATIONS

The first generation of server-side technologies is common gateway interface (CGI), usually developed in Perl programming language. This technology employs server-side applications to complete a predefined task when a client's request is received. This may be far from efficient since the server-side application is heavy-weighted and consumes many server resources. The next generation of server-side technologies is server-side scripts such as Active Server Pages, Java Servlet/Java Server Pages, PHP, etc. These technologies allow the server to generate a thread to handle a client's request. The thread is light-weighted and consumes much less resources than the application generated by CGI. Hence, this is more efficient than the CGI technology.

As opposed to server-side technologies, client-side technologies, which are discussed in detail in the next section, execute tasks at the client-side. Why are server-side technologies more commonly utilized in existing web-based applications? Many factors contribute to this scenario. However, there is a necessary condition for server-side technology to be feasible: when the client requests the server to process a task, the response must be fast enough to avoid a web-browser time-out. The default time-out for most web-browser is a few minutes and proxies may have an even lower time-out threshold. This implies that if the server response time is longer than a few minutes, it is not recommended to keep connections between the server and the client. Since there is generally, no guarantee for networks to be secure and anything could happen at anytime.

It is highly possible for some server-side web-based applications to meet the response time threshold due to their computational features. With the assumption that the network delay is minimal, the response time depends on the actual running time at the server. The running time is relatively small for most popular web-based applications, since generally the task is not computationally intensive. For example, many e-Commerce and on-line banking systems are heavily involved with database queries and the execution time depends on the nature of the queries. Since current enterprise database drivers can complete a query in milliseconds at a 1GHz desktop computer, there is certainly no problem for a more powerful server to complete a database query almost instantly.[1]

Common Gateway Interface (CGI)

The common gateway interface (CGI) is a standard for interfacing external applications with information servers, such as HTTP or web servers. A plain HTML document that the web server **retrieves** is **static**, which means it exists in a constant state: a text file that doesn't change. A CGI program, on the other hand, is **executed** in real-time, so that it can output **dynamic** information.

For example, let's say that you wanted to "hook up" your Unix database to the World Wide Web, to allow people from all over the world to query it. You need to create a CGI program that the web server will execute to transmit information to the database engine, and receive the results back again and display them to the client. This is an example of a *gateway*, and this is where CGI got its origins.

The database example is a simple idea, but most of the time rather difficult to implement. There really is no limit as to what you can hook up to the web. The only thing you need to remember is that whatever your CGI program does it should not take too long to process. Otherwise, the user will just be staring at their browser waiting for something to happen.

Since a CGI program is executable, it is the equivalent of letting the world run a program on your system, which isn't the safest thing to do. Therefore, there are some security precautions that need to be implemented when it comes to using CGI programs. Probably the one that will affect the typical web user the most is the fact that CGI programs need to reside in a special directory. Then the web server knows to execute the program rather than just display it to the browser. This directory is usually under direct control of the webmaster, prohibiting the average user from creating CGI programs. There are other ways to allow access to CGI scripts, but it is up to your webmaster to set these up for you.

In your web server folders, you will see a directory called /cgi-bin. This is the special directory mentioned above where all of your CGI programs currently reside. A CGI program can be written in any language that allows it to be executed on the system, such as:

- C/C++
- Fortran
- PERL
- TCL
- Any Unix shell
- Visual Basic
- AppleScript

It just depends what you have available on your system. If you use a *programming language* like C or Fortran, you know that you must compile the program before it will run. If, however, you use one of the *scripting languages* instead, such as PERL, TCL, or a Unix shell, the script itself only needs to reside in the /cgi-bin directory, since there is no associated source code. Many people prefer to write CGI scripts instead of programs, since they are easier to debug, modify, and maintain than a typical compiled program.[11]

The output of any CGI script contains two basic parts. The first is a set of message headers, followed by a blank line. The second, optional, part is the actual message body. A script can output just a header with a Location: line, providing the URL of a new document that the client should retrieve. In either case, the script must send the client a series of message headers, followed by a blank line, then the (optional) message body. To instruct the client browser to retrieve a new document as the result of the CGI script, the program must send the client a Location: line giving the complete URL, like this:

```
Location: http://www.utilityreporting.com/EISdemo/
index.htm
```

On most servers, if you want to return a reference to another document on the same server, you can specify just the changed part of the path (a relative URL), as shown:

```
Location:/EISdemo/index.htm
```

NOTE: This requires not one, but two blank lines after Location:

When the client browser receives this output from the CGI script, it will go to

www.utilityreporting.com and retrieve/EISdemo/index.htm.

More usefully, the CGI script can return an actual document, image, sound file, or other MIME file. To send a full document to the client, the script must send a Content-type: header with a valid MIME type such as text/html followed by any other header lines, a blank line, then the document body. For example, a script might return the following data:

Example Script #1

```
Content-type: text/html

<HTML> <HEAD><TITLE>This is my output</
TITLE></HEAD> <BODY> <H1>Welcome to
CGI!</H1> <P> This is the output of a simple
CGI script. </BODY> </HTML>
```

When the client receives this output it will strip off the HTTP headers (everything before the first blank line), and use the 'Content-type' to determine the type of file it has received. In this case it will see that it is 'text/html' and render the HTML to display the results to the user. The web browser behaves just as if it had retrieved a static HTML document from the web server.

The basics of a CGI script do not require much effort to create. A simple program, which prints the current date, requires a few simple lines of PERL:

Example Script #2

```
#!/usr/local/bin/perl5
# # This tells the client to expect a HTML document
#
print "Content-type: text/html,""\n\n";
$today='date'; # Get the date
chop $today;
# Remove the newline on the end
print "Today's date is $today. \n";
$this_server=$ENV{'SERVER_NAME'};
print "You can <A HREF=\"http://$this_server/\">re-
turn to the server</A>";
exit(0);
```

See what it looks like in the browser at:

[http://utilityreporting.com/cgi-bin/cgi-example2.pl]

When the server launches a CGI program, it passes important information to the program in two main ways.

Environment Variables

Information about the web server and the current client is provided in the Unix environment variables available to the client. The following environment variables are specific to CGI, as documented by NCSA[http://hoohoo.ncsa.uiuc.edu/cgi/env.html].

The way these environment variables are accessed varies depending on the language used for the CGI program. For example, the variable SERVER_NAME contains the hostname of the server, such as www.utilityreporting.com. Here are some examples of how this value is accessed in various languages:

LANGUAGE	ACCESSED
'C'	getenv("SERVER_NAME")
PERL	$ENV{'SERVER_NAME'}
SH	$SERVER_NAME
TCL	$env(SERVER_NAME)

Input Stream

The form contents from POST requests are provided as a stream of characters on the Unix *stdin*, or 'standard input'. The total number of bytes made available is provided in the CONTENT_LENGTH environment variable. For example, here is a simple PERL script that accepts the input from a POST form and prints the stream of bytes:

```
#!/usr/local/bin/perl5
$methodused = $ENV{'REQUEST_METHOD'};
print "Content-type: text/html\n\n";
print <<ONE;
<HTML>
<HEAD>
<TITLE>This page generated by Perl</TITLE>
</HEAD>
<BODY>
<h3>Thank you for submitting this form.</h3>
<p>The method used was: <b>$methodused</b>
<p>The path info is:
<p><b>$ENV{'PATH_INFO'}</b>
ONE
if ($methodused eq "GET") {
```

```
$whatwassent= $ENV{'QUERY_STRING'};
}else{
$contentlength = $ENV{'CONTENT_LENGTH'};
read(STDIN,$whatwassent,$contentlength);
}
print "<p>Before translation, this was sent:";
print "<p><b>$whatwassent</b>";
$whatwassent =~ tr/+//;
$whatwassent =~ s/%([a-fA-F0-9][a-fA-F0-9])/
pack("C," hex($1))/eg;
print <<TWO;
<p>The translated query string or posted data
are:
<p><b>$whatwassent</b>
</BODY>
</HTML>
TWO
```

You can see how this works at:

http://utilityreporting.com/cgi/example1.htm]

Some Other Common Mistakes and Their Remedies:

1) Missing access permissions
All directories in which CGI scripts reside should allow you to read, write, and execute files, and everybody else to only execute the files from the directory. If you 'cd' to the cgi-bin directory, you can set the permissions with the command:

```
chmod u=rwx,go=x
```

If the actual CGI programs are compiled binaries (e.g. from C source code) then only world-execute access is needed (chmod u=rwx,go=x file.cgi), if they are interpreted scripts such as a PERL program, then they need world read access as well, the chmod command to set the permissions in that case is:

```
chmod u=rwx,go=rx file.cgi
```

And of course, HTML files and images don't need execute program, so their permissions can be set to:

```
chmod u=rw,go=r file.html
```

2) File in wrong location, or has wrong extension
On most web servers, only files located in or under the cgi-bin directory can be used as cgi scripts, and generally the script must have an extension of .cgi or .pl to be executed.

3) Server returns an error
When the server returns an error message to the client, the error is often intentionally vague. The server error_log file, or equivalent, usually contains detailed information on the exact reason that the request failed.

There are numerous resources for information on the Common Gateway Interfaces available on the web. Some of the more valuable include:[12]

- NCSA[http://hoohoo.ncsa.uiuc.edu/cgi/interface.html] has the canonical copy of the CGI specifications as well as a simplified primer [http://hoohoo.ncsa.uiuc.edu/cgi/primer.html].
- Thomas Boutell's *Web FAQ* [http://www.boutell.com/faq]
- O'Reilly produces a book, *CGI Programming on the World Wide Web* [http://www.oreilly.com/catalog/cgi/].

Active Server Pages (ASP)

Files created with Active Server Pages have the extension .ASP. With ASP files, you can activate your web site using any combination of HTML, scripting, such as JavaScript or Visual Basic® Scripting Edition (VBScript), and components written in any language. This means your ASP file is simply a file that can contain any combination of HTML, scripting, and calls to components. When you make a change on the ASP file on the server, you need only save the changes to the file, the next time the web page is loaded, the script will automatically be compiled. How does this happen? It works because ASP technology is built directly into Microsoft Web servers, and is thus supported on all Microsoft Web servers.

ASP runs as a service of the web server and is optimized for multiple threads and multiple users. This means that it's fast, and it's easy to implement. If you use ASP, you can separate the design of your web page from the nitty-gritty details of programming access to databases and applications. It all works together via scripting.

Here's how it works. When a browser requests an ASP file from your web server, your web server calls Active Server Pages to read through the ASP file, executing any of the commands contained within and sending the resulting HTML page to the browser. An ASP file can contain any combination of HTML, script, or commands. The script can assign values to variables, request information from the server, or combine any set of commands into procedures.

ASP uses the delimiters "<%" and "%>" to enclose script commands.

For example, the code below sets the value of the variable "MyFavTVShow" in the user cookie to "I Dream of Jeannie."

<%Response.Cookies("MyFavTVShow")=
"I Dream of Jeannie"%>

The scripting languages supported by ASP in turn support use of the If-Then-Else construct (something that will undoubtedly warm the hearts of all coders out there). Finally, you can embed some real logic into your HTML. For example, the following code from the IIS documentation shows how you can set the greeting shown based upon the time of day.

<FONT COLOR="GREEN"> <%If Time >= #12:00:00 AM# And Time < #12:00:00 PM# Then%> Good Morning! <%Else%> Hello! <%End If%> </FONT>

Built-in Objects

ASP includes five standard objects for global use:

- Request—to get information from the user
- Response—to send information to the user
- Server—to control the internet information server
- Session—to store information about and change settings for the user's current web-server session
- Application—to share application-level information and control settings for the lifetime of the application

The request and response objects contain *collections* (bits of information that are accessed in the same way). Objects use *methods* to do some type of procedure (if you know any object-oriented programming language, you know already what a method is) and *properties* to store any of the object's attributes (such as color, font, or size).

The Request Object

The request object is used to get information from the user that is passed along in an HTTP request. As mentioned earlier, the request and response objects support collections:

- ClientCertificate—to get the certification fields from the request issued by the web browser. The fields that you can request are specified in the X.509 standard
- QueryString—to get text such as a name, such as my favorite TV sitcom above
- Form—to get data from an HTML form
- Cookies—to get the value of application-defined cookie
- ServerVariables—to get HTTP information such as the server name

The Response Object

The response object is used to send information to the user. The response object supports only cookies as a *collection* (to set cookie values). The response object also supports a number of properties and methods. *Properties* currently supported are:

- Buffer—set to buffer page output at the server. When this is set to true, the server will not send a response until all of the server scripts on the current page have been processed, or until the Flush or End method has been called.
- ContentType—to set the type of content (i.e: text/HTML, Excel, etc.)
- Expires—sets the expiration (when the data in the user's cache for this web page is considered invalid) based on minutes (i.e.: expires in 10 minutes).
- ExpiresAbsolute—allows you to set the expiration date to an absolute date and time.
- Status—returns the status line

The following *methods* are supported by the response object:

- AddHeader—Adds an HTML header with a specified value
- AppendToLog—Appends a string to the end of the web server log file
- BinaryWrite—writes binary data (i.e, Excel spreadsheet data)
- Clear—clears any buffered HTML output.
- End—stops processing of the script.
- Flush—sends all of the information in the buffer.
- Redirect—to redirect the user to a different URL
- Write—to write into the HTML stream. This can be done by using the construct
- Response.write("hello")

or the shortcut command
<%="hello"%>

The Server Object

The server object supports one property,

ScriptTimeout, which allows you to set the value for when the script processing will time out, and the following methods:

- CreateObject—to create an instance of a server component. This component can be any component that you have installed on your server (such as an ActiveX).
- HTMLEncode—to encode the specified string in HTML.
- MapPath—to map the current virtual path to a physical directory structure. You can then pass that path to a component that creates the specified directory or file on the server.
- URLEncode—applies URL encoding to a specified string.

The Session Object

The session object is used to store information about the current user's web-server session. Variables stored with this object exist as long as the user's session is active, even if more than one application is used. This object supports one method, **Abandon**, which abandons the current web-server session, destroying any objects, and supports two properties, **SessionID**, containing the identifier for the current session, and **Timeout**, specifying a time-out value for the session. One thing to bear in mind about the session identifier: It's not a GUID. It's only good as long as the current web-server session is running. If you shut down the web-server service, the identifiers will start all over again. So don't use it to create logon IDs, or you'll have a bunch of duplicates and one heck of a headache.

The Application Object

The application object can store information that persists for the entire lifetime of an application (a group of pages with a common root). Generally, the IIS server is running this whole time. This makes it a great place to store information that has to exist for more than one user (such as a page counter). The downside of this is that since this object isn't created anew for each user, errors that may not show up when the code is called once may show up when it is called 10,000 times in a row. In addition, because the Application object is shared by all the users, threading can be a nightmare to implement.

To create a simple ASP page you need a text editor such as Microsoft's note pad, which comes pre-installed in most all versions of Windows. Microsoft Front Page 98/2000[http://www.microsoft.com/frontpage/] also has built in support for ASP editing with the "HTML" tab which allows the viewing and editing of straight HTML page source, which is where your ASP code goes.

Now choose your tool, and start by cut and pasting the following code into your text/HTML editor.

Source file:

```
<html> <head><title>My first ASP page</title> </head> <body> <p>The current date is <%=Date%>.</p> </body> </html>
```

Save this file as test.asp.

Save this on a location on your hard drive such as x:\ASP\ (replace "x" with the letter of your hard drive).

In order to run your test ASP file (test.asp); you need a web server that supports ASP. Here is a list of web servers known to support ASP.

- IIS (Internet Information Services)
- PWS (Personal Web Server)

IIS ships standard with Windows 2000 Server and Windows 2000 Advanced Server. Windows 2000 also comes with a limited version of IIS, which allows a maximum of five simultaneous connections at any one time. This means that it is not designed to serve to more than a few visitors if any.

Hosting services are another main service provided to run your ASP files.

Free ASP hosting services

- Brinkster.com[http://www.brinkster.com]

Step 1. Upload your file **test.asp** to the hosting service provider base folder.

Step 2. Enter the URL to the file "**test.asp**" in your web browser and you should receive the following page content:
[http://www.yourhost.com/test.asp]

The current date is 3/24/2003.

If you see the following page, congratulations! You have made your first ASP page.

The current date is 3/24/2003.

Or try this URL:
[http://www26.brinkster.com/dcgreen/test.asp]
to view the results.

JavaTM Servlet

JavaTM servlet technology provides web developers with a simple, consistent mechanism for extending the functionality of a web server and for accessing existing business systems. A servlet can almost be thought of as an applet that runs on the server side. JavaTM servlets have made many web applications possible. Servlets are the Java platform technology of choice for extending and enhancing web servers. Servlets provide a component-based, platform-independent method for building web-based applications, without the performance limitations of CGI programs. In addition, unlike proprietary server extension mechanisms (such as the Netscape Server API or Apache modules), servlets are server- and platform-independent. This leaves you free to select a "best of breed" strategy for your servers, platforms, and tools. Servlets have access to the entire family of Java APIs, including the JDBCTM API to access enterprise databases. Servlets can also access a library of HTTP-specific calls and receive all the benefits of the mature Java language, including portability, performance, reusability, and crash protection. Today, servlets are a popular choice for building interactive web applications. Third-party servlet containers are available for Apache Web Server, Microsoft IIS, and others. Servlet containers are usually a component of web and application servers, such as BEA WebLogic Application Server, IBM WebSphere, Sun ONE Web Server, Sun ONE Application Server, and others.[13]

JavaServer Pages

JavaServer pages technology is an extension of the JavaTM Servlet technology. Servlets are platform-independent, 100% pure Java server-side modules that fit seamlessly into a web server framework and can be used to extend the capabilities of a web server with minimal overhead, maintenance, and support. Unlike other scripting languages, servlets involve no platform-specific consideration or modifications; they are Java application components that are downloaded, on demand, to the part of the system that needs them. Together, JSP technology and servlets provide an attractive alternative to other types of dynamic web scripting and programming. It offers platform independence, enhanced performance, separation of logic from display, ease of administration, extensibility into the enterprise and most importantly, ease of use.

Web developers and designers use JavaServer Pages technology to rapidly develop and easily maintain information-rich, dynamic web pages that leverage existing business systems. HTML-savvy web page developers and designers can use JSP technology without needing to learn how to write Java scriptlets. Although scriptlets are no longer required to generate dynamic content, they are still supported to provide backward compatibility.

JavaServer Pages technology uses XML-like tags that encapsulate the logic that generates the content for the page. Additionally, the application logic can reside in server-based resources (such as JavaBeansTM component architecture) that the page accesses with these tags. All formatting (HTML or XML) tags are passed directly back to the response page. By separating the page logic from its design and display and supporting a reusable component-based design, JSP technology makes it faster and easier than ever to build web-based applications.

The JSP specification is the product of industry-wide collaboration with industry leaders in the enterprise software and tools markets, led by Sun Microsystems. Sun has made the JSP specification freely available to the development community, with the goal that every web server and application server will support the JSP interface. JSP pages share the "Write Once, Run AnywhereTM" characteristics of Java technology. JSP technology is an essential component in the Java 2 Platform, Enterprise Edition, Sun's highly scalable architecture for enterprise applications.[14]

PHP

PHP is a language for easily building dynamic web pages. It provides an easier way to accomplish web related programming tasks. These tasks are accomplished only with difficulty in more complex and powerful languages, such as Perl or C. It is ideally suited to the web because PHP scripts live inside web pages right along with the HTML tags and content. For that reason, PHP is called an embedded scripting language. Developers can embed programs in their web pages, making them dynamic. They can treat programs just like web pages. PHP pages can contain both regular HTML and PHP code. This allows you to develop web applications quickly. However, unlike some web scripting languages, PHP makes a clear distinction between sections of PHP code and sections of the HTML document. When the web server fills a request for a PHP enabled page, it first looks through the page content for sections of PHP code and executes any it finds. Any normal HTML sections are passed to the browser without any changes. This means that you can freely mix snippets of program into a web page anywhere.

Here are a few good reasons to choose PHP for enabling interactive content on your web site:

- It is open source
- It uses similar syntax and constructs, knowledge of PHP can help you in learning the C language
- The data types and structures of PHP are easy to use and understand, PHP knows what you mean and can convert types automatically
- You don't have to know any special commands to compile your program, it runs right in your web browser
- PHP serves as a "wrapper" for many standard C libraries, which are easily compiled into the language giving it the flexibility to respond more rapidly to changes in web technology or trends
- PHP enabled web sites can be deployed with amazing rapidity, due to its being tuned for dynamic pages and database backends

You don't have to know everything there is to know about PHP before you can write useful programs. Therefore, we will start with a few simple examples.

Embedding snippets of code into a web page has many uses. For example, in an otherwise static web page you may obtain the value of a variable and later use it to dynamically change the content of the page. This PHP example displays the text identifying the web browser on a web page.

```
<?php $browser = getenv("HTTP_USER_AGENT"); ?>
<P>You are using the <?php echo($browser);?> web
browser. </P>
```

The quickest way to learn PHP is to start using it and see what happens. We'll start by diving right into the good old "Hello World" script you may have seen when learning other languages. It simply prints "Hello World" to the browser screen using the **echo( )** function. Each line of code requires a semicolon at the end. Perl syntax is very different from PHP, which is more formal with its C influenced functional syntax. For example in Perl you can pretty much drop a regular expression into a statement anywhere, but in PHP you must explicitly make a function call to a regular expression function (ereg( ) or the Perl-like preg( ). What might be natural for a Perl aficionado to code right into their program without thinking requires putting the regular expression into a function in PHP.

PHP code is delimited by left < and right > brackets. This gives a one-line PHP program the appearance of an HTML tag. One liners are often used for outputting content mixed right into HTML.

```
<?php echo("Hello World\n"); ?> <?php print "Hello
World\n"; ?> <?php print "Hello World<br>"; ?>
```

Most people favor the print function over echo( ), I think for clarity. "echo" is a term that might cause confusion while "print" is obvious.

The **echo( )** function sends one or more strings of text to the browser for display. Actually, echo is not function but a "language construct" but you do not have to worry about that. It acts like a function. However, it does mean that you can save some typing. Therefore, the parenthesis is not required.

Now, let's look at some of the typical PHP syntax in the script. The beginning and end of a PHP code section is marked by angle brackets (less than and greater than signs) followed or preceded by a question mark. The end of each line is marked by a semicolon. The "\n" is a special "escape code," meaning print a new line (or "newline") that represents a line break in plain text.

First, PHP outputs a content type header stating the following output is to be HTML. It then sends the P tag and "Hello" to the browser untouched. Once encountering the PHP start tag, it then executes the code until it reaches the PHP end tag.

Since PHP 4.0.3 the track_vars configuration directive has been turned off by default. This is a security measure to stop people setting random vars by simply passing them in the URL string. If you didn't implicitly set a variable within your script this could cause problems. For this reason, track_vars should be turned off.

The way to access the form variables is by the arrays $_GET and $_POST depending on what your form is using. So if you have a text input called "name" in your form, just do:

```
$name = $_POST['name'];
```

This is the basic structure of many browser-based applications. If the form variable does not exist in the CGI environment, we know to display the form asking the user for their name.

```
<?php
$frmName = $_POST['frmName'];
if(!$frmName)
{

//Switch to HTML mode to display form
?>
<form action="hello-web.php" method="post">
Name:
```

```
<input type="text" name="frmName" size="24">
<input type="submit" value="Submit">
<?php
}else{
?>
Hello <?php echo($frmName); ?>
!
<?php
}
?>
```

At the very end, we switch back to PHP mode just to close the else clause. This may be overkill in this small application, but this kind of mode switching part of the php-way. You could use echo or print commands and forget the mode switching.[15]

Here is a PHP 4 database example.

```
<?php
$dbname="php.dbf";
if (!$fp = dbase_open($dbname,0)) {
    echo "Cannot open $dbname\n";
    exit;
}
$nr = dbase_numrecords($fp);//Number of
    records.
echo "<Br>";
echo "Data Table<Br><Br>";
for ($i=1; $i <= $nr; $i++) {//From 1 to $nr as
    you know.
    $temp = dbase_get_record($fp,$i);
//if (chop($temp[0]) == $frmName) {//
    $frmName comes from FORM via WEB
echo "$temp[0], $temp[1], $temp[2] <Br>";
$data_array["column1"][$i-1] = $temp[0];
$data_array["column2"][$i-1] = $temp[1];
$data_array["column3"][$i-1] = $temp[2];
//}
}
array_multisort($data_array["column1"],
    SORT_STRING, SORT_ASC,
    $data_array["column2"],$data_array["column3"]);

echo "<Br>";
echo "Sorted by Column # 1<Br><Br>";
echo $data_array["column1"][0];
echo ," ";
echo $data_array["column2"][0];
echo ," ";
echo $data_array["column3"][0];
echo "<Br>";
echo $data_array["column1"][1];
echo ," ";
echo $data_array["column2"][1];
echo ," ";
echo $data_array["column3"][1];
echo "<Br>";
echo $data_array["column1"][2];
echo ," ";
echo $data_array["column2"][2];
echo ," ";
echo $data_array["column3"][2];
echo "<Br>";
echo $data_array["column1"][3];
echo ," ";
echo $data_array["column2"][3];
echo ," ";
echo $data_array["column3"][3];
dbase_close($fp);
exit; ?>
```

You can view the results of the example above at:

[http://www.utilityreporting.com/php/dbase.php4]

FoxWeb

FoxWeb is another technology that enables developers to create dynamic web pages, especially if the pages involve underlying FoxPro databases. Like the ASP script engine, the FoxWeb script engine can handle complicated database manipulations, especially for FoxPro databases. This makes FoxWeb very attractive to developers who need to convert legacy desktop applications powered by FoxPro databases to web-based, FoxPro-driven applications. Similar to an ASP web page, a FoxWeb web page is essentially a text file with extension of *.fwx that contains HTML code as well as FoxWeb scripts. Also, the latest FoxWeb scripting object is compatible with Microsoft's Active Server Pages (ASP) objects. This makes it easier for developers who are already familiar with ASP to become familiar with FoxWeb.

CLIENT-SIDE TECHNOLOGIES

Applications for power systems analysis tend to have different features from e-Commerce applications. Power system applications are by nature computation-oriented. It may be neither efficient nor practical to web-enable them with server-side technology. However, client-side technology may be worthwhile for consideration in this case.[1]

There are a number of popular client-side technologies available to implement extra functionality on top of HTML. This section discusses three standard client-side technologies, client-side scripts, Java applets, and ActiveX controls.

Client-side scripts are probably the simplest client-side technology. A client-side script is just a collection of commands interpreted, rather than compiled, by the web-browser. The browser interprets and executes the script to create more dynamic HTML pages, when certain events occur. Client-side scripts are relatively easy to implement, but limited in functionality compared with full-featured programming languages. Another drawback is that the code can be viewed from the browser. Hence, any proprietary knowledge should not be implemented in client-side scripts. The most popular client-side script language is JavaScript. However, it is not the only one. VBScript can also be used for client-side script programming.

JavaScript

JavaScript is used in millions of web pages to improve the design, validate forms, and much more. JavaScript was developed by Netscape and is the most popular scripting language on the internet. JavaScript works in all major browsers that are version 3.0 or higher.

- JavaScript was designed to add interactivity to HTML pages
- JavaScript is a scripting language—a scripting language is a lightweight programming language
- A JavaScript is lines of executable computer code
- A JavaScript is usually embedded directly in HTML pages
- JavaScript is an interpreted language (means that scripts execute without preliminary compilation)
- Everyone can use JavaScript without purchasing a license
- JavaScript is supported by all major browsers, like Netscape and Internet Explorer

This example:

```
<html>
<body>

<script type="text/javascript">
document.write("Hello World!")
</script>

</body>
</html>
```

produces this output:

Hello World!

VBScript

VBScript is a scripting language. A scripting language is a lightweight programming language. VBScript is a light version of Microsoft's programming language Visual Basic

When a VBScript is inserted into an HTML document, the internet browser will read the HTML and interpret the VBScript. The VBScript can be executed immediately, or at a later event.

This example:

```
<html>
<body>

<script type="text/vbscript">
document.write("Scripts in the body section are
    executed when the page is loading")
</script>

</body>
</html>
```

produces this output:

Scripts in the body section are executed when the page is loading.

Java Applets

A Java applet is a Java program downloaded from a web server into a client browser and runs inside a Java Virtual Machine (JVM) that is integrated into the web browser. Since they are written in Java, applets have all the features of Java such as platform independence, built-in security mechanism, memory management, error handling, etc. In addition, because Java is a full-featured programming language, applets are nearly as powerful as stand-alone desktop applications. Applets are well able to locally process intensive user interactions like drawings, which are often difficult or impossible to implement with HTML and scripts.

Browsers like Internet Explorer (IE) may provide their own JVM, but Sun provides a standard JVM, Java Plugin, that has some significant advantages. The biggest one is probably the Swing classes in Java Plugin, as opposed to the AWT classes in the browser

JVM. Both Swing and AWT are used for GUI development, but Swing is much faster and consumes less memory than AWT. In addition, from the point of view of developers, Swing is probably the cleanest design and implementation of a windowing system. It has well-defined relationships between containers, components, and UI elements. Swing's architecture is based on the Model-View-Controller (MVC) design pattern, which separates data from presentation and the manipulation of that data. The price for this is to initially download a Java Plugin into the client machine. Since the download is just a mouse-click and only performed the first time, it may be worthwhile if the web-based application requires a complicated GUI. Figures 24-2a and 24-2b show different code segments for tags used in a web page to invoke the browser JVM and Sun's Java Plugin, respectively.

Security is built into Java applets to prevent any writing operation on users' disks. If write access is desired, applets can be digitally signed by vendors and authenticated by third-party authorities to ensure security.

Java Web Start (JWS) is an emerging approach to creating web-based applications. From the user's point of view, JWS is similar in nature to Java Plugin, but differs mostly in the first step. JWS currently requires more involvement from administrators and users alike, which is more tedious than the browser's automatic installation of the Java Plugin. In addition, from the developer's point of view, JWS is more like a web-deployment solution. After the initial launch from an HTML link, a JWS application has little connection or interaction with the

```
<HTML>
<HEAD>
<TITLE>Applet page : Browser's JVM</TITLE>
</HEAD>
<BODY>
<APPLET CODE=HelloWorld.class ARCHIVE=HelloWorld.jar>
Your browser may not support Java applets
</APPLET>
</BODY>
</HTML>
```

Figure 24-2a. Using Browser's JVM

```
<HTML>
<HEAD>
<TITLE>Applet page : Sun's Java Plugin</TITLE>
</HEAD>
<BODY>
<OBJECT classid="clsid:8AD9C840-044E-11D1-B3E9-
00805F499D93"
width=100% height=100
codebase="./j2re-1_3_0_02-win.exe#Version=1,3,0,2">
   <para name="code" value="HelloWorld.class">
   <para name="archive" value="HelloWorld.jar">
   <para name="cache_archive" value="HelloWorld.jar">
   <para name="cache_option" value="Plugin">
</OBJECT>
</BODY>
</HTML>
```

Figure 24-2b. Using Sun's Java Plugin

parent browser, while an applet can interact more with the parent browser.

ActiveX controls

ActiveX technology is an alternative to Java technology for complex web-based UI development. The term ActiveX generally refers to ActiveX controls that are software modules based on Microsoft's Component Object Model (COM) architecture. ActiveX enables a program to add functionality by calling ready-made components that blend in and appear as normal parts of the program, such as toolbars, calculators, and even spreadsheets. The power of ActiveX controls in web development is that they can be linked to a web page and downloaded by any ActiveX-compliant web browser. This virtually turns web pages into software pages that can perform operations just like any desktop application. An ActiveX control can be written in a variety of programming languages from Microsoft's suite of Visual tools, including Visual Studio.Net.

ActiveX controls can be quite dangerous and even lethal to client machines because they make extensive use of the windows operating environment. To add some security, most browsers can be set to reject the download of ActiveX controls that are not marked safe for scripting and that are not digitally signed by the developer with a certificate issued by a third party verifier such as VeriSign.

However, if the application is used in a secure environment such as an intranet then ActiveX can be quite safe. Here's an example of how the Tabular Data Control works. Tabular Data Control is a Microsoft ActiveX control that comes pre-installed with all versions of IE4+.

The data file mydata.txt looks like this:

```
Utility | Units
~Electric ~ | ~KWH~
~Water ~ | ~KGAL~
~GAS ~ | ~BTUS~
```

The HTML page example1.htm looks like this:

```
<OBJECT ID="data2" CLASSID="CLSID:333C7BC4-
    460F-11D0-BC04-0080C7055A83">
        <PARAM NAME="DataURL"
        VALUE="mydata.txt">
        <PARAM NAME="UseHeader"
        VALUE="TRUE">
        <PARAM NAME="TextQualifier"
        VALUE="~">
        <PARAM NAME="FieldDelim" VALUE="|">
</OBJECT>
<TABLE DATASRC="#data2" BORDER="2">
<THEAD>
        <TH>Utility</TH>
        <TH>Units</TH>
</THEAD>
<TR>
        <TD><SPAN DATAFLD="Utility"></SPAN></
TD>
        <TD><SPAN DATAFLD="Units"></SPAN></
TD>
</TR>
</TABLE>
```

The following page will display the results:

[http://utilityreporting.com/activex/example1.htm]

Utility	**Units**
Electric	kWh
Water	kgal
GAS	Btus

You can read more about the Tabular Data Control at:

[http://wsabstract.com/javatutors/tdc.shtml]

In summary, different client-side technologies have different features and fit different tasks. Client-side scripts may serve as glue to connect HTML, applets, ActiveX controls, and other server-side technologies. While client-side scripts may be the simplest in functionality and easiest to program, applets and ActiveX controls are more complicated and more capable. With the help of Swing, Java applet is very good choice for interactive GUI development and offers other advantages such as platform independence, built-in security mechanism, etc. ActiveX controls in C++ may be the most capable to handle intensive engineering computation such as domain logic, since the performance of C++ is the best among popular languages. In the final analysis, an engineering web-based application is usually a combination of all these technologies taking advantage of the situational strengths of each.[16]

CONCLUSION

Not all of the technologies discussed in this chapter are required to build effective utility data web pages. Also, there are certainly other technologies not discussed here that can be helpful. Pure HTML is the most reliable method of designing web pages. XHTML is the latest version of HTML providing the syntax necessary to incorporate XML data into web pages. XML provides portability of data between database formats. CSS makes web pages less complex, more standardized and produces smaller, faster loading pages. DHTML adds some features that make web pages more intuitive but add some risk that different web browsers may interpret the page differently. Scripts like JavaScript, VBScript, PHP and ASP add the functionality of a subset of higher level programming languages into web pages. Server-side scripts are more secure but place the processing burden all on the server. Therefore, response times may be slow. Client-side scripts are less secure and limited in functionality but distribute the processing burden across the client machines leading to faster response times. Java applets are most useful for graphing trends in data.

References:

[1] Fangxing Li, Lavelle A.A Freeman and Richard E. Brown, "Web-enable Applications for Outsourced Computing," IEEE *Power and Energy Magazine*, vol.1, no.1, pp53-57, January-February 2003.

[2] "XHTML.ORG"[article on-line] (August 4, 2000, accessed 6 May 2003); available from http://www.xhtml.org/; Internet.

[3] Infinite Software Solutions, Inc.,"DevGuru XHTML Introduction"[article on-line] (accessed 6 May 2003); available from http://www.devguru.com/Technologies/xhtml/quickref/xhtml_intro.html; Internet.

[4] Richmond, Alan,"DevGuru XHTML Introduction"[article on-line] (February 2, 2000, accessed 6 May 2003); available from http://www.wdvl.com/Authoring/Languages/XML/XHTML/; Internet.

[5] Refsnes Data,"XHTML Tutorial"[article on-line] (accessed 6 May 2003); available from http://www.w3schools.com/xhtml/; Internet.

[6] W3C,"W3C Markup Validation Service"[article on-line] (December 5, 2002, accessed 6 May 2003); available from http://validator.w3.org/; Internet.

[7] Western Civilisation Pty. Ltd.,"The Complete CSS Guide"[article on-line] (accessed 6 May 2003); available from http://www.westciv.com/style_master/academy/css_tutorial/toc.html; Internet.

[8] Traversa, Eddie,"Intro to DHTML"[article on-line] (September 13, 2000, accessed 6 May 2003); available from http://www.dhtmlshock.com/articles.asp?ArticleID=1; Internet.

[9] " Rough Guide To The DOM"[article on-line] (accessed 14 May 2003); available from http://www.devshed.com/Client_Side/DHTML/DOM/DOMpart1/page1.html; Internet.

[10] W3C,"XML-Data"[article on-line] (January 5, 1998, accessed 6 May 2003); available from http://www.w3.org/TR/1998/NOTE-XML-data-0105/; Internet.

[11] NCSA,"Common Gateway Interface"[article on-line] (accessed 6 May 2003); available from http://hoohoo.ncsa.uiuc.edu/cgi/intro.html; Internet.

[12] Kadow, Kevin,"Common Gateway Interface, Quick Reference Guide"[article on-line] (accessed 6 May 2003); available from http://www.msg.net/tutorial/cgi/; Internet.

[13] Sun Microsystems, Inc.,"Java Servlet Technology, The Power Behind the Server"[article on-line] (accessed 6 May 2003); available from http://java.sun.com/products/servlet/; Internet.

[14] Sun Microsystems, Inc.,"JavaServer Pages, Dynamically Generated Web Content"[article on-line] (accessed 6 May 2003); available from http://java.sun.com/products/jsp/; Internet.

[15] Knoblock, Steve,"Getting Started With PHP"[article on-line] (accessed 6 May 2003); available from http://www.phphelp.com/article/1p1.php; Internet.

[16] Fangxing Li, Lavelle A. A. Freeman, "Using Client-side Technologies to Develop Web-based Applications for Power Distribution Analysis," Distributech 2003 Conference, February 4-6, 2003, Las Vegas, Nevada.

Chapter 25

Utility Data Web Page Design—Presenting the Data

David C. Green
Green Management Services, Inc.

PRESENTING UTILITY DATA in a meaningful way may be an art or it may be a science. However, one thing is for sure, the design of the presentation depends entirely on those who will be using it. Once the general specification for the utility information is created, it is the responsibility of the developer to make the pages informative, intuitive, reliable and robust. Using web technologies, a developer can make large amounts of utility data come alive with informative "vision" into utility consumption. Web pages full of utility data can be transformed instantaneously from one report to another. Trends can show critical values that may have been missed otherwise. Data can be compared and analyzed immediately rather than sifting through reams of paper reports and spreadsheets. The user of the data becomes the master of it rather than the other way around.

INTRODUCTION

The two most important things to remember about displaying utility data on a web page are to display as much data as will fit on the page and to use as many links as possible. Links are both informative and functional. They display data as well as provide a means of changing the display. Database filtering options are lists or text fields used to define dates, locations, type of utility or other variables. They should be visible and available to change at all times to make the application robust. These options are used to narrow the scope of the data returned from the database query and displayed in the reports. Reports provide both the means of displaying the information found in the database and the links that allow the user to navigate the application.

Links are used to navigate to graphs or help pages. They connect to filter routines or sort routines to change the data display. "Onmouseover" is a DHTML feature that helps make the links more intuitive by displaying a description of the actions in the message line of the browser. Toggle links are great for options that only have two or three choices and do not change often. The user can browse through the option choices very quickly. Browser arrows are used to change a date, day, month, year or any other value, incrementally up or down.

Options are usually made available in a separate frame from the data display. The option values are stored in a "user table" uniquely named for each user. This insures the selected options will remain until the next time the user accesses the application. HTML forms will be used most of the time to select and submit values for options. Drill down lists are used to change levels and provide lists of options specific to the selected level.

Reports containing both data and links are displayed in a frame of their own. Reports must always display the values of options so that if the report is printed it is well qualified. Filtering or sorting changes the reports most dramatically. Filter and sort options can also be stored. Data that falls outside a specified range is highlighted in the report. Highlighted data are called a flag since it stands out in the report. Flags help the user focus on problem areas. Page widths must be limited for viewing and printing the data.

Graphs are an important way of conducting trend analysis on utility data. Line graphs and perhaps bar graphs are useful for creating comparison graphs from month to month or year to year. It is important to include legends and all qualifying information on the graph. Dual-axis graphs are useful for comparing utility values to some other value such as outside temperature data.

It is quite helpful to provide a means of creating an image file of the graph for emailing to others. Some graph applications are "server-side" applications that create the images first and then use them to display in a web page. Other methods use Java applets or Active-X to generate the images on the client browser in real-time.

LINKS TO GRAPHS, HELP SCREENS, FILTERS AND SORTS

HTML links are what bring utility data alive. They provide the path through the data that give the application its robust look and feel. They make the application intuitive for the user by providing a quick response to easy requests for change. Changing the presentation of the data is essential to providing the information in a way that is meaningful to a wide range of users. Links provide instant access to information not directly related to the data itself, such as help instructions or contact information. Options can be quickly changed using links rather than pull down lists or text fields. Links should be used as first choice for changing the web page content.

The "onmouseover" event method provides information about links to help the user determine the effect of its action before clicking it. Holding the mouse cursor over a link will cause a custom message, pertinent to the link, to appear in the browser status field at the bottom of the window. The message can include data retrieved dynamically from form fields or databases as well as text. The messages add to the help features of the application by letting the user anticipate the functionality of the application with very little effort. Using the "onmouseover" event adds some overhead to the processing required to load the web page in the browser window. However, their helpfulness far outweighs the cost of implementing them.

The anchor element to create the link with the "onmouseover" event method looks as follows:

```
<a href=http://localhost/scripts/foxweb.exe/a01@udp/
programs/view?sortby=school~
onmouseover="window.status='Select This Link to
Sort the List by School'; return true" ><font
size=1>SCHOOL </font></a>
```

Toggles are an especially quick way to change options related to a particular page design or its content. A toggle is a link that changes between two or three choices at the most. Each time the user clicks the link a new choice is displayed as the link. This makes it easy for the user to scroll through choices rather than picking from a small list. Toggles are best used for options that do not change frequently. The following example shows a toggle for turning the auxiliary axis of a graph on or off.

Aux Axis Off ⟵⟶ Aux Axis On

Browse arrows are great for sequential options such as dates for integers. A set of arrows, one on either side (or top and bottom) of the value can be linked to the next value in the sequence. This lets the user quickly scroll through a large list of option values in either direction if desired. Often times the number of date options are too numerous to list. Typing in the dates requires too many keystrokes. Calendars are nice for large jumps in values, but require at least a few mouse clicks for each change, whereas browse links can change values with a single click. For dates, the combination of a calendar and browse buttons provides the best access to those option values. Varieties of arrow images are available for free download on many of the image internet sites. In the example below, clicking on June 2002 will bring up the relevant data for that month. Selecting the arrows on the left will change the month to the previous month—May 2002. Selecting the arrows on the right will change the month to the next month—July 2002.

<<June 2002>>

OPTIONS

Options are the heart of the data presentation. Options describe which data to display and they describe how to display the data. Options are unique to each user concurrently. A user enters and changes options by using links or forms. Sequential lists of links can be helpful in "drilling" down to a particular option such as a location. Frames are sometimes helpful to separate the options display from the content display. Common options are dates, utility types, etc.

Obviously each user requires their own particular options and it would be nice if those options would be retained as long as the user desires. This is best done by using a user data table to store the current options for a particular user. User data tables are stored on the server side of the web application and therefore are quite reliable. The table can be named by a unique identifier associated with the user like the user id (if the application

LABEL	USERID	MONTH	YEAR	AUXAXIS	DISTRICT	SORTBY
A01	GREEND	6	2002	ON	GREEN	PCTBTU/SF
A01	PJA	7	2002	OFF	GREEN	COST

requires a log in). This method provides a "state" for the user that remains indefinitely.

HTML forms are used to select and store options. They contain pull down lists, input text fields, radio buttons or check boxes. All forms require a "submit" button to execute some process that sends the form field values to their storage location. Forms should only be used if necessary since they tend to distract the user from the data itself. Chances are the form will have to temporarily occupy the same space on the page as the data. Forms are ok for configuration items that the user will rarely change after the initial setting. Sometimes options simply require many choices, and forms are needed.

Figure 25-1. List of School Districts

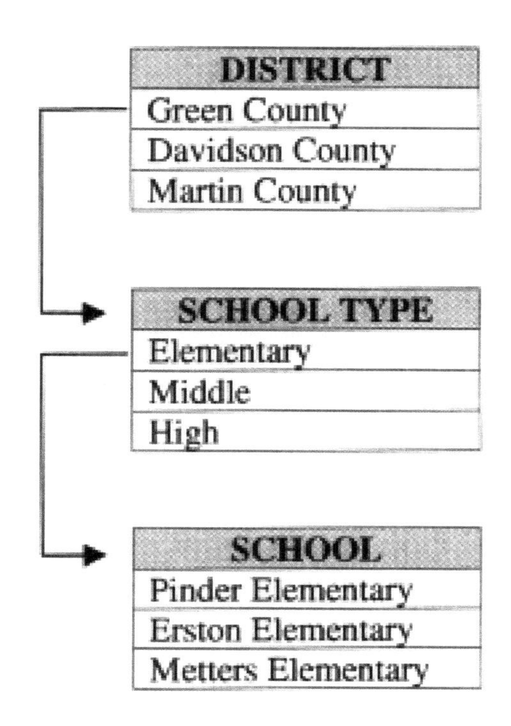

Drill down lists are a great way to lead the user to selecting from a group of hierarchical values such as building locations within large complexes that are part of a corporation with facilities in several cities. A list of links associated with each level and linked to the level below would allow the user to quickly move from the top of the hierarchical tree down to whichever level is desired. Once the appropriate level is reached a list of options and data specific to that level could be displayed.

Examples for options in energy use reports are location levels such as complex, facility, building or meter. Then there are the values for each of the levels such as specifically which complex, facility, building or meter. An option for which utility to focus in on is most definitely needed (electric, water, gas, etc.). The date is also a common option. Date options can take many forms as in month/year combinations, month only, year only, month/year range, date only (single day), date range, etc. It's useful to provide an option to define comparison data to be shown along with the primary data on a graph. An example is outside temperature data. However, any data may be needed as comparison data along with the primary data set.

REPORTS

Reports are best displayed in HTML frames. This allows the option choices to be separated from the database information itself. However, option values displayed in the report provide background information to describe the data. Links are used to navigate throughout the frames. Filtering links allow the user to "zoom" in on specific areas of interest: sorting links make evaluating the data much easier. Links to graphs allow the user a quick look at trends in the data. Flags highlight data with colors or markers that make it stand out. The report HTML inside frame may need to be sized appropriately to fit most printers. Columns and rows can be arranged to best suit the purpose of the report. Use a font that is clearly readable in a web browser. Adding a link to a file containing the data in some other format such as an Excel file, a Word document, or an Acrobat PDF file is quite valuable.

HTML frames are helpful to separate the option choices from the data display so that they are always visible and available to change. This also makes the data display less cluttered and more like a report that can be printed if desired. The database information itself is displayed in the largest frame. Typically, the option frame is located along the left side or the top of the browser window, or both. Frames are frequently referred to by name within HTML and the Document Object Model (DOM) so it is helpful to pick names that are easy to remember. For example, title frame, options frame and display frame, etc. Option values should be redisplayed in the report so that if the report is printed the background information necessary to qualify the data is there. Most of this information can go into the title of the report or the headings of columns.

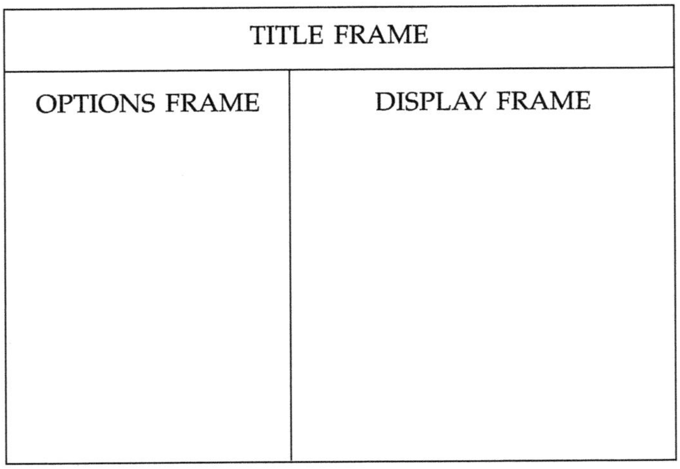

TITLE FRAME	
OPTIONS FRAME	DISPLAY FRAME

Figure 25-2. Layout of a Web Page with Frames

Links allow the user to navigate through the page and frames to change options, filter, sort and graph. Links associated with some items in the report may be used to filter by location, utility type, etc. This gives the user the ability to "drill" down into the report and look specifically at areas of interest or print a report that is not cluttered with irrelevant information. Links on headings are used to sort the columns by that heading. This quickly shows how values in the list compare allowing easy evaluation. Sorted columns should be denoted in some way. Links to graphs show trend analysis quickly without re-selecting options. Pre-defined graphing routines take the data to a larger or smaller scope such as the last 30 days, last 12 months or a 24-hour period. Graphs can be used to compare one period to another or one data set to another.

Links change the sort or filtering by passing a parameter in the HTTP request similar to the sort link shown below. Graph links work in a similar manner. Below is the address used in a link, as it would show up in the browser address field, to sort the data display by "cost."

http://localhost/scripts/foxweb.exe/a01@udp/programs/view?sortby=cost~

Flags are used to make certain data stand out from all the rest. Highlighting the values or marking them in some manner when they breach a certain threshold will quickly show the user which data are critical. For example, kWhs that increased from one year to the next would show a positive percent change and could trigger a flag so the viewer could easily see which locations increased the kWh consumption. Configuration settings can pre-define which data and values will trigger the flags.

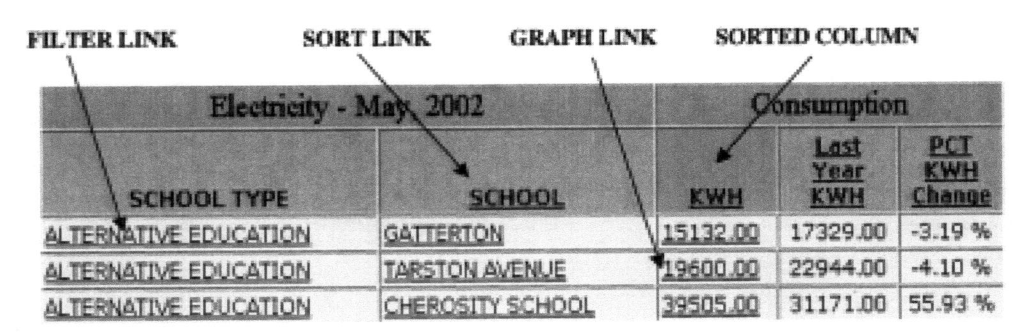

Electricity - May 2002		Consumption		
SCHOOL TYPE	SCHOOL	KWH	Last Year KWH	PCT KWH Change
ALTERNATIVE EDUCATION	GATTERTON	15132.00	17329.00	-3.19 %
ALTERNATIVE EDUCATION	TARSTON AVENUE	19600.00	22944.00	-4.10 %
ALTERNATIVE EDUCATION	CHEROSITY SCHOOL	39505.00	31171.00	55.93 %

Figure 25-3. Data Display Frame with Links

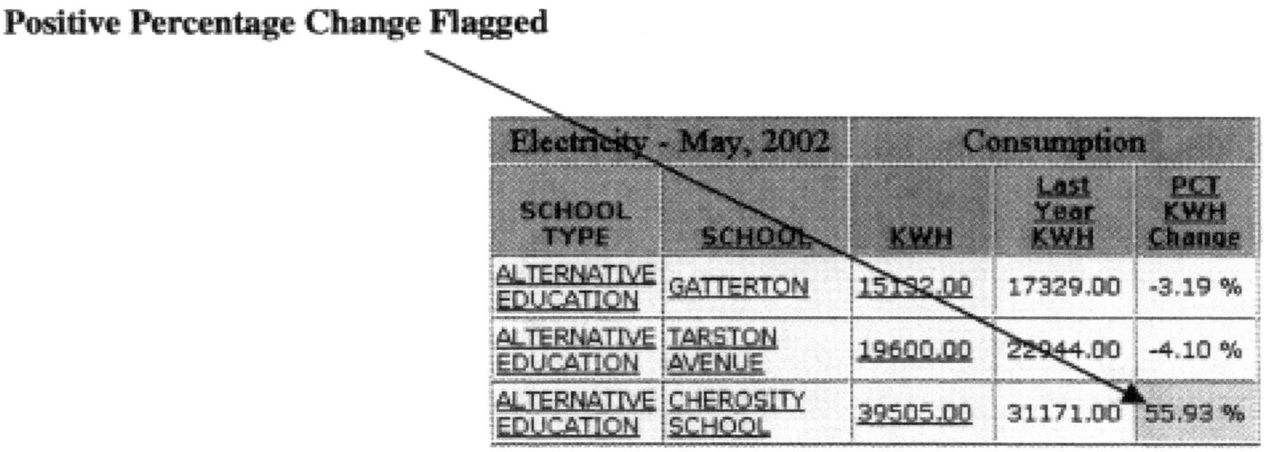

Positive Percentage Change Flagged

Electricity - May, 2002		Consumption		
SCHOOL TYPE	SCHOOL	KWH	Last Year KWH	PCT KWH Change
ALTERNATIVE EDUCATION	GATTERTON	15132.00	17329.00	-3.19 %
ALTERNATIVE EDUCATION	TARSTON AVENUE	19600.00	22944.00	-4.10 %
ALTERNATIVE EDUCATION	CHEROSITY SCHOOL	39505.00	31171.00	55.93 %

The display frame may need to be sized appropriately to view or print the data. Portrait printouts typically need to be no more then 670 pixels wide. Landscape printouts can be larger, up to 970 pixels, but require a note to remind the user to adjust their page settings to landscape mode before printing. Use the **TABLE** HTML element to control the page width as shown below.

<table border=1 width="670" CELLSPACING=0 CELLPADDING=2>

Sometimes the reports are more complicated than just rows and columns. Some columns may need to be split into sub-columns to further define a different set of data. The same goes for the rows. Rows can be combined to organize the data into groups that standout by highlighting in different colors or separating with a blank row. A good font is Verdana or Arial. It is recommended to use a few alternate fonts in case the user's browser does not accommodate the primary one. Using the FONT HTML element you can specify alternate fonts as shown below.

<font face=Verdana, Arial, Helvetica, sans-serif size=1>

It is a good idea to provide the data in other formats for compatibility with other programs. A simple link at the bottom of the report or options frame can lead to a report in another format such as MS Excel or XML.

GRAPHING

Graphs allow for trend analysis of the data on a regular basis. Line graphs are the most useful since data sets can easily be compared. Graph titles, axes, and legends should contain enough of the information found in the options to qualify the data. Graphs can contain multiple data sets for comparison along the same axis or opposite axis. Outside temperature is a good item to place on an opposing axis. Graphs are created in the browser window or frame using a variety of methods. Java applets or Active-X make use the of client processor to display the graphs but often lack the ability to create an image for emailing or saving. Server-side graphing applications provide the graph as a GIF or JPEG image to begin with and then display it within the HTML of the frame or window.

Trend analysis gives the user the "big picture" about what the data are doing over time. It is a lot easier to evaluate a graph than to look at a list of numbers even if they are sorted appropriately. It makes communicating the issues much easier as well.

Line graphs seem to be best suited for showing incremental data spread out over time. In addition, they easily accommodate comparison data sets without cluttering up the graph to the point that it is unreadable. Graph titles need to contain the same information that a report would contain. The remainder of the qualifying information needs to be present in the axis titles or the legend. Legends are placed immediately adjacent to the graph. Most graphing applications allow for flexible configuration of the fonts and placement of the legends. Comparison graphs are used to evaluate two or more data sets to see if a trend is consistent across locations or to compare data sets to values of some other nature such as weather data.

When compared data sets differ in values to the point that the axis will be stretched too far to show accurate results, then those data sets that are outside the normal range need to be placed on an opposing axis. In the graph below, two data sets comparing values of elec-

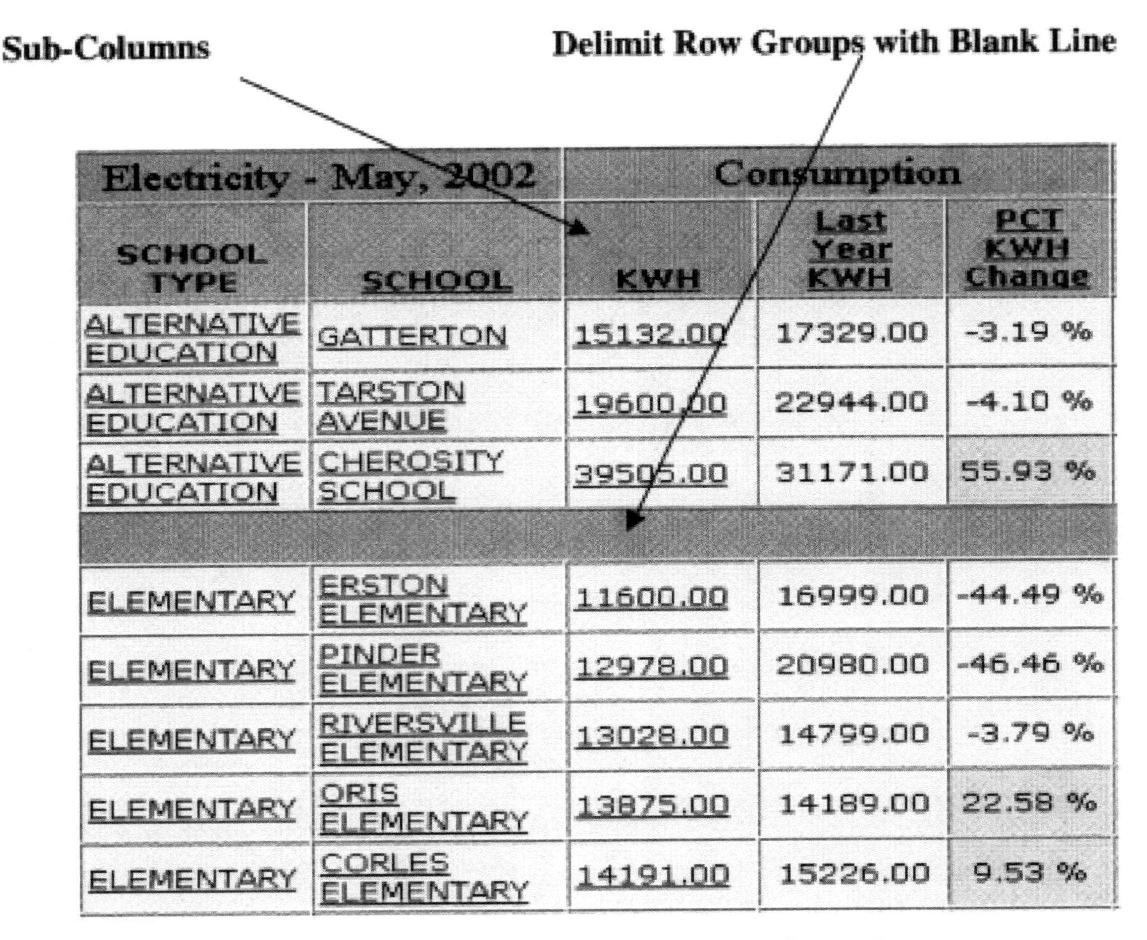

Electricity - May, 2002		Consumption		
SCHOOL TYPE	SCHOOL	KWH	Last Year KWH	PCT KWH Change
ALTERNATIVE EDUCATION	GATTERTON	15132.00	17329.00	-3.19 %
ALTERNATIVE EDUCATION	TARSTON AVENUE	19600.00	22944.00	-4.10 %
ALTERNATIVE EDUCATION	CHEROSITY SCHOOL	39505.00	31171.00	55.93 %
ELEMENTARY	ERSTON ELEMENTARY	11600.00	16999.00	-44.49 %
ELEMENTARY	PINDER ELEMENTARY	12978.00	20980.00	-46.46 %
ELEMENTARY	RIVERSVILLE ELEMENTARY	13028.00	14799.00	-3.79 %
ELEMENTARY	ORIS ELEMENTARY	13875.00	14189.00	22.58 %
ELEMENTARY	CORLES ELEMENTARY	14191.00	15226.00	9.53 %

Figure 25-4. Data Display with Sub-Columns and Row Groups

tric consumption for two different years are on the left-hand axis. The average outside temperature is on the right hand axis. This makes the graph very readable; the change in energy use from one year to the next is easily viewed, showing a reduction in use over the previous year. The use for each year can be compared against average temperatures to see what the weather-related effects on energy use are.

Client-side applications such as Java Applets and Active-X generate graphs using the client machine. Java applets have to be downloaded the first time but are stored on the client machine and respond quickly afterward.

The downside to client-side applications is that they do not easily create images of the graph. The graph results can easily be communicated to others by simply emailing an image or inserting it in a document. Server-side graphing applications create images first and then embed them into the resulting web page or frame. This makes it easy to transfer the information elsewhere and it doesn't require client processor time to generate the graph.

CONCLUSION

Utility data are complex. It is directly related to time, temperature, economic conditions and perhaps even political variables. It's important to see the "big picture" and analyze the data in an objective light. However, many people are involved in energy decision making, and they each desire a unique angle to examine the data and turn it into meaningful information.

Web technologies provide the flexibility and portability required to query and present utility data to many different people in many different ways. These web technologies merge together into a mosaic of databases, programming scripts, links, forms, reports and graphs that tell the story of the data collected. A robust ability to select options quickly and easily allows the

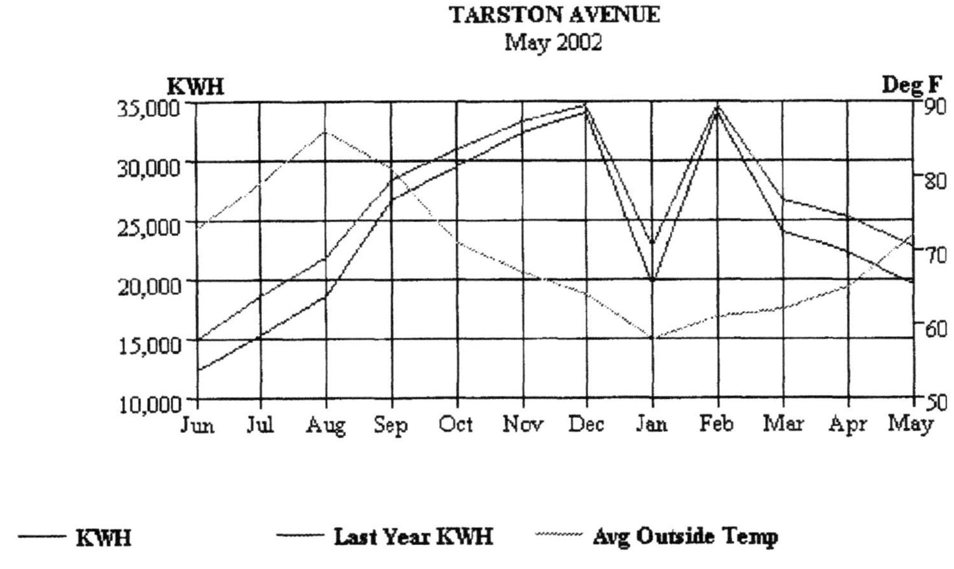

Figure 25-5. Comparison Graph with Different Left and Right Axis Values

user to ask questions of the data and get answers. Web enabled database engines provide data on demand per the requests of the users. Reports provide a familiar basis for understanding the data while the graphs show trends over time. Links to sorting, filtering and graphing routines provide an endless number of "views" for analysis of the data. However, those views are manageable. Created on demand, from logical choices, the data are streamlined to fit the need of the analyst. Visit UtilityReporting.com to try out a utility data presentation demo and explore links to information on many web technologies.

Acknowledgments

Thanks to Bill Bragg for showing me what "reams of data" really look like and confirming that there is a need for us to find some way to "master" the huge amount of data that surrounds us in this Information Age. His short, but meaningful, discussions on the subject were a great inspiration.

Section Nine

Integration and Communication Technology Issues in Implementing EIS and ECS Systems

Chapter 26

BACS Industry Shaped by Standardization and IT Technology*

*Hans R. Kranz***
Marketing and Standardization
Siemens Building Technologies Ltd
Building Automation, Zug, Switzerland

A SHORT OVERVIEW of today's trends in BACS (building automation and control systems) is presented. The influence of standard protocols and web technology is examined, and some of the most promising ones, in the author's view, are analyzed in more detail. The actual status of the different standard protocols by international standard bodies is outlined. Finally a prediction is given as to how the future in the building automation world could look in the next few years.

INTRODUCTION

Facilities must meet increasingly stringent business requirements, while reducing operating and maintenance costs. Leaner budgets and environmental considerations demand advanced energy management to help reduce energy and utility costs. Growing expectations for comfort and service require advanced managerial instruments. Since every enterprise has a unique pattern of building management requirements, choosing the right solution to meet specific needs is crucial.

It's time to change the way we look at facility and building management. Only a coherent automation and control framework, within which many different systems can evolve, can remove the barriers to progress. A common interface, an open system, normative system standards, web technology and advanced integration are prerequisites for success.

An integrated management system enables a building to be operated more easily and with greater efficiency, and helps to reduce operating costs and increase profitability. Instead of using a variety of dedicated devices for operation, the facility manager's team can perform all activities from operator stations with the same visual display and operating characteristics, keeping training and operating costs to a minimum and reducing the likelihood of incorrect operations. Dedicated security systems, of course, send their alarms to the security guards and the fire brigade. However, if information for all systems in the building is brought together to one or more networked operator stations with a unique human interface, faults and alarms can be identified and dealt with immediately, which can provide better protection of people and property.

Stable and reliable communication is one precondition for system integration. Protocols and functions must be described exactly if successful performance and integration are to take place. In the case of multi vendor systems integration, the right contracts, mutual coordination and agreed-upon responsibility are indispensable. Nevertheless, there are practical limits to implementation:

- The fusion of different concepts is difficult and demands considerable investment of time and energy. This is hardly justifiable for a single application!
- The delimitation of responsibility at the interfaces and uncoordinated alterations in the subsystems represent almost intolerable complications for integration.

To overcome these limits the BACS industry has agreed on some communication standards, which de-

*Derived from a web publication for "www.AutomatedBuildings.com" (January 02) and used with their permission.

scribe the basic functions in the areas of HVAC, lighting, safety, etc. These basic functions, defined as objects, and their values can then be easily exchanged by different services, disciplines or management systems and with a minimum of effort via communication networks, routers, gateways etc.

Integration plays a key role in bringing down operating costs and allows us to look at the building as one single process. Integration will only be really successful, however, when standard protocols are implemented internally within the individual systems. This eliminates complex conversion and, above all, too much loss of functionality. The standardization of engineering data and the engineering and commissioning workflow in the future will further help to lower costs and to increase the reliability and functionality of integrated systems.

Typical applications in a building that need to be controlled are: air, water, lighting and sun-blinds, heating, cooling, ventilation, energy distribution, safety and security, refrigeration, transportation and ancillary systems. These could all be controlled successfully and more easily with an integrated energy management system. Multi-facility systems also lend themselves to integrated controls.

Below is a list of actual customer requirements.

- Reduction of initial investments
- Building automation using existing IT and CT infrastructure
- A unique user interface for all building disciplines (HVAC, Lighting, Elevators, Fire, Access, Intrusion, ...)
- Higher supplier independence in case of later system enhancement
- Reduction of operating, energy and maintenance costs
- Ability of mobile operators to operate the building at any time, from everywhere, immediately
- "Obvious" operating ability needed for less trained operator
- Technical competence center to serve many buildings in a defined geographical region
- Growing expectations for comfort and service
- Reduced energy consumption

The answer: System and Communication Standards and Web-Technology

IT & WEB TECHNOLOGY STATE-OF-THE-ART

IT and web technology is widely used by BACS. This allows smooth data-exchange with the customer's office automation system and business process. Existing communication infrastructures such as LAN/WAN (wide area networks) can be shared—which reduces cabling cost and allows flexible solutions. Web technology brings additional benefits such as easier and centralized software updates and maintenance and local operation using thin, low cost clients (only a browser is needed). Some of the most important technologies from the IT world are: ethernet, TCP/IP, internet, web, OLE/COM, XML, and SOAP.

THE TECHNOLOGY BEHIND...

System Standards for BACS

The main expenditure for using BACS systems is in the efforts for the complete engineering of the systems. The costs are in direct relation to the type and number of functions necessary. The input-output points just provide—or use—the signals. To avoid additional costs at tendering, execution and billing, the use of standardized terms is necessary. Free and fair competition, including the potential for innovations, is only possible by specifications using a unique language for all parties involved. The CEN/TC247WG3 (Committee for European Standardization) has developed a set of standards for a common understanding of hardware and BACS functions. In the case of system integration, this definition of functions specifically helps to determine the tasks of each involved system to avoid double engineering and to calculate the proper costs for the system engineering. This work has been adopted by the relevant ISO Technical Committee TC 205.

Communication Standards for BACS

Standardized communication protocols such as BACnet™ EIB/KNX, LONMARK®, etc. and interfacing standards such as OPC (OLE for Process Control) allow exchange of information between different manufactures devices, systems and programs in an agreed way. Communication standards are the prerequisite for cost efficient integration of various manufacturers' subsystems.

WHAT HAS TO BE COVERED BY OPEN COMMUNICATION STANDARDS FOR BUILDING AUTOMATION AND CONTROL SYSTEMS?

There are several important concepts which must be considered when deciding which protocol is the right solution for BACS. The following issues have to be addressed:

Data Has to provide complex and simple data structures to accomplish daily functions like:

- Data exchange between devices
- Monitoring and operation of inputs, outputs, setpoints, alarms
- Time Scheduling
- On-line Grouping/Regrouping
- Trend/History
- Backup/Restore

The data approach has to be object oriented.

Services Has to support services for system start-up, network management and backup/restore.

Media Has to provide independent media access to support modern IT & CT networking standards and cabling systems.

Extension Has to allow extensions to provide the possibilities of future innovations.

The European standardization committee had to weigh over 100 issues for the selection of the present pre-standard protocols.

BACNET VERSUS LONMARK

If we apply the above criteria to BACnet™ and LonMark® we get the following picture.

LonMark®

The LonTalk® protocol is a proprietary product from Echelon Inc. and belongs to their LonWorks® product family. The defined LonMark templates cover the functionality needed for open interactions at the field level:

- Data exchange and interactions between devices on the same bus
- Monitoring of values such as temperatures, operating states
- Simple control functions e.g. lights on/off, etc.

The field device functions are normally implemented in one single chip (the Neuron®). The

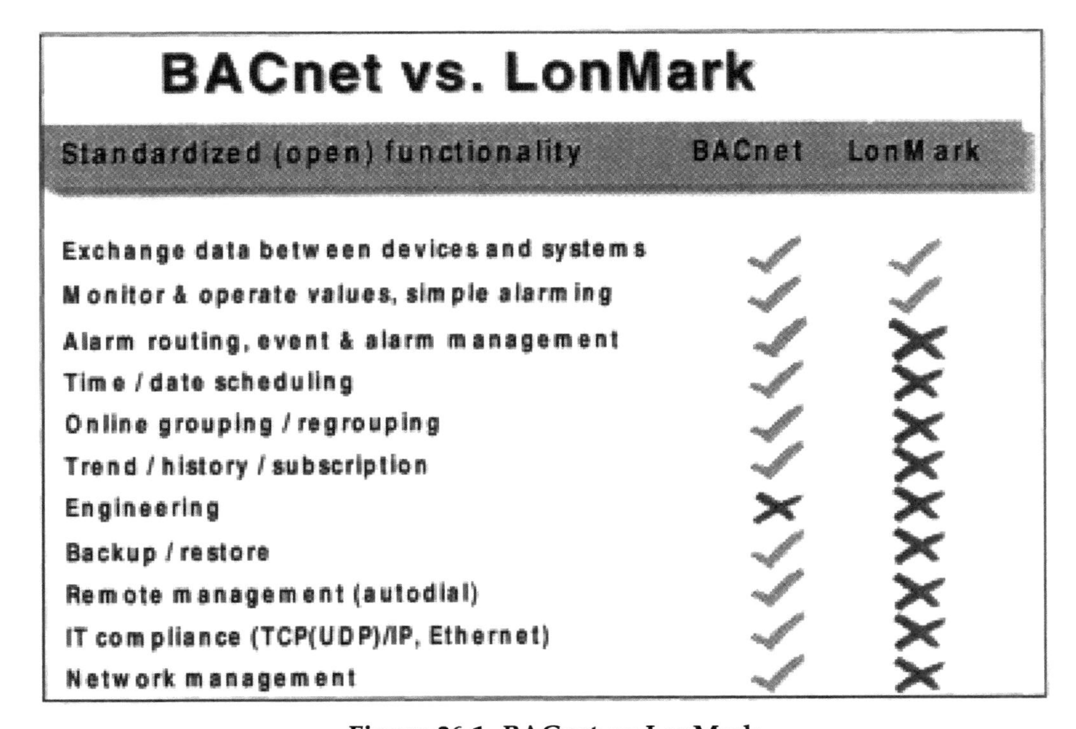

BACnet vs. LonMark

Standardized (open) functionality	BACnet	LonMark
Exchange data between devices and systems	✓	✓
Monitor & operate values, simple alarming	✓	✓
Alarm routing, event & alarm management	✓	✗
Time / date scheduling	✓	✗
Online grouping / regrouping	✓	✗
Trend / history / subscription	✓	✗
Engineering	✗	✗
Backup / restore	✓	✗
Remote management (autodial)	✓	✗
IT compliance (TCP(UDP)/IP, Ethernet)	✓	✗
Network management	✓	✗

Figure 26-1. BACnet vs LonMark

LonWorks® transceivers support a range of transmission media, e.g. the inexpensive Cat. 4 cables. In addition there is a relatively large (>180) number of manufacturers who make devices for building applications, but the product's interoperability has to be checked by the system integrator. All this may give the LonWorks based products a future potential to network at the field level. But "normal" sensors and actuators cost less—if you don't take the overall system costs and the engineering into account.

EIB/KNX

The application of a standardized and internationally accepted protocol for specific applications is still no guarantee that a device by manufacturer A can be exchanged for one supplied by manufacturer B. For this purpose, further standardization of design and of the exact functionality of a device is necessary. One step in this direction has been made by the EIB/KNX concept for building applications. Each of the more than 7,000 available devices for the Konnex bus is supplied with its relevant parameterization data stored on a floppy disc that can be used—together with the unique and neutral EIBA Tool Software ETS—by the installer for combination. (EIBA = European Installation Bus Association, Brussels/Belgium). There are 65 certified training organizations and 15 national EIBA Organizations. More than 80.000 projects have been finished all over the world (2002).

In a convergence project the three European Associations EIBA, BCI, EHSA, have been migrated to the Konnex Association (March 2001) and the protocols "EIB," "BatiBus" and "EHS" merged into the "KNX" (Konnex) protocol. (http://www.konnex-knx.org) The EIB/KNX Communications Protocol is part of the ANSI EIA 776.1-.5 standard and incorporates a CEBus—EIB Router.

Today's market provides such an extreme diversity of applications that an all-comprehensive testing procedure is hardly economically feasible. Testing is less arduous and expensive for less sophisticated devices (e.g. devices for the EIB) whose range of functions has been defined through "interworking standards." The EIB functional blocks describe not only the semantics of a function in English language, but also define how to access the services associated with that function. This is done on the basis of Data Points.

The EIBA has been assigned a unique BACnet vendor ID (74, decimal) by the ASHRAE organization. This vendor ID must be used for the mappings of EIB/KNX applications to BACnet. The document ISO/TC205 WG3 BACnet draft addendum H.5 describes how Konnex Functional Blocks, which are part of Object Interface Specifications (ObIS) of the Interworking Model EIS (EIBA Interworking Standard), are mapped to the corresponding building automation and control networks (BACnet) object model definitions and services.

BACnet

In most other protocols, some of the basic BACS functionalities needed for open interoperability are missing. Services and functions which are needed for mature building management are: trend/history, time scheduling, back-up/restore, remote management, alarm distribution, and many others. BACnet covers all these requirements. In addition, BACnet supports, in a flexible way, modern IT and networking technology such as ethernet and IP (internet protocol) as this is needed in today's modern IT environments. BACnet is therefore the best choice for the upper system levels where a broader functionality and full IT compliance is needed.

EUROPEAN STANDARDIZATION

After having analyzed these main topics in order to decide which standard protocols are preferred for our BACS system, we should also look at the standardization situation in Europe, because the open communication was started earlier there than anywhere else. Implementation began in Germany in 1984 with IBM's FACN protocol for GPAX and Series/1 computers, followed by the FND and Profibus in 1987.

For each functional "system level," there is a choice of various standards with different functionalities and benefits. Since there were over 20 protocols in the beginning of the standardization process (1990), the decision was to limit the number to three per functional level. Fortunately the experimental standards may only stay for 5 years, then a new decision must be made. The FND was withdrawn as a standard in 2001, but it still is there as a "private" public document. The experts within CEN decided to treat the other protocols in relation to their acceptance in the marketplace also.

BACnet is the only standard covering management, automation and field level functions in an open standardized way, supporting various data transportation methods. This brings additional benefits for customers and system integrators: Since no protocol conversion is needed we get higher functionality at lower cost (no gateways needed)—the upper two system levels can be collapsed and even more complex field devices can be integrated.

On the field level both EIB/KNX and LonMark have future potential. EIB, BATIbus and EHS are about to evolve to one new standard: KONNEX (KNX), the EIBA has migrated to the Konnex Association covering over one hundred producers with over 7,000 products.

- LonMark is well suited for complex field level applications where a certain engineerable flexibility is required.
- KONNEX (KNX) will be optimal for standard solutions—cost effective, simple and easy to implement.

INTERNATIONAL STANDARDIZATION

The ISO/TC205 "Building Environment Design" WG 3 committee treats the building automation and control system. This covers hardware, functions, applications, communication, conformance testing, project and system implementation. There is only one protocol without any level dedication: BACnet as ANSI/ASHRAE 135, in 2003 it became the International and European Standard EN ISO 16484-5 "BACS—Protocol."

The EIB/KONNEX Functional Blocks, which are part of the EIB Object Interface Specification, are a part of the BACnet specification by mapping them to the BACnet objects (Addendum H).

All inquiries for the BACS standards will be done in parallel among European and International countries. Parts 2, 3 and 5 came out with no negative votes, so we can expect to have one international system standard.

BACNET, EIB/KNX, LONMARK, ETC.—WHERE TO USE WHAT

So where do we use what? After considering all the above the picture gets clearer. On the floors, for Room Automation, where a large number of devices have to be linked together and to interact, use EIB/KNX for real open craftsmen solutions and LonMark for more complex engineered solutions. Both offer good cost-effective products covering the functionality needed.

In the plant rooms, more functionality is needed. This functionality is currently best covered in a standard way by BACnet. For BACnet the most effective physical media has to be selected, e.g. the same as for the floors.

For management applications we need both high functionality and IT compliance, so BACnet on ethernet/IP will cover the requirements.

OPC—A NEW ALTERNATIVE?

OPC stands for "OLE for Process Control," a communication standard based on OLE/COM technology that brings the same benefits to industrial hardware and software that standardized printer drivers brought to

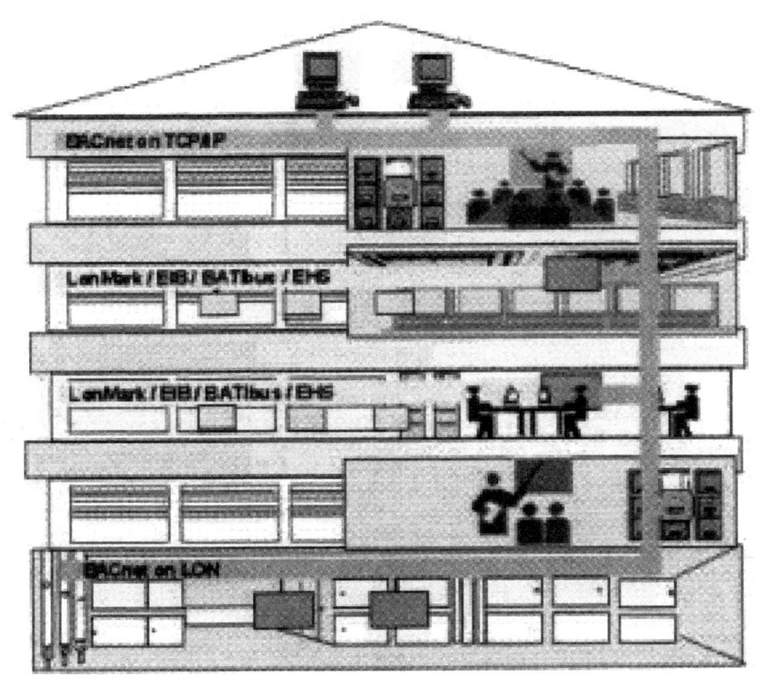

Figure 26-2. Standard BACS Communication Protocols

word processing. The OLE/COM technologies define how individual standard components can interact and share data. Based on Microsoft's OLE, COM (Component Object Model) and DCOM (Distributed Object Model) technologies, OPC consists of standard interfaces, properties and methods for use in process control and manufacturing-automation applications. OPC provides an interfacing standard for factory automation where every system and every communication driver can freely connect and communicate. Having such a standard, the communication and interactions between different applications, from the plant to the MIS (Management Information System), become easier and allow truly open enterprise communication.

OPC... What it is Not?

OPC is not a new standard bus or a universal protocol, but it is an interface definition defined by different industrial automation companies and Microsoft. Complex data structures are not yet fully defined in OPC, leaving them dependent on the specific application. The driver is still vendor specific.

SOME ACTUAL CUSTOMER NEEDS COVERED BY WEB-TECHNOLOGY

Web technology is found almost everywhere. The following attributes of the internet are also valid for web applications within BACS:

- Global connectivity
- Global companies need global BMS information
- Easy to operate
- Access at any time, from anywhere, immediately
- Use of mainstream technology
- IT-world compliance
- Thin, inexpensive clients

In addition, the web technology facilitates and centralizes software maintenance, updates and support. Last but not least, the web offers new services and opens up new business opportunities.

EMBEDDED WEB-SERVER VERSUS APPLICATION SERVER

Web technology can be implemented in almost any device from small embedded systems like a DDC controller to large application servers. In the future, will we still need conventional management stations or can they be fully replaced by embedded systems?

BACS functionality can be realized using web technology; embedded web servers can and will fully cover the area of small and simple applications where functionalities such as graphical plant operation, alarming, simple trending and reports are needed. A central application server will be needed for larger and/or distributed installations where centralized functions such as navigation between various sites, alarm dispatching, data evaluation, billing, statistics, maintenance management, etc. are required. Web-embedded systems and application servers are complementary. Functionality and data management is distributed on autonomous servers within an intranet/internet.

APPLICATION SERVER—SOLUTIONS

The application server will primarily support the following technical and technological solutions:

Operating
- Reporting
- Alarm handling
- Plant optimization
- Security
- Lighting
- HVAC
- Media access
- Room Automation

Data Management
- Reporting
- Energy data evaluation
- Consumption Control
- Process data evaluation
- Trends
- Billing
- Error and System Log

Service and Maintenance
- Application libraries
- Customer specific solutions and programs
- On-line documentation
- Knowledge database
- Preventive maintenance information
- Support/Hotline

Integration Platform
- Integration on operating level (e.g. using ActiveX)

- Subsystem integration via standards communication protocols and interfaces (BACnet, EIB/KNX, LONMARK, OPC, ...)
- Data exchange with customers business processes, office environment, etc.

GENERAL SYSTEM ARCHITECTURE

In the traditional Windows-based management stations, business logic and presentation (using e.g. MS Windows) are running on the same HW platform, normally a personal computer. With state-of-the-art web technology, the presentation layer and the business logic are separated. To view and operate the system requires only a simple web browser, which runs on a large variety of clients. The core part, the business logic, where functions such as alarm routing, trend/history data evaluation, etc. are realized, remains on a dedicated machine, the application server. Real-time process data are handled by dedicated DDC controllers optimized for the various building management disciplines. A standard database, preferably the leading IT standards SQL/MSDE, is used to store engineering and process data on the management level.

WILL WEB-TECHNOLOGY REPLACE BACNET?

The communication protocol between web-server and clients is based on leading edge internet technology such as HTML, DHTML, XML, SOAP, NET-framework. These standards from the IT world are optimized for their specific purposes: to transmit data needed for presentation by thin to fat clients through firewalls over the intranet/internet.

For communication between the subsystems and the application server, the specific BACS functionality such as trending/history, alarming, scheduling etc. has to be covered. Here BACnet is the right choice. Only BACnet currently provides this functionality within an open standard format—to assure interoperability between different systems in an open way. So we see both BACnet & the web are needed!

THE FUTURE OF BUILDING AUTOMATION

What will building automation look like in the next few years? Here are some possibilities:

- Neutral, functions oriented BACS specifications based on International Standards will increase, as a prerequisite for true open systems.
- Proprietary communication protocols will lose importance.
- In the future there will be a limited number of communication and interfacing standards.
- BACnet, EIB/KNX, LonMark have good future potential as protocols and OPC as the method,
- Newer ultra low cost technologies, as e.g. Motorola's "digital dna" LIN family (local interconnect network) may change the transportation protocols, but not the standardized BACS communication objects semantic.
- Web technology will be used on the operating level.
- Both embedded servers and applications servers will play important roles and allow distributed functionality and data management in geographical networks.
- Technologies with very good future potential are: HTML, DHTML, XML, SOAP, NET-framework.
- BACnet, OPC and web are complementary technologies.
- Unbundling—best of breed solution for every level/application.
- More and more integration of all technical applications for building services.
- Turnkey solutions from one supplier, but with standardized interfaces.
- Application service providers will offer new ways to operate, optimize and manage a building.

And do not forget: We have tried to explain in which direction the building automation and control industry is moving. As usual, however, the marketplace will make the final decision!

Acknowledgment

I would like to thank Othmar Giesler, Product Marketing, Siemens Building Technologies Ltd., Zug, Switzerland, for providing me with the idea, figures and English expressions for this chapter.

Chapter 27

BCS Integration Technologies—Open Communications Networking*

Tom Webster, P.E.
Center for the Built Environment
University of California, Berkeley CA
Lawrence Berkeley National Laboratory
Berkeley, CA

THE PURPOSE OF this chapter is to provide energy managers with some basic informational tools to assist their decision making process relative to energy management systems design, specification, procurement, and energy savings potential. It is important for energy practitioners to have a high level of knowledge and understanding of these complex systems.

This chapter will focus on building control system (BCS) networking fundamentals and an assessment of current approaches to open communications protocols. Networking is a complex subject and the networks form the basic infrastructure for energy management functions and for integrating a wide variety of OEM equipment into a complete EMCIS.

Note: Please refer to the Glossary of this book for a complete listing of all acronyms and their definitions.

INTRODUCTION

The two primary driving forces behind the vast changes that are occurring in BCS communications networking technology are: 1) technological change, and 2) the open systems movement. Technological change was discussed in Chapter 8; this chapter will focus on open systems.

In general, open systems embrace three major concepts: open source software, open communications protocols, and open data exchange. Open data exchange includes standardization of databases, data objects, and data presentation software (e.g., browsers). While a major driver for open systems derives from the user's desire for simplification, interoperability, and low cost, one of the strongest drivers comes from the internet and the move to "web-enable" virtually all applications. While all of these categories of open systems will have an impact on BCS development, we will focus here on open protocols and open data exchange. Furthermore, since a discussion of open systems cannot be divorced from a discussion of standards, standards issues will be interwoven into the open systems discussions.

The central focus of open protocol efforts in the BCS industry is the standardization efforts by BACnet and LonMark, and the corresponding changes the BCS vendors are making in their proprietary offerings to accommodate openness. A similar process is occurring in other industries, most notably industrial process control and information systems (IS), which will also affect BCS development.

In assembling this assessment, we have relied heavily upon building control system experts, product literature, white papers, technical papers, and news and journal articles. Our intent is to provide an impartial and accurate portrayal of the state of practice with emphasis on evolutionary trends and emerging technologies.

NETWORKING FUNDAMENTALS

It is essential to have a grasp of basic networking concepts to be able to clearly understand the issues as well as to interpret the information being provided by vendors and consultants. This chapter covers the following concepts: (1) network architectures as they apply to the BCS industry, (2) networking fundamentals, includ-

*This chapter was originally published as a report for the Federal Energy Management Program, New Technology Demonstration Program, Report No. LBNL—47358

ing a short primer on protocols, and (3) the contending approaches to open protocols.

Network Architectures

Although product and technical literature contains descriptions of various BCS network architectures, it is the evolution of these architectures and how they fit into a broader perspective that is most important. The ongoing convergence of technologies (voice, data, video) and the increasing internetworking of communications infrastructures are the hallmarks of the information age.

Figure 27-1 shows a generalized view of the interrelationship between network types and how BCS networks relate to others. This diagram illustrates the most common arrangements being developed today but does not indicate the vast array of legacy systems that make up the bulk of the installed base.

Networks can be broken down into two fundamental types:

1. **Point-to-point**—store and forward, or switched WANs that pass messages through a network node by node, and
2. **Broadcast**—multiple access or multi-drop LANs that (typically) use baseband* signaling with various access methods to share a single channel; i.e., each node sees packets sent to other nodes.[1]

BCSs can be characterized by the four-level architecture shown in Figure 27-1. While this hierarchical structure predominates in the buildings industry, it is being "flattened" by merging levels together as the technology evolves; e.g., sensor and terminal bus merged together so sensors and terminal controllers co-exist on the same sub-network, or field panels and terminal unit controllers on the same bus.** However, in general, the sensor bus is not yet implemented as a discrete layer, although it is being developed in the industrial process industry. For BCS networks the trend is toward more internetworking just as it is for IS networks. In addition, as IS protocols become more standardized, and components less expensive, they are migrating further down the network hierarchy (see discussion below about Opto 22). One reason this is important is because of the impact it is having on the development of EMCIS standards.

As indicated in Figure 27-1, the sensor bus connects sensors and actuators together and interfaces to the *terminal bus* level (referring to HVAC terminal equipment unit controllers such as VAV boxes, fan coils, etc.), which in turn connects to field panels at the BCS *backbone* level. Above the backbone level are various levels of EIS networks that ultimately connect to WANs. Typically EISs are client-server based, which distinguishes them from real-time peer-to-peer control networks. Client-server functions are contained in a set of protocols that sit on top of the networking protocols that are most familiar to the building control industry. Integration with the EIS is important in distributing control system information to higher-level EIS applications. In fact, one of the major drivers for change in BCS technology is the ability to port control system information into the EIS environment and thus service a much more diverse set of enterprise information needs than previously was possible.

Table 27-1 is a "roadmap" showing the relationship between the typical BCS network architecture and the various protocols used. This list includes common protocols that are or will be candidates for the architectural level shown. These are included to provide the reader with a basic framework for tracking the evolution of protocol development efforts that affect the BCS industry.

Networking Protocols & Standards, Basic Concepts

Protocol

A protocol is a detailed, structured method of communicating certain types of information. When we use the term "protocol" in reality we are referring to a "protocol stack," i.e., a series of protocols that are used to send messages between devices (nodes) on a network or between different networks. When discussing protocols it is useful to use the OSI multi-layer model for a protocol stack that was created by ISO and is outlined in Table 27-2. The message, or packet, consists of various headers that are significant to the services being performed at each layer as well as for interfacing between the layers (i.e., addressing, contention resolution, etc.), and a data element that contains the application information. The lower four layers are considered "connectivity" layers, and the upper three "application" layers. For BCS applications, only the top application layer is generally used.

The workhorse protocol for WANs has been the TCP/IP stack that was originally developed for the internet but now is used routinely for intranets (LANs within enterprises) as well. Of course, there are many

*Baseband refers to digital signaling whereas broadband refers to analog signaling. LANs and WANs use both of these techniques.

**In actual practice flattening depends more on the functions of a system, its size, and the desire to segregate communications traffic than on technology availability. Merging these networks lowers costs and allows more direct integration with enterprise wide applications and WAN access and is changing the administration of these infrastructures away from the facility departments to the MIS departments.

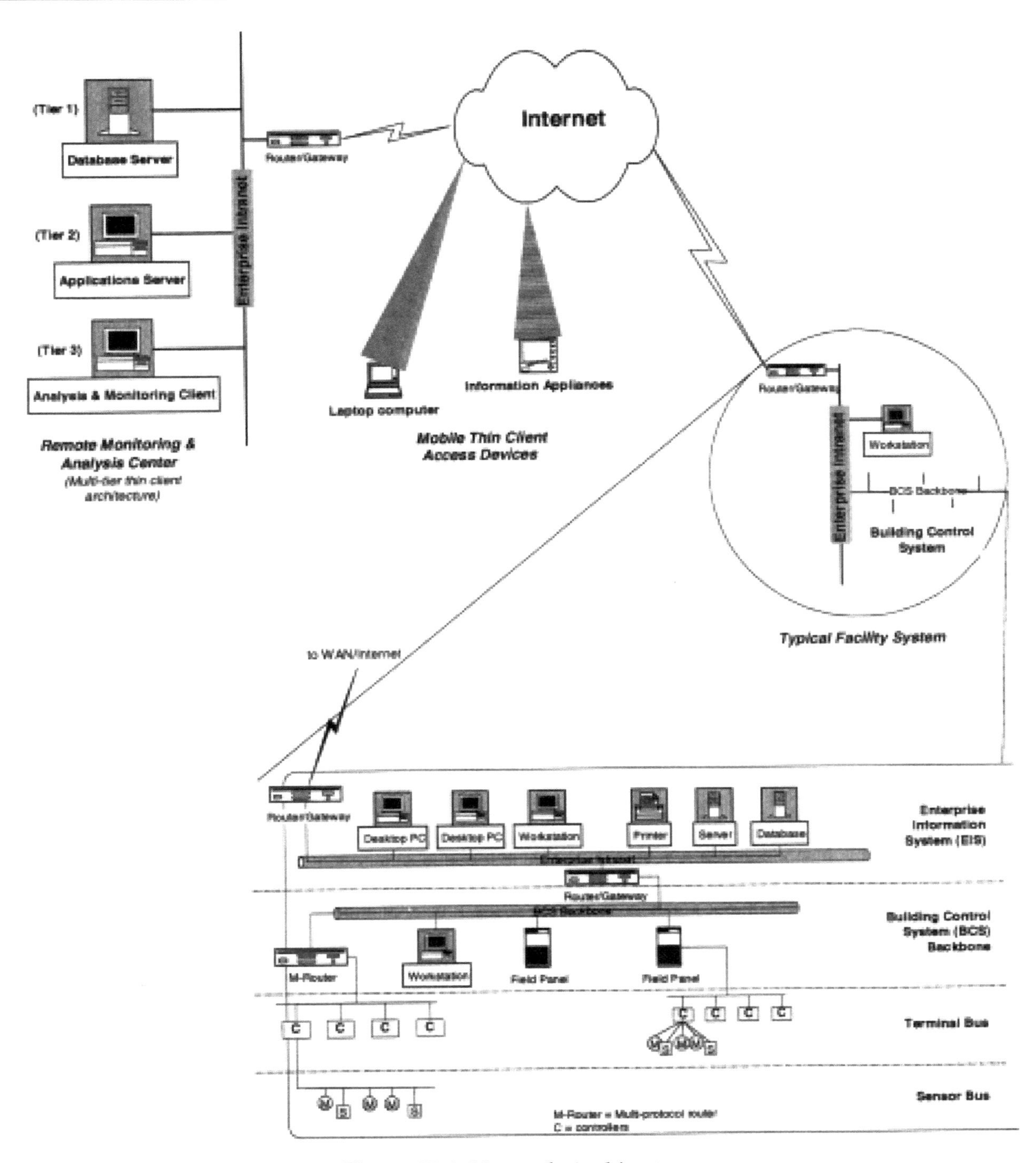

Figure 27-1. Network Architectures

other protocols being used in various types of LANs and WANs but there appears to be a steady migration in newer LANs towards the TCP/IP/ethernet* standards. Although this makes internetworking easier it still does not make it possible to connect LANs and WANs together without some sort of interface device. For TCP/IP intranets it is usually a "multi-protocol router" since the signaling protocols for the two networks are different (see discussion below about networking protocols). All of the networks can be classified under the terms "distributed processing" or "distributed intelligence" networks, as opposed to the central processing systems of old.

*The slash between acronyms is synonymous with "over" and denotes the hierarchical relationship of the OSI model; i.e., TCP/IP means TCP (Layer 4) over IP (Layer 3).

Table 27-1. Roadmap of Networking Protocol Options*

Network Level	Protocol Suite Options* Note: / indicates 'on top of' or 'over' data link/physical layers.	Standard or Specification Sponsor**	Remarks
Enterprise/ Intranet	CORBA IIOP	OMG	Client-server object standards via consortium//SIG of 800 companies
	COM, DCOM DDE OLE,ActiveX,OPC	Microsoft	Microsoft's client-server object model specifications that compete with CORBA. OPC is the initiative to adapt ActiveX to industrial real-time control.
	Java RMI	Sun	Java based C-S object model that competes with COM and CORBA
	TCP-IP/Ethernet	Internet Society RFCs/ IEEE 803.2	
	Novell IPX SNA	Novell IBM	Two examples of early computer networking technologies developed by Novel and IBM that still have large installed bases
BCS Backbone - Field panels (Control)	TCP,UDP-IP/Ethernet	Internet Society RFCs/ IEEE 803.2	Protocol suite of choice for intranet LANs
	Proprietary/Ethernet, ARCNET	Various control and HVAC OEMs	
	BACnet® /Ethernet, ARCNET BACnet-IP/Ethernet	ANSI/ASHRAE –135-1995, IEEE-803.2	
Terminal bus - unit controllers (Fieldbus)	Proprietary/EIA-485	Various control and HVAC OEMs	Typically proprietary master-slave or token passing low speed, low cost protocols used ubiquitously by BCS vendors since 1980's.
	Proprietary/Ethernet, ARCNET	Various control and HVAC OEMs	
	BACnet MS,TP/EIA-485	ANSI/ASHRAE –135-1995, IEEE-803.2	
	LonTalk/TP, PLC, RF	ANSI/EIA-709.1 and 709.2,.3	Adoption of Echelon's LonTalk. Requires a license from Echelon. Twisted pair, RF, and powerline carrier standards for are contained in EIA-709.2,.3
	Profibus	EN 50 170	European fieldbus bus standard
	DeviceNet	Allen Bradley	Short distance industrial sensor bus derivative of CAN that competes with other field & sensor buses. Uses CSMA.
	Modbus/Modbus+//TCP//IP	Modicon	Token passing industrial process protocol for PLCs now supported by several BCS vendors
Sensor Bus – sensors & actuators	DeviceNet, ControlNet	Allen Bradley, Rockwell	Short distance industrial sensor bus derivative of CAN that competes with other sensor buses. Uses CSMA.
	LonTalk/TP, PLC, RF	EIA-709.1	
Others (Emerging protocols to watch)	Bluetooth	Bluetooth SIG	Short-range self-configuring wireless networking protocol for information appliances (e.g., PDAs, etc.) being developed under auspices of 1200 member Bluetooth SIG
	Wireless Ethernet	IEEE 803.11	Wireless Ethernet LAN protocols being developed by IEEE.
	Firewire	IEEE 1394	High speed data bus primarily for PC peripherals
	Fieldbus	Fieldbus Foundation	Fieldbus Foundation works to implement the ISA SP50 specifications to develop an interoperable sensor bus for distributed process control.
	IEEE 1451	IEEE 1451.1-.4	Industrial process sensor/actuator interoperable interface standards.

* We do not attempt to distinguish individual OSI layers here, only indicate the various protocol suites that are used or under development for the particular architecture level shown. Parentheses indicate typical industrial process level designations.

** In this table we include both *de jure* standards as well as "specifications" being developed by SIGs and private companies.

Table 27-2. OSI Protocol Model

An easy way to conceptualize protocols is by using a layered model. Each layer performs specific functions and interfaces to the layers above and below. While the generalized model used by ISO (called the OSI model) consists of 7 layers, most HVAC protocols utilize only 3-5 layers (typically Layers 1-3,4 and 7).

Layer	Description	Function	Examples
Layer 7	Application	Interface to applications logic; data object interfaces	HTTP (HTML/XML web pages); SMTP, SNMP (email), Telnet, FTP BACnet
Layer 6	Presentation		
Layer 5	Session		
Layer 4	Transport	End to end security; assembly and ordering of message fragments	TCP UDP
Layer 3	Network	Addressing and Routing	IP BACnet LonTalk
Layer 2*	Data Link (LLC+MAC**)	Contention resolution, medium access, channel allocation	TDM FDM WDM (DWDM) ATM CDMA (wireless) CSMA (Ethernet) PPP, SLIP (for IP protocols) Token passing (ARCNET) Master/Slave
Layer 1*	Physical	Signaling, topology and media	EIA 232,485 10,100BaseT SONET (fiber) RF,FHSS, DSSS (wireless) ISDN

*Note that many LAN standards combine these two layers into one specification; i.e, Layer 2 and Layer 1 are not necessarily "mix and match."
** The MAC sub-layer is a very important component that specifies the packet structure and the rules for how a device accesses the network.

Standard

More care should be exercised in the industry when this term is used. We reserve this term for "true" or *de jure* standards, i.e., those promulgated by recognized standards organizations such as ANSI, ASHRAE, IEEE, EIA, ISO, and ITU. To apply this term to proprietary or *de facto* protocols tends to confuse and mislead and usually results from overzealous marketing efforts. Just as important as standardization is the availability of multiple suppliers; a standard that does not capture enough of a market to warrant multiple suppliers fails to achieve one of the primary goals of standardization—low cost from competition and thus easy access to a uniform implementation.

Even worse is the overuse of the designation "open system." Open can have just about any meaning one wants to put on it. As generally applied in the BCS industry, "open system" usually refers to those systems that allow connection by alternative protocols used by competing vendors. This definition has little to do with standards, proprietary systems can be interconnected by virtue of an agreement between two parties (see discussion below). "Open" implies the willingness to cooperate but the devil is in the details. For example, when a license is required to use a protocol, it represents a barrier to use and may compromise the concept of "free and open access to any and all parties." The author believes that a protocol cannot be truly "open" unless it is a standard as defined here. The process of standardization makes the specifications available to all comers without restriction or preconditions at virtually no cost. Even this definition is not completely adequate, since for example, implementing ANSI/EIA-709.1* requires a license from Echelon Corp.**

LAN Characteristics and Performance

A detailed discussion of networking technology is beyond the scope of this chapter but a basic understanding of some key characteristics of LANs typically used in BCSs is important to frame the following discussions about BACnet and LonMark. There are three data link protocol technologies that are important to BCS networking: Master-slave (M-S), token passing (TP), and carrier sense multiple access (CSMA).

*ANSI/EIA-709.1-A-1999 is the ANSI/EIA standard that incorporates Echelon's LonTalk protocol.[3]

**When a license is required there is another critical concern—who determines interoperability. To preserve openness, a separate, independent certification organization is required to allow competing vendors a certification method that does not require the "blessing" of the company granting the license.

The master-slave protocol has been the most popular choice for lower levels in the network due to its simplicity and flexibility. This technique relies on a master node to orchestrate traffic on the network; the master initiates all transactions usually by polling the slaves. Although this is not technically a peer-to-peer solution, in practice it can be made to work like one.

Token passing schemes are "multi-master" and operate by passing a token from node to node. Only when a node holds the token can it transmit a message.

CSMA is the technique used in ethernet (IEEE 802.3) and LonTalk. CSMA methods are fundamentally different than the two above. M-S and TP are orderly non-contention schemes while CSMA is a contention method where each node begins to transmit a message at will unless it detects* a collision with a message being sent at the same time by another node. Various types of probabilistic backoff algorithms are used to schedule retries.

Each of these has advantages and drawbacks. There is no perfect solution, and each was invented to solve certain problems and/or to overcome limitations from previous developments. The pros and cons can be evaluated to some extent from the following key factors:**

- **Efficiency**—ratio of message data bits to total bits in a packet; accounts for overhead in the protocol.
- **Throughput**—ratio of actual packet transmission rate relative to line speed signaling rate; depends on how efficiently network traffic is managed.
- **Response time**—the time it takes to respond to a request for data; important for time critical operations.
- **Determinacy**—the consistency in delivery of message packets; are there varying or known delays.***
- **Peer-to-peer**—the ability of nodes to talk directly to one another without a master being involved.
- **Signaling method**—baseband vs. broadband; i.e., digital vs. analog signaling; also determines cost of implementation.
- **Data rate**—raw signaling rate, i.e., bandwidth.
- **Complexity**—relates to difficulty of implementation in terms of initial cost, memory size, and maintenance.

*Technically this protocol is called CSMA, CD where the CD denotes collision detection..

*Most of these factors result from the MAC and physical layer specifications, but upper layers also will have an affect.

***This issue is being overcome for CSMA schemes by advances in the technology; see the discussion under the BACnet section.

- **Topology**—the geometric layout of the nodes and wiring. Bus, star, and ring are the configurations of most importance to BCS networks.
- **Bus length and number of nodes**—the combination of length of wire and number of nodes that can be supported without repeaters.

All of these factors influence one another in determining overall performance. For example, for CSMA networks, speed and bus length determines contention slot timing and therefore the minimum packet size required. Bus length is inversely proportional to speed; the lower the speed the longer the bus can be. Table 27-3 provides an overview of the three data link protocols of interest in terms of these key factors.

When considering the pros and cons it is good to bear in mind Tannenbaum's view. He points out that numerous studies of these different access schemes have been conducted and have not identified any clear winner in terms of performance (i.e., throughput, efficiency, etc); they all work reasonably well so frequently factors other than performance drive the choice of one over the other.[3]

Interoperability

This is a complex subject. However, for our present purposes, we assume that the Holy Grail of interoperability is "plug and play"—the ability to substitute devices of equivalent *functionality* for one another without special tools and configuration. This allows end users to enjoy the benefits of easily interchanging devices from multiple vendors that work the same way as one another.

Communications connectivity (meaning the lower four layers of the protocol stack) is assumed to be compatible within this concept. However, compatible connectivity itself is no assurance of interoperability. Interoperability is facilitated by the adherence to stan-

	CSMA	TP	M/S
Peer-to-peer	Yes	Yes	No
Complexity (relative)	Medium	Medium -High	Low
Signaling method (typical for BCS industry)	Baseband IEEE 802.3, EIA-485 (and derivatives)	Baseband IEEE 802.4,5 or EIA-485	Baseband EIA-485
Data rates (typical bandwidths for BCS)	39 kps – 100 Mbps	19.4 Kbps – 10 Mbps	9.6 kps – 78 Kbps
Determinate	No	Yes	Yes
Efficiency	Decreases with speed increase since minimum packet size increases. Low (10-15%) for short messages due to minimum packet size.	Medium to high - Increases with load and message size	Medium to high - Increases with message size
Throughput	Excellent at low to medium loads, saturates at high loads due to increase in number of collisions	Excellent at high load	Low to medium
Response time	Fast for low traffic loads, degrades with load	Slow relative to CSMA for low loads due to token passing time, but no degradation for high loads	Slow
Topology	Star, bus	Bus, Ring, Star	Bus
Bus Length	500 m – 2.5 km	~600 m	~1200 m
Other	Variations in implementation have potential to improve on determinacy and performance	Susceptible to lost token which complicates protocol	Requires a master node which makes the network susceptible to its failure and compromises response time and throughput.

Table 27-3. Data Link Protocol Comparison

dards in the upper layers, primarily the application layer (Layer 7). Object standards, are application level methods for facilitating interoperability. Interoperability requires that all disparate applications adhere to the same object standards. Furthermore, use of these objects may be necessary for interoperability, but in and of themselves they do not guarantee equivalent *functionality* which is in the domain of the control logic.

Integration

Closely allied with interoperability, integration connotes the interfacing of multiple systems of distinct functionality such as HVAC, lighting, security, access, fire and life safety. While interoperability is useful to accomplish system integration it is not sufficient; many other issues must be resolved for these complex system to interact in a seamless and synergistic way. For example, deciding which data to exchange between the systems and what an appropriate system response should be to data input from other systems.

Networking Components

Gateway, Routers, Bridges and Repeaters

There is much confusion (as well as marketing hype) generated around these concepts. In an attempt to avoid being bogged down in technical nuances inherent in these concepts, we believe HVAC practitioners and energy managers will benefit from using the conceptual framework illustrated in Table 27-4. All of the devices listed in Table 27-4 are considered at minimum two-port devices that support interfacing functions between two message streams. Remember that there is a "protocol" associated with each layer of the OSI "protocol stack" and a message (or packet) uses all of the available layers.

The five component types listed in Table 27-4 perform a service (action or "translation") of some kind between the two message streams at the protocol layer(s) indicated.* It is only the gateway, however, that actually performs a translation of the *data* portion of the packet. This is why the gateway generally has limited functionality, is a customized device, and requires support and maintenance.** As such it can be an expensive undertaking. Implementation of a robust gateway is estimated to cost $20-50K and take 3-6 months to design, program and test. However, for simple data objects it can be considerably cheaper and in fact has been routinely done in the form of "drivers" in the industrial process industry for a number of years at costs as low as $2-5K.

Likewise in the BCS industry this gateway function is also routinely supported in field panel level controllers that support sub-networks of various other controllers (see Figure 27-1). Some vendors such as Johnson Controls (JCI), have made gateways a fundamental part of their business model; e.g. JCI's Metasys Connectivity Partners program which claims to support protocols from over 100 other companies. Newer web-based product companies such as Silicon Energy also rely heavily on gateways to allow access to legacy networks to support their energy monitoring and analysis software products. An excellent discussion of gateways can be found in[4].

Tunneling routers obviate the need for a gateway to pass messages through different types of networks; e.g., using the internet to collect data from remote buildings. In tunneling, the entire message (data, addressing, etc.) of a given protocol is "wrapped" (i.e., contained in the *data* portion of the packet) in a secondary protocol as opposed to making a translation. The message travels between two nodes of the tunneling (secondary) network that are each in turn connected to primary networks. At the destination the message is unwrapped and placed on the primary network port untouched. The interface devices between the two network types are called "multi-protocol routers" since they are performing a routing function between two different networks but also generally include support for different data link and signaling formats. Tunneling is being used extensively to support interconnection between networks via the internet, and within enterprise ethernet based intranet LANs.

Driver

Driver is a term frequently used as a catchall phrase for code used to interface a protocol to a device. It is similar to an application-programming interface (API) in that it provides a means for interfacing the protocol to a platform's computing resources/OS and thereby performing a gateway like function of translation into the platform's native schema.

*Layers 5 and 6, (Session and Presentation) are not listed here because they are generally not used in BCS protocol stacks.

**Gateways can become complex because they must link domains that may not share the same ideal of what objects are and how their associated methods perform; the gateway has to know a lot about both object domains to successfully bridge between them. To simplify this process, the object translations can be done at low levels in the network thereby reducing the burden on higher level objects.

OPEN COMMUNICATIONS APPROACHES

Proprietary Networks

Proprietary protocols have been the workhorses of earlier generations of BCS networks. They have become

Table 27-4. Networking Components

Protocol			Gateway	Tunneling or Multi-protocol Router	"Pure" Router	Bridge	Repeater
Layer	Description	Function					
		Device Function	Route/forward and translates data frames of packets between dissimilar networks	Route/forward packets between dissimilar networks	Route/forward packets between similar networks	Store and forward data link frames	Copy bits between cable segments; regenerate weak signals
7	Application	Data object interfaces	X*				
4	Transport	End to end reliability	X				
3	Network	Addressing and routing	X	X	X		
2	Data Link	Contention resolution	X	X		X	
1	Physical	Signaling	X	X			X
		Examples	• LonMark to BACnet • Proprietary to BACnet or LonTalk	• LAN-WAN router • LonTalk to IP/Ethernet • BACnet/Ethernet to IP/LAN, WAN (Annex H tunneling) • LonTalk iLON** • LonTalk to BACnet	• BACnet routers • LonTalk Routers • LAN-LAN and WAN-WAN routers		• Media extenders • LonTalk repeaters • EIA-485 repeaters

X indicates which layers are involved in providing the services of the indicated device; the functionality of the other layers is the same for both ports.

* Although generally a gateway device supports two entirely different protocols and therefore supports routing functions, it's the application layer "translations" that are key to the gateway concept. A special form of gateway may involve only the applications layer; where all other functions remain unchanged except a translation is made for the application data objects. This could occur, for example, if BACnet objects were used in a LonTalk network.

** The iLON now being developed by Echelon performs both tunneling router functions and gateway functions. The latter is inherent in supporting a web-server thus requiring a translation between a BCS protocol and HTTP for HTML support.

robust due to continual upgrading over long periods of time and because the vendors had a vested interest in ensuring a reliable infrastructure for their control devices. Since one vendor provided virtually everything, users had a single point of responsibility to address problems. Many users prefer to work with a single providers system simply to reduce complexity.

The Present

Proprietary networks use the same protocol layers and techniques described earlier. Generally, they have evolved from early generations that used simple collapsed three layer structures. For the lower layers, many vendors have used specifications very similar to one another and/or used older *defacto* standards such as Modbus, Opto22, or simple EIA-485 master-slave protocols running at 1.2 to 9.6 kbps. Most of these protocols have been upgraded to higher speeds due to the availability of better transceivers and embedded processors. Also, IT developments have fostered the upgrading of the lower layers of these proprietary networks to more modern LAN protocols such as ethernet and therefore provide much better support for greater data transfer demands. However, the proprietary nature of these protocols is derived not so much from the lower layer specifications but from the Applications layer implementations that were and still are custom solutions.

Openness with this approach is based on vendor-supported access; i.e., agreements between vendors to support each other's needs. This has worked well and can be considered an alternative to a standards-based approach. As mentioned previously, this approach follows the model that the industrial process industry has used for many years.

There are four basic ways that proprietary networks support openness:

1. **Open access protocol**—Equipment OEMs provide specifications for an access protocol to their equipment controllers that allows network providers to integrate the equipment into a BCS network usually using a gateway device or integration module of some sort that is interfaced to the proprietary network. JCI supports at least 75 third-party vendors using this approach.
2. **Open network protocol**—BCS providers supply open access via published bus protocol specifications (usually upon request and at lower levels of the network hierarchy) to allow implementation by others directly into the BCS network. This then becomes a third party vendor supported gateway to third party devices. JCI's Open N2 offering is a good example of this approach.
3. **Gateways**—Gateway support for the BACnet standard or LonMark (see section below). Most of the major BCS vendors now support both technologies.
4. **Front ends**—System integration provided by support of third party protocols at the front-end workstation. JCIs Unity workstation product is an example of this approach.

The Future

Complications have arisen with the proprietary openness approach because users are now demanding that these traditional networks support integration with other vendors' equipment, legacy systems, and other vendor sub-networks in a seamless way. Since the rate of adoption of new standards is slow and there is a large installed base of proprietary networks, and because the major BCS providers still want to retain control over the supply of BCSs, there will always be a mix of proprietary and standards-based approaches to networking similar to what exists today. However, the lower layers are rapidly becoming standardized either via IT developments or by HVAC industry efforts such as BACnet. The Application layers however, will most likely remain proprietary in the core of the major vendors networks until there is more widespread adoption of either BACnet or LonMark. However, support for these two approaches alongside proprietary offerings has, and will continue to, become more widespread. It is also likely that BCS vendors will continue to provide an alternative integration mechanism by supporting a wide number of third party protocols.

BACnet and LonMark*

BACnet, the standard protocol suite that ASHRAE has developed, and LonMark, a protocol suite developed by Echelon Corp. based on their LonTalk protocol, currently represent the two main contenders for BCS standardization. Although there is competition between these two technologies, one thing is clear—there will be no "winner."[5] These two technologies (and others) will share in being options for EMCIS specifications. This competition is most intense at the terminal and sensor bus levels of the network where the control devices are located. BACnet and LonMark devices located on the

*Henceforth in this chapter when we refer to LonMark we assume that it represents the LON protocol suite that is comprised of LonMark application level objects and the LonTalk protocol. See Appendix A for a more complete description of the LON Technology.

same bus are incompatible with one another. Thus the more established one protocol becomes, the greater the potential for revenues based on it. Of course both of these protocols are competing with established proprietary offerings by BCS vendors.

This is somewhat analogous to the situation in other industries where newly developed protocols and standards are continually competing with older ones that make up a large installed base. In fact we contend that all of the BCS protocols will be impacted to a large extent by ongoing advancements in information technology. The market for IT is orders of magnitude larger than those for BCS and industrial processes so component prices are low, standardization efforts are greater, and capabilities are ever increasing. However, although IT dominates the landscape, each industry still needs its own standard objects and services to integrate with IT standards in order to support industry specific requirements. It is this set of industry specific objects and services that defines the real value to efforts such as BACnet.

Major efforts have been mounted in the industrial process arena to adapt IT to real-time control applications. PC based SCADA, and OPC, are examples of these efforts, but the most notable is the continuing development of ethernet. Vast changes in the ethernet standards are being made to make it more suitable for real-time control and thus a candidate for the lower layers for virtually all levels of the network.[7-11] Opto 22 offers a digital I/O product today that uses TCP/UDP-IP over ethernet for transmitting sensor data. It even includes a web server so that this device can be accessed over the internet with a web browser. Cisco and GE have recently formed an alliance to pursue factory automation based primarily on the realization that ethernet is now ready for widespread use in real-time networks.*[13]

BACnet

BACnet is a good example of a "true" or *de jure* standards based technology. The development of BACnet has been long and difficult, but significant progress has been made. Most BCS vendors now support BACnet to a greater or lesser degree**, but only a few such as Alerton, Automated Logic, and Delta Controls have complete, native implementations at all levels of the network. However, there is still considerable confusion about BACnet's usefulness and impact. Some of the issues are discussed in the following comments.

The Present

BACnet was developed by a recognized standards body, ASHRAE, under a consensus process and is truly open and non-proprietary. Ignoring all arguments about technical issues and innovation impacts, this process is the best method for ensuring open standardization. This process has been, is being, and will be increasingly used throughout the world in virtually all industries in order to level the playing field and ensure broad and uniform implementation of and access to technology.

In the original standard, BACnet developed Layer 7 applications objects and Layer 3 addressing conventions. For Layers 1 and 2, existing standards were specified; e.g., ethernet and ARCNET. Although not a standard at the time, LonTalk was also included as another data link option. BACnet also developed a version of the commonly employed master-slave and token passing schemes called MS/TP used over the popular EIA-485 signaling protocol. These latter protocols have long been the workhorses for terminal level devices in older systems. The BACnet specification also includes a point-to-point (PTP) protocol based on the ubiquitous EIA-232 physical layer. PTP is the basis for accessing networks over modems or direct connections of BACnet gateways to workstations.

With the approval of Addendum 135*a* in January 1999, BACnet fully conforms to IP standards. Technically BACnet/IP is a version of BACnet that consists of BACnet Layer 7 objects, UDP for transport, IP for addressing, and choices (typically ethernet) for Layer 2 data links. It also includes support for broadcast messages. This is a significant development in that BACnet/IP devices can operate on standard TCP/IP networks using widely available IT networking components. It also greatly reduces the need for gateways to access BACnet networks. BACnet internetworking options are summarized in Figure 27-2.

The Future

There are those that argue that TCP/IP is not appropriate for real-time control applications.[14] The argument is that since TCP/IP protocols were primarily devised for client-server networks that do not require the robust and deterministic two-way peer-to-peer communications capabilities of control networks, they are fundamentally unsuitable for control applications. Client-server applications generally are dedicated to trans-

*For the upper layers, the trend in IT is for the applications to talk to databases and web servers using IT object standards (i.e., XML) and protocols thus obviating the need for a separate set of object standards.[12]

**Forty-four companies that offer one or more product types are listed on the BACnet web site.

1. **Basic BACnet** - Original specification for LANs that uses BACnet specifications for Layers 7 and 3, and choices for data link and physical layers.

2. **Tunneling BACnet (Annex H)** - A part of the original specification that supports BACnet messages over WANs via tunneling with TCP/IP, Routers carry the burden of managing IP.

3. **BACnet/IP (Annex J, Addendum 135*a*)** - Added in January 1999, this capability allows BACnet to support both LANs and WANs using "true" TCP/IP; i.e., network nodes are IP addressable.

Figure 27-2. Internetworking with BACnet

actions that are large, bursty and not time critical. Control applications are just the opposite. Furthermore, TCP (Layer 4) is a "connection oriented" protocol that establishes a virtual circuit between nodes during transactions and uses many sub-transactions for acknowledgments, packet sequencing, etc. that compromise the ability to perform real-time control. UDP on the other hand is connectionless with few accouterments and thus has become the basis for adaptations for real-time applications.

Many of these arguments are being overcome by advancements in IT and control technologies. This situation is analogous to the early arguments about using ethernet for control networks in that ethernet timing is not deterministic (due to the use of contention techniques). This issue has largely been overcome by the brute force of high speeds (fast ethernet at 100 Mbps and, in the near future, gigabit ethernet at 1 Gps), switching technology (allows for full duplex transmissions and private channel communications), segmented network design, and advancements in ethernet technology (e.g., prioritize messages).[10,15,16] As a result, ethernet has supplanted ARCNET (ANSI Standard 878.1, a deterministic token passing protocol) as the data link of choice in today's BCS networks. Some BCS manufacturers have migrated to supporting ethernet after basing their backbone network on ARCNET for many years. In addition, it is likely that advances in industrial process technology will, as usual, filter down to the BCS industry. The industry is pushing the development of ethernet very hard, as noted previously.

As more emphasis is placed on open object based systems and on integrating control and enterprise information systems, it would appear that mapping BACnet objects in a way that is compatible with these trends is the preferred path. BACnet object services might even be augmented by more advanced and robust client-server implementations such as CORBA, DCOM, and Java RMI.* (Only CORBA, however, is an industry consensus specification.) This trend is currently being fueled by an alliance between Tridium, Inc. and Sun Microsystems with the development of the Building Automation Java Architecture (BAJA) standard. This effort is an attempt to standardize interoperability at the enterprise level using JAVA, XML and other standard internet protocols.[17] In any event, developments like these leave the industry-specific object definitions and services themselves as the primary elements of significance that BACnet brings to the table.

The availability of BACnet/IP is a major step in facilitating the wider use of BACnet but the following additional efforts also will have a major impact:

- Conformance classes are being replaced by new BIBB specifications.** A BIBB is a collection of BACnet services that support functions such as data sharing, alarm/event management, trending, scheduling, and device and network management. These BIBBS are in turn used in standard BACnet application profiles. All of the functionality of a BACnet device (both standard and proprietary) is required to be reported in the device PICS.

- There is much discussion in BACnet circles about developing high level objects similar to LonMark profiles that would simplify configuring and programming. However, progress has been very slow

*Since higher level applications will most likely rely on IT object standards, it is important that BACnet objects do not interfere with implementation of services at this higher level; i.e., the focus should be on the behavior of objects, not on the ultimate implementation of them.[12]

**Conformance classes may have been a good idea but the particular way they were implemented was confusing.

and no clear consensus has emerged as to how to proceed.*

- Conformance testing tools, procedures and testing agents are being established. Currently this capability is embodied in the open source VTS tools and procedures developed by NIST, which are the basis for a companion standard (Standard 135.1) to BACnet currently (4/2001) under public review. These tools have formed the basis for current conformance testing activities by the BMA.[18]

- NIST has developed a DDC guide specification for BACnet systems.[19]

- Revisions and improvements in the standard are continually being made (e.g., Addendum 135*b* contains 17 changes and additions to the standard). The BACnet committee has been proactive in tracking and adapting to new technologies, as they become available.

Adoption

At first BACnet was being adopted slowly, but now it seems to be gaining momentum. The BMA has compiled the statistics shown in Table 27-5 reflecting the state of deployment of BACnet devices as of late 2000.[20]

Table 27-5. BACnet Deployment

Item	*Number*
Installations	19,054
Gateways	2,410
Devices Network Type	
• ARCNET	95,567
• Ethernet	11,920
• MS/TP	248,500
• PTP	1,549
Workstations	15,807
Large controllers	53,391
Unit controllers	299,600

*In this regard LonMark might be considered to be ahead of BACnet; i.e., BACnet has only recently (1999) started attempting to develop higher-level objects similar to the LonMark profiles. And, in fact, there has been some discussion of using the LonMark profiles as a model for BACnet profiles.

Note that these numbers are based on reports from only six BACnet vendors but represent about 90% of the production of BACnet devices. Note also the small number of gateways and the large number of MS/TP devices relative to the others. It should also be pointed out, however, that one manufacturer who reports 15,000 installations dominates the number of installations. The change in these statistics over time will be of key importance in assessing the ultimate penetration of BACnet.

LonMark

As opposed to BACnet, LonMark exemplifies a *de facto* standards-based technology. The strategy with this approach is to create such a presence in the marketplace that users will be compelled to use it simply because everybody else does and ultimately to have it adopted by a standards setting body. Most BCS vendors offer LON support and a few support it exclusively (e.g., ESUSA, Circon). Even more so than with BACnet, there is considerable confusion and controversy about the overall efficacy of this technology. Appendix A contains a detailed analysis of LON technology and its attributes and limitations. The material in this section is largely excerpted from the more complete analysis that appears in Appendix A.

The Present

The LonTalk communications protocol stack (a part of the LonMark protocol suite) is modeled after the full seven layer OSI stack contrary to many other BCS systems that use much simpler 3-4 layer structures. LonTalk consists of new protocols for each layer rather than implementing existing standards. In fact the lower layers are a derivative of the CSMA technique that ethernet uses. This approach was taken so LonTalk could address a wide variety of applications in various industries and operate over various media. To some extent these changes improve on low-load efficiency and high load saturation characteristics of IEEE 802.3 protocols. It also resulted in a maximum data rate of 1.25 Mbps, although most systems seem to use 78 Kbps.* While this scheme suffers from the same issues of non-determinacy as ethernet, it seems to work well for the lower levels in the BCS architecture as long as appropriate network design is followed.

These features (plus the "packaged" LonWorks technology) denote the major innovation that Echelon has brought to the BCS industry: peer-to-peer network-

*As of 1999 the maximum rate was increased to 2.5 Mbps.

ing technology at the terminal and sensor bus level, using twisted pair (EIA-485 type) signaling.* The FTT polarity-free, twisted pair transceivers that Echelon has developed represent a major improvement over other EIA-485 implementations.

Another key feature of this technology is that it is *hardware based* in that the technology is imbedded in proprietary Neuron chips as opposed to software based solutions that can be used on any suitable hardware platform. LON technology originally derived its "openness" from the fact that multiple vendors implemented Echelon's proprietary technology.

LonTalk (not LonMark) is now a *standard* due to its adoption by ANSI and EIA. It is still not a *de jure* standard as is BACnet since it was not created by a standard setting body using a consensus process. The support of LonTalk by ASHRAE and EIA are fundamentally different. ASHRAE's BACnet adopts LonTalk as a *data link* specification only; none of LonTalk's upper layers are specified (nor any of LonMark's application level objects). Specification of LonTalk does not ensure BACnet conformance; it represents only one part of BACnet conformance—only the data-link and physical layers much like the MS/TP and ethernet specifications. For true compliance, BACnet objects and networking need to be implemented. LonMark's Functional Profiles are a competing object model to BACnet's Layer 7 objects; they are not compatible with one another.

EIA, on the other hand, has adopted LonTalk layers 2-7 in EIA-709.1 and Layer 1 options in EIA-709.2 and .3 but also does not include LonMark Functional Profiles in the standard: *EIA-709 standardizes LonTalk—not LonMark—profiles*. LonTalk does not support Layer 7 applications services other than the rudimentary SNVTs that can be used to facilitate sharing of variables over a network. The LonMark Profiles use these Layer 7 SNVTs to implement the interoperability guideline conventions. Thus the LON standardization effort falls short of being a complete standard since it is still missing an essential element—a full application layer object specification.

The LonMark organization was created in 1994 to further the cause of creating interoperable LonTalk based products for various applications. This was necessary to address the deficiencies in the LonTalk application layer for supporting interoperability. LonMark is a trade association sponsored and controlled by Echelon (i.e., Echelon owns the LonMark trademark) and therefore LonMark lacks the autonomy and neutrality of an independent industry organization or standards body. Furthermore, the LonMark guidelines are not subjected to public review, as is the BACnet standard. A degree of interoperability is obtained by the voluntary adherence of LonMark members to the LonMark guidelines (i.e., implementers' agreements). Vendors that do not have products certified are unlikely to be compatible with LonMark certified devices, despite having compatible connectivity.

Conformance is based on a review of conformance documentation submitted by the product manufacturer (.ixf interface files) for adherence to mandatory and optional variable definitions; it is not necessary to submit the product itself. "Testing" in the LonMark conformance process refers to the review process, not actual vendor-to-vendor compatibility testing. A new process was under development (slated for release in the second quarter of 2000) to allow self-certification using special testing devices that were to be used in-situ on each type of device offered by a vendor. As of October 2000 there was been no mention of this on the LonMark website.

A number of special tools and technologies have been developed to address the deficiencies inherent in LonTalk to service emerging requirements (i.e., LonMark for interoperability, LNS for client-server support, iLON and LNS for internet access). Although these are important for broader integration, it results in a cumbersome development, installation and maintenance process for what is ultimately a sub-network of a larger BCS network. For smaller systems that are solely LonTalk based, LonWorks technology may make more sense. For larger systems, offerings from providers such as Tridium that have well integrated support for LonMark products (as well as BACnet) in their web-enabled architecture obviate the need for LonWorks accouterments.

The Future

On the one hand LON technology has become well established in the buildings industry as evidenced by its wide support by BCS vendors. On the other hand the future potential is mixed as summarized in the following comments.

- The packaged concept of LonWorks as opposed to the protocol itself appears to be the most compelling reason for using LonMark devices. The LON technology is a fairly complete set of tools to build products around that includes most of the necessary micro-controller, programming, and networking components as well as network management and interfacing tools. The design and development

*The other media supported by LonTalk have not seen significant use in the BCS networks.

tools were built around the "one size fits all" concept to offer developers a "universal" platform for control devices. This was to obviate the need to develop low level micro-controller capabilities from scratch for each new application; a basic micro-controller platform was made available that supported "typical" functions with communications built-in from inception. On the face of it, this "black box" concept allows designers a relatively easy path to build products without having to develop low level aspects from scratch. On the other hand, this approach results in some significant compromises as discussed in the *Other Limitations* section of Appendix A that limit its future potential.

- The adoption of LonTalk by EIA (EIA-709), has resulted in all layers of the LonTalk protocol now being "opened" so that the protocol can be implemented on alternative platforms. Although a license is still required from Echelon, developers are no longer required to buy Neurons or Echelon based workstation software to use the protocol. Although opened in 1996 via EIA-709 (and via ANSI acceptance in October 1999), very few alternative implementations can be found today. This suggests that there is not great incentive to "port" the protocol to other platforms most likely because it is so wedded to the Neuron processor structure and/or there is not enough market incentive to do so.*

- Neurons are computationally slow and relatively expensive. A better option might have been to develop a chip that implements the connectivity layers in firmware without the applications layers. In any event, even this approach would be challenged by the imminent rise of ethernet as a universal connectivity standard for all levels of the network.* In 1999 Toshiba introduced upgraded versions of the Neuron that included a 20 MHz clock speed (allowing 2.5 Mbps communications bit rates) and more on-board memory. This improves the raw processing capabilities but does nothing to improve the relatively old fundamental processor technology upon which the Neuron is based, or expand its computing capabilities; e.g., although promised years ago, there are still no 16-bit versions of the Neuron.

  Since ethernet is now undergoing fundamental changes to improve its real-time performance, there is less and less reason to use other data link protocols such as LonTalk. On the other hand, third parties have provided ethernet support for LonTalk allowing for its use on high-speed networks. In addition, Echelon's iLON product, LNS network operating system software, network management tools, and development systems make up a complete development and operating suite of tools that cover at least the basic requirements for web-enabled systems.

- Echelon claims that they are finally on the verge of major cost reductions due to new integrated chips being made by Cypress Semiconductor that combine the Neuron with the FTT transceiver and because of large orders derived from the adoption of LON technology by ENEL, the Italian utility as well as other non-building industries.* It remains to be seen if this in fact comes to pass.

- Although they are *de facto* standards, LonMark functional profiles appear to be the only high level objects available since BACnet has yet to develop them.

- The conformance process is weak and appears to be unfinished as discussed above. Also, there is no assurance in the current process that products of different types can be made compatible. This was supposed to be addressed by a new set of "system" certification procedures being developed by LonMark but there is no indication that these procedures have been adopted. In addition, there is no explicit control over future changes in the profiles since the modifications are voted on only by a select set of preferred members, i.e., "sponsors" that pay the greatest membership fees, of which Echelon is one. A further limitation is that development, configuration, and network management tools are based on proprietary technology that is

---

*Although the LonTalk reference implementation available from Echelon allows access to the protocol, the license agreement governing its use restricts commercial development. Commercial uses of LonTalk on other platforms are subject to additional license agreements governed by Echelon.

**The real competition to LonMark is not BACnet, but TCP-IP/ethernet based products that are likely to be the focus for the future.

*Pricing levels of $2 per Neuron long promised by Echelon have never been achieved.

only available from Echelon and a few select vendors.

Despite the drawbacks noted herein, it appears to be a significant ramp-up in vendor acceptance of LonMark technology that may ultimately have a major impact on the overall BCS market.

Adoption

Echelon has long claimed that LonTalk was a *de facto* protocol standard even before adoption by EIA. This, however, is questionable if the test is ubiquitous installation in the buildings industry (e.g., Windows OS is truly ubiquitous and therefore a *de facto* standard in the business environment; although some would argue that it is still proprietary because Microsoft drives the specification process). For example, Echelon estimates the following breakdown of Neuron uses as of June 2000.

- 13 million nodes sold
- 45% used for BCS, 25% for industrial process,* 20% transportation, and 10% in miscellaneous products.
- The split is roughly 50% US and 50% non-US.

Thus it appears that approximately 2.3 million Neurons were used in the US buildings industry over the past decade. Based on the analysis contained in Appendix A it appears that most of the ~2 million nodes are dedicated to a mixture of lighting, access, residential applications, and BCS vendor offerings.** This number of nodes is a small fraction of the BCS installed base.** Since the major equipment OEMs currently offer very few or no LonMark based products most of the volume is provided by BCS vendors.

Echelon also claims that worldwide 3500 companies are involved in developing products and that 1400 products now exist. These numbers depend heavily on how the counting is done. Echelon literature suggests that 3500 represents the number of development systems sold, not the number of products being developed for sale. Moreover, many companies produce slightly different versions of the same product. If we use LonMark listed products as an example, we find that Leviton offers 7 types of occupancy sensors, and Siemens offers 29 versions of their DESIGO RX controller for fan coils and radiant heating and cooling systems.* If *distinct* product types were counted, the number is likely to be far less as indicated by our estimates in Table A6;** which shows that the total distinct products is about one-half of the total products listed on the LonMark website. Likewise, LonMark claims to have over 200 member companies. However, only about 50*** companies are listed on the LonMark product list.

*A study conducted by Venture Development Corp. disclosed the following facts about the industrial market and LonTalk penetration of it. (1) In 1998 the total *annual* device market consisted of 24 million control devices; Echelon's estimated 3 million nodes produced over 15 years is a very small fraction of the total. (2) Ethernet is estimated to increase to 22% of the industrial market in 2003 from 8.4% in 1998. Over 75% of the market is projected to be divided between just four basic protocols. Although not explicitly mentioned, LonTalk is assumed to be included in the "others" category that accounts for 24% market share.[7]*Lack of detailed and reliable data prevents a finer breakdown.

**If market growth had matched expectations projected at the 1995 LonUsers conference where annual volumes of 100 million (downgraded to 85 million at 1996 conference) Neurons were anticipated by year 2000, then claims of being a *de facto* standard may have been legitimized. In fact only a total of 10 million chips were sold in 12-15 years.[22]

SUMMARY AND CONCLUSIONS

Our major conclusions regarding the evolution of open systems networks are summarized in the following:

General

- Information technology will drive the development of EMCIS and BCS communications networks; these technologies will augment and possibly displace elements of current protocol and/or object standards.
- Ethernet is likely to become the standard for the lower layers in all levels of the network.
- Networks will be "flatter" and less hierarchical.
- Despite the increased influence of new IT based technologies, there will still be a need for the industry-specific, application-level objects and services that LonMark and standards like BACnet provide.

Proprietary Networks

- Proprietary solutions have adapted to the demands for integration and interoperability by supporting third party protocols and emerging standards.
- Proprietary communications network offerings

*Ironically DESIGIO systems use BACnet for the BCS backbone.

**This product list grew by about 10% in a one-year period, mostly in lighting and I/O products categories.

***As of mid-2000.

represent an alternative to the pure standards based approach.

- Proprietary networks will continue to be an important part of the mix of solutions for the foreseeable future.

BACnet

BACnet is still very much a work in progress. However, BACnet has a number of attractive features:

- BACnet is truly an open and complete *de jure* standard allowing implementation on virtually any computing platform of choice without licensing requirements.
- BACnet/IP will facilitate the use of BACnet in TCP-IP/ethernet networks, and the emerging standard for higher levels in the BCS architecture. It will also foster integration with IS networks and the internet.
- The imminent approval of Standard 135.1p and the advent of conformance testing by the BMA will significantly improve the conformance certification process.*
- BACnet will continue migrating to lower levels in the network.
- BACnet is having difficulty moving beyond the primitive object level to create higher-level applications objects similar to LonMark's.
- Due to its inherent flexibility, software vs. hardware orientation, and scalability, BACnet is well suited to sophisticated solutions and adaptation to technological change.

LonMark

In terms of current availability, LonMark has an apparent edge over BACnet because it has a more complete offering including support and development tools and hardware components supplemented with the LonMark conformance certification procedure.

- Hardware dependence on Neurons will limit their long-term usefulness.
- The attractiveness of the LonTalk protocol will be challenged by the imminent rise of ethernet as a universal connectivity standard for all levels of the network and will compete with LonTalk and other similar protocols.*
- LonMark is a technology best suited for low-end applications for small systems (light commercial and residential) or for lower levels in large EMCIS networks.
- The LonMark profiles represent a significant contribution toward simplifying implementation of interoperability.
- For EMCIS specifiers (and developers), *caveat emptor* should be exercised when reviewing LonWorks marketing and promotional materials (see Appendix A).
- LonMark products are distributed broadly across low-end applications, are supported by many large BCS vendors, and the installed base is growing.

Federal Facilities Perspective

It is inevitable that energy practitioners will be drawn into the controversy surrounding protocols. This is especially true with regard to BACnet and LonMark because Federal practitioners place greater emphasis on adherence to standards than their commercial counterparts.

However, one should bear in mind that the primary impact of standards will be on the configuration, procurement, and integration of systems and components. Although there is tremendous interest, lack of consensus, and even controversy surrounding protocol options, energy and O&M savings are derived primarily from *applications* and not communications technology or its infrastructure (except in so far as it might improve control dynamics). Ultimately, the applications are where the true intelligence of these systems resides.** None of the options (proprietary, BACnet nor LonMark) are total solutions or panaceas. If the goal is true interoperability and vendor independence, then BACnet and LonMark can be seen as one step in the process toward this goal, but they share the solutions landscape with proprietary offerings for the foreseeable future.

Although these protocols will have no significant direct impact on operations, control, and energy use, there may be an impact on reliability and on first cost (higher initially, lower later). The protocols represent

---

*Critics have pointed to the BACnet conformance issue as evidence that BACnet is not really interoperable. However, most BACnet providers have either thoroughly tested their products with other vendors on their own initiative or through the NIST conformance testing standard development project. In addition, about 60% of BACnet vendors use the Cimetrics BACnet protocol stack which Cimetrics claims has been rigorously tested for interoperability.[31]

*The real competition to LonMark is not BACnet, but TCP-IP/ethernet based products that are likely to be the focus for the future.

**Considerable controversy surrounds projects such as 450 Golden Gate as to the impact that implementing a multi-vendor BACnet network has had on increasing energy savings. Any savings that have resulted have been due to the changes in control logic and equipment rather than overtly to the protocols themselves.[30]

esoteric details that manufacturers and implementers might care about, but end users are primarily interested in the functionality of the system and good reliability at low overall cost (including maintenance and upgrade cost).

Thus multi-vendor interoperability and interchangeability are important issues for the end user, over and above the subtleties of how they are achieved.

Standards help because they tend to cultivate uniformity, longevity, broad support, and reliability. However, standards need to be supplemented by an appropriate conformance certification and testing process and attention to equivalent functionality.

Trademark Notices:

- LON, LonTalk, LonWorks, LonMark, SNVT, Echelon, and Neuron are trademarks of Echelon Corp.
- ARCNET is a trademark of ARCNET Trade Association
- BACnet is a trademark of ASHRAE

All other products, trademarks, company or service names used are the property of their respective owners.

Acknowledgments

The author would like to acknowledge the dedicated support and guidance provided by Bill Carroll of Lawrence Berkeley National Laboratory (LBNL) and DOE/FEMP/NTDP for providing funding for this work.

APPENDIX A

Part of the material in this appendix has been excerpted in the preceding chapter. However, several topics are covered here in much greater detail, and the editors of this book decided to include the appendix for the reader who desired more complete information on LON technology.

LON Technology

Definitions

One must distinguish carefully between some of the major elements of Echelon's technology and related terminology in order to avoid confusion.

- **LonWorks** refers to the overall technology developed by Echelon Corp.; it includes an array of hardware and software components and tools to develop and operate LonWorks based systems.

- **LonTalk** on the other hand, refers to the communications protocol part of the LonWorks technology; it is the only part that is standardized.

- **LonMark** refers to the trade organization that Echelon formed to develop implementers agreements to promote interoperability efforts. The LonMark organization has developed a series of Functional Profiles that represent the application level object definitions that promote interoperability between LonMark devices.

- **LON** stands for local operating network, an Echelon coinage of their LAN technology.

- A **neuron** is the fundamental building block of the LonWorks technology; it is a custom micro-controller now being manufactured by Toshiba and Cypress Semiconductor (Motorola, the original maker of Neurons, has ceased production of these chips).

The LonTalk communications protocol stack is modeled after the full seven layer OSI stack contrary to many other BCS systems that use much simpler 3-4 layer structures. This was done so LonTalk could address a wide variety of applications in various industries and operate over various media. Unfortunately, this also introduces extra complexity and overhead not generally required in BCS control systems.* LonTalk consists of new protocols for each layer rather than implementing existing standards. In fact the lower layers are a derivative of the CSMA technique that ethernet uses. This was done to optimize its performance for lower speed networking typically found in low-end applications and to allow for consistent operation over multiple media. To some extent these changes improve on low-load efficiency and high-load saturation characteristics of IEEE 802.3 protocols. It also results in maximum data rates of 1.25 Mbps although most systems seem to use 78 Kbps.** Although this scheme suffers from the same issues of non-determinacy as ethernet, it still seems to work well for the lower levels in the BCS architecture as long as appropriate network design is followed.

---

*The interested reader may want to review the reasons why the full OSI protocol stack is not particularity good for actually implementing communications; it has been more useful as a model for discussing layered communications protocols (see Tannenbaum[1], Section 1.4.4 for an excellent discussion of this point).

*In 1999 Toshiba introduced upgraded versions of the Neuron that included a 20 MHz clock speed (allowing 2.5 Mbps communications bit rates) and more on-board memory.

These features (plus the "packaged" LonWorks technology) denote the major innovation that Echelon has brought to the BCS industry: peer-to-peer networking technology at the controller and sensor bus level, using twisted pair (EIA-485 type) signaling.* The FTT polarity-free, twisted-pair transceivers that Echelon has developed represent a major improvement over other EIA-485 implementations. What is less clear, however, is if there is any significant advantage to using LonTalk or CSMA for that matter at these levels of the network. As shown in Table 27-3 there are many tradeoffs that must be considered to determine whether there is a significant advantage for the particular applications being addressed. Moreover, since ethernet is now undergoing fundamental changes to improve its real-time performance, there is less and less reason to use other technologies like LonTalk. The choice to use LonTalk frequently boils down to the "other factors" that Tannebaum denotes.

The packaged concept of LonWorks appears to be the most compelling reason for using LonMark devices. The LON technology is a fairly complete set of tools to build products around that includes most of the necessary micro-controller, programming, and networking components as well as network management and interfacing tools. The design and development tools were built around the "one size fits all" concept to offer developers a "universal" platform for control devices. This was to obviate the need to develop low level micro-controller capabilities from scratch for each new application; a basic micro-controller platform was made available that supported "typical" functions with communications built-in from inception.

On the face of it this "black box" concept allows designers a relatively easy path to build products without having to develop low-level aspects from scratch. On the other hand, this approach results in some significant compromises as indicated in the Other Limitations section below.

Standards

LonTalk (not LonMark) is now a standard due to its adoption by ANSI and EIA. It is still not a *de jure* standard as is BACnet since it was not created by a standard setting body using a consensus process. Echelon is attempting to follow the path of other proprietary protocol developments such as ethernet and more specifically ARCNET in becoming a standard, which is first to try to become a *de facto* standard by shear volume in the market.*

The real significance of the adoption of LonTalk by EIA (EIA-709), however, is the fact that all layers of the LonTalk protocol have now been "opened" and can be implemented on alternative platforms. Although a license is still required from Echelon, one is no longer required to buy Neurons or Echelon based workstation software to use the protocol. However, prior to EIA adoption, none of LonTalk protocol layers were standards; they were all proprietary. Another key feature of this technology is that it is hardware based in that the technology is imbedded in proprietary Neuron chips as opposed to software based solutions that can be used on any suitable hardware platform. LON technology derived its "openness" from the fact that multiple vendors have implemented Echelon's proprietary technology. Although opened in 1996 via EIA-709 (and via ANSI acceptance in October 1999), very few alternative implementations can be found today. This suggests that there is not great incentive to "port" the protocol to other platforms most likely because it is so wedded to the Neuron processor structure and/or there is not enough market motivation to do so.**

The support of LonTalk by ASHRAE and EIA are fundamentally different. ASHRAE's BACnet adopts LonTalk as a *data link* specification only; none of LonTalk's upper layers are specified (nor any of LonMark's application level objects). EIA, on the other hand, has adopted LonTalk layers 2-7 in EIA-709.1 and Layer 1 options in EIA-709.2 and .3.

Specification of LonTalk does not ensure BACnet conformance; it represents only one part of BACnet conformance—only the data-link and physical layers much like the MS/TP and ethernet specifications. For true compliance BACnet objects and networking need to be implemented. LonMark's Functional Profiles are a competing object model to BACnet's Layer 7 objects; they are not compatible with one another. *EIA-709 standardizes LonTalk—not LonMark—profiles.* LonTalk does not support Layer 7 applications services other than the rudi-

*The other media supported by LonTalk have not seen significant use in the BCS networks.

*Ethernet was created by Xerox/DEC/Intel and later adopted by IEEE as IEEE 802.3. ARCNET was a tightly controlled proprietary protocol (similar to LonTalk) for almost 20 years, finally standardized in 1992 but still has only two suppliers.

**Although the LonTalk reference implementation available from Echelon allows access to the protocol, the license agreement governing its use restricts commercial development. Commercial uses of LonTalk on other platforms are subject to additional license agreements governed by Echelon.

mentary SNVTs that can be used to facilitate sharing of variables over a network. The LonMark Profiles use these Layer 7 SNVTs to implement the interoperability guideline conventions. Thus the LON standardization effort falls short of being a complete standard since it is still missing an essential element—a full application layer object specification.

LonMark Products Penetration

Echelon has long claimed that LonTalk was a *de facto* protocol standard. This, however, is questionable if the test is ubiquitous installation in the buildings industry (e.g., Windows OS is truly ubiquitous and therefore a *de facto* standard in business the environment; although some would argue that it is still proprietary because Microsoft drives the specification process). For example, Echelon estimates the following breakdown of Neuron uses as of June 2000:*

- 13 Million nodes sold.
- 45% used for BCS, 25% for industrial process,** 20% transportation, and
- 10% in miscellaneous products.
- The split is roughly 50% US and 50% non-US.

Thus it appears that approximately 2.3 million Neurons were used in the US buildings industry over the past decade. This installed base is made up primarily of OEM factory-mounted control products and field-supplied products by BCS vendors. On the OEM side, we have estimated that a total of almost 9 million units of various types of commercial HVAC equipment*** have been produced by equipment OEMs over the 10 year period of 1990 to 2000. Of these we estimate that about 3 million have digital controls. Roughly 150,000 of these units are likely to have LON based controls. Based on this analysis it appears that most of the 2 million nodes are dedicated to a mixture of lighting, access, residential applications, and BCS vendor offerings.* This number of nodes is a small fraction of the BCS installed base.**

This analysis suggests two things; 1) a substantial fraction of the HVAC equipment production is still sold *without* factory mounted controls, and 2) BCS vendors are the primary purveyors of LonMark products. Since the trend is for more factory mounting of controls, it remains to be seen how this might change over time since the major equipment OEMs currently offer very few or no LonMark based products.

Echelon also claims that worldwide 3500 companies are involved in developing products and that 1400 products now exist. These numbers depend heavily on how the counting is done. Echelon literature suggests that 3500 represents the number of development systems sold, not the number of products being developed for sale. Moreover, many companies produce slightly different versions of the same product. If we use LonMark listed products as an example we find that Leviton offers 7 types of occupancy sensors, and Siemens offers 29 versions of their DESIGO RX controller for fan coils and radiant heating and cooling systems.*** If *distinct* product types were counted, the number is likely to be far less as indicated by our estimates in Table 27A-1.† This list totals to about one-half of the total products listed on the LonMark website.

LonMark

The LonMark organization was created in 1994 to further the cause of creating interoperable LonTalk based products for various applications. This was necessary to address the deficiencies in the LonTalk application layer for supporting interoperability. LonMark is a trade association, sponsored and controlled by Echelon (i.e., Echelon owns the LonMark trademark); therefore LonMark lacks the autonomy and neutrality of an independent industry organization or standards body. Furthermore, the LonMark guidelines are not subjected to public review, as is the BACnet standard. A degree of interoperability is obtained by the voluntary adherence of LonMark members to the LonMark guidelines (i.e.,

*As of June 2000 these numbers have changed somewhat: 13 million nodes worldwide, 40/60% US/other, 25% industrial.[21]

**A study conducted by Venture Development Corp. disclosed the following facts about the industrial market and LonTalk penetration of it. (1) In 1998 the total annual device market consisted of 24M control devices; Echelon's estimated 3M nodes produced over 15 years is a very small fraction of the total. (2) Ethernet is estimated to increase to 22% of the industrial market in 2003 from 8.4% in 1998. Over 75% of the market is projected to be divided between just four basic protocols. Although not explicitly mentioned, LonTalk is assumed to be included in the "others" category that accounts for 24% market share.[7]

***These consist primarily of VAV boxes, medium to large rooftops, water source heat pumps, fan coils, and packaged terminal air conditioners (used primarily in hotel and motels).

*Lack of detailed and reliable data prevents a finer breakdown.

**If market growth had matched expectations projected at the 1995 LonUsers conference where annual volumes of 100 million(downgraded to 85 million at 1996 conference) Neurons were anticipated by year 2000, then claims of being a *de facto* standard may have been legitimized. In fact only a total of 10 million chips were sold in 12-15 years.[22]

***Ironically DESIGIO systems use BACnet for the BCS backbone.

†This product list grew by about 10% in a one-year period, mostly in lighting and I/O products categories.

Table 27A-1. LonMark Products

Product class	Distinct Products*
Access	1
Energy management	3
Fire	1
HVAC	
• Chilled Ceiling	2
• Fan coil	6
• Heat pump	2
• Damper actuator	5
• Equipment controller (e.g., AHU)	5
• Roof tops	4
• Thermostat	1
• Vav Box Controller	9
• I/O products	21
Industrial	4
Lighting	17
Motor controls	4
Networking	3
Sensors	16
Other	6

* Distinct products per company times the number of companies; e.g., two companies that make the same device are counted once each, but the same company that makes variants of a product for essentially the same application gets counted once only for each application, not each variant. [24]

implementers' agreements). Vendors that do not have products certified are unlikely to be compatible with LonMark certified devices, despite having compatible connectivity.

Conformance is based on a review of conformance documentation submitted by the product manufacturer (.ixf interface files) for adherence to mandatory and optional variable definitions; it is not necessary to submit the product itself. "Testing" in the LonMark conformance process refers to the review process, not actual vendor-to-vendor compatibility testing. A new process was under development (due for release in the second quarter of 2000) to allow self-certification using special testing devices that were to be used in-situ on each type of device offered by a vendor.* However, there is no assurance in the current process that products of different types can be made compatible. This is being addressed by a new set of "system" certification procedures being developed by LonMark. In addition, there is no explicit control over future changes in the profiles since the modifications are voted on only by a select set of preferred members, i.e., "sponsors" that pay the greatest membership fees, of which Echelon is one. A further limitation is that development, configuration and network management tools are based on proprietary technology that is only available from Echelon and a few select vendors.

Moreover, interoperability between devices in legacy LonTalk networks is not assured since pre-LonMark systems (prior to 1994) relied heavily on the technique of "foreign frames" in which a proprietary protocol was embedded into a LonTalk frame (see the tunneling discussion above). This means that communications with these systems is essentially proprietary and incompatible with newer LonMark based devices. Furthermore, these are virtually inaccessible to newer remote access technologies without a gateway.[23]

Another issue is device complexity. Apparently, LonMark objects are somewhat weak in terms of being able to inherit properties of other objects. Therefore, as complexity grows new objects have to be created rather than being a separate instance of a more robust single object. This could explain why there are so many versions of basically similar devices in the LonMark list. On the other hand, LonMark has succeeded in developing higher-level objects that make implementation easier, something that BACnet is still struggling with.[12]

LonMark Acceptance

LonMark claims to have over 200 member companies. However, only about 50** companies are listed on the LonMark product list. This plus the arguments presented earlier about distinct LonMark products suggests the following possibilities:

---

*As of October 2000 there was been no mention of this on the LonMark website.

**As of mid-2000.

1. There are many LonMark products still in the pipeline awaiting agreements.
2. There is a lag in the commitment to interoperability in general.
3. LonMark is not being broadly accepted in the buildings industry.

Other Limitations

LonMark nodes are generally used for terminal and ancillary devices since there are limitations on the number of variables that can be shared on the network (64 network variables), in the amount of memory that can be supported with Neuron chips, and the bandwidth of the bus and therefore the amount of traffic that can be supported. Large applications need to be supported by a number of nodes with continuous interaction, a solution that has not been wholeheartedly embraced by the industry, or by using the Neuron as a communications coprocessor with another processor for applications—a better solution but one that increases cost.* Given these limitations, LonTalk has not been used as the sole EMCIS network protocol or as a backbone to any large extent in large building systems.** The vast majority of large system BCS vendors include LonTalk networks as sub-nets of larger systems for terminal or sensor bus level devices. Thus gateways are required at some point, usually at the field panels, in the network.

Some developers contend that it takes significant effort to get around the built-in limitations and roadblocks inherent in the LonWorks approach. For example, the SNVTs are actually quite limited resources that are mostly committed (i.e., bound) during configuration. If during later monitoring one wants to acquire unbound SNVTs data, one must resort to other more arcane methods to access them. These types of limitations are inherent in "packaged" or generic solutions. Packaging results in many tradeoffs and compromises and the broader the scope of applications to be addressed by a package the more compromises there are for any given application. Added to this is the fact that the basic Neuron processor technology is old—the basic design is now over 15 years old.*** Many of the limitations arise from the need to protect against overwhelming the processor's capabilities. Unfortunately, the very high level of software integration that made the Neuron so attractive in the first place now makes it immune from improvement. Neuron software is not upgradeable, so the installed base of Neurons represents a non-upgradeable legacy product. Most current technologies such as system-on-a-chip solutions incorporate flash memory that allows quick, remote upgrades of OS, protocol and application codes.

The lack of effective network management and support tools has been a major impediment to easy deployment of LonMark systems. There are some alternative platforms available for network tools, including Echelon's LonWorks Network Services (LNS) and IEC's Peak Components. Performance (such as speed of discovery of networks) is an issue, as well as other features such as platforms that they can operate on, industry standard interfaces that they support, etc. For example, LNS is designed to operate on a PC platform, while Peak was designed to operate on smaller embedded platforms as well as PC platforms. LNS has it's own plug-in interface, which it markets as a *"de facto"* standard. The Consumer Electronics Association (CEA) has recently created an open device plug-in standard for network tools (EIA/CEA-860) that is independent of network management platforms like LNS and Peak. At the time of this writing, it is unknown how widely adopted EIA/CEA-860 will become in the future.[25]

These limitations result in LonWorks technology, being relegated to lower end, simpler applications developed by lower skilled developers—precisely the way it is being played out in the market. LonWorks is not capable enough for high-end, custom, robust, high complexity systems. Going forward it will be at an increasing disadvantage compared to newer processing and communications technology currently being developed.[12,26] Implementation costs are another issue. Anecdotal comments suggest that building a product on LonWorks technology is not as simple as Echelon portrays—several projects required significantly more time and money than originally anticipated. These appear to result primarily from having to find ways around some of the limitations built into the technology as alluded to earlier.

Marketing and Promotion

A discussion of LonMark would not be complete without a comment about marketing and promotion.

In what may be a response to a competitive environment, Echelon's marketing strategies and methods appear to control information availability, and thus can make it difficult for a prospective client or specifier to

*Some practitioners believe that LonWorks does not scale well in large applications using multi-layer architecture due to the namespace limitations of the network variables.[12]

**One example of this might be the Dirksen Courthouse in Chicago that was originally specified with a LonTalk backbone but was subsequently changed to ethernet apparently due to performance problems.

****See footnote 32.

make well-informed purchasing decisions.* Many of Echelon's marketing materials regarding their own products are heavily promotional in nature. From these materials it is often difficult to obtain a clear appraisal of the potential of LonWorks, LonTalk and LonMark in terms of acceptance, use, and capabilities, as we have indicated in the discussions above. This tends to cause confusion among developers, specifiers, and users. It also engenders lowered confidence about the overall merits of the technology in general. We therefore caution energy practitioners about accepting at face value statements in literature of this type. Additionally, Echelon statements regarding the merits of alternative approaches such as BACnet (see statements about BACnet in[24, 29]) should be viewed with caution and evaluated with care. We recommend using independent information sources and trusted, unbiased experts to evaluate functionality and performance claims before making purchasing decisions.

Because it is important to ensure availability of a sufficiently wide range of compatible products for future extensibility, practitioners should also be cautious when it comes to evaluating the penetration of both LonMark and BACnet technologies in the market. There is very little solid data to back up claims being made. Data about types of products or, better yet, sales volumes of products by type are the only reasonable way to make definitive statements about penetration and growth rates. In terms of number of nodes being sold it appears the market is somewhat balanced between BACnet and LonMark with Honeywell and Siebe (Invensys) leading with LON devices and Alerton, Automated Logic, Delta Controls and Trane leading with BACnet products. It would be better if the types of products and their volumes were known.**

References

[1] Tannebaum, A.S., *Computer Networks*, 3rd ed., Prentice Hall, New Jersey, 1996.

[2] "Control Network Protocol Specification," EIA-709.1, Electronic Industries Alliance, April 1999.

[3] Tannebaum, p. 301.

[4] Bushby, S.T., "Communications Gateways: Friend or Foe," *ASHRAE Journal*, April 1998, p. 50.

[5] Hull, G.G., "Myths of LonWorks and BACnet," *ASHRAE Journal*, April 1999, p. 22.

[6] Internet Protocols[Website]. Cisco, November 8, 1999. Available from: www.cisco.com/univercd/cc/td/doc/cisintwk/ito_doc/ip.htm.

[7] "Control Network Shuffle Continues," *Control*, June/July 1999, p. 12.

[8] Caro, D., and R. Mullen, "Ethernet as a Control Network," *Control*, February 1998, p. 38.

[9] Merritt, R., "Technology Trends," *Control*, April 2000, p. 40.

[10] Waterbury, B., "Ethernet Ready to Strike," *Control*, August 1999.

[11] "Profibus Supporters Ponder Life with Ethernet," *Control Design*, October/November 1999, p. 14.

[12] McParland, C., Personal Communication, January 2001, *Computer Scientist*, Lawrence Berkeley National Laboratory.

[13] "GE, Cisco Systems Form Networks Unit for Factories," *Wall Street Journal*, June 6 2000.

[14] Knollman, R., Personal Communication, November 1999, Applications Engineer, CTI Products.

[15] Thomas, G., "Looking Deeper into Ethernet," *The Industrial Ethernet Book*, Spring 2000, p. 16.

[16] Feeley, J., R. Merritt, T. Ogden, P. Studebaker, and B. Waterbury, "100 Years of Porcess Automation," *Control*, December 1999, p. 17.

[17] Tridium, "White Paper: Baja: A Java-based Architecture Standard for the Building Automation Industry," Tridium Inc., 2000.

[18] "Method of Testing Conformance to BACnet," ASHRAE 135.1P, ASHRAE, 2000.

[29] Bushby, S.T., Newman, H.M., Applebaum, M.A., November, "GSA Guide to Specifying Interoperable Building Automation and Control Systems Using ANSI/ASHRAE Standard 135-1995, BACnet," NISTIR 6392, National Institute of Standards and Technology (NIST), 1999.

[20] BMA, "BMA 2000 Annual Report," BACnet Manufacturers Association (BMA), December 2000.

[21] Tennifoss, M., Personal Communication, February 27, 2001, Vice President, Marketing, Echelon Inc.

[22] "LonWorks Product Update," presented at LonUsers International, San Jose, CA, May 1995.

[23] Chervet, A., Personal Communication, January 2000, FAE, Echelon Corp.

[24] Current Listing of All LonMark

*For example, IEC Intelligent Technologies has been repeatedly denied access to LonWorld to show their products that compete with Echelon's[28]

**Accumulating these numbers is a valuable contribution that an organization such as the BMA could make.

Products[Website]. LonMark Interoperability Association, November 8 2000. Available from: www.lonmark.org/products/prod_list.cfm.

[25] Wittkowske, C., Personal Communication, May 31, 1999, Systems Integration Engineer, Trane Co.

[26] Robin, D., Personal Communication, February 2000, Senior Research Engineer, Automated Logic.

[27] DesBiens, D. "NOOO PEAKING!" at LonWorld®—AGAIN!

[28] IEC Denied Admission[Website]. IEC Intelligent Technologies, 2000. Available from: http://www.ieclon.com/News/20000928.html.

[29] LonTour99[Powerpoint]. Echelon Corp., November 1999.

[30] Diamond, R., T. Salsbury, G. Bell, J. Huang, and O. Sezgen, R. Mazzucchi, J. Romberger, "EMCS Reftofit Analysis Interim Report," LBNL-43256, Lawrence Berkeley National Laboratory, March 1999.

[31] Lee, J., Personal Communication, April 2001, President, Cimetrics.

Bibliography

Allen, E., Bishop, J., "The Niagara Framework," Tridium, Inc., 1998.

"BACnet Test Standard Recommended for Review," *ASHRAE Journal*, March 2000, p. 22.

"LonMark Product Conformance Review," LonMark Interoperability Association, December 1999.

"LonWorks Product Update," presented at LonUsers International, San Jose, CA, May 1995.

"The Next Wave," *Business Week*, August 31 1998, p. 80.

"Potential for Reducing Peak Demand with Energy Management and Control Systems," WCDSR-137-1, Wisconsin Center for Demand Side Research, 1995.

"Standard 135-1995—BACnet™—A Data Communication Protocol for Building Automation and Control Networks," ANSI/ASHRAE 135-1995, ASHRAE, 1995.

Bushby, S. T., Personal Communication, November 1999, Senior Program Manager, NIST.

Elyashiv, T., "Beneath the Surface: BACnet Data Link and Physical Layer Options," *ASHRAE Journal*, November 1994, p. 33.

Falk, H. S., "Overview of Plant Floor Protocols and their Impact on Enterprise Integration," presented at Autofact '94, Detroit, MI, 1994.

Feeley, J., "What Network Works for You," *Control Design*, February/March 2000, p. 30.

Hartman, T., "Practical Considerations for Protocol Standards," *Heating/Piping/Air Conditioning*, August 1994, p. 45.

Hougland, B., "Ethernet Advantages at I/O Level," *The Industrial Ethernet Book*, Spring 2000, p. 9.

Hess, M., Personal Communication, December 2000, Systems Marketing Engineer, Trane Co.

Kinsella, T., R. Hirschmann, "Ethernet in Industrial Automation—Today and Tomorrow," *The Industrial Ethernet Book*, Spring 2000, p. 29.

Klien, J., "Battle of the Protocols," *Energy User News*, February 1999, p. 1.

LeBlanc, C., "The Future of Industrial Networking and Connectivity," *The Industrial Ethernet Book*, Spring 2000, p. 6.

Merritt, R., "DCS Dead Yet?," *Control*, July 1999, p. 34.

Middaugh, K.M., "Industrial LANs: Sorting out your communications options," *I&CS*, November 1993, p. 45.

Mostia, W.L., "What's New in the Wireless World," *Control*, June 1999, p. 77.

Naylor, T., Personal Communication, December 1999, Application Engineer, Coactive Networks.

Newman, H.M., "BACnet Goes to Europe," *Heating/Piping/Air Conditioning*, October 1998, p. 49.

Newman, H.M., Direct Digital Control of Building Systems, *Theory and Practice*, 1st ed., John Wiley & Sons, New York, 1994.

Newman, H.M., "Integrating Building Automation and Control Products Using the BACnet Protocol," *ASHRAE Journal*, November 1996, p. 26.

Putnam, A., Personal Communication, November 30, 1999, Development Director, Cimetrics Technology.

Roberts, J., Personal Communication, January 19, 2000, Principal Engineer, LonMark Interoperability Association.

Rosenbush, S., "Charge of the Light Brigade," *Business Week*, January 31, 2000, p. 62.

Sakamar, G., "FireWire Gets Ready for Control," *Control Design*, April/May 2000, p. 74.

Studebaker, P., "Object Technology Targets Process Control," *Control*, June 1998, p. 57.

Tatum, R., "Interoperability Marks new Generation of Commercial Building Products," Building Operating Management, July 1996, p. 23.

Turpin, J.R., "Getting to the BAS Point (and Counterpoint)," *Engineered Systems*, August 2000, p. 48.

Waterbury, B., "Network Computers are Dead: Long Live Thin Clients," *Control Design*, April/May 1999, p. 34.

Waterbury, B., "Platforms for Decentralized Intelligence," *Control*, March 2000, p. 97.

Waterbury, B., "Web Closes the Loop on Real-time Information," *Control Design*, April/May 2000, p. 37.

Weiss, G., "Smiling as Highfliers Blow Up," *Business Week*, June 26 2000, p. 223.

Chapter 28

ANSI/EIA 709.1, IP, and Web Services: The Keys to Open, Interoperable Building Control Systems

Michael R. Tennefoss
Alex Chervet
Echelon Corporation

THE ADVENT OF THE ANSI/EIA 709 control standard, internet protocol (IP) based networks, and SOAP/XML web services has opened the door to a new generation of open, interoperable control systems. These systems leverage LANs, WANS, and the internet to deliver higher reliability, more vendor choices, lower life-cycle costs, and more flexibility. Seamlessly merging the worlds of data and control allows facility owners to gain deeper insights into the operation of their facilities, better analyze and manage their buildings, and better leverage their capital investments.

INTRODUCTION

Before the advent of solid-state electronics, control systems consisted of pneumatic controls or wire bundles connected to relays, switches, potentiometers, and actuators. Cabling was installed point-to-point between electrical panels, sensor inputs and actuator outputs. The functionality of these control systems was relatively rudimentary and inflexible, and changes often required the assistance of a controls engineer.

The arrival of the transistor provided a way to replace relays and pneumatics with logic circuits in direct digital controllers (DDCs), which were programmed or configured with a data terminal. As increasingly powerful microprocessors and more sophisticated control algorithms were made available, the control systems grew more complex as did the issues associated with making adds, moves, and changes. The system controllers and their associated software became more expensive, and susceptibility to a single point of failure grew as reliance on controllers increased (Figure 28-1). In order to ensure a future revenue stream, manufacturers of DDCs developed them using proprietary internal architectures: the expansion of a DDC system required the use of components from the original manufacturer, making competitive bidding difficult if not impossible.

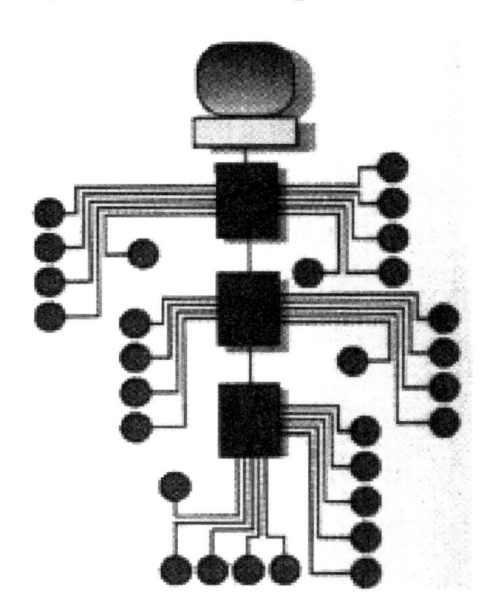

Figure 28-1.

The incompatibilities between products from different manufacturers were highlighted when customers attempted to interconnect DDC systems or devices from different manufacturers. The use of incompatible communication protocols, data formats, and electrical interconnections made it very difficult to exchange information. Seeking the communication equivalent of the "least common denominator," systems integrators and manufacturers turned to the use of gateways at the workstation level to tie together subsystems from different manufacturers (Figure 28-2).

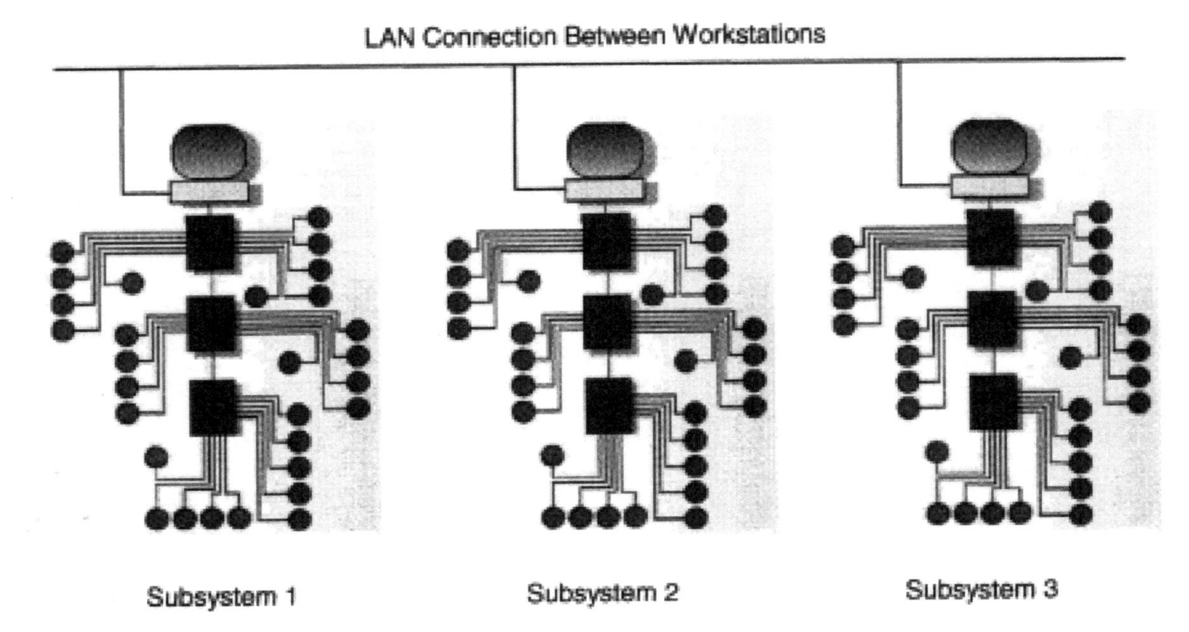

Figure 28-2. Using Gateways in Workstations to Link Subsystems from Different Manufacturers

These gateways didn't provide a detailed, seamless view into the different systems to which they were connected. They allowed only limited status and control information to be passed between the different subsystems. Fault status information couldn't be shared, information from different sensors wasn't accessible for combinatorial logic programs, and systems couldn't adapt in real-time based on direct device-to-device communications. Furthermore, the gateways needed to be changed whenever one of the subsystems was modified, creating an open-ended development and support problem for integrators and facility owners alike.

INTEROPERABILITY AND OPEN SYSTEMS

Creating a seamlessly integrated control system requires interoperability among the components of that system, as well as other related systems that must exchange information (Figure 28-3). Interoperability is the process by which products from different manufacturers, including those in different industries, exchange information without the use of gateways, protocol converters, or other ancillary devices. Achieving interoperability requires a standardized means of communicating between the different devices and managing device commissioning and maintenance; it depends on a system level approach that includes a common communication protocol, communication transceivers of different types, media routing, object models, and management and troubleshooting tools.

The benefits made possible by interoperability are many. Since one sensor or control device can be shared among many different systems, fewer sensors/controls are needed and the overall cost of the control system drops appreciably. For example, in a building automation system, one interoperable motion sensor can share its status with the zone heating system for occupancy

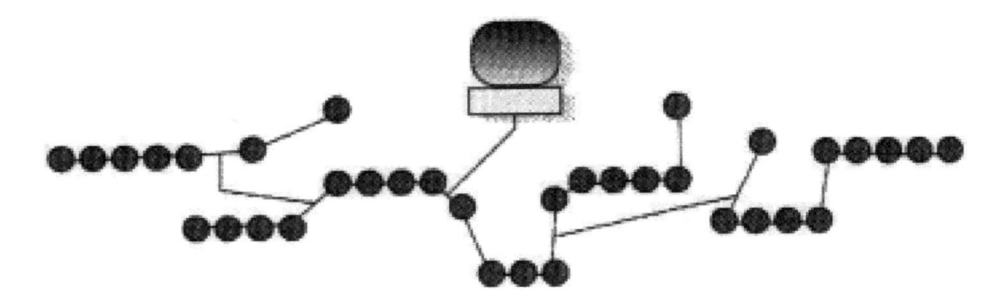

Figure 28-3. Open, Interoperable Control Network with Minimal Wiring

sensing, the access control system for request-to-exit purposes, the security system for intrusion detection, and the fire alarm system for occupancy sensing. The motion sensor still performs the same task—detecting motion—but it can share the information with the many subsystems that can make use of its status.

The ability to share more information between systems makes possible many long sought-after applications, including integrated energy control systems. For example, in response to access control reader data and daylight illumination sensors, the HVAC and lighting systems can automatically adjust the comfort and illumination levels in pertinent work areas based on individual preferences and energy costs. Lighting can be adjusted on a cubicle-by-cubicle basis for computer operators and occupants near windows—either automatically or through commands entered from a user's PC via the corporate LAN. Heating and air conditioning can be similarly tailored. Or, based on signals from smoke detectors, the HVAC system can create positive or negative air pressure of select areas to cause a fire to move away from occupied areas while the lighting system leads the way to the closest exit. The possibilities are limited only by the creativity of the designers.

For a facility owner, interoperable products offer the advantage that devices can be selected from among different manufacturers; the owner is no longer tied to any one manufacturer's closed technology. Aside from the cost savings achieved by open competition, the facility owner is safe in the knowledge that replacement products will be available if any one manufacturer goes out of business or discontinues products. Service contracts can be openly bid since no proprietary devices will be used, thereby avoiding single source service contracts.

Interoperability also benefits equipment manufacturers because their products will be assessed based on their quality and functionality—not on their ability to meet a closed, proprietary specification. Interoperability levels the playing field and increases competition, insuring that better devices will be built and the best devices for the job will win.

Interoperability has been a driver in the growth of the internet, too. TCP/IP, the connectivity standard for internet communications, provided a common transport mechanism for connecting far-flung governmental and institutional computers. The availability of HTML and web browsers enabled the creation of interlinked pages that allowed anyone with a few basic tools to use the internet, fueling both its popularity and the development of new applications.

The next level of growth will be driven by web services, which are positioned to drive expansion in network connectivity by leveraging the huge infrastructure investment that created the World Wide Web. Web services (XML/SOAP) provide a standard means of allowing disparate computing systems to exchange information. Just as a web browser uses the internet infrastructure to find a specific computer URL, and the corresponding server responds with a stream of formatted text, a web service consumer allows a computer to request data via the internet and receive an intelligible response (Figure 28-4). By providing a standardized, interoperable platform for system-to-system communications, web services have the potential to drive new applications in the area of web interaction. The question is how web services can best be leveraged in the world of building controls.

THE DAWN OF BACNET

In an effort to create a standardized method of interconnecting heating, ventilation, and air conditioning (HVAC) subsystems from different manufacturers, the American Society of Heating Refrigeration and Air Conditioning Engineers (ASHRAE) set about to create an open standard called building automation and Control NETwork (BACnet). BACnet was originally intended to eliminate the need for proprietary gateways between workstations by defining a standardized means of communicating over a local area network (LAN) to which

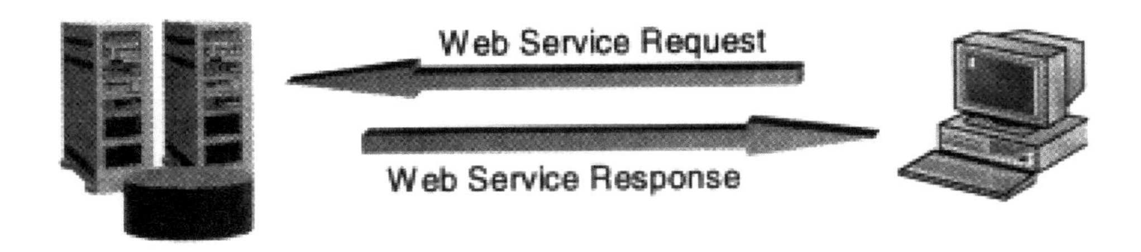

Figure 28-4. Web Services Communication Model

the workstations were connected. The workstations would, in turn, be connected to their respective control systems. Several different LANs were defined including point-to-point, master slave/token passing, ANSI/ATA 878.1, ethernet, and LonTalk® (an open control standard also known as ANSI/EIA 709.1).

One of the key features of BACnet is the use of an ethernet backbone running the BACnet protocol that was intended to improve overall system performance relative to the comparatively slow speed buses used by commercial control system vendors. The ethernet backbone was expected to be especially important in large systems at times of high network activity, such as an automatic restart following a power outage.

BACnet defined a messaging format that used objects (a logical representation of an input, output, or functional grouping of inputs and/or outputs), properties (the characteristics of an object through which it is monitored and controlled), and services (the means by which BACnet devices obtain information from one another). Since under BACnet, devices could have varying levels of functionality, even if they performed the same task from an end user's perspective, BACnet defined conformance classes that categorize the capabilities and functionality of devices.

Devices within a conformance class needed a minimum set of features, but optional features were permissible. All of the features of a device were presented in a device's protocol implementation conformance statement (PICS). A specifying engineer needed to know which objects and services were supported by which devices, since this varied from device to device, and the PICS provided most of this information. The PICS represented a point at which manufacturers could diverge in their implementation of BACnet: products which appeared to perform identically could vary considerably in terms of their functionality and the accessibility of data.

While the BACnet standard had the potential to be open and non-proprietary, its implementation varied considerably from manufacturer to manufacturer. This variability undermined the fundamental precepts of what BACnet was intended to be and do, and resulted in the creation of closed, proprietary devices flying the BACnet banner. Facility owners have the right to expect that BACnet workstations, sensors and actuators from different manufacturers may be used seamlessly in a common control network, and that devices from one manufacturer can be replaced by devices from another manufacturer without assistance from the manufacturer or the redesign of the control system. Due to variations in the implementation of BACnet PICS by different manufacturers, however, neither scenario has been realized. In short, BACnet devices from different manufacturers are neither interoperable nor interchangeable. The truth of this point is driven home by the dearth of interoperable, multi-vendor BACnet systems that support device-to-device, peer-to-peer control.

In an effort to bridge the gap between existing control networks and BACnet networks, and to appease engineers who are specifying BACnet because of its promise, some manufacturers turned to BACnet gateways. A BACnet gateway converts data from the format used by one control network into the BACnet format. This allows the manufacturer to state that a product supports BACnet because BACnet packets can be received by the gateway and forwarded to the non-BACnet system, and vice versa.

The problem is that the use of a gateway violates the spirit of BACnet and fails to deliver interoperability to control systems. Why? As discussed earlier, in the process of converting information between two networks a gateway discards information, thereby limiting the scope of the tasks that can be performed across the gateway. Diagnostic information from nodes, network traffic statistics, network management messages—all are affected by the insertion of a gateway. While gateway manufacturers may have had the best of intentions in mind, gateways do little to realize the promise offered by BACnet and instead merely extend the life of closed, proprietary systems.

The vision of a unified BACnet network was transformed into the reality of islands of proprietary networks linked by workstations and gateways running the BACnet protocol over IP-based LANs. BACnet workstations and gateways impede the free flow of network information, making impossible peer-to-peer communication between sensors and actuators situated within different proprietary islands. BACnet did not deliver on its promise and has been sullied by hype that hasn't been matched in the execution of BACnet compliant products or systems.

ANSI/EIA 709.1, IP TUNNELING, AND WEB SERVICES

One of the control networks specified within BACnet for communications between sensors and actuators is the ANSI/EIA 709.1 protocol. The ANSI/EIA 709.1 protocol allows all manner of control devices to communicate with one another through a commonly shared protocol. Communication transceivers and transport mechanisms (ANSI/EIA 709.2 power line signaling, ANSI/EIA 709.3 twisted pair signaling, ANSI/EIA 709.4 fiber optics) are standardized, as are object models and programming/troubleshooting tools to enable the rapid design and implementation of interoperable, ANSI/EIA 709.1-based devices. Network management software, protocol analyzers, internet protocol (IP) servers, network interface cards, and development tools are available off-the-shelf to speed development and reduce time to market. In short, ANSI/EIA 709.x offers a system level approach to interoperability, and comprises a complete set of tools and products. Roughly 30,000,000 ANSI/EIA 709.1 compliant devices have been shipped to date by thousands of product manufacturers.

Ensuring the interoperability of these network communications is the responsibility of an organization called the LONMARK® Interoperability Association. Funded through member dues, the LONMARK Association defines the interoperability guidelines for devices based on the ANSI/EIA 709.1 protocol, including communication transceivers and object models. Products that bear the LONMARK logo are certified to adhere to the LONMARK interoperability guidelines and can be used with confidence in integrated control systems.

One of the underlying tenets of ANSI/EIA 709.1 is that every device in a network should have the ability to send packets to, and receive packets from, any other device in the network without an intermediate gateway that filters and modifies information. This capability is one of the cornerstones of any open, interoperable, peer-to-peer network.

Additionally, the ANSI/EIA 709.1 protocol is routable, and an internet server/tunneling router may be used with any IP-based network (including LANs, WANS, and the internet) as a seamless pathway to communicate on a peer-to-peer basis between ANSI/EIA 709.1-enabled workstations, sensors, actuators, and displays. Where BACnet workstations and gateways filter information, ANSI/EIA 709.1 compatible internet servers intelligently tunnel packets through the IP network on a peer-to-peer basis, making these packets available to any sensor, actuator, or workstation that needs them.

By bridging the data and control worlds, tunneling routers allow facility owners to leverage high speed information technology (IT) infrastructure as an extension of the control network. Peer-to-peer networking, network management, device diagnostics, and software downloading—all features of ANSI/EIA 709.1 are retained, but the overall installation costs are reduced because one common IT infrastructure supports from the control and data networks. From the IT manager's perspective, the two systems are seamlessly linked because the internet server behaves like typical IT gear (and includes a programmable packet aggregation feature which throttles the rate at which control packets are broadcast on the IP network, ensuring that the IP network is not overwhelmed with control-related packets). From the building automation System manager's perspective, the two systems are seamlessly linked because the internet server behaves like an ANSI/EIA 709.1-based device.

The ability to tunnel ANSI/EIA 709.1 packets through an IP network allows devices to communicate on a peer-to-peer basis across a LAN that connects floors of a building, a WAN that connects buildings within a city, or over the internet to facilities spread around the world. Eliminating the need to install a separate and dedicated data network for the control system, and instead using the existing LAN or WAN data infrastructure for control networking, saves both installation and on-going maintenance costs.

The use of an ethernet backbone was one of the claimed advantages of BACnet over other control systems. Today, however, ANSI/EIA 709.1-based systems have a significant advantage over BACnet because they support peer-to-peer networking, they share the IT infrastructure with existing data systems, and ANSI/EIA 709.1-based internet servers behave like standard IT equipment.

ANSI/EIA 709.1-based networks have one other significant advantage over BACnet—they can leverage the newest open web standard, web services. Using an internet server that exposes a SOAP/XML web services interface, enterprise applications from Oracle, PeopleSoft, Siebel, SAP, and others can draw data from, and send messages to, any ANSI/EIA 709.1-based networks. By leveraging existing enterprise applications and IT infrastructure, web services allow companies to derive greater utility from their existing software and capital investments, while providing unparalleled access to device-level information gathered from remote facilities.

From the perspective of a facility owner, web services can shorten the payback period of existing enterprise software by expanding its use into new areas such as remote facility monitoring, like-facility comparisons, preventive maintenance, predictive failure analysis, and support logistics optimization. These new tasks can be accomplished without replacing ANSI/EIA 709.1-based control devices since an internet server with SOAP/XML web services interface will be compatible with existing

ANSI/EIA 709.1 devices. This is perhaps the best example of the power of using open, interoperable standards for building control—existing software and existing devices can perform new functions through the addition of a simple internet server operating over an existing IP network.

SUMMARY

The world of building control systems has come a long way technologically since the days of relay and pneumatic controls. The availability of ANSI/EIA 709.1-based devices has opened the door to a new generation of open, interoperable control systems that leverage the internet and SOAP/XML web services. These systems deliver on the benefits promised by those who espouse the benefits of open systems—higher reliability, greater vendor choices, lower life-cycle costs, great flexibility. Seamlessly merging the worlds of data and control allows facility owners to gain deeper insights into the operation of their facilities, better analyze and manage their buildings, and better leverage their capital investments. ANSI/EIA 709.1, IP, and web services are truly the wave of the future, so hitch a ride and enjoy the benefits.

Chapter 29

Real-time System Integration*

John J. "Jack" McGowan, CEM
Energy Controls, Inc.

THE BUILDING AUTOMATION business is overrun with new terms that try to describe "state-of-the-art" in controls. *Real-time system integration* may simply be one more such term, but it does capture the challenge and the opportunity for automation, while hinting at the internet's role for the future. There are a wealth of new control products, technologies, and buzzwords, yet there is a simple evolution underway toward the internet. No, it isn't likely that the internet will ever control buildings at the equipment level, but it is the ultimate vehicle to enhance control and to dramatically expand the information management capabilities of automation systems.

AUTOMATION AND CONTROL

It is important to start by predicting that the future of automation systems will not be determined by a small group of control manufacturers. The Industry now includes companies offering many types of software and hardware including controller interfaces and equally importantly those that offer internet technology to expand the power of these systems. Industry leaders seem to agree that automation and information technology are converging. As a result many different players, not just name brand control companies, may introduce new advances in automation. Therefore it is important to consider each industry participants in examining the state of the automation business. The primary industry participants are control manufacturers that develop products, which have evolved from simple mechanical and electrical devices to sophisticated electronic networked systems. It is easy to identify important milestones in the migration of DDC from process control to heating, ventilation and air conditioning (HVAC) systems. The movement from stand-alone controllers to distributed DDC was spurred by user demand for energy management, along with cost reduction and comfort. The next major milestone was evolution to open systems or communication standards. DDC open systems were "the price of admission" to leverage off-the-shelf technology and the internet, in integrated system projects like one on the Albuquerque Academy campus and Heidelberg Web Systems plant.

The question remains: what new innovation will have the most far-reaching impact on the control discipline? Ten years ago, answers to that question might have been: BACnet™, other communication standards, pre-integrated microprocessor-based HVAC controls or new controller platforms. Yet automation architecture, including hardware and software, remains fairly stable, though there have been ongoing advances to simplify engineering and programming of control sequences, with the exception of internet control which is worthy of some discussion.

Automatedbuildings.com is a focal point for discussion of control technologies in general, and by extension related internet technology due to the e-zines unique vantage point on the industry. Having recently returned from the AHR show in Chicago it seems appropriate to address this juxtaposition of automation and the internet, as it seemed to be a significant factor every booth. Native TCP/IP, internet-ready, IP enabled and internet control are among the related buzzwords that were used, along with a host of references to web browsers and other internet based features. At the same time, many attendees were asking; what does it mean and how do you cut through the hype to see if it brings any value? The logical expansion of building automation has been to move from heating ventilation and air conditioning (HVAC) control to direct digital control (DDC) and integration of fire and security, and potentially more. The next step for DDC was internet access, but web-based automation or "internet control" goes beyond simple access. *Internet control* automates facilities, HVAC and processes, while expanding the scope of control to

*This chapter originally appeared on the web site AutomatedBuildings.com, April, 2003, and is used with their permission.

the enterprise level, thus using the internet to convert a control system into a *management information system.*

It is worth exploring the development of internet control. The controls business has been evolving toward system integration for more than two decades, as a result of the demand for standard communication and the desire for interoperability between systems. Professionals began combining legacy DDC systems together, or with new standard-based systems, for interface via a single front end and sometimes, limited control integration. This process of standardization and integration was a precursor to internet control. Software and hardware based approaches, such as gateways, to leverage "drivers," are available to implement these systems. Some drivers were developed by companies that shared protocols, and others have been backward-engineered to achieve some level of integration. Integration usually means communication with multiple systems from one computer, but in some case sequences have integrated Fire / Life Safety or Security systems with HVAC control.

Data Communication Standards

Standardization of data communication for direct digital control was instrumental to provide the foundation for internet control. Yet it is possible that TCP / IP could be the true standard for data communication, which may make BACnet™ and LON™ moot points. Of course, these standards will always be important for communication at lower levels of the building architecture. That background covers some common ground, but it still does not explain the term internet control. Internet control automates facilities though an architecture of hard-wired and web-enabled devices. The key is that digital control is still executed in a distributed processing fashion at equipment level by controllers. In the illustration below, typical control system architectures are shown that contain DDC controllers, which execute sequences and directly enable, disable, modulate and monitor equipment. An interesting semantic variation being used by some companies is "Infrastructure" to refer to the networking of cabling and devices that supports data communication. In addition to industry specific DDC devices, "Web Appliances" for HVAC control are also available that speak "native" TCP / IP, but the internet would not control a VAV box. Web-based technologies make it possible for systems to integrate information from internal hardwired points and external sources, and that is the beginning of a definition for internet control. Yet there is much more in the illustration, and therefore energy managers will need to become network managers in the world of internet control. Ultimately the result is an energy information and control system that introduces new types of data and the opportunity for real time sequences, providing an unparalleled management tool. Of course there are also risks that must be managed. The illustration introduces some components, like the firewall, that are necessary for security. Given that this expanded system is a combination of the energy and information systems, the firewall and other equipment are generally already in place. The key is working with information systems staff to accomplish the interface.

Internet Control Terms

Driver: software to convert data from one protocol to another

Firewall: Network protection to protect against viruses, etc.

VPN: Virtual Private Network is basically a wide area network

TCPIP: Language of the Internet Transmission Control Protocol/Internet Protocol

IP Tunnel: Secure communication between two IP addresses.

Infrastructure or Architecture by Any Other Name

Internet control requires the same devices as legacy system integration, and some more for expansion to the enterprise level. As the illustration shows, the lower level architecture remains in tact, and many of the companies showing devices at AHR depicted BACnet MS / TP (Master Slave Token Passing RS-485) or LON Talk at this level. The diagram shows a router connecting that level to a server or PC, but in most cases these functions were included with another port on the server. The ultimate potential for internet control is to allow full-scale integration of building systems including HVAC, fire and security for access control. Security via IP addressable cameras for web-based video surveillance is also possible, though this requires larger scale integration. Without question this is clearly the future of building automation, but of course it will require further education, such as reading this book.

The second set of automation industry participants includes companies that develop independent software and hardware devices including Internet Tools. This includes *routers* that interface between BACnet™, LON™ and other automation local area networks (LANs) and facility ethernet LANs. Routers are hardware devices, common in data communication networks that accomplish this transition between networks. There is merit to clarifying the role of routers compared to other network components that are referenced in the control industry, particularly servers and gateways. The first area of con-

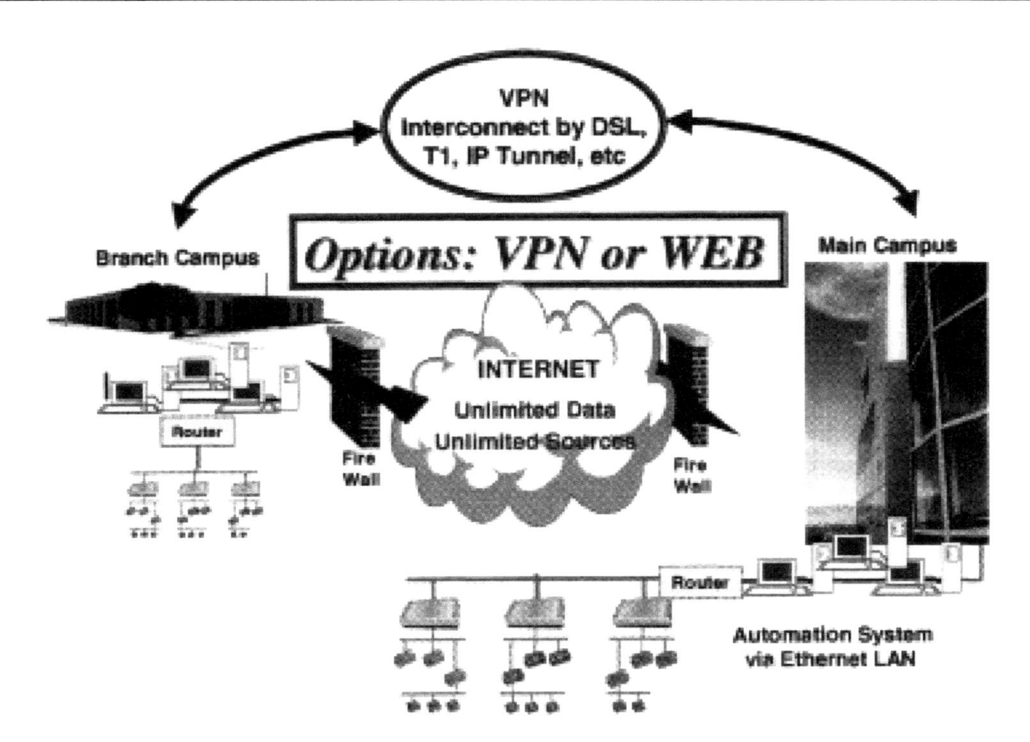

fusion is that these functions may be built into a single device. Communication standards break up the process of data interchange into a series of discrete steps, which define requirements for everything from the wire to the application or job being done by a computer. As the highest level are *servers*, sophisticated computer based devices, concerned with everything from control, and even office applications, to managing access to the network. The *gateway* is implemented when two devices on a network must communicate but do not speak the same language or protocol. *Routers* are devices, which allow two dissimilar networks, i.e. BACnet™ and ethernet, to be connected. A router is only concerned with the transportation of data via a network of devices, but converting to an ethernet network provides the added benefit of making data available from the LAN to TCP/IP, the internet protocol.

TAC Americas uses a device called the NetPlus Router to connect commercial LAN/WAN ethernet networks and control system networks. Using independent routers in automation is beneficial because they are dedicated data communication devices, which eliminates overhead on the control network. Routers look like any other node on the LAN and can provide simple fast ethernet communication. TAC's router was applied at the Albuquerque Academy project discussed here to achieve campus-wide communications for the TAC automation system using the schools' ethernet LAN. Another example is Delta System Manager module, which is a modular communication unit that is able to integrate all the networking requirements for an automation system. In the Heidelberg plant, Jim Evers of ibcontrols, inc. completed an integrated system for HVAC, lighting, access control and CCTV using the Delta System. There are also generic routers available from a number of sources that can integrate any piece of equipment with a serial port an ethernet LAN and allow interface to the internet.

At the January 2003 AHR Show four free automation sessions were offered and the dominant themes were integration and the internet. Other developments were discussed such as wireless sensors and communication buses and self-configuring controls, but the real excitement is the integration that is possible outside of the controllers themselves. This comes as no surprise to Anto Budiarjo, creator of "BuilConn," the first industry conference and trade show targeted specifically at system integration. Budiarjo says, "BuilConn is about all of the systems found in buildings." Beyond traditional automation Budiarjo says this includes security, specifically: access control, CCTV, intrusion, and fire. Builcon's target audience is management and technical staff with integration and contracting organizations, and conference tracks will cover IT, open systems, internet, integration, energy, facility management and contracting issues. Integration is clearly where the leading edge in automation may to be found.

System Integration

The term "system integration" has evolved in definition over time, much like the term direct digital control. Early definitions of system integration actually referred to what we now call building automation. Integrated systems combined HVAC control with fire/life safety and security systems. Actual integration of control sequences ranged from extremely limited, to complex approaches for smoke evacuation and after hours access to selected areas of a building. Over the years a new breed of controls contractor, or system integrator, has evolved. A new study published in *SDM Magazine* identified the top 100 system integrators and presented system integration as a business not a building automation system feature. SDM targets *security system integrators*, but many of the companies listed are clearly in the building automation business. The study reported that the Top 100 generated $2.4 billion in integration during 2001! This is a dramatic piece of information, especially when the article also points out 33% growth from 2000 to 2001.

Of particular interest here is to analyze what type of business is being done by system integrators. There are actually three different types of integrator:

- companies that specialize in combining legacy DDC systems together or with new standards based systems,
- companies that combine HVAC control with fire/life safety and security systems and
- a very small number of companies that have started implementing real-time system integration.

Combining legacy or standard-based DDC systems is not the focus here. This form of integration has been common for over a decade, and there is software and hardware available to implement such systems. This is a well-defined market, but interestingly it made up little, if any, of the revenue reported in the SDM study. The vast majority of the revenue reported appears to have been generated from companies that combine Fire and Security systems with some potential for HVAC control. It is possible that some of this income could have been derived from integrating Legacy Systems, but that is not clear. Of significance is that integrators saw a major increase in CCTV security and access control, and that those surveyed attributed 42% of their revenue to integrated systems.

The third type of integrator specializes in real-time system integration, which is of greater interest for building automation. Real-time system integration is more sophistication than interfacing Fire and Security with HVAC control for energy management. The simplest level of this technology combines the systems above while expanding their benefit by integrating the internet, offering a completely new type of integration, in real-time. This is clearly the future of building automation and it takes system integration to its logical conclusion. It also demands that system integrators become network integrators. An excellent example is a project at Albuquerque Academy that includes: central plant retrofits, 19 building campus-wide automation with interface to an ethernet LAN, access control, web-based video surveillance for security and fire annunciation. This $2 million capital project won a 2002 *Energy User News* Efficient Buildings Award for Best Education Project. The project integrated HVAC control with security systems for access control, web-based CCTV video surveillance and Fire Annunciation to optimize overall system performance. This was a rare opportunity to implement full-scale integration as part of a campus central plant upgrade with 5 new pulse boilers, a new chiller, and replacement of existing automation systems from several manufacturers. This integration centralizes interface for all campus systems through an internet-based Real-time EnterpriseDashboard™ (RED™). This powerful approach to facility automation interface expands the scope of control to the enterprise level, using the internet to convert a control system into a *management information system*.

Real-time Enterprise Dashboard™ (RED™) and Internet Tools

The Real-time EnterpriseDashboard™ expands web browsers for automation interface to the next level. RED™ opens the door for true *energy* web services companies to offer DDC, fire and security, with interface to an ethernet LAN, and maybe even become facility internet service providers. Access to real-time information anywhere, anytime through an internet-enabled automation system is the value of "real-time system integration." Please note that showing ECI RED™ is not intended to be an advertisement, but this concept is so new that it is easiest to describe it by showing a real example. The dashboards concept has been growing in popularity for several years in the information technology business, and now with RED™ an energy dashboard becomes a *campus home page*. In most organizations data are trapped, and sometimes hoarded, in various departments. Facility managers have data that is normally only available to them, and the same situation exists for accounting, finance, purchasing, etc. In most cases these groups are happy to share their information, but this requires someone to become proficient with special software and usually go to someone's office.

The dashboard concept is based on the idea of web services, which uses the internet to share information between many different software and hardware systems, that was previously not shared or available.\

Combining the concepts of web services and dashboards leads to the genesis for a truly valuable tool. RED™ is even more exciting when it is designed to make information available via the internet in real-time. The Dashboard becomes a tool that brings together information so that managers can make effective decisions. So what is a Real-time EnterpriseDashboard™ or RED™? It is a portal that allows the manager to get access to real-time information about building operations, and to link out to other systems, like building automation or access security, for interrogation and update. The distinction between a traditional "front end" and RED™ is that the home page allows the manager to link to other sites like the local utility or to internal software like the budget. A screen capture from a RED™ that is under development is shown below. On the left-hand side, the screen provides energy costing data from the local utility and internet energy resellers, as well as a break-even calculation to trigger on-site generation. Also on the left is a link to the system integrators support site to email a work order and to see a service history database showing the status of work orders currently in process and completed. This section also shows data from one of the web-based video surveillance cameras integrated with the system. Below the banner are hypertext hot links to launch from RED™ to building automation and other systems, as well as a link to the customers operating budget. On the right-hand side of the screen there is current local weather and links to the customers web site including a message from the headmaster. Finally in the center there is a quick snapshot of key events on campus and the user can link to a web page for more information and a real-time measurement and verification report.

Other new internet tools are also being introduced. Again automation companies no longer dominate the industry, and developments will many sources. For example, consider the wealth of LON-based products available for every application from HVAC to lighting and access control. These products make it possible to integrate functions that in the past would have required completely dissimilar systems. Another example is the development of web appliances, often called "embedded devices," which are internet-enabled and perform a variety of functions. Computrols has also introduced a suite of internet ready controllers. According to Mike Dolan, Computrols analyzed automation trends to design a product with the best possible "headroom into the future." The Computrols team concluded that TCP/IP was clearly the future for controller communication. So

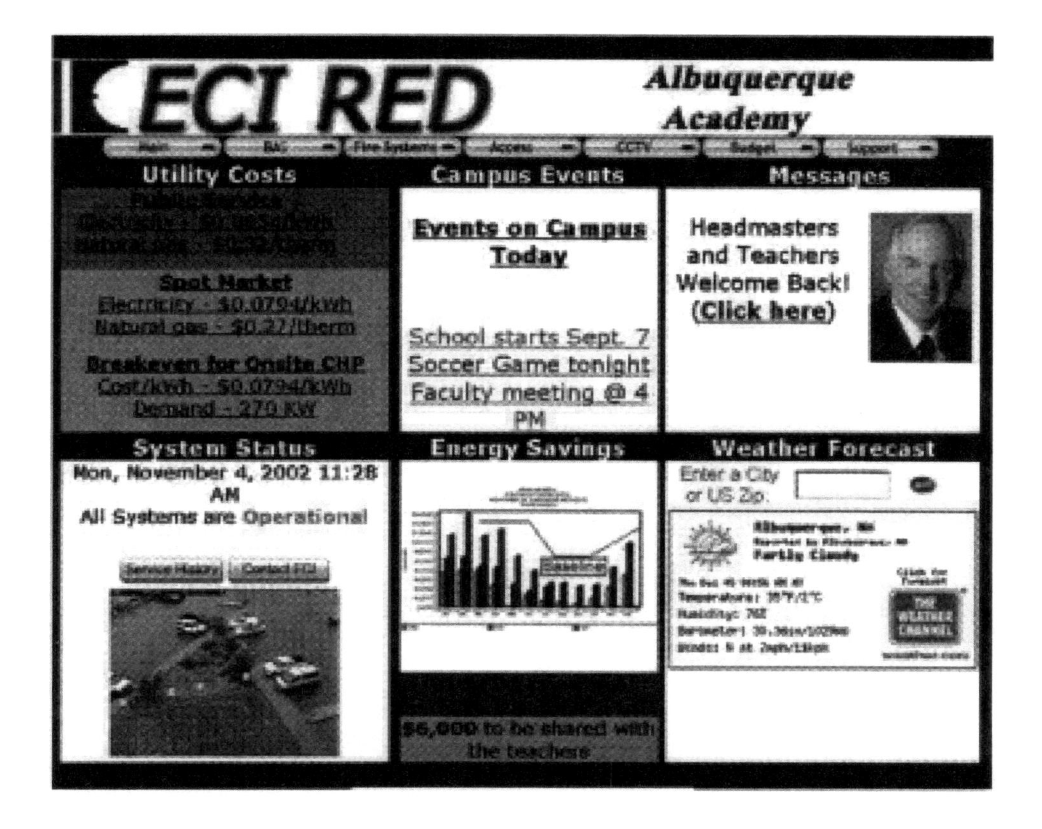

they decided to develop internet ready controllers speak "Native TCP/IP," and have a range of point density options. The controllers offer pre-programmed routines for standard HVAC applications or integrators can write custom routines. Of course the real story is the internet capability, plus built-in ethernet connections and an onboard web server. Web services features are under development to allow controllers to access internet data in XML for use in control sequences. Devices can be programmed at commissioning with a zip code and the controller pulls weather data for that location from the web instead of an outside air sensor.

This technology is exciting, but it also requires integrators to learn new technology. Just in time, a number of options are presenting themselves for professionals that need to be trained.

Real-time System Integration Training

The expanding technologies that comprise integrated systems require industry participants to revisit training needs. Knowledge necessary to be successful in building automation starts with an understanding of DDC and HVAC control, and that now includes fire/life safety and security integration. That sentence covers a tremendous amount of ground and, in a world that demands specialization, there are two or three areas to which a person could devote an entire career. In the real world however, the industry demands more, so after mastering building automation the integrator must understand data communication. The dramatic industry effort to develop standards for networking and communication has resulted in a host BACnet™ and LON™ compatible products, but it has also been the price of entry to embrace the Internet. So the third key area of knowledge for real-time system integration is to understand the internet including the underlying technology and software that make it possible.

Sources for this training are the next question and there are some exciting opportunities. For those who are just starting in the industry and need a core level training, Penn State's Penn College of Technology offers a Bachelor of Science degree in building automation technology. Most students begin with some experience and possibly an associate degree in HVAC, electrical technology, electronics or architectural technology. The program includes courses in DDC equipment control, legacy systems and control theory, building automation programming, control networks, chiller and boiler control, interoperability, energy management and building commissioning. Two valuable elements of the program, says Professor Phil Henning, are that students do a building automation summer internship between the junior and senior year, and complete a senior project that requires solving a real world problem or opportunity in building automation. Penn College is also planning a new building automation lab to support the program. More information is available from Henning at phenning@pct.edu.

There are also industry sources for seminars and short courses covering specific topics in integration and internet-based control. Automatedbuildings.com has launched several on-line courses and offers these in an asynchronous training environment that allows students to learn at their own pace, while accessing real-time information from other participants via virtual seminar meetings. The Association of Energy Engineers (AEE) offers similar programs, and is launching an on-line university that will offer synchronous seminars using the internet. These seminars meet at a specific time and the instructor broadcasts audio and video images wile offering real-time information.

The Future

The major industry effort to standardize automation system data communication provided the foundation for real-time system integration and web services for buildings. Web-based technologies now make it possible for systems to integrate information from internal hardwired points and internet sources. Accomplishing this in real-time is the next generation for building automation. This opportunity is not without challenges however and there are risks that must be managed. Data security is an ongoing concern, but the benefit of combining automation and information systems is that integrators can leverage existing firewalls and other security equipment. Data reliability is another question, particularly if the control information is completely reliant upon the web. These obstacles can overcome as they have in other industries like on-line banking. If customers trust the internet with money, how much of a leap of faith is it to trust it with data? There is little question that building automation on-line is the future and to accomplish that goal real-time system integration is required. The key is to track the developments and trends in automation hardware and data communications technology, as well as internet-enabled building services. Without question integrators and astute owners will succeed in creating cost effective building environments by synthesizing industry information to leverage technology and optimize system control and facility management.

Section Ten

Web-based EIS and ECS Applications and Case Studies

Chapter 30

Specifying, Selecting, and Evaluating Web-accessible Control Systems*

Steve Tom
Automated Logic Corp.

WEB-ACCESSIBLE CONTROL systems (WACS) are making dramatic changes in the way building control systems can be operated. No longer are operators tied to their desks; instead they can access the building with PDAs, cell phones, or laptops computers as well as the traditional workstations. This chapter explains the benefits of WACS and describes what to look for when specifying a system. The final section includes lessons learned from actual systems installed around the world.

INTRODUCTION

Recently a new breed of building automation System has appeared on the market. Instead of relying on a proprietary software program to serve as the "front end," these systems present building information in the form of web pages that can be viewed through a standard web browser such as Netscape Navigator or Microsoft Internet Explorer. These systems are called either web-based control systems or web accessible control systems (WACS).

The overarching benefit of WACS technology is that it dramatically improves an operator's ability to access a building control system. Instead of being limited to a few central workstations or laptop computers running expensive proprietary software, the operator can use any computer or web-enabled device to access the building. This means that any technician, engineer, manager, service contractor, or other authorized individual can access the building from any location.

This chapter was originally published as a series of white papers on the web site www.automatedlogic.com, which are used with permission of Automated Logic Corporation.

BENEFITS OF A WEB-ACCESSIBLE CONTROL SYSTEM

Some of the benefits of WACS are:

- Access to the BAS using workstation browsers, cell phones, and portable computers.
- Complete control of system operating parameters, such as schedules and setpoints, from virtually any modern communications device inside or outside of the building.
- The ability to use secure socket layers (SSL),[1] wireless-access-protocol devices,[2] digital subscriber lines,[3] virtual private networks,[4] and other emerging network architectures and technologies that are being developed for the entire web community. Because the research-and-development effort devoted to new web technologies dwarfs anything that could be undertaken by the building-automation industry, the capabilities of WACS will expand at a dizzying pace.

Globalization and multi-lingual support are made easier by the inherent flexibility of web standards.

Of course, WACS can also provide all of the benefits you would expect from a well-designed conventional BAS, such as:

[1]SSL is a 128-bit encryption standard that safeguards critical data being communicated over the web.
[2]WAP devices such as cell phones and personal digital assistants provide wireless access to the web.
[3]DSL is a telecommunications standard that allows communication with a WACS at a speed up to 50 times faster than is possible with a modem.
[4]VPN is a communications technology that allows the use of the internet to connect buildings while still maintaining the security of a private intranet.

- An intuitive user interface that shows you the status of your facilities at a glance and makes managing your building systems easy.
- A sophisticated alarming package that provides immediate notification of problems.
- The integration of multiple building systems into a cohesive user interface.
- Powerful tools that allow you to quickly and easily adjust equipment schedules to meet the needs of occupants.
- Energy-saving features that reduce overall demand and lower peak-demand charges.

WACS also provide secondary or indirect benefits, including the attraction and retention of talented engineers and technicians who want to learn about and use state-of-the-art technologies to ensure themselves an upwardly mobile and satisfying career.

CONSIDERATIONS WHEN SPECIFYING A WACS

As with any building automation system, it is essential that designers and owner-engineers participate in the specification, procurement, and commissioning of a WACS. With few exceptions, their involvement is much the same as it would be for any networked BAS. The WACS itself should not significantly impact the construction project's scope, schedule, or budget. The installation costs and billable hours of a WACS are comparable to those of a conventional building automation system.

In terms of hardware and software, the primary difference between a WACS and a conventional automation system is that a WACS has a server, a computer that generates web pages on demand and makes them available to the browser on the user's workstation. This replaces the dedicated workstation on a conventional BAS, so there should be no additional cost for the server. There are, however, a few design decisions to be made which are unique to a WACS:

What is Actually on the Web

How much of a WACS actually is on the web is up to the user. A WACS can be restricted to a local building network and still provide the flexibility and interoperability benefits described above. If people outside of the local building need to access the system, the WACS can be put on a company intranet, or it can be put on the internet so it can be accessed from anywhere on the planet. This does not mean that anyone and everyone *has* access to critical facility data. Security must be a major concern in any specification of a WACS.

Security

Theoretically, any system can be hacked; however, the security measures already available for web communications make a properly designed WACS more secure than most previous-generation systems. SSL encryption, for example, is widely used to protect credit-card transactions and other financial communications on the web. That same level of protection is available for WACS communications, but only if the WACS utilizes SSL.

Product Design

There are significant differences in the designs of the web-accessible control systems in use today. A detailed description of these differences would fill many pages, but there are three major areas where different design philosophies have shaped the products: (1) how much control is provided via the browser interface; (2) how the web pages are generated; and (3) how well the product supports open protocols.

Browser Interface Control

The first area is key to everything that follows. Many manufacturers view web access as an "accessory" to their workstation software. Their system will display key sensor readings on a web page, and maybe let the operator adjust a setpoint or two, but for anything more complex the operator needs to use the proprietary workstation software. System configuration, schedules, trending, alarms, reports, and other "normal" features are not available through their web interface. When a manufacturer takes this approach, each new feature that is added to the web interface represents a significant additional workload on the developers. It soon becomes natural for them to ask "Why do users need to do this through the web?" Since their proprietary workstation software already provides this tool it's easy for them to decide it isn't necessary to provide this feature over the web. As a result, their web system's feature set tends to remain very limited.

Alternatively, a few manufacturers have developed web systems that *replace* their proprietary workstation software. Like the Spanish explorer, Cortez, who burned his ships off South America to show his troops there was no turning back, these manufacturers have forced themselves to develop a complete toolkit for their web-based system. For them, the question "Does the user *have* to do this through a browser?" becomes an edict "The user *will* be able to do this though a browser!" Not surprisingly, the web-based control systems offered by these manufacturers provide a much richer environment than those that are merely intended as an adjunct to a traditional workstation-based system. Trending, scheduling, alarming, reports, program changes, memory downloads—all can be accomplished through a browser. It also follows that these systems provide a better platform to support the new generation of "gadgets," such as WAP cell phones, because these devices are also based upon a browser interface.

Generating Web Pages

The second major area where the different manufacturers' philosophies become apparent is in the methods used to generate the web pages. There are many web authoring tools on the market today that can be used to generate web pages for a building automation system. The problem is that hand crafting individual web pages for even a moderately sized control system is a very time consuming process. The customer is ultimately the one who pays for the labor required to create these pages, whether the control system manufacturer creates these pages as part of the control system design or merely gives the tools to the user so he can create his own pages. Maintenance can also become a costly proposition when hand crafted pages are used. If every change to a sequence of control requires manual editing of the associated web pages, costs can skyrocket. Also, there is always the danger of having the web pages get out of sync with the actual control system.

A better approach is to develop a system where the web pages are automatically generated by the same tools that are used to engineer the control system. Write a control program for, say, a VAV Air Handling Unit and the web page that provides the user interface to this unit is created automatically. Make a change to this control program and the web page is automatically updated to reflect this change. Here again, systems in which the browser serves as the primary user interface may have an advantage because these systems are designed from the ground up to be web-based. For these systems, the web pages are not an afterthought—they are an integral part of the system design. Since the manufacturers of these systems must provide a way to access every single element of their system through a browser, it is in their own best interest to automate the creation of web pages.

Support for Standard Protocols

The third major area in which the manufacturer's philosophy shows itself is in the support for standard protocols. Web-based control systems aren't any different than other systems in this regard. The degree to which a control system supports standard protocols has a dramatic effect upon its ability to interact with a control system made by a different manufacturer, and it also affects the freedom the customer has to compare manufacturers when the system needs to be modified or expanded. Today the concept of interoperability has become a "motherhood" issue in the marketing of control systems. You'd be hard pressed to find a single manufacturer who says they're against interoperability. However, the definition of "interoperability" varies greatly from one manufacturer to another.

BACnet, LonWorks, MODBUS, and other protocols are often touted as the "one true path" to interoperability. At the controller level, there is not a clear winner today. The most interoperable system may be the one that supports all of the popular protocols. At the workstation to controller level, however, there is a clear winner. That winner is BACnet. This is because BACnet is the only open protocol that defines a standard way to handle all of the higher workstation level functions, such as trending, alarming, scheduling, and file transfers. A workstation that uses BACnet to implement these functions will be able to interoperate with other systems that support these BACnet services. This is true whether the workstation is web-based or not. A web-based system may have a slight advantage in this regard because it will of necessity support internet standards such as TCP/IP, HTML, and XML, but it is important to realize that these standards do not in themselves guarantee interoperability.

TCP/IP, for example, is the standard used by the internet for packaging, addressing, and delivering information "packets." TCP/IP imposes no standard upon the information contained in these packets. It could be a BACnet object or Japanese Haiku poetry. The content makes no difference to TCP/IP, but it definitely makes a difference to the control system that receives this packet! Similarly, HTML simply defines the way information will be presented on your computer screen. While this may provide a degree of interoperability if you are simply viewing a graphic page from another system, it doesn't help much if your system is receiving an alarm from another system and needs to decide what actions to take in response.

XML is a system for creating and documenting data structures. It provides an excellent platform upon which to create an interoperable system, and ASHRAE's BACnet committee is in the process of defining an XML-based structure called "BML." By itself, however, XML does not define a standard way of communicating alarms, schedules, setpoints, or any of the other data required for a true interoperable system. These internet standards facilitate the transfer of data between control systems, but if the data being communicated needs to be interpreted by the other system it has to be packaged in a standard building automation protocol like BACnet.

FEATURES ACCESSIBLE THROUGH THE WACS

Any system which provides some level of access through a web browser can be classified as a WACS; however, the features which are offered through the web browser differ significantly from one manufacturer to another. Some provide a limited set of tools to view and adjust selected building parameters such as setpoints. Others use the WACS as the primary user interface, enabling the user to perform all commissioning, monitoring, troubleshooting, and adjustments through a web browser. The functionality available through the browser interface, the way web pages are generated, the interoperability of the systems, the ease of use, the security, the control of operator privileges, and many other features of the products should be compared. When specifying and selecting a WACS it is important to decide what features you will want to access through a web browser and ensure the system you purchase will provide that functionality. The following list of questions can help guide you in the specification and selection of a WACS. We suggest that you weight the answers according to the importance each feature holds for you. You may wish to add your own questions to this list, as well.

Features

1. Are all of the operating features that are available through the host workstation software also available through a browser interface?
2. Is the ability to schedule any segment of a building for "occupied," "holiday," or "override" mode through a browser a standard feature?
3. Through a browser:
   - Can you create groups of building segments for scheduling purposes?
   - Does the system include trend logs on all points?
   - Can you display multiple variables on a trend graph?
   - Can you change the variables to be trended?
   - Does the trend graph provide zoom, pan, and data-point annotation?
   - Does the web-based solution include an alarm-management screen that enables you to view and acknowledge alarms through a browser?
   - Can the alarm-management screen restrict viewing to those alarms associated with specific geographic segments of a building?
   - Can you sort and filter alarms based on their source, type, and criticalness through a browser?
   - Does the web-based solution allow you to configure alarms and reporting actions through a browser?

Ease of Use

1. How would you rate (from 0 to 10, with 10 being the easiest) the ease of operation through the browser?

Accessibility

1. Does the web-based solution work through a browser without any special software, including plug-ins?
2. Is the web-based solution designed with a platform-specific technology (such as ActiveX) that would prevent it from operating on a variety of internet appliances?
3. Can operators speaking different languages access the same information at the same time?

Subsystems

- Can the web-based solution monitor and control a wide variety of third-party HVAC and electrical equipment through a browser?

Open Standards

1. Can the server communicate with BACnet, LonWorks, and Modbus-dedicated hardware at the same time over the same wire?
2. Does the server-software component of the web-based solution run on all major server platforms, including Windows®, Linux, and Sun® Solaris™?
3. Does the server-software component of the Web-based solution support all major database servers, including MS SQL Server, Oracle®, and IBM® DB2?

Security

1. Does the Web-based solution include SSL at no additional charge?

2. Can every operator be password-restricted from:
   — Configuring the system?
   — Changing parameters?
   — Viewing certain areas of the building?
   — Acknowledging alarms by type?
   — Any combination of the above?

Technology

1. Under worst-case conditions, how quickly does the web-based solution provide end-to-end alarm delivery and annunciation for critical equipment, including central-plant and primary air-handling units?
2. Does all of your field hardware use Flash memory for upgrades to the firmware through a browser?
3. Does the web-based solution come with a comprehensive interactive computer or web-based training program?
4. Can the sequence of operation algorithms be completely developed using graphical methods and without line programming?
5. For troubleshooting purposes, are the algorithms used in controlling your equipment's sequenced operations graphically viewable through a browser, and do they include dynamic display of variables?

Configuration

1. Does the system come with all the tools needed to create or edit control programs and graphics?
2. Can graphic screens be created with no knowledge of or skills in HTML programming?
3. Can one generic graphic template be used to represent multiple occurrences of the same equipment type?

WEB ACCESSIBLE CONTROL SYSTEMS—LESSONS LEARNED

When web accessible control systems first became available, many articles were written that focused on what might be possible with these systems. As the first HVAC controls company to actually bring a web-based system to market, Automated Logic has gained considerable experience with these systems and this section will present some of the lessons we have learned by installing these systems around the world. The following four categories summarize these lessons.

Lesson 1: Web-based Systems Work

From Shanghai to Newark, in hundreds of offices, schools, museums, data centers, and other buildings, web-based systems are taking care of day-to-day building operations while giving their owners unparalleled access in the process. These systems provide all the functions of a conventional building automation system, including direct digital control of equipment, trending, scheduling, alarming, reporting, and interfacing with other building systems. Unlike conventional systems, however, the user interface is not limited to a few dedicated workstations running proprietary software. Web-based systems generate conventional web pages as the user interface, allowing users to log into the system and perform any required function from any computer on the network. If the network includes the World Wide Web, this means virtually any computer in the world can serve as the operator workstation. If the network includes a wireless service provider, WAP-enabled cell phones and other wireless devices can also be used to access the building system.

Obviously these systems also provide a high degree of security to prevent unauthorized access, but authorized users can gain significant productivity improvements from this increased access. Since these systems provide all the functions of a conventional automation system, there is no drawback to this increased access.

The custodian of a small public school which has no internet access can use a dial-up connection to access his system from his home computer, checking on alarms and making needed adjustments without leaving the house. For him, the advantage of a web-based system is that it does not require him to buy multiple copies of proprietary software from the system manufacturer to install on his work computer, his home computer, his neighbor's computer, his wife's laptop, or any other computer he may be near when an alarm comes in on his pager.

At the other extreme, customers with world-wide operations can use these systems to access multiple facilities scattered around the globe. Not only does the web access allow them to bring all their facilities together into a unified system, it allows energy managers and consultants located anywhere in the world to access these buildings remotely to study energy usage and "fine tune" the building operation.

Lesson 2: Bring Your IT Staff into the Picture Early

Conventional building automation systems have been network based for many years now, and it is not unusual for them to utilize the existing IT (information technology) network for at least part of their structure.

This has not caused any major problems in the past, but web-based systems introduce at least two new wrinkles into this situation:

- They have the potential, at least, to provide outside access to the existing IT network, and
- They utilize technologies and devices such as web servers, routers, and IP addressing, which are traditionally the province of the IT department.

Neither of these issues is insolvable, and in fact the "solution" is often simply a matter of providing the IT department with the proper information. They are, however, reasons to involve the IT staff when planning to install a web-based system, and it's best to bring them into the picture as early as possible.

Conventional control servers tended to use proprietary hardware and proprietary terminology, which was of little interest to the IT staff. If you wanted to install an operator workstation and a few primary controllers on their network, you were no different than the hundreds of other users installing PCs and network printers. Replace that operator workstation with something called a web server however, and it's an entirely different situation. Now you're treading on IT territory and they are intensely interested. Their interest is understandable. Professionally speaking, the IT network is their sole reason to exist, and they live or die based on how dependable that network is. Anything that has the potential to disrupt that network is of great interest to them.

In actuality, a properly configured web-based system does not pose a threat to their network. There are many ways to provide security for a web-based system, as will be explained in the following section, but it is important to involve the IT staff in the planning process early so their concerns can be addressed. If you "surprise" them by hanging a web server on their network without consulting them in advance, they are bound to get defensive and create all kinds of obstacles. (The reaction of a mother bear with cubs that is surprised by the appearance of hikers is nothing compared to an IT staff that is surprised by the appearance of a web server.)

Lesson 3: Address Security Concerns Early

Whenever a system can be accessed by "outsiders," security needs to be addressed. There is nothing unique about web-based control systems in this regard. Indeed, conventional control systems with dial-up access also have the potential to be "hacked" by outsiders. In many ways, web-based systems have an advantage over conventional systems in that they utilize internet technologies and can take advantage of the security systems that have been developed to protect bank transactions, personal information databases, and other sensitive data that is routinely transmitted over the internet. The list of customers who use our web-based system includes universities, military bases, financial institutions, telecommunication centers, and others who have reason to be especially careful about security. Again, the key is to involve the IT security personnel in the planning process, and address their security needs up front. Factors to consider include:

- Use a "fine-grained" password protection scheme. Passwords have long been used as the first line of defense for conventional building automation systems, and web-based systems are no different. A "fine-grained" password protection scheme allows you to adjust the access levels for each individual operator to give them access to the equipment and properties they need, and to prevent them from accessing anything else. This type of system will allow you to control access to setpoints, schedules, tuning parameters, hardware configuration, and other sensitive functions. Operator privileges can be controlled from "view only" to full system administrator rights, depending on the needs and experience of the individual operator. Figure 30-5 shows a partial list of operator privileges available through a "fine grained" password control.

- Use a web server which is inherently limited. There are "general purpose" web servers available on the market which are designed to provide web access to a wide range of network functions, often as an "add-on" to an existing database. Unfortunately, the widespread use and wide range of features of these servers also makes them very popular with hackers. We use a dedicated web server which is built into our product and which only provides access to the building automation functions. Additional security is provided by only activating the port needed to access web pages. (Typically port 80.) By not activating the ports needed for Telnet, NetBios, FTP and other network functions, the potential for unauthorized access is greatly reduced.

- Use encrypted transmissions. Secure socket layer (SSL) connections can be used to provide 128-bit encryption on all transmissions between the browser and the web server. This effectively prevents anyone from tapping into your communications to capture passwords and other sensitive data. SSL is commonly used to protect internet credit card transactions, and this same level of protection should be available on your web-based control system.

- Use firewalls and/or virtual private networks. The WACS server should be compatible with external firewalls and virtual private networks (VPN) to provide additional security. VPN can be configured to restrict access to specific computers which have been individually configured as a member of the VPN. Computers which are not a designated member of the VPN cannot connect to the network, regardless of any passwords or "inside knowledge" the operator may possess.

- Restrict your system to an intranet. While the term "web accessible control system" implies a connection to the World Wide Web, this connection is not required in all cases. While not connecting your system to the internet does reduce the "access from anywhere" advantage of a WACS, the system is still available from any computer on the internal network. Universities, data centers, and other customers with a 24/7 facilities staff may find this is an acceptable trade-off since their on-duty staff still has full access to the control system.

The security arrangements needed for any particular installation will depend on the security needs of the customer. Not all of the provisions outlined above need to be implemented in all cases. A careful review of the trade-offs involved with each option needs to be reviewed when planning for a particular installation, but the bottom line is that web-based systems can be made as secure as needed.

Lesson 4: The Use of Open Protocols Enhances the Interoperability of WACS

Web-based systems typically utilize internet standards such as HTML for generating and transmitting web pages. These internet standards are designed to render computer data as text and graphics which can be understood by humans. While this allows operators to access multiple systems through a single web browser, these internet standards do not provide interoperability between different pieces of equipment within the building automation system.

Fortunately, WACS can be fully compatible with existing standard protocols for equipment, such as BACnet, LonWorks, or Modbus. Using these protocols within the WACS greatly simplifies the integration of equipment made by different manufacturers into a unified system, and the WACS interface makes this system available through a web browser. Status values, setpoints, PID gains, and other commonly accessed values can be presented in a unified front end that shields the operator from having to learn a different interface for each vendor. BACnet adds higher level functions such as scheduling, trending, and alarming to this mix, making it even easier to manage the entire system from a single front end. While all of these functions can be integrated without the use of an open protocol, the use of open protocols eliminates the need for custom programming and greatly simplifies integration. A few examples:

- A government facility manager uses BACnet to integrate eleven buildings in three states using control systems made by four different manufacturers into a single energy management system. Five of the buildings were controlled by native BACnet systems to begin with, while six use proprietary systems which are translated into BACnet through a gateway. The net result is a unified interface which allows operators and consultants across the US to view and manage energy data through their web browsers. (Interestingly, the WACS system that unifies this network was made by a fifth manufacturer, further illustrating the interoperability provided by open protocols.)

- The Shanghai ScienceLand Museum, in Shanghai, China, uses BACnet to integrate 3,200 control points and 1,000 monitoring points into a single WACS. AHUs, chillers, pumps, lighting, elevators, fire, security, and other building systems supplied by 9 different vendors are monitored and controlled by a unified operator interface.

- The Houston Bush Intercontinental Airport is using a WACS to integrate two native BACnet systems plus multiple Modbus systems into a single, unified system. With construction nearing completion, it is safe to say this integration could not have been completed within the time and budget available without the use of open protocols.

CONCLUSION

Web-based systems are not just a concept to be shown at trade shows and discussed in journal articles.

They're real, they work, and they are providing unparalleled access to facilities around the world. Assuming good design and quality components, building automation systems that use standard protocols at all levels of the system architecture will have the lowest life-cycle cost because they can be maintained more inexpensively than can proprietary solutions. WACS integrate building systems and operators with the greatest efficiency because they are broadly familiar, inherently visual, and entirely flexible. Implementing a BAS with web accessibility should not add cost, extend the project schedule, or involve any additional risk. In fact, web accessibility can expand capabilities, tighten security, reduce a building's energy and service costs, help attract and retain quality staff, and provide opportunities for owners to give their occupants more monitoring and control functions to raise the occupants' level of satisfaction. The planning and implementation suggestions described in this chapter can make the installation of these systems go more smoothly, and can help you decide if a web-based system is right for your operation.

Chapter 31

Custom Programs Enhance Building Tune-up Process

Paul Allen, Walt Disney World
Rich Remke, Carrier Corporation
David Green, Green Management Services Inc.

THE BUILDING TUNE-UP (BTU) is a process in which building heating, ventilation and air conditioning (HVAC) system and the energy management system (EMS) are analyzed and optimized for improved performance and energy efficiency. The BTU process is also known in the industry as re-commissioning or continuous commissioning (1). Although difficult to predict for a single building, facility energy savings resulting from a BTU can be expected to be between 5-15%. Thus, the BTU is one of the most cost-effective energy conservation projects available to an energy manager.

This chapter describes the basics of the BTU process and showcases two custom programs that aid in that process: (1) the facility time schedule (FTS) program and (2) the Building tune-up System (BTUS).

BUILDING TUNE-UP PROCESS BASICS

The building tune-up process is one of the most cost-effective energy management projects available to an energy manager. The actions taken are generally low-cost or no-cost adjustments to an existing EMS and will not only minimize current operating costs but will also lower future maintenance costs. The building tune-up is stage 2 of the five-stage Energy Star Buildings program (2). The BTU does not necessarily involve the purchase and installation of new equipment or technology. Instead it requires an investigative-style approach to ensure that the EMS controls are working and controlling the HVAC and lighting systems optimally.

Goals

The BTU process has the following goals:

- Reduce utility consumption by optimizing air conditioning and lighting time schedules/setpoints
- Improve EMS performance by improving control algorithms & documentation
- Identify corrective action items by monitoring HVAC system operation
- Measure and track utility savings

BTU Team

The most effective way to perform a BTU is to organize a BTU Team. Initially, a core team is pulled together from maintenance/engineering services and the energy management team. The purpose of the core team is to get the BTU process rolling. The first steps that are taken include:

1. Determine who is responsible for day-to-day operation of the EMS?
2. Make initial assessment of the condition of the EMS.
3. Agree on the time/place for a weekly meeting.
4. Determine additional technical resources required.

Initial BTU Stage

This initial stage is designed to get the EMS into good working order and improve the EMS documentation. Shown below is the scope of work for the EMS assessment:

On-Site EMS Checkout

- Set all control valve and damper actuators to design ranges
- Ensure control valves shut completely when commanded off
- Check that supply/return/exhaust fans turn off when commanded off
- Replace sensors that have malfunctioned
- Check operation of Variable Frequency Drives (VFD)

- Check outside air damper operation and proper setting
- Check HVAC system air and water side balance vs. design
- Check EMS communications to all controllers
- Check that EMS panel is neat and clean

EMS Programming & Documentation Check-up
- Review all control algorithms and sequences for proper configuration
- Create/update wire list to match existing EMS Panel configuration
- Update EMS graphics as-needed
- Add time and setpoints to Facility Time Schedule database
- Add HVAC/EMS data to the Building Tune-up System (BTUS)

Weekly Meeting

The other important part of this process is establishment of a weekly meeting. It is best to set up a one-hour meeting and use the same time and place so everyone gets accustomed to the routine. This doesn't seem like much time, but a slow and steady pace is best for this type of work. The weekly meeting provides a way to recap progress and to plan remaining tasks that can be documented in a weekly status report/email. Most of the detailed work actually is spent between the weekly meetings checking out the EMS controls for proper operation and documenting the EMS programming and setpoints. This process might last weeks or months, depending on the state of the HVAC/EMS and the number of buildings being studied.

Review System Schedules

Once the detailed EMS tune-up is completed, the next step is to review the HVAC and lighting systems time/setpoint schedules with the "Building Owners." This is best accomplished in a meeting with the appropriate representatives of Operations, Food & Beverage, Merchandise and Custodial/Housekeeping. The time schedules and setpoints for each building and HVAC system are established until all of the systems have been reviewed.

BUILDING TUNE-UP PROGRAMS

Two programs were developed to aid in the BTU process. The facility time schedule (FTS) program manages the equipment time schedules and temperature setpoints using a server-side program and automatically resets time schedules and temperature setpoints on a daily basis. The Building Tune-up System (BTUS) is a web-based program that was developed to provide the "Building Owners" a view into the EMS control settings, without actually having access to the EMS.

The Facility Time Schedule Program

The Facility time schedule (FTS) program is a custom client/server program that interfaces with the Carrier ComfortWORKS energy management system. The purpose of the FTS program is to provide the energy manager a method to manage the time and setpoint schedules for a large campus facility in a master schedule database. Time schedules can be set up as "relative schedules" that incorporate the facility opening/closing times, dusk/dawn times and by day of week. Each day, the FTS program determines the appropriate open/close/dusk/dawn times and calculates actual time schedules that are broadcast to the Carrier EMS controllers on the Carrier ComfortWORKS Network (CCN). Without this automatic reset feature that is created when the master schedules are downloaded each night, the time schedules and setpoints eventually get changed from their optimal settings.

The FTS program can also handle special events that occur after the normal open/close time schedules by sending additional time schedules that effectively increase the HVAC/Lighting equipment run-time to accommodate the special event. The FTS Program provides some additional operational features:

- Users can make local EMS panel adjustments to both time and setpoint schedules to respond to building conditions without worrying that the changes would be permanent. All schedule changes will revert to the master schedule at the programmed download time the next day.

- Setpoint schedules can be grouped into common areas or types and can be programmed with a bias offset. This offset can be used during loadshed conditions to change the setpoint low and high values to a user adjustable level, reducing energy consumption. Schedule groups can also be used to pre-cool or pre-heat an area during special functions. The setpoint bias will be removed from the setpoints after the next automatic download.

- These master schedules are sent automatically to each Carrier EMS controller on a daily basis. For each schedule, the system administrator can choose whether and when to send either or both time and setpoint schedules.

- If the facility open/close times are changed during a given day, the new time schedules are re-calculated and downloaded again to the EMS controllers to reflect these changes.

FTS Program Details

The FTS program is comprised of three main parts: TSServer, TSClient and TSserver Database. The TSServer and TSClient programs are custom Microsoft Visual Basic Programs. Both programs are compiled into executable (exe) format and installed using a setup installation program. The TSServer database is based on Microsoft Access and contains the master schedules and other important tables that link the TSServer and TSClient programs. Figure 31-1 shows a schematic diagram for the FTS program/database relationship.

TSServer runs on the ComfortWORKS server. TSServer uses ODBC (Open Database Connectivity) to connect to both the ComfortWORKS SQL Database and to the FTS Microsoft Access database. The TSServer is set up on the server as a service to insure that it operates even if no one is logged into the ComfortWORKS server. This requires the use of Service Mill from ActivePlus software (http:\\www.activeplus.com). TSServer will automatically calculate the time schedules, special event schedules and setpoints and broadcast them to the EMS controllers at a specified time each day.

TSClient is the user interface for the FTS program. It is installed on the user client PC and interfaces with the Microsoft Access database on the server. TSClient controls the following FTS program features:

- **Program Setup Screen**: This includes most of the program defaults as well as what rights this client installation has. The setup screen is password protected and is only accessed by the system administrator.
- **Import/Update**: Import time and setpoint schedules from Carrier ComfortWORKS database.
- **Time/Setpoint Schedule Changes**: EMS controllers are picked from a tree-type menu interface to allow changes to each individual time/setpoint schedule.
- **Open/Close Times**: The facility open/close times are entered for each day.
- **Time Sched–ule Groups**: Individual time schedules can be grouped together for use with special events.
- **Special Events**: Time schedule groups can be combined into a special event and assigned a date and time to accommodate functions that occur after normal operating hours.
- **Open/Close Time Change**: If the open and closed times change on the current day, the revised time schedules are re-calculated and re-broadcast immediately.

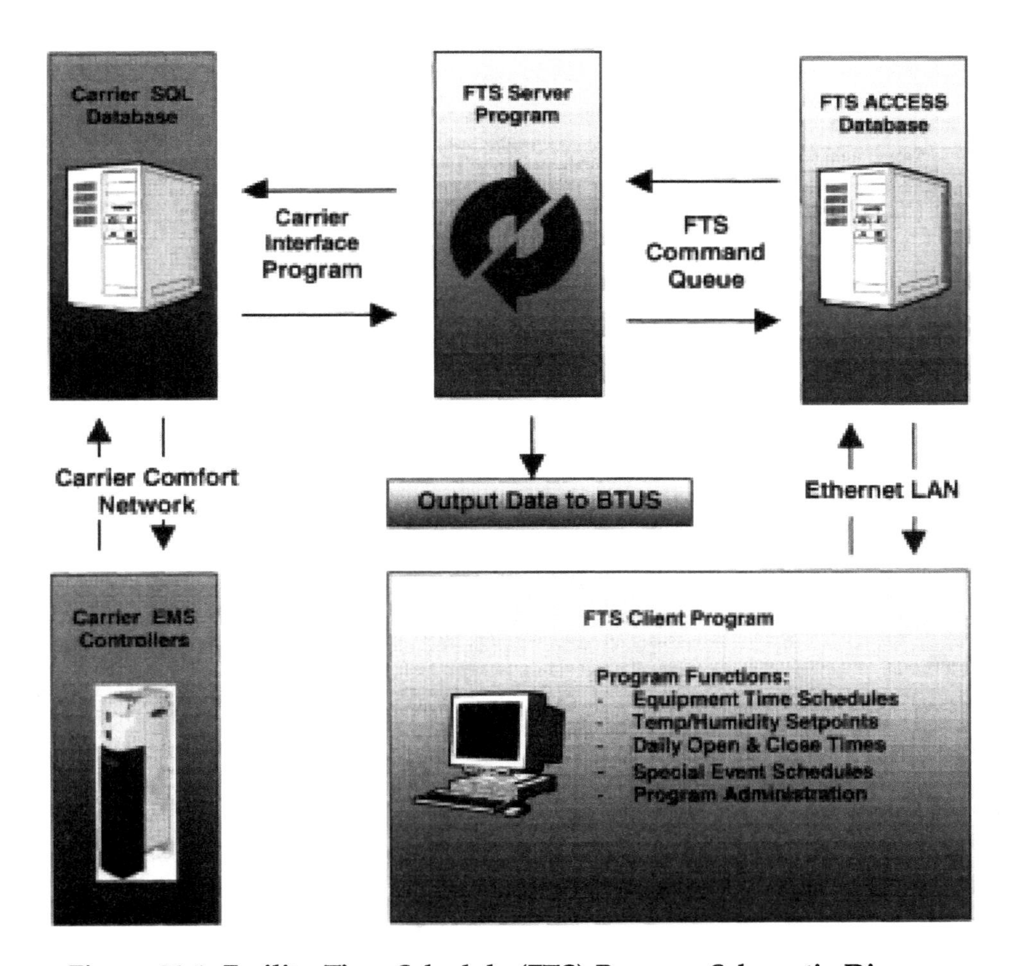

Figure 31-1. Facility Time Schedule (FTS) Program Schematic Diagram

- **Program Administration**: Program operation functions, including system log, failed downloads, password access and FTS Server command queue tables.

The Building Tune-up System

The building tune-up system (BTUS) is a web-based program that provides information on each air conditioning system and shows the time and setpoint schedules that are in effect. The purpose of the BTUS is to give building owners a view into their buildings HVAC and lighting systems without accessing the EMS directly. The building owners provide the information on how their HVAC systems are controlled by establishing the time and setpoint schedules. Because it is web-based, it also allows a broader audience to access it via its web browser interface.

The BTUS shows the following information for each HVAC system:

- Description of area serviced. A color-coded floor plan can be displayed if available.
- Time and setpoint schedules. Includes both desired schedules and a link to look at the most recent schedules broadcast by the FTS program.
- Shows equipment in need of repair.
- HVAC temperature, humidity and status trends can be graphically displayed if available

The main control on the BTUS is a drop-down menu that allows the user to select the area desired. The user is then presented with a list of buildings from which to pick. Once the user selects an individual building, the detailed data for each HVAC system is displayed. Figure 31-2 shows a screen shot that shows a list of the HVAC systems in the building PAVILION1. Links are available to show the detailed information on each HVAC system.

The BTUS displays the HVAC system time schedules and setpoint schedules for each HVAC system. Clicking on the link for these schedules displays the latest actual schedules that were downloaded the previous night to the EMS controllers by the FTS program. Figures 31-3 and 31-4 show screen shots for these displays.

Another useful feature of the BTUS is to display a graphical floor plan that shows what each HVAC system covers. The floor plan is color coded to show the coverage areas for each HVAC system.

If an EMS trend report is available for a particular HVAC system, a link under the ID# will be highlighted. This makes an easy-to-use method to display trend data graphically. This feature is further explained in the chapter "Creating Web-Based Information Systems from Energy Management System Data."

CONCLUSION

The building tune-up process is a systematic approach to fine tuning an energy management system for optimal performance. This effort can be considered one of those proverbial "low hanging fruit" energy projects that all energy managers should focus on.

The non-technical side of the building tune-up focuses on the development of the building tune-up team. It allows both technical and non-technical staff to work together in an on-going continuous improvement process to lower utility costs with minimal capital outlay. This provides an excellent venue to organize energy conservation efforts within an organization. Every team member can play an important role and can contribute to the overall team success.

The technical side of the building tune-up focuses on computer programs that keep the energy management system settings at their optimal state. The FTS program automatically resets the energy management system time schedules and setpoints to their optimal valves. The BTUS program allows all users to view the time schedules and setpoints from their own PCs using web browser software.

The building tune-up process helps users understand how their HVAC system operates and provides a method to prevent EMS degradation by auto-resetting time and setpoint schedules on a daily basis. Keeping the control values optimal and enabling users to view the control settings from their own desktops makes this a continuous improvement process.

References

Continuous Commissioning SM in Energy Conservation Programs, W. Dan Turner, Ph.D., P.E., Energy Systems Lab, Texas A&M University, 409-862-8480, e-mail: dturner@esl.tamu.edu

ENERGY STAR Buildings Upgrade Manual—Stage 2 Building Tune-Up, US EPA Office of Air and Radiation, 6202J EPA 430-B-97-024B, May 1998

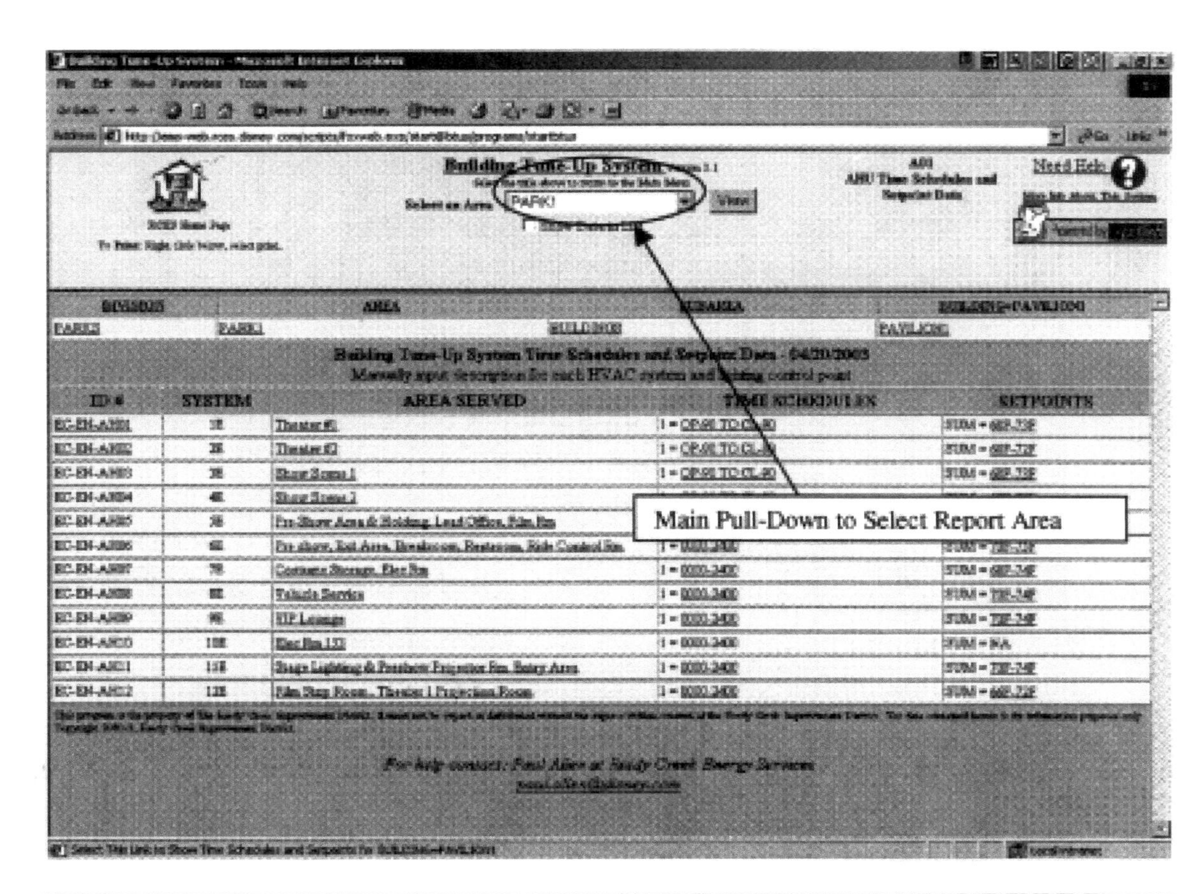

Figure 31-2. BTUS Main Display Screen

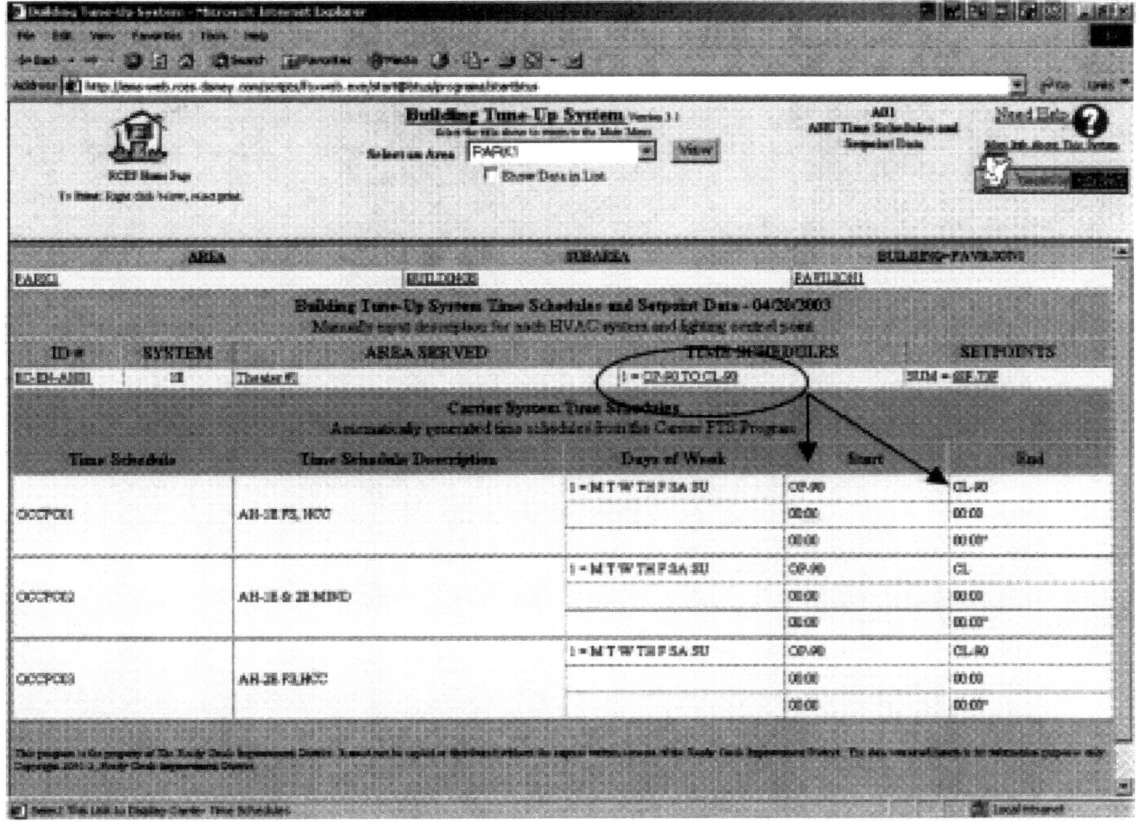

Figure 31-3. BTUS Display After Clicking on Time Schedule Link

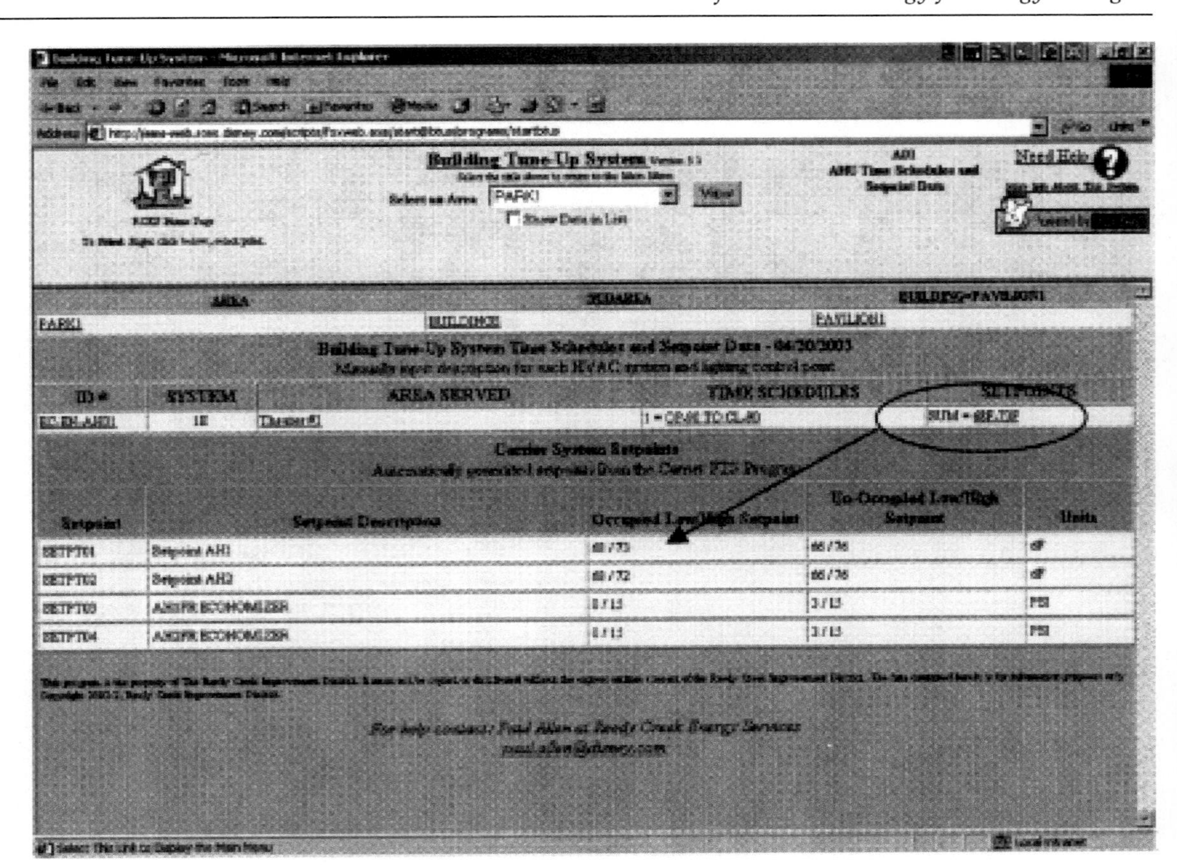

Figure 31-4. BTUS Display after Clicking on Setpoint Schedule Link

Figure 31-5. BTUS Display Showing EMS Trend Reports and Floor Plan Links

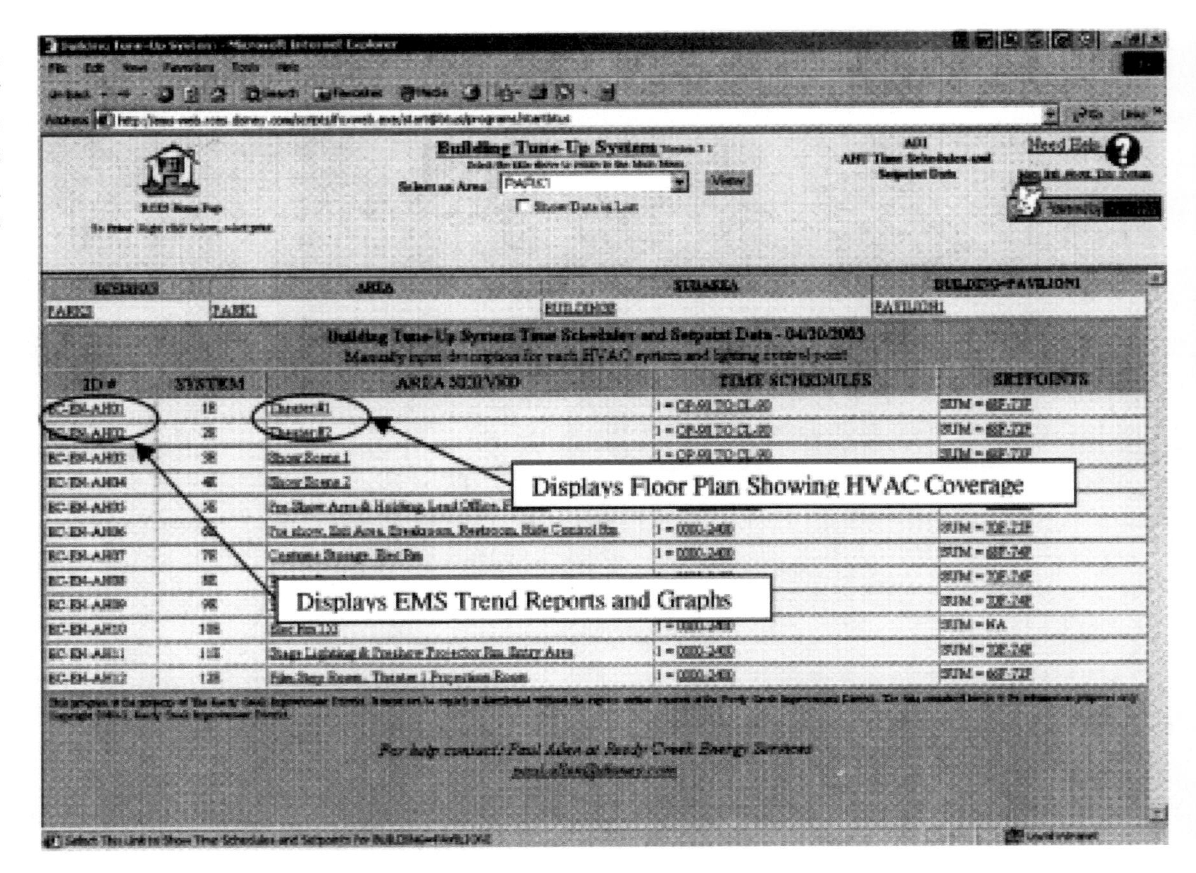

Chapter 32

Application Examples of Energy Information Systems

Jim Lewis
Obvius Corporation

THIS CHAPTER CONTAINS five examples of uses for various features of Energy Information Systems. The first example explains the use of meters and sensors to measure and verify energy savings from energy retrofits. The second example involves the use of meters and data acquisition servers to allocate energy costs within a facility. Example three describes data acquisition systems that are installed in multiple buildings and provide energy use information to a remote database for load aggregation. Example four illustrates the use of a data acquisition server to monitor the electric loads and environmental conditions in internet data centers. The fifth example shows the use of sensors with a data acquisition server to monitor indoor air quality for commercial and industrial buildings.

EXAMPLE ONE: MEASUREMENT AND VERIFICATION OF ENERGY RETROFIT PROJECTS

Meters and sensors are installed in a facility to monitor the energy savings from energy retrofit projects on systems such as lighting and HVAC (heating, ventilating and air conditioning). The sensors are connected to a data acquisition server (DAS) or building control system (BCS) to measure the energy consumption of the monitored equipment. These values are compared to energy usage prior to the installation.

Background

Facility managers in commercial and retail buildings are charged with operating the buildings as efficiently as possible in order to maintain (or improve) the operating margins and profits for the owners. In many industries, profit margins tend to be small. Savings in energy can use translate into significant bottom line impacts, but the energy savings have to be realized with little or no impact on temperatures, lighting levels, etc. There is no shortage of potentially energy reducing technologies such as higher efficiency lights, variable speed motors, high efficiency compressors, etc., but it is very challenging for facility managers in a dynamic environment to determine the benefits of these technologies.

Consider the following case: a retail store facility manager attends a trade show and learns about a new, low-cost technology for lighting in areas such as storage and warehouse space. Savings of 30% and more are projected and it looks like a logical program for all the manager's 100 stores, but he wisely decides to do a pilot program in one store to evaluate the results. He hires a lighting contractor who does the retrofit and after a commissioning session and a training program for local staff, he waits for the next month's lower electric bill. When the bill arrives, he finds that the bill is actually higher than the previous month and the prior year. Certain that the contractor made a mistake, he calls the lighting contractor who goes to the job site and assures that after significant review and testing, the system is working as specified and should be producing savings.

What went wrong? The facility manager calls the local store manager to find out if there were any changes to operations or other factors that might have influenced energy usage. He finds out the following from the local store manager:

- "It seemed a lot hotter this year than last year."
- Sales were up 3% from last year and the receiving area was "busier than usual."
- "Our 'Midnight Madness' sales promotion on a couple of weekends added a fair number of hours to our normal operation."

To make a long story short, it's virtually impossible for the facility manager to determine whether the lighting retrofit fell short, met or exceeded the energy savings expectations. It's budget time, and the facility manager needs to decide whether to plug in a capital item to do all 100 stores, but he has no way of knowing whether the test site was a success or not. A clearly defined measurement and verification (M&V) program that provides definitive proof of the savings from this retrofit would have prevented this dilemma. This application example provides some guidelines for getting a solid M&V program for energy retrofits.

How Does it Work?

The owner and the energy management contractor must agree in advance to measure and verify the energy savings because it is important to establish a "benchmark" for performance before changes are made. The M&V system may have to be in place and operational for a time period of several days to several months in advance of the retrofit to establish a sound baseline for comparison. The key to the ultimate success or failure of the M&V program is the ability to isolate the specific systems being modified (e.g., lighting) from the rest of the energy-consuming equipment in the building. The steps required in a typical M&V program are:

Pre-retrofit

Before any systems changes are made, data sensors and the data acquisition system are installed to determine how much energy is being used by the system to be modified. The amount of time required to establish a baseline will depend on a number of factors, such as seasonal changes, operational complexity, etc. If the building has been part of an accountability metering program (see Chapter 10—Example One-Accountability Metering), this baseline data may already be available. The purpose of this benchmarking exercise is to verify the expected savings and to adjust the payback estimates to reflect the actual usage.

During Installation

If possible, the sensors and data acquisition server (DAS) installed during the baseline period should remain in place during and after the installation to provide consistency in measurement. The information from the DAS can be valuable in commissioning the new system as this data provides 24 hour, 7 day a week measurement to ensure that systems are operating to the proper setpoint and times. Many commissioning and acceptance programs focus on a "snapshot" view of system performance that may or may not be reflective of the longer-term operation. Reports at this period are invaluable for fine-tuning system operation for the maximum return on investment for the contractor and owner.

Post Retrofit

Many energy retrofits perform well initially, but the ROI is never realized because system efficiency degrades over time due to calibration drift, manual overrides, etc. Post-retrofit M&V will not only serve as a warning system for loss of efficiency, but can also provide valuable insight to the total cost of installing and maintaining energy efficiency. For example, assume that a package unit retrofit costs $10,000 and saves $1,000 per month, providing a payback period of 10 months. If post-retrofit monitoring shows that the new system requires quarterly calibration visits (at $250 per quarter) to maintain efficiency, the actual payback period is 11 months, or 10% longer. If this particular retrofit is being used as a model project for multiple locations, determining the total cost of the project is crucial prior to a wider rollout.

Benefits

The facility manager evaluating the success of a particular project can move forward with a much greater degree of certainty about the energy and dollars saved from the project if he or she is looking at actual data from the data acquisition system rather than projections and calculations. In addition, the manager actually improves the likelihood of a successful project if the proper tools are in place for M&V as real time feedback becomes available to the manager and the contractor.

Drawbacks

Adding real M&V to an energy retrofit project adds costs in the form of hardware and software and the manpower needed to review the reports and track the project's success. Establishment of a baseline for measurement can delay the implementation of the project for a few weeks, stretching the payback time.

Installation Requirements

The specific hardware required for M&V will, obviously, be dependent on the systems being retrofitted and the expected outcome. Consider the following examples for general guidelines:

Chiller retrofits

Chiller retrofits generally are designed to improve the efficiency of the chiller system in producing cooling (typically measured in kW/ton). The benefits and savings from most chiller retrofits can be determined using the following:

- *AcquiSuite™ DAS*—gathers the data on user-specified intervals and stores it till it is sent to the remote server

- *Electrical sub-meter*—provides data on the power (kW) consumed by the chiller system to produce cooling (tons)
- *Chilled water supply and return temperature sensors*—used to measure the temperature differential between supply and return temps (ΔT)
- *Flowmeter*—used to determine how many gallons of chilled water move through the chiller; when combined with ΔT, the amount of cooling produced (in tons) can be calculated
- *Software* to calculate chiller efficiency (in kW/ton) to provide comparison to before and after retrofit and to fine tune performance

Variable Speed Drives

Great savings can be realized from converting constant volume fans and pumps to variable speed as the load on the fan or pump can be reduced during non-peak load periods, but if the system is not properly sized and calibrated, the savings can be lost. There are several options for monitoring the performance of VSD's:

- *AcquiSuite™ DAS*—gathers the data on user-specified intervals and stores it till it is sent to the remote server
- [Option 1] *Electrical sub-meter*—provides data on the power (kW) consumed by the motor; most accurate and most expensive
- [Option 2] *Analog current sensor*—connected to one leg of the motor electrical supply, the current sensor measures amps which can be used as a measure of the load on the motor
- [Option 3] *Analog output from the drive*—many VSD's provide an analog output signal proportional to the load. For example, a 4 to 20 mA output signal would read 12 mA at 50% of full load. This is usually the least expensive option if the analog signal is available from the drive

Lighting Retrofits

Most simple lighting retrofits (e.g., ballast and tube replacements) are the easiest to measure since the amount of energy saved (in watts) per fixture is nearly constant and the only variable that needs to be monitored is the run time. Typical requirements are:

- *AcquiSuite™ DAS*—measures the total runtime on user-specified intervals and stores it till it is sent to the remote server
- *On/off current sensor*—this sensor is similar to the analog current sensor described above, but the current sensor simply uses current flow to determine whether the lights are on and the DAS calculates the run time to verify that the hours of operation are the same as the baseline period
- [Optional] *Analog current sensor*—the analog sensor can be used to measure the actual current draw and verify not only that the hours of operation are consistent, but that the equipment is functioning properly
- [Optional] *Ambient light sensor*—If there is concern that the retrofit will have a negative effect on light levels, one or more ambient light sensors can be installed to verify that the lighting levels are acceptable

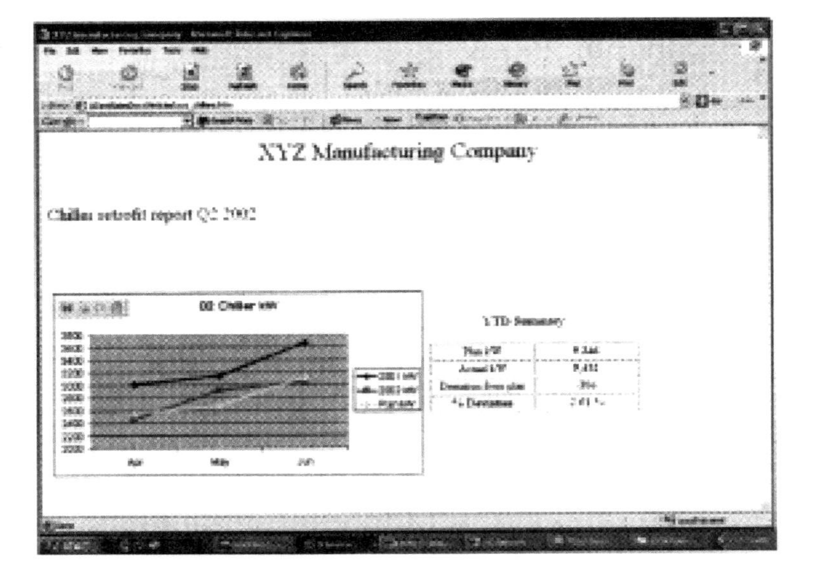

Figure 32-1. Planned vs. Actual Savings for a Chiller Retrofit

Reports

The reports for M&V can be relatively simple (baseline energy use vs. actual) or can involve a great deal of post-processing. For example, analysis of a chiller retrofit would likely have to include some "normalization" of energy use to include temperature differences from the baseline period to the M&V period to account for higher temperatures that might impact the load on the chiller.

A simple example showing plan vs. actual savings for a chiller retrofit might look like Figure 32-1.

Analysis/Actions

A targeted M&V program can greatly accelerate an energy program because the facility manager (and contractor) can:

- React more quickly to correct problems that are identified in the M&V audit. Many retrofits are installed or operated incorrectly and the problems are not identified until weeks or months later when the utility bills are analyzed. In many cases, it is difficult if not impossible to determine the root cause of poor performance months later.
- Quickly determine the success (or lack thereof) of a particular energy retrofit and make better decisions about wider implementations to more locations.
- Provide measurable proof for owners and executives of the impact of energy retrofits on the bottom line, greatly improving the chances of getting additional program support.

Costs

As with all the application examples in this chapter, it is very difficult to estimate costs due to a variety of factors (wiring distances, communications issues, scheduled shutdowns, etc.), but some general guidelines for costs (hardware and installation) are:

- AcquiSuite™ data acquisition server—$1,200 to $1,800
- Electrical sub-meter (3 phase)—$600 to $1,000
- Current sensor (digital or analog)—$100 to $200
- Data storage and reports—$20 per month per AcquiSuite™

Notes/miscellaneous

Historically, energy contractors and building owners have been reluctant to add equipment for measurement and verification, choosing instead to rely on calculations (a.k.a., stipulated savings), snapshot studies and utility bill analysis to verify savings from energy retrofits. This view has resulted in many projects that may not have any return being accepted as successful and has also almost certainly caused projects with good returns to be deemed as failures because extraneous factors masked real savings. The simple fact is that a good M&V program (using new cost-effective technologies from companies like Obvius) pays for itself both in the near term and over the long haul as managers are assured that the retrofits they invest in continue to provide the expected returns for many years to come.

EXAMPLE TWO: COST ALLOCATION FOR CAMPUS AND INDUSTRIAL FACILITIES

Cost allocation refers to the use of meters and data acquisition servers (DAS) to submeter and allocate energy usage by department or cost center within a campus or industrial facility. Allocating these costs provides accountability for energy use in campuses and allows businesses to accurately determine the cost of products and services.

Background

Many, if not most, campus facilities have primary metering for energy consumption (electricity, gas and water), but have little capacity for determining how energy is used "behind the meter." For the energy manager responsible for operations, this typical structure poses several major problems:

- Internal users have little or no incentive to reduce energy costs.
- Outside users (food services providers, community groups, etc.) are often subsidized by internal users.
- Prioritizing and evaluating energy conservation efforts is virtually impossible
- Isolating and correcting operational problems can be difficult.
- Determining the actual cost of manufacturing products is difficult and standard cost allocation methods (e.g., as % of revenue) may lead to poor decisions on product mix and pricing.

Submetering to allocate costs alleviates these problems and provides the means to measure performance.

How Does it Work?

For each building or department, the facility manager installs the necessary sensors and meters to isolate and measure the energy used by that department. On a monthly basis, the data from the DAS is analyzed to produce a cost for energy assignable to the department.

Benefits

For most energy and facility managers, there are benefits and payback to investing in the hardware and software to submeter campus facilities:

Energy Savings from Internal Users

"If you can't measure it, it won't happen…" is a phrase often used in business texts to stress the importance of being able to measure the success at meeting objectives. The same applies to energy conservation and to modifying the behavior of building users to manage energy wisely. The governing board may set an overall objective to reduce energy costs, but if the responsibility and accountability for the reduction is not measured at the department or facility level, no one owns the goal and it is unlikely to be realized.

Recently, a large international hospitality company adopted a corporate wide goal of reducing energy usage by 5% in all its facilities. One of the largest facilities was served by a few primary meters, and there was no way to assign responsibility for energy usage to individual operations. The energy manager chose to begin the energy saving process by sub-metering individual operations within various divisions of the campus. Data gathered from these sub-meters was used to benchmark current consumption and to provide information about operations within the campus. This benchmark data was used to compare the relative success of each group within the campus and was published monthly on the company's intranet site with rankings of the groups and their success at meeting the corporate energy savings goals.

With the exposure of the monthly reports, it is hardly surprising that managers began to be aware of energy usage and requested additional sub-metering to allow them to drill down into the usage within their organizations. The campus not only met, but even exceeded the corporate goal for energy savings simply through making users accountable with cost allocation and monitoring of energy usage. The company was able to track with a high degree of accuracy such things as signage left on, motors left running and HVAC systems in operation during unoccupied periods.

There are also occasions where a department or other internal user group may have increased operating hours (e.g., final exams, special holiday events) for which charges may be assigned on a per square foot or per hour basis that may not reflect the actual cost of providing energy.

Cost Recovery from External Users

Many campus facilities offer space for rent by non-affiliated users (e.g., food service vendors, retail outlets, community groups). The usage by these non-affiliates is often on a different schedule than the normal operating hours and the energy cost is allocated to these users using either a fixed monthly cost or a formula for allocation based on square footage or per hour costs. In some cases, the external users may have a much higher energy density (energy usage per square foot) than the typical affiliated user and this higher usage is not allocated to the users appropriately in a fixed per square foot or per hour allocation.

Sub-metering of non-affiliated users is the fairest means of allocating the costs of energy as it provides an actual representation of the usage and costs for all users. The non-affiliates are charged fairly for their share and the affiliated users are not expected to subsidize the costs for these non-affiliates.

Drawbacks

The primary drawback to implementing a cost allocation program is the expense involved in installing the hardware and the ongoing cost of managing the data and producing useful reports.

Installation Requirements

The hardware required to complete a cost allocation system will obviously vary depending on the layout of the campus and the method of operation for the physical plant (i.e., central plant vs. distributed systems), but the following general guidelines will apply to each building:

- *AcquiSuite data acquisition server (DAS)*—a standalone web server located on the building site that communicates with the sensor(s), stores interval information and communicates with the remote server
- *Electrical submeters* to monitor electricity usage (kWh) and demand (kW) for the building or department being monitored
- *Gas meters* if the natural gas is supplied directly to each building rather than supplied from a central plant
- *Water meters*
- *Steam meters* if the steam is supplied from a central plant
- *Btu meters* if chilled water is supplied from a central plant to the buildings
- *Phone line or local area network (LAN) connection* for communication with the remote server

Reports

The typical report for cost allocation will be prepared monthly and will provide a breakdown of the usage for each department by utility type. This report can then be provided to the accounting department for use in assigning costs to various groups on the campus. In the simplest scenario, each department is charged for its consumption based on a "blended" rate for the total facility from a primary meter. There can also be more sophisticated reports that allocate demand and time-of-use charges if appropriate. A sample report for a campus might look like Figure 32-3.

Analysis/Actions

On a monthly basis, the accounting department would take the information from the report and assign

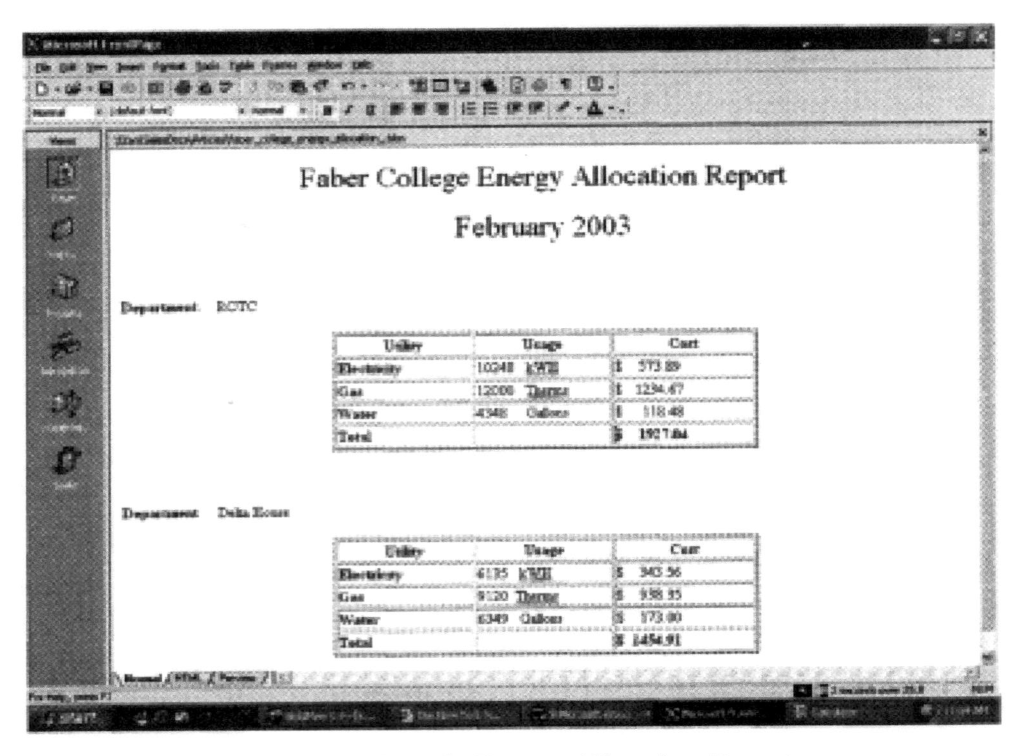

Figure 32-2. Sample Energy Allocation Report

the costs to the appropriate department. These costs can be used to develop budgets for succeeding years and to compare year-to-year changes in operations or energy usage.

Individual managers responsible for the departments being billed will likely be inspired to reduce the energy usage (Chapter 10, Example One: Accountability Metering) to minimize the budget impact and free resources for better use.

A side benefit to cost allocation is that the campus facility manager can use the data to determine if there are problems in the system (e.g., leaking steam traps). If the total cost allocated to all of the buildings is significantly less than the primary meter, it could indicate that system losses of steam or water due to leaks is much higher than expected and could provide a valuable tool in determining the priority for locating and repairing the leaks.

Costs

As with all the application notes in this series, it is very difficult to estimate costs due to a variety of factors (wiring distances, communications issues, scheduled shutdowns, etc.), but some general guidelines for costs (hardware and installation) are:

- *AcquiSuite™ data acquisition server*—$1,200 to $1,800
- *Electrical sub-meter (3 phase)*—$600 to $1,000
- *Flow meter for gas or water*—$500 to several thousand
- *Btu meter for chilled or hot water*—$300 to $1,000
- *Data storage and reports*—$20 per month per AcquiSuite™

Notes/miscellaneous

Cost allocation can be an important tool for the facility manager of a campus or industrial facility as it provides not only accountability for energy use, but also can be a valuable asset in determining priorities for energy retrofits. Identifying the major energy users is the first step in deciding where to allocate budgets for physical plant upgrades, lighting retrofits, control system upgrades, etc.

EXAMPLE THREE: LOAD AGGREGATION OF MULTIPLE FACILITIES

Data acquisition systems installed in multiple buildings provide energy usage information to a remote database. The information from these buildings is aggregated to produce a single electrical bill as if the individual buildings were one single facility.

Background

During the days when deregulation of the electric industry was all the rage, there were a number of factors cited as benefits to deregulation: lower costs, better service, and a wider range of offerings for the consumer. Among the biggest projected winners from deregulation were the many companies with multiple smaller locations scattered across one or more regions in the U.S. In theory, these companies (e.g., convenience stores, big box retail stores and restaurants) would be able to dramatically reduce their electricity bills by combining the consumption of some or all of their locations into a single account. Thus, a company with 100 stores of 10,000 square feet would have a similar rate structure to a single location with 1,000,000 square feet and would benefit from significantly lower rates. In addition, these companies would be able to get bids from multiple competitors to supply the power and would gain even more benefit from competition.

As just about everyone knows, the pace of deregulation has slowed considerably, to say the least. Disastrous outcomes such as those experienced in California in the first summer after deregulation have led other states to reconsider deregulation (really "reregulation") and to attempt to apply different models to achieve the benefits of a competitive power grid with a minimum of volatility in the supply and cost of electrical energy. In anticipation of coming deregulation, some utilities have begun to offer existing customers the option of aggregating some buildings. It is also fairly safe to say that some form(s) of deregulation will come into the market over the next few years and it will be beneficial for owners of multiple facilities to build a database of energy usage.

The actual price a utility is willing to charge for electricity depends on a number of factors, among them:

- Geographic location and number of locations to be supplied
- Total consumption (in kWh) of electricity for a given period of time (typically one month)
- The load profile (or load shape) of the electrical demand for a particular interval (typically 15 minutes). This load profile is particularly important to the utility as it helps to determine just how much power (in kW) the utility must have on-line at any given time to meet the needs of its customers. Customers who consume electricity during off-peak times are likely to receive lower rates than those who use electricity during peak demand periods.

How Does it Work?

In its simplest form, the utility simply adds the consumption for all locations and also compares the load profiles to determine peak demand intervals for the potential customer. A typical load profile is shown in Figure 32-3.

Using the data from load profiles such as those shown above, the utility can determine the total load (the heavy line in the graph above) that the utility will experience if both of these stores are on the grid. If, for example, Stores A and B had higher load requirements in the non-peak hours (say 3:00 AM), the total package to be offered by the utility is likely to be far more attractive than if the profile is similar to that shown above.

Benefits

As with most products, consumers who purchase more electricity from a single supplier tend to get lower

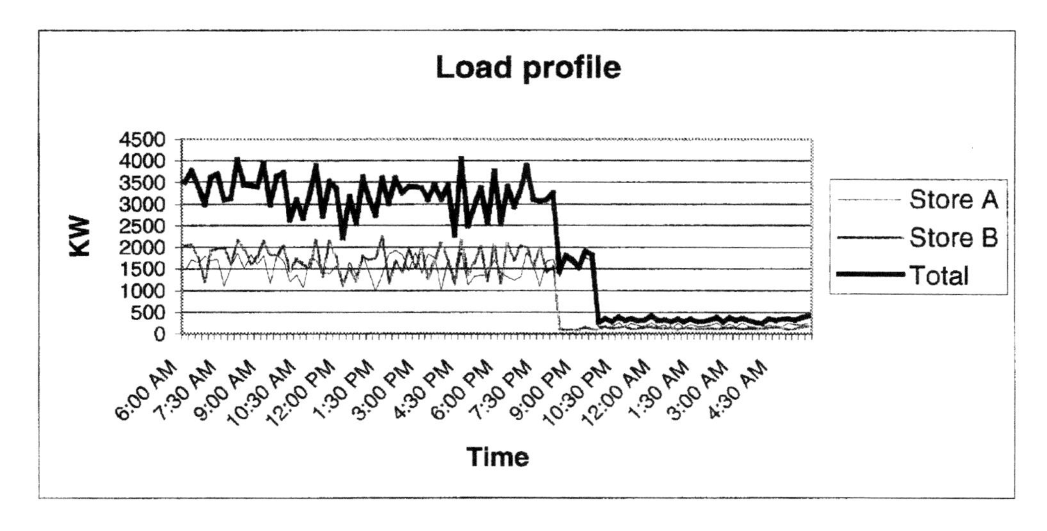

Figure 32-3. Load Profile for Two Stores

prices and this is reflected in the different rate structures offered by utilities to commercial and industrial (C&I) customers. C&I customers who are able to aggregate multiple locations benefit from being able to purchase electricity at a much more attractive rate schedule than they would when buying power for each location separately. This can also result in lower costs for handling the purchasing as the bills will come from a single source and the costs of handling and paying multiple vendors is greatly reduced.

Customers who have installed metering systems to monitor operational performance (see "Accountability metering" Example One, Chapter 10) have the added benefit of developing a load profile as part of the metering program. Since the information necessary for developing load profiles is the same as that gathered for operational monitoring, the customer has the required information available immediately and thus saves the time and expense of assembling hundreds of utility bills for utilities to bid on.

The utility benefits from adding customers (preferably those with attractive load profiles) to the grid and from the lower cost of billing a single customer for multiple locations, minimizing the need for billing multiple locations. The utility can also afford to spend money on automating the gathering of energy information as these costs can be spread across multiple stores for a single customer.

Drawbacks

The biggest issues for C&I customers considering load aggregation are:

- Gathering the load profile information necessary for utilities to put together proposals
- Evaluating a potentially wide variance in rate structures from different utilities
- Evaluating and managing the costs of service from a single supplier to a variety of locations
- Making commitments for longer term contracts

For the utility, the major hurdles are:

- Potentially losing an existing customer to a competitor
- Having to provide support for geographically dispursed locations
- Installing and maintaining a system for reading and billing from multiple locations to a single client
- Accepting a lower margin on each location in exchange for getting or keeping the customer

Installation Requirements

Depending on the requirements for additional information beyond energy consumption and demand, the typical installation for load aggregation at each location are:

- *AquiLite™ data acquisition server* (DAS) from Obvius to collect pulses from new or existing electrical meters
- *Pulse outputs from new or existing meters*
- *Phone line or LAN connection* for communication to remote server

Reports

Reports for load aggregation are relatively simple and consist primarily of a summary bill each month, backed by graphical or tabular presentations of detailed load profile information for some or all of the locations.

Analysis/Actions

Unless there are concerns or disputes over the bills, there is little analysis or action required for the billing of the aggregated loads. The customer and/or the utility may, however, use the load profile data to provide additional benefits such as ranking the different facilities ("racking and stacking") or using the load profile information to establish priorities for energy studies and retrofits.

Costs

For a simple load aggregation project, the typical cost for each location would be:

- *AcquiLite™ DAS*—$500 to $600
- *Pulse from existing meter*—depends on the utility (may be anywhere from free to $1,000 +
- *Pulse from new shadow meter* [Optional if no existing pulse]—$600 to $800
- *Data storage and reports*—typically included in the billing and rate structure

Notes/miscellaneous

As mentioned above, the slow pace of deregulation has limited the opportunity for facility managers to take full advantage of load aggregation, as most utilities have not been forced to make aggregation available. It is important to note, however, that C&I customers who adopt an energy information plan in order to take advantage of operational savings will benefit from having load profile information readily available in the future when aggregation does become a viable option in more locations.

EXAMPLE FOUR: DATA CENTER MONITORING

The AcquiSuite™ data acquisition server (DAS) from Obvius is used to monitor the electrical loads and environmental conditions in internet data centers. Specialized multiple circuit monitors from Veris Industries provide a cost effective means of monitoring loads on individual circuits in both new and retrofit installations.

Background

The dramatic increase in the use of the internet as the backbone for corporate and personal communications, there is a growing demand for data centers that serve as centralized locations for hosting servers and ancillary equipment. These centers may be developed exclusively for a single client (e.g., banks or financial institutions) or may serve multiple clients by providing hosting services for servers. A key element to all data centers is the need to monitor the electrical loading for each server to ensure that the current load does not exceed the breaker capacity and cause the unit to "trip" or go off-line.

Monitoring of individual branch circuits allows the data center manager to determine when circuits are approaching high loading levels. This information can be monitored locally or using the Obvius' web site: www.buildingmanageronline.com (BMO). If the BMO service is used, the operator has the option to set alarm levels for each branch circuit and to receive email or pager notification if any of the alarm limits are exceeded.

How Does it Work?

The data center manager installs H704 branch circuit monitors (BCMs) or H6238 multi-circuit monitors (MCMs) from Veris Industries to monitor the loads on each panel. Current transformers are installed on each circuit within the panel (using either solid core or split core CT's) which are then connected to the main board of the BCM or MCM. The main board provides a serial RS 485 output (using Modbus RTU protocol). The serial output from one or more circuit monitors is connected to the RS485 port on the AcquiSuite™, which automatically recognizes the devices and loads the appropriate drivers. The user then connects the AcquiSuite to a local LAN and completes the configuration using a web browser such as Internet Explorer. The AcquiSuite now monitors the circuit monitor(s) continuously and logs the data on a user-selected interval (from 1 to 60 minutes).

If alarm levels are not exceeded, the AcquiSuite simply stores the interval data in non-volatile memory

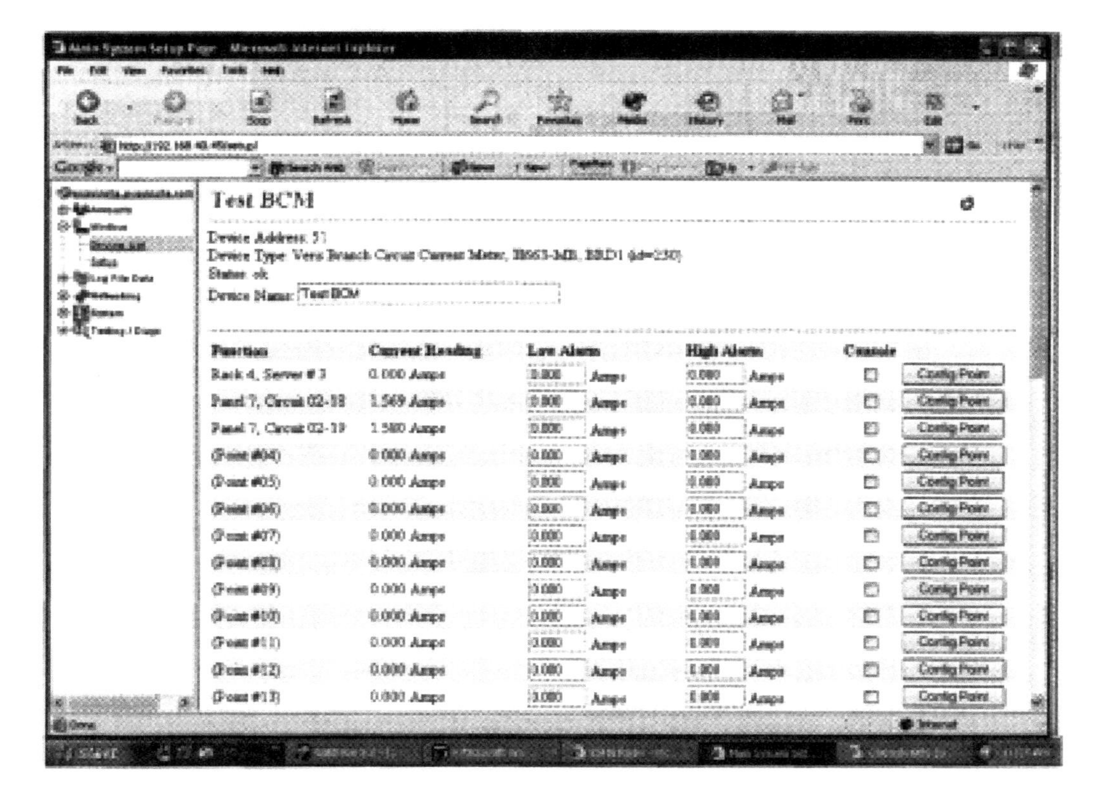

Figure 32-4. Setup Screen for Alarms on BCM

until the next scheduled time for the DAS to upload the data to the BMO web site. The DAS initiates an HTTP request using a local LAN connection and pushes the interval data from the local site to the BMO database server where it is available for display to authorized users with internet browsers (see "Reports" below).

If an alarm is received, the DAS initiates an immediate upload to the BMO server and the BMO server sends out email or pager notification to the users who have requested to be notified on alarms. The alarm message indicates which point is alarm, what the value of the monitored load is and the time the alarm was received. The alarm notice can be sent to multiple users.

Benefits

The data center operations manager benefits in many ways from the monitoring program, with primary benefits being:

Load Management

By tracking the loads on each circuit on a 24/7 basis, the operations team is able to determine which circuits are nearing capacity and which can accept additional loads. This information can be very useful when new capacity is required to avoid overloading branch circuits with new servers or ancillary equipment

Alarm Notification

Alarming at levels below the trip point of the circuit breaker allows the operator to detect potential problems before they result in costly downtime of equipment and web sites. For example, if the operator sets an alarm point at 90% of the capacity of each circuit, the BMO server will notify the operator in time to make adjustments in the loading and keep the systems up and running

Customer Assurance

Most data center operators provide QoS (Quality of Service) guarantees to customers that address issues such as uptime. Data centers that provide continuous monitoring of electrical loads can provide additional assurance to new and existing customers of their ability to ensure high reliability

Drawbacks

The major drawback to the installation of monitoring equipment is in the initial cost of hardware and labor to install the equipment (see "Costs" below). In addition, if alarm levels are not properly set, the operator may receive "nuisance" alarms that do not reflect real problems and it may take some time to establish proper alarm limits for each circuit.

Installation Requirements

A typical installation for monitoring data center loads and environmental conditions includes:

- *AcquiSuite A8811-1 data acquisition server (DAS)*—a standalone web server located on the building site that communicates with the sensor(s), stores interval information and communicates with the remote server.

- *H704 BCM from Veris Industries*—a device that monitors the current draw on up to 42 branch circuits. The H704 is a solid core CT unit, the H663 provides the same functionality with split core CT's for ease of installation in retrofit projects.

- *H8238 MCM from Veris Industries*—monitors the power (kW) from up to eight individual circuits.

- Power quality (at the primary meter and throughout the facility) can be monitored using *power quality sub-meters* such as the ION 6200, 7300 and 7500 from Power Measurement Ltd. These meters allow the operations manager to monitor the quality of the electrical supply (e.g., harmonics, power factor, sags and surges) in addition to the quantity of the energy used.

Optional Features Include:

- *Environmental sensors* (e.g., temperature, humidity, dewpoint, indoor air quality) that provide 4 to 20 mA or 0 to 10 Vdc outputs can be connected to the AcquiSuite to track indoor conditions in addition to loads
- *On/off current sensors* (such as the H800 or H900 from Veris Industries) can be connected to the AcquiSuite to provide alarm outputs from HVAC equipment to detect failures before they impact operations

Reports

For evaluation of load profiles on each circuit, the simplest report is a graphical indication of the loading on each circuit (Figure 32-5).

The device list screen on the BMO site provides immediate visual indication of the status of all points on the device (Figure 32-6.)

Custom reports can be developed using standard

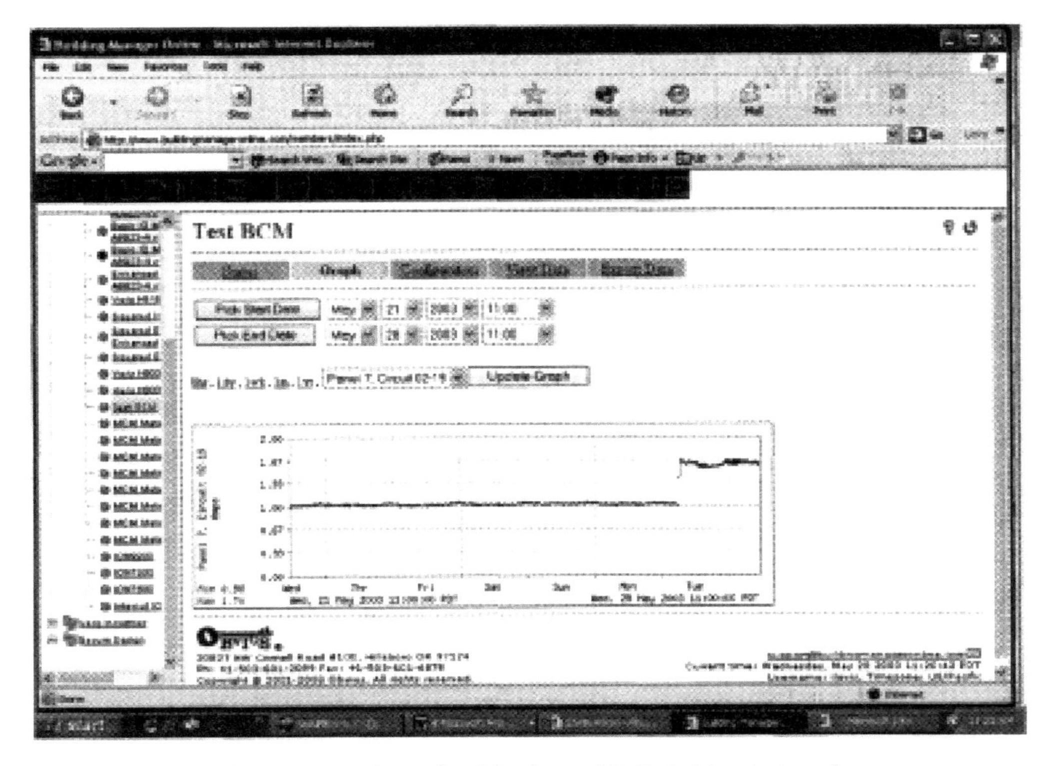

Figure 32-5. Sample Display of BCM Circuit Load

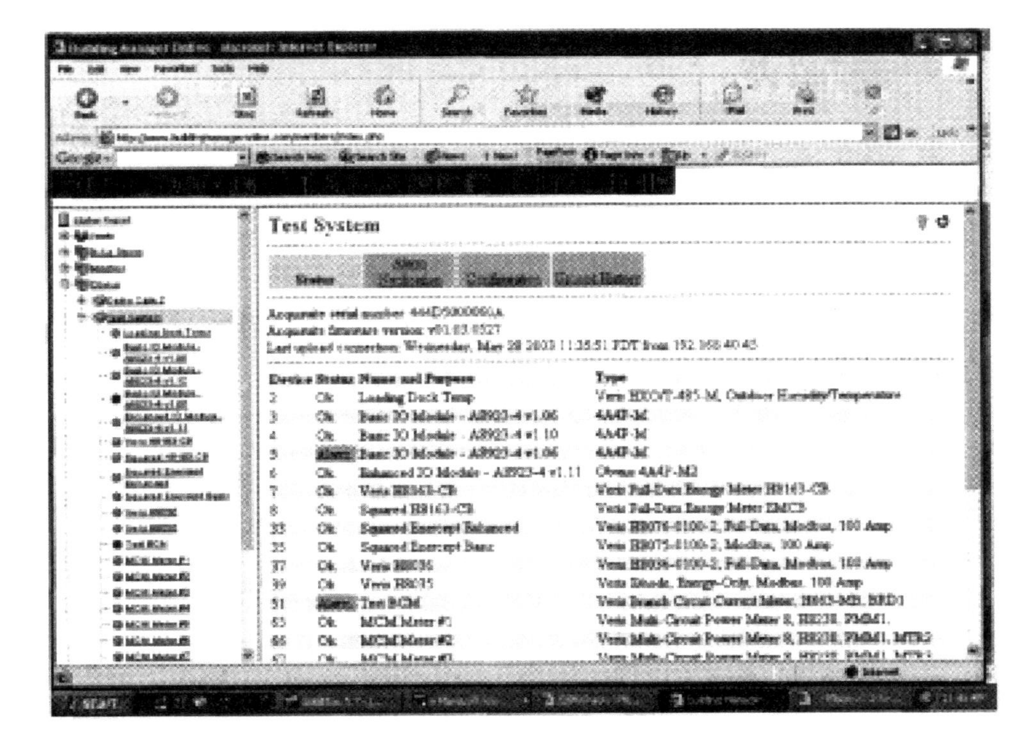

Figure 32-6. Alarm Indication

tools such as Microsoft Excel or Access by downloading files from the BMO web site. These reports can be developed to meet the specific needs of individual centers or customers.

Analysis/Actions

Alarm conditions will obviously dictate the actions as the operations group will need to respond and shift loads to balance the electrical loads on each circuit.

Load monitoring curves can be used to determine where additional equipment can be added without overload conditions and power quality meters can be used to ensure that the quality of the power delivered to the facility is adequate.

Costs

Installation costs will vary widely depending on the number of circuits, the distance of the wiring runs, etc., but in general terms the cost of installing monitoring equipment for a typical panel would be:

- *AcquiSuite™data acquisition server*—$1,200 to $1,800
- *H704 Branch circuit monitor*—$1,000 to $1,500 (per panel)
- *H8238 Multi-circuit monitor*—$1,500 to $1,800 (per panel)
- *IAQ sensor(s)*—$100 to $1,000
- *Data storage and reports*—$20 per month per AcquiSuite™

Notes/miscellaneous

As noted earlier, data from the BMO web site is available from anywhere in the world using a web browser and each authorized user has access to a variety of AcquiSuites. These features allow monitoring of one or multiple sites by a single operator or many operators in remote locations, producing a substantial reduction in the cost of operating and maintaining a data center.

EXAMPLE FIVE: MONITORING INDOOR AIR QUALITY

A variety of sensors is used with a data acquisition server (DAS) such as the AcquiSuite from Obvius to continuously monitor the indoor air quality (IAQ) of commercial and industrial buildings. Sensors from a variety of suppliers can be used to tailor the monitoring system to meet the needs of the individual building and its occupants.

Background

One of the many challenges facing the owners of commercial buildings today is the need for ensuring that the indoor air quality (IAQ) within their facilities meets established guidelines. The combination of new materials and energy conserving ventilation systems has generated concern among building occupants that the quality of the air they are breathing falls below minimum standards. Commercial building owners need to monitor the indoor environment to identify potential problems or to verify that guidelines are being met. Among the potential problem pollutants that can be monitored:

- CO_x
- NO_x
- Formaldehyde
- Ozone
- Particulate matter
- CO_x
- Water vapor
- Radon

In addition to the items listed here, virtually any gas or particulate that can be measured can be added to this list. Typical application areas include:

- Garages
- HVAC ventilation
- Conference rooms or other occupied spaces with variable loads
- Outdoor air ducts near heavily traveled streets or garage access points
- Bars and restaurants with smoking and non-smoking areas

How Does it Work?

The building owner or contractor selects the sensors that will detect the pollutants of concern and provide an industry standard (4 to 20 mA or 0 to 10 Vdc) output. The sensors are then installed either in the occupied space or in the HVAC system (e.g., in a return air duct) and connected to one of the 4 analog input terminals on the AcquiSuite. If there are more than 4 analog inputs, up to 32 auxiliary I/O modules (A8923) can be used with up to 4 analog inputs each (a total of 128 inputs).

The DAS is setup to read the value from the sensor on user-specified intervals and convert the analog signal to appropriate engineering units (e.g., parts per million

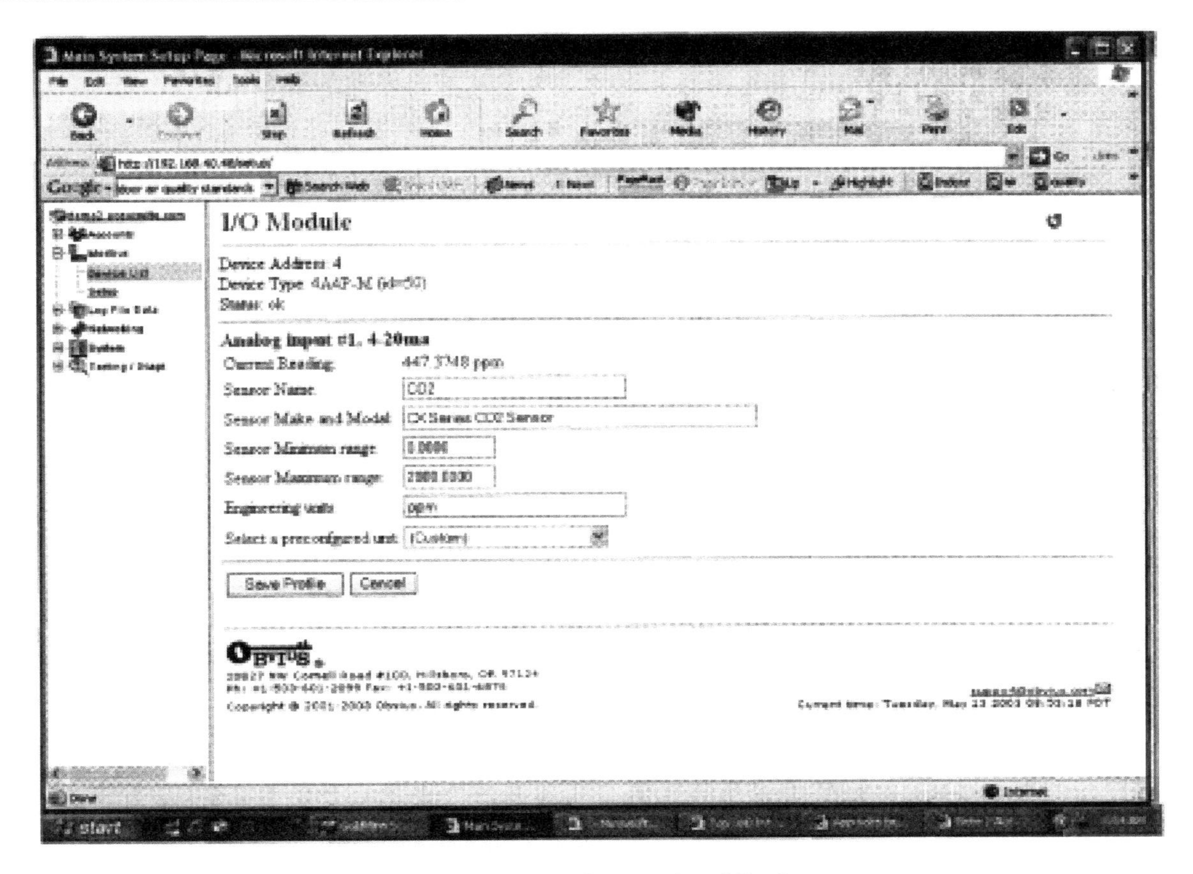

Figure 32-7. Setup Screen for CO_2 Sensor

(PPM)). A typical setup screen is shown in Figure 32-7.

The AcquiSuite provides the installer with the ability to set up any sensor that provides a measurable output and also provides preconfigured setup options for many of the most common inputs such as indoor and outdoor temperature and humidity.

Data from the DAS is recorded continuously and then uploaded to a database server (http://www.obvius.com) where it is stored and can be displayed using a typical web browser such as Internet Explorer (see the "Reports" section below for examples). The data upload can be accomplished either via phone lines or an existing LAN connection.

There is also an option to import the data from the AcquiSuite in a comma- or text-delimited format to a local PC for display and analysis using Microsoft Excel or Access.

Benefits

There are two primary benefits for the building owner in monitoring IAQ levels:

Identification of Problem Areas

Providing continuous monitoring of IAQ can help the building owner or manager to identify and correct potential problems before they become a health hazard for occupants. For example, in many HVAC systems the amount of fresh air introduced into the system can be adjusted to ensure that the level of CO_2 in the space does not exceed recommended levels.

Documentation History

In many instances, claims of poor air quality can be transient (e.g., new carpet installation) or seasonal. The only way to be sure that recommended guidelines are met is to provide continuous monitoring of the areas involved and to be able to document the levels. This can be critical in defending claims of harm to building occupants as the owner is able to document the IAQ levels within the building.

Drawbacks

There are costs involved in installing the hardware on-site to provide the monitoring as well as ongoing costs for storing and displaying the information via the web (see Figure 32-8 below). There is also a potential risk for the building owner in that if the monitoring program uncovers problems that need to be addressed, the owner will either have to address these issues or face potential legal action from tenants.

Installation Requirements

In a typical installation, the following components would be required:

- AcquiSuite (A8811-1) data acquisition server
- I/O module (A8923) if more than 4 analog inputs are required
- Sensor(s) with 4 to 20 mA or 0 to 10 Vdc outputs
- Communications connection (either phone or LAN) if the information is to be uploaded to a remote site for viewing (e.g., www.obvius.com).

Reports

The reports for an IAQ monitoring project are fairly simple and will provide either a display or table of the levels of the monitored values for a user-specified interval.

Analysis/Actions

The AcquiSuite provides the option for the owner to set alarm levels for any points being monitored, allowing the owner to receive notification of excessive levels via email or page. A typical setup page is show in Figure 32-10.

If the alarm level set in the AcquiSuite is exceeded, the server at www.obvius.com will send either an email or page to the owner or his representative to allow the problem to be corrected. As an example, if an exhaust fan on an underground garage failed, the CO level would rise and the building facility manager would get an alarm and get the fan fixed or replaced.

Monitoring of gas levels (particularly CO_2) can be very useful in saving energy as the level of outside air introduced into the HVAC system can be adjusted to save energy as long as the CO_2 levels remain acceptable. This can be automated using a building automation system (BAS) that continuously monitors and adjusts the outside air damper based on the IAQ levels.

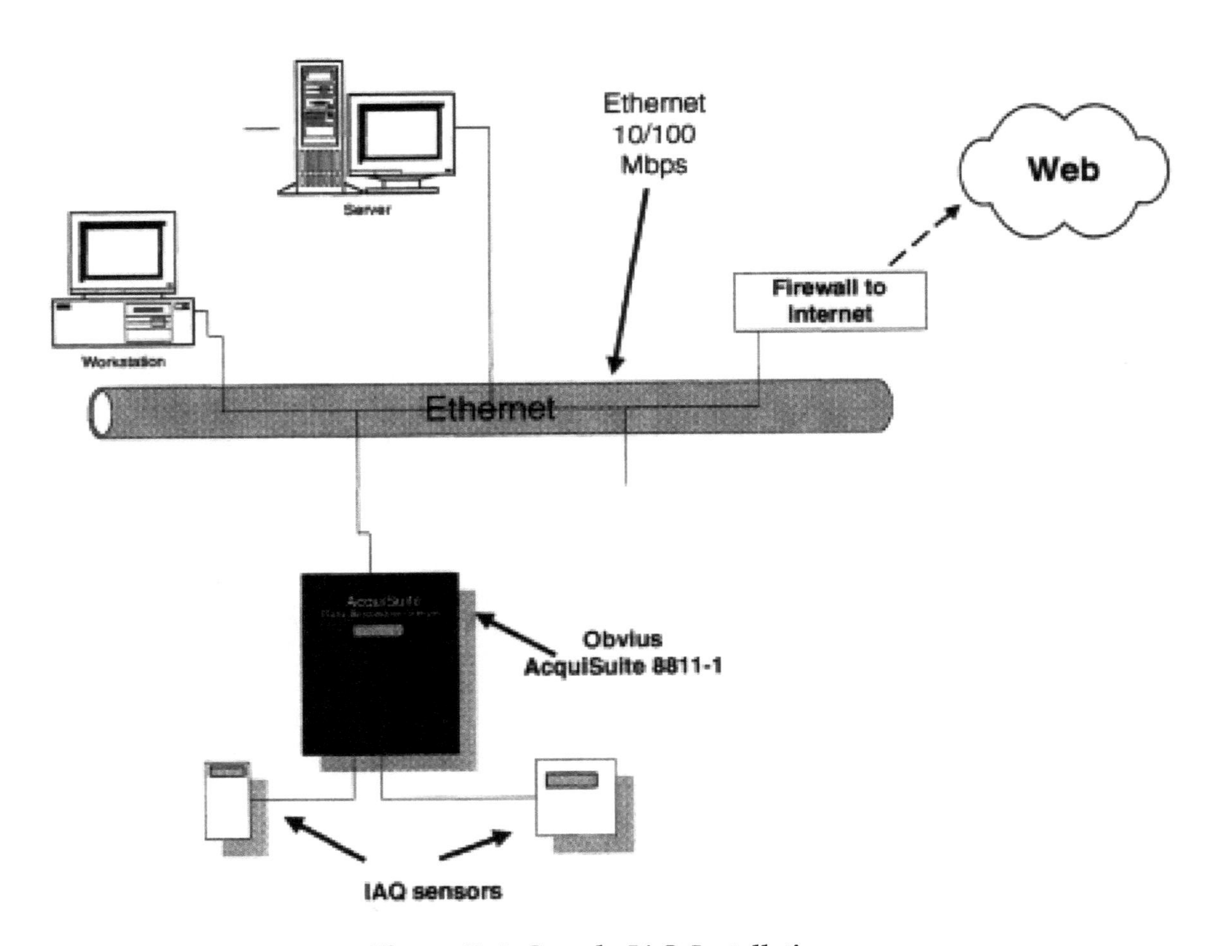

Figure 32-8. Sample IAQ Installation

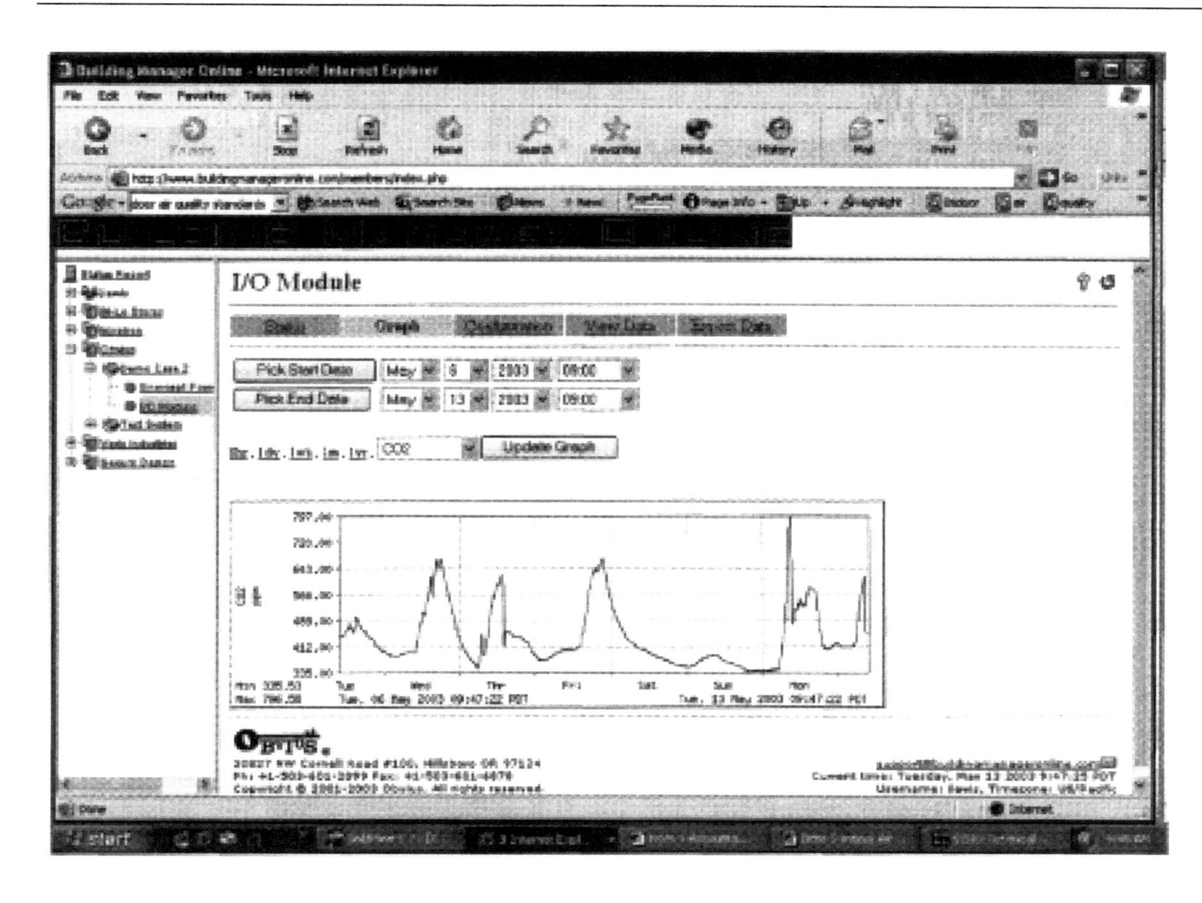

Figure 32-9. One Week Graph of CO_2 Levels

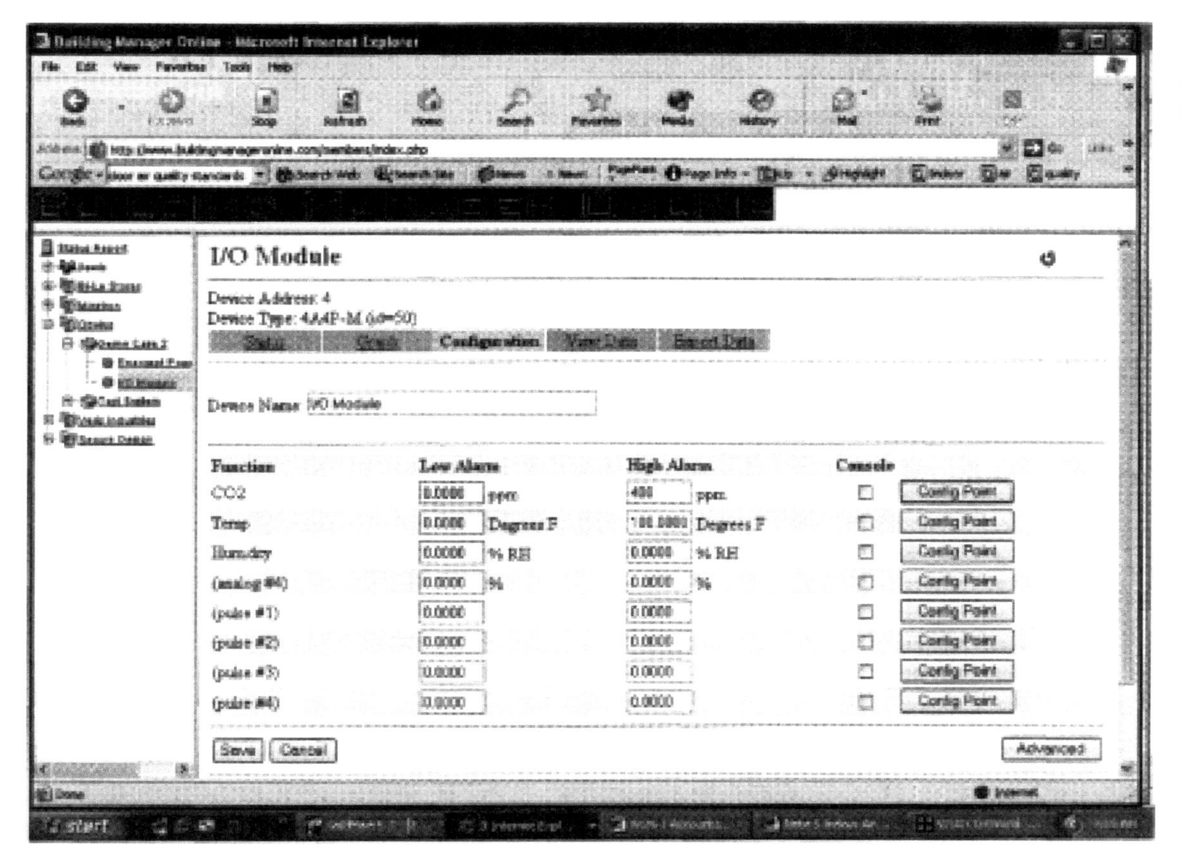

Figure 32-10. Alarm Setup for CO_2 Sensor

Costs

The actual installation costs for IAQ monitoring will vary greatly depending on factors such as the environmental factors being monitored and the costs involved in wiring the devices and the communications, but the following costs can be expected:

- AcquiSuite™ data acquisition server—$1,200 to $1,800
- IAQ sensor(s)—$100 to $1,000
- Data storage and reports—$20 per month per AcquiSuite™

Notes/miscellaneous

As IAQ concerns (and litigation) increase, the building owner needs to stay ahead of the game and take proactive steps to manage and document the air quality within his or her buildings. An effective monitoring program is the first step to gaining control of IAQ issues and heading off complaints and legal action.

For more information or a demonstration of these features, you can contact Obvius at: www.obvius.com, (503) 601-2099, or toll free at (866) 204-8134

Chapter 33

Leveraging IP Technology for Remote Monitoring in the Energy Industry: Pipeline Operations and Facilities Management*

Deepak Wanner, President
Precidia Technologies

THE ENERGY INDUSTRY needs a low-cost way to remotely manage and monitor serial devices simply and cost-effectively. Equipment that meets operating requirements lacks network connectivity to link to the company's LAN/WAN. Upgrading or replacing existing serial equipment is costly and the incremental functionality does not justify the large price tag. One way to access equipment that is so widely dispersed is by using internet protocol (IP) technology that seamlessly connects serial devices to an IP network. This chapter presents several scenarios that have some elements in common: the need to access information from a central location, the importance of reducing downtime, and the ability to reduce staff time in checking and monitoring equipment at many locations.

INTRODUCTION

Ethernet. Addressing. Twisted pair. Data packets. The protocol stack. TCP/IP is not easy. It demands recognized experts to design, implement and manage a TCP/IP network. Yet, TCP/IP has become the de facto networking standard and can be found in virtually every corporate network structure of most office environments across North America.

The need for a low-cost way to remotely manage and monitor serial devices simply and cost-effectively has become a demand in the energy industry, from pipeline monitoring, to building automation. Much of the energy sector's serial equipment meets operating requirements but lacks network connectivity to link to the company's LAN/WAN. It is costly to upgrade or replace existing serial equipment and the incremental functionality does not justify the large price tag. The question is—how do you access equipment that is so widely dispersed? One such solution is internet protocol (IP) technology that seamlessly connects serial devices to an IP network.

In a world where companies are driven by profit, and schools and other similar facilities face scarce resources, there is a limit on how much can be spent on developing remote management. There has to be a solution that satisfies budget requirements and technical needs. This chapter will outline how IP technology enables companies to remotely monitor and manage their mission critical serial equipment.

Today, with the emerging network-enabled world and the evolution of the internet and company intranets, there are cost-effective solutions to securely monitor equipment, using IP via ethernet, point-to-point protocol (PPP) or a wireless network. IP technology will allow energy companies to keep their existing serial equipment and connect it to a device that will IP enable the equipment.

THE OBSTACLES TO INTEGRATING EXISTING EQUIPMENT

Monitoring remote pipelines, utility plants, and other facilities has been a constant struggle and an expensive challenge for the energy industry. Most of the energy sector's equipment operates effectively but because they utilize standard RS-232, RS-422 and RS-485 serial connections, they do not link to corporate LAN/WANs.

*Portions of this chapter have appeared on the website, AutomatedBuildings.com and are used with their permission.

IP technology will provide bi-directional communications between the equipment and server, providing remote network management capabilities and increasing IT management efficiency. An IP based solution also offers enhanced security, and reduces overall maintenance costs.

IP TECHNOLOGY: BACKGROUND

TCP/IP was initially developed to deliver basic services that everyone needs (file transfer, electronic mail, remote log-on) across a very large number of client and server systems. Several computers in a small department can use TCP/IP (along with other protocols) on a single LAN. The IP component provides routing from the department to the enterprise network, then to regional networks, and finally to the global internet. TCP/IP was designed to be robust and automatically recover from any node or phone line failure. This design allows the construction of very large networks with less central management.

As with all other communications protocol, TCP/IP is composed of a series of layers. Within each layer are other protocols. The key elements of the TCP/IP architecture that provide flexibility, convenience, and real-time information for the user are discussed below.

Session Layer

Simple Network Management Protocol (SNMP)

This is a communication protocol that has gained widespread acceptance as a method of managing TCP/IP networks. SNMP is used to manage devices supporting the protocol from any number of hosts or PCs running SNMP software, including individual network devices, and devices in aggregate.

SNMP defines the client/server relationship between the SNMP network manager (the client) and the SNMP agent (the server). The manager opens virtual connections to an agent on a remote network device, which serves information to the manager regarding the device's status. With SNMP, you can gather statistics or configure the IP device. Users can gather statistics with get-request and get-next-request messages, and configure them with set-request messages. Each of these SNMP messages has a community string that is a clear-text password sent in every packet between a management station and the Precidia device (which contains an SNMP agent). The SNMP community string is used to authenticate messages sent between the manager and agent. Only when the manager sends a message with the correct community string will the agent respond.

Secure Sockets Layer (SSL)

This protocol was developed for transmitting private documents via the internet. SSL works by using a public key to encrypt data that's transferred over the SSL connection. Another protocol for transmitting data securely over the World Wide Web is Secure HTTP (S-HTTP). Whereas SSL creates a secure connection between a client and a server, over which any amount of data can be sent securely, S-HTTP is designed to transmit individual messages securely. SSL and S-HTTP, therefore, can be seen as complementary rather than competing technologies.

Dynamic Host Configuration Protocol (DHCP)

This is a protocol for assigning dynamic IP addresses to devices on a network. With dynamic addressing, a device can have a different IP address every time it connects to the network. Dynamic addressing simplifies network administration because the software keeps track of IP addresses rather than requiring an administrator to manage the task.

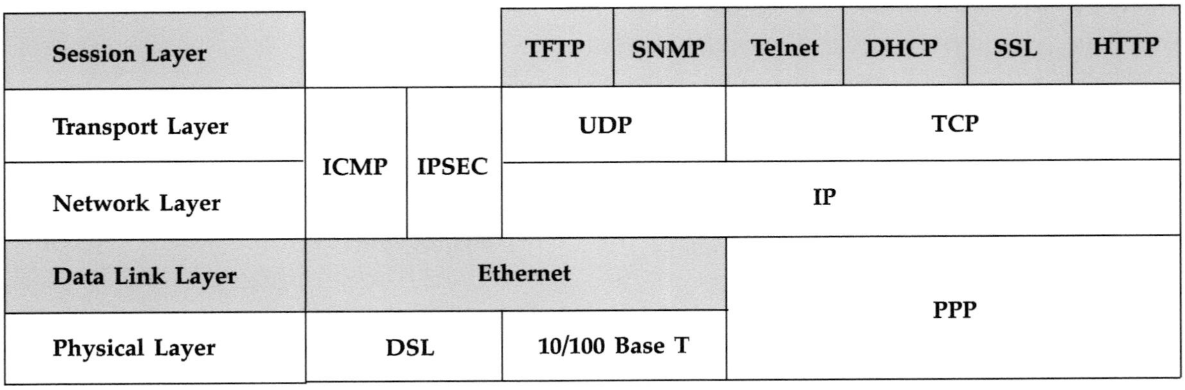

<table>
<tr><td>Session Layer</td><td colspan="2"></td><td>TFTP</td><td>SNMP</td><td>Telnet</td><td>DHCP</td><td>SSL</td><td>HTTP</td></tr>
<tr><td>Transport Layer</td><td rowspan="2">ICMP</td><td rowspan="2">IPSEC</td><td colspan="2">UDP</td><td colspan="4">TCP</td></tr>
<tr><td>Network Layer</td><td colspan="6">IP</td></tr>
<tr><td>Data Link Layer</td><td colspan="4">Ethernet</td><td colspan="4" rowspan="2">PPP</td></tr>
<tr><td>Physical Layer</td><td colspan="2">DSL</td><td colspan="2">10/100 Base T</td></tr>
</table>

Figure 33-1. TCP/IP Architecture

Telnet

Telnet is not a protocol. It is a program that runs on a computer and opens a "Telnet Session" that connects a PC to a server on the network, enabling the user to control the server and communicate with other servers on the network.

Transport Layer

Transmission Control Protocol (TCP)

TCP is a transport protocol responsible for verifying the correct delivery of data from client to server. With TCP/IP, it establishes a connection between two hosts to communicate back and forth. The connection provides a reliable communication stream to send data over packet-switched networks. Data can be lost in the intermediate network. TCP guarantees delivery of data and also guarantees that packets will be delivered in the same order in which they were sent. TCP adds support to detect errors or lost data and to trigger retransmission until the data are correctly and completely received.

With high-speed networking, serial data can continually retrieve with only milliseconds of latency. The cost of bandwidth is dropping so fast that not only is it economical to increase the frequency of data capture, but it is also possible to increase the density of sensors.

The primary reason why data bandwidth has become so cheap is the advent of standardization. In the past, there were many incompatible protocols that were optimized for specific applications. But largely due to a government initiative, one protocol dominated. That protocol, internet protocol (IP) networks, found its origins in a defense project, but is now so prevalent that most people have long forgotten that it was originally designed to allow Military officials in Washington to have a reliable link to the missile silos in the Midwest. With standardization comes commoditization. And with every commodity comes pricing pressure. Thus, data bandwidth today is significantly cheaper than it was a decade ago.

Network Layer

Internet Protocol (IP)

This protocol is responsible for moving packets of data based on a four-byte destination address (the IP number). The internet authorities assign ranges of numbers to different organizations. The organizations assign groups of their numbers to departments. IP operates on gateway machines that move data from department to organization to region and then around the world.

Data Link Layer

Point-to-Point Protocol (PPP)

Dial-up IP devices access an IP network using a series of protocols called Point-to-Point Protocol (PPP). One stage of the connection is the authentication protocol that sends a "username/password" to verify that the device is allowed to access the network.

TYPES OF NETWORKS

The ubiquitous internet that allows us to do web surfing and emails is an example of an IP network. However, when it comes to transmitting sensitive information, it is highly recommended that either the data be encrypted over secure IPSec tunnels over the public internet or moved to a private internet generally known as a Virtual Private Network (VPN). Access to a VPN can be through one of two primary methods, wired or wireless.

Wired data networks began with low-speed, dial-up networks. Frame relay introduced the always-on, reliable link that has become widely available today. The cost however reflected the highly reliable, industrial nature of the media. The newcomer to the scene is DSL or Cable Modems. This uses the existing phone line or Cable and overlays a high frequency signal that can carry large amounts of data. Initially, this underdog technology grew up with a fairly bad reputation of being difficult to install and notoriously unreliable, but with

Table 33-1. Comparison of Types of Network Access

DSL	*Frame Pervasive*	*Wireless*	*Satellite*
❒ Inexpensive	❒ Costly	❒ Limited security	❒ Very reliable
❒ Increasing reach	❒ Very reliable	❒ Least expensive	❒ Expensive/costly
❒ Secure	❒ Limited flexibility	❒ Growing reach	❒ Maximum reach
❒ Reliable	❒ Secure	❒ Reliability issues	❒ Secure
❒ Service levels	❒ High service levels	❒ Low bandwidth	

ever advancing technology and support organizations, you will be pleasantly surprised with the advances that have been made. But the real driver is cost. In most cases, you will find that even though they carry over 10 times the data of a Frame Relay line, they can cost less than half the price.

Wireless networks were limited to short range solutions, but satellite has introduced lower cost options that are highly competitive. Personal communications service (PCS)/global system for mobile communications (GSM) is also gaining a foothold in the market. Using cellular digital packet data (CDPD) and/or iDEN wireless coverage provide an "always-on" connection, unlike dial-up, which must re-establish a connection each time data are transmitted.

REMOTE ACCESS DEVICES

With the growth of the internet and company intranets it is the ideal time to leverage existing infrastructures and create connections with remote locations. There are numerous products on the market that connect serial devices to IP. These can connect via ethernet, point-to-point protocol (PPP) or wireless networks.

Precidia Technologies' ethernet products, the Ether232, Ether422, and Ether485 are examples of available products for serial-to-IP conversion. These network-enabling devices allow serial devices to migrate seamlessly to a company's LAN/WAN. They facilitate the upgrade of legacy equipment and consolidate stand-alone equipment.

Precidia's IP232 enables the user to remotely manage any RS232 serial device, without being connected to a local LAN. While other solutions have an ethernet interface for connection to a LAN or WAN, the IP232 has an embedded v.34 modem and offers a dial connection access to IP networks. Once the connection is established, the IP232 receives raw serial data from its RS232 connection, converts it to IP packets, and sends the packets to the internet or private IP/VPN via a Point-to-Point Protocol (PPP) dial connection. This allows the user to leverage the low cost and global reach of the internet to seamlessly connect serial devices such as pumps, valves and data acquisition devices, to IP networks.

The Cell232*Plus* wireless IP solution connects any stand-alone asynchronous serial device to a wireless network. It provides access to a low cost, wireless VPN without changing the user's existing serial equipment. It is ideal for connecting utility meters and other serial equipment remote from land lines.

BENEFITS OF THE IP SOLUTION

IP technology will allow users to keep their existing serial equipment and connect it to a device that will IP-enable the equipment. It will provide bi-directional communications between the equipment and server, thus providing remote network management capabilities, increasing IT management efficiency, and reducing overall maintenance costs.

Cost Savings

The devices seamlessly install into the existing infrastructure and/or equipment, and require no upgrade or replacement of older operational devices. They enable older equipment to communicate from remote sites to the LAN/WAN or wireless network, eliminating the need to purchase additional equipment and wiring. It allows organizations to monitor energy usage, device statistics, equipment performance and react to problems quickly with real-time information

Remote Administration

With remote access devices, the IT and/or engineering departments are able to remotely configure the devices to access the equipment without having to leave the office—saving hours in support. The devices also have the capability to perform automatic routing of traffic to a backup system during a disaster or accident.

Remote Diagnostics

The IP-enabling devices facilitate remote diagnostics. Very large cost savings are realized by reducing the need to dispatch personnel on service calls, and reducing personnel required to provide on-site monitoring at remote locations. The devices can simultaneously route data to multiple destinations. They also allow for data and diagnostics to be handled simultaneously, in real-time.

APPLICATIONS

Pipeline Application

Each day, hundreds of millions of gallons of petroleum products are shipped all over the world by pipeline companies. Oil and natural gas consumption is expected to grow, primarily due to higher projected energy demand in the commercial and transportation sectors. Natural gas, viewed as an environmentally friendly fuel, will see significant increase in consumption by the electric utilities. "Petroleum product consumption in

North America is projected to increase by 10.3 million barrels per day from 1999 to 2020, at an average annual growth rate of 1.8 percent."[1]

The growth in the movement of petroleum products has resulted in a vast underground and aboveground pipeline network. "Two thirds of all domestic deliveries flow through 160,000 miles of oil pipelines in the United States."[2] This huge infrastructure will continue to expand to meet the projected demand. In President George W. Bush's National Energy Policy Speech, in May 2001, he said, "The U.S. will need newer, cleaner and safer pipes to move larger quantities of natural gas—up to 38,000 new miles of pipe and 263,000 miles of distribution lines."[3]

Safety concerns in the pipeline industry focus on the possibilities for damage from terrorism as well as from natural occurrences and erosion of pipeline equipment. Damage to a pipeline can be extensive even with a small incidence. For example, last year a stray bullet hit an Alaska pipeline. An estimated total of 285,600 gallons spilled, of which 108,600 gallons were recovered.

According to the Office Of Pipeline Safety, it was reported that, between 1986 and 2001, there were 3,035 accidents with hazardous liquid pipeline operators, 285 injuries & fatalities, and over $764,198,000 in property damage. Pipeline operators are looking for new ways to reduce spills, because they affect the environment and are very expensive to clean up. "Spilling one 42-gallon barrel of oil from a pipeline typically costs the pipeline company $50,000 in lost revenue and clean-up expenses."[5] Over the years spills have been declining, however there are still thousands and thousands of gallons that spill throughout the year.

To date, a physical presence is one of the few ways available to have active deterrence of any terrorist incidence or to minimize damage from a natural occurrence. The safest way to protect the asset from terrorism or vandalism is with armed guards. To have them stationed along the length of the pipeline, as well as a guard stationed at each pumping station would be the optimal solution, but this alternative is cost prohibitive to any pipeline company.

Pipeline companies already have to deal with the increase in administrative responsibilities resulting from the additional and aging pipelines. More technicians are required to monitor the large and growing network and to inspect this equipment, which is so essential for the economic well being of the country. "Interstate pipelines deliver over 12.9 billion barrels of petroleum each year."[4]

In most cases, such as the Alaskan pipeline spill, it would have been difficult to avoid the situation, but it is possible to reduce the impact through quick action. This brings up the concept of "real time" monitoring. Having real-time access to data provides the ability to report on unusual signals or messages that would imply a problem with one of the pipelines.

While it may be cost-prohibitive to have guards stationed along the pipelines, remote monitoring will allow you to respond to alarms in real-time and make quick management decisions to re-route the oil, avoiding serious damage to local farmers, the environment and surrounding suburb life.

Today, only a small percentage of field equipment can be easily monitored or remotely configured under current conditions. Every time there is a new pipeline connection, new network or maintenance concern, it requires IT personnel, technicians and/or engineers to venture out to remote areas to install, configure or inspect the equipment. Network managers are under pressure to maintain and improve the current performance of the network, regardless of size. This has serious cost implications for energy companies.

There is pressure to find new solutions that will provide "remote eyes" on the network without hindering the existing equipment or infrastructure. With the lack of network connectivity it is difficult and costly to improve response time to security alarms, or to remotely monitor perimeter and exterior sensors, cameras, access control and lights.

Many pipeline companies are realizing the benefits of IP-enabling equipment. For example, one pipeline company, a leader in the petroleum products industry, was looking for a way to connect serial equipment to their network for remote monitoring capabilities. They could install a phone line to communicate with their Omni Flow Computers, but that would incur installation fees and the monthly expenses associated with a dedicated phone line. They wanted a solution that could leverage their existing LAN/WAN infrastructure, saving them time, money, and IT resources.

The company wanted a cost-effective way to connect and remotely monitor data from their Omni Flow Computers, which control valves, measure pipeline flow, and count barrels. Precidia Technologies serial-to-IP connectivity product, the Ether232, was chosen because it easily and cost-effectively connects stand-alone equipment to IP networks via ethernet.

Using Precidia's device, the company was able to reduce the personnel costs associated with troubleshooting, configuring, and programming field equipment, as the device monitors equipment in real time from a standard PC. This IP solution saves the company days of work each month.

This cost-effective solution has extended the service life of the company's expensive pipeline equipment, and simplified the management and maintenance of the network. The data from the sensors can now be connected to the network cost-effectively for troubleshooting, configuring, programming, and real-time data collection.

Building Automation/Metering

Schools Application

Every year school boards invest hundreds of thousands of dollars maintaining their school facilities. Whether in a school's boiler room, or the thermostats in each classroom, serial equipment is the foundation of school facilities. These serial devices can include HVAC systems, lighting, security alarms, UPS systems and data loggers. With the volume of this equipment growing, and school budgets tightening, there is a pressing need to manage these widely dispersed devices more efficiently. Today, the only means of managing these devices is through the serial port; the challenge is to find a way for the serial ports of these devices to 'talk' to the LAN or WAN. Otherwise,, these devices must be physically monitored on-site by an administrator or technician.

The Limestone District School Board, located in Kingston, Ontario, is comprised of 59 elementary schools. Recently, Limestone installed a Precidia access device in each school. This allowed each of the schools, which were already on the WAN and connected to the school board's server, to be connected to each school's Andover[2] Control Panel. Limestone can now remotely view and monitor each school's HVAC system.

Dwight Allport, Mechanical Maintenance Supervisor, Limestone District School Board describes the benefits of connecting the Andover Control Panel to a TCP/IP network:

> "The Precidia device solved a long standing issue for Limestone: how to gather information from all of our remote locations within the school district, monitor the data and alarms from our Andover Building Control equipment and feed it into a central PC. Precidia's Ether232 provided the most cost-effective solution.
>
> "By connecting the Ether232 to the Andover device in each school's boiler/mechanical rooms, the data can now flow through our WAN to a central location. Solving this problem has been a key step in deploying our mission-critical HVAC monitoring system."

Limestone has demonstrated the benefits of Precidia's solution. The devices allow metering equipment to be connected to a central location which facilitates remote monitoring, reduces the resources necessary for meter monitoring, and helps decrease their energy costs. In addition, the school board can now control a range of security access systems through a centralized location over a TCP/IP network. Several door access control products including door controllers, magnetic swipe card readers, lift controllers and alarm controllers have an RS232, RS422 or RS485 serial connection and can now be connected to TCP/IP. HVAC systems can be monitored from a central location with a Precidia device; school boiler systems are controlled from a central location, sending out alarms to custodial staff when problems occur. Even electronic signs for indoor and outdoor applications can now be connected to a network and programmed from a centralized location

This is just one example of the benefits of remote monitoring via an ethernet LAN. The ability to remotely monitor stand-alone serial devices in real time is beneficial in any environment where access to information and reducing downtime is critical.

Hospital Application

Monitoring energy usage is a critical element of any large organization's facility management mandate. Consider the case of hospital organizations with multiple facilities. Specializing in commercial metering solutions for large clients, Metering Sales & Service needed a solution to help a large Philadelphia area hospital group better manage their electricity consumption. Power monitoring is used to measure consumption of electrical energy, usually in kilowatt hours (kWh).

Metering Sales & Service attached an SHM[3] data logger to each electricity meter throughout the hospital group's numerous local and satellite locations, many of which had been added to the group through acquisitions. Several of these satellite locations had their own LANs. While the data loggers can collect vital data on energy consumption, the customer needed a means to bring this data onto the hospital's LAN. By connecting each data logger to a Precidia Ether232, electricity consumption data captured in the data logger could be accessed from a central location on the LAN, which was then connected to the larger city-wide WAN.

Steve Helzner, Meter Systems Engineer at Metering Sales and Service, spearheaded this ambitious project:

"We needed a solution which would provide the 'missing link' for our customer: the ability to access the data retrieved from the data loggers in real time from their LAN. With the Ether232, they can now retrieve real-time data on the hospital's power consumption from a remote location. With this solution, our client is able to easily identify and eliminate energy waste." The Precidia solution has been a tremendous success; the hospital group has enjoyed a return on their investment in under a month, reducing energy costs significantly, and gaining new control over their expenditures. Building upon this success, Metering Sales & Service will soon expand the application of Precidia's device, bringing steam, water and gas meter data onto the LAN for an even greater cost savings.

CONCLUSION

The demand on network and facilities managers to reduce network downtime and improve operating efficiency is greater than ever; even a few hours of downtime can lead to millions of dollars in lost revenue for some organizations. The challenge is formidable, keeping old equipment functioning well enough to avoid major facility damage such as flooding or freezing, without any significant capital expenditures. When you consider the security mandate of pipeline operators, this mission-critical challenge becomes even more significant.

All of the situations discussed in this chapter have some elements in common: the need to access information from a central location, the importance of reducing downtime, and the ability to reduce staff time in checking and monitoring equipment at many locations. When these challenges are faced by network or facilities managers, IP access is the answer.

Increasing energy efficiency, manageability and security are part of the facility manager's professional mandate. The extent to which much of this equipment is geographically dispersed, across countries in the case of pipelines, makes them difficult to monitor, but advances in telecommunications have led to dramatic price reductions in network communications service, opening the door to tremendous gains from real-time monitoring. However, to exploit these lower network costs, companies need to move away from non-standardized protocols to the more ubiquitous TCP/IP standard. TCP/IP technology has been available for years, but unfortunately, energy companies have been slow to adopt the technology in their own operations, until just the last few years. With the ongoing evolution of IP networks and the continuing price decreases expected from this technology, it's an ideal time to consider migrating.

With the capability to remotely monitor equipment, users will be able to detect any problems or remove poorly performing equipment, because they will have real-time access to data 24 hours/7 days a week. Real-time data logging is exciting because it will present a new opportunity to enable quick decision-making and dynamic facility management. In the past, technicians and engineers spent a lot of valuable time on-site performing troubleshooting, configuration, and general support. Now through the use of IP enabling access devices, they can spend time on other valuable tasks and save money at the same time.

References

[1] Energy Information Administration, Annual Energy Review 2000, August 2001
[2] American Petroleum Institute, A Report from Petroleum's Oil Pipeline Industry
[3] President George W. Bush's National Energy Policy Speech, May 17, 2001 in St. Paul, Minnesota
[4] Association of Oil Pipelines
[5] American Petroleum Institute, A Report from Petroleum's Oil Pipeline Industry

Chapter 34

Creating Web-Based Information Systems From Energy Management System Data

Paul Allen, Walt Disney World
David Green, Green Management Services Inc.

THE DATA THAT ARE reported from an EMS (energy management system) can be very useful in managing system operations. This chapter describes some methods of presenting this voluminous amount of data in such a way to make it easy to manipulate, using a web-based information system. Typical EMS reports include data trending, alarm reports and EMS system activity. Most EMS on the market today do an adequate job of collecting the data but a poor job of publishing and presenting the data to a wide audience. In the past, the EMS was the only system that could access and display the data collected. Only a few people in any one organization could turn this raw data into something meaningful. Today, using the internet/intranet to publish data or its derivatives is becoming more popular. Sharing EMS data with others, perhaps across the globe, is a reasonable expectation. Simply put, the internet has changed the way we share data

This chapter will describe how to get the most out of EMS data and turn it into a meaningful and useful information system.

USING THE EMS TO CREATE REPORTS

An EMS is used to control functions of a heating, ventilating, and air conditioning (HVAC) system. An EMS is also commonly referred to as EMCS (energy management control system), BMS (building management system), or BAS (building automation system).

Most EMS's on the market today have some method for collecting data from sensors, control devices and meters attached to the system. The EMS text files produced from these trend reports are huge. There is generally no easy way to graph the data from these reports. However, if report files are recorded so they can be read by a spreadsheet, the data can then be further sorted or graphed by the User. Alarm logs and system activity event reports can also provide insight into EMS operation and system administration. Again, if these reports are formatted so they are easily read by a spreadsheet, then they can be easily sorted and filtered. Pulling the data into spreadsheets is probably fine for a small EMS with only a few users, but for large campus-wide EMS systems with numerous users, this quickly becomes a very inefficient and time-consuming process.

A better approach would be to read these EMS text reports into a relational database. If the report files are formatted so they can be read into a relational database, then a custom web-based program can be written to display, search, sort, filter and graph the data using a standard web-browser interface. Access to this data becomes quicker and easier than ever and is also available to everyone on the company intranet instead of the limited few that are EMS users.

There are three types of reports that are particularly useful for management of an EMS: (1) trend and consumable reports; (2) alarm history reports; and (3) system activity reports. An EMS user must first understand the process for creating EMS text file reports using their EMS. The EMS text reports should be automatically output to data text files once a day on the EMS front-end computer. The user should look over the raw data using a text editor or spreadsheet program to verify the data integrity and accuracy.

The next step in the process is to copy the EMS report files to the web-server PC for post-processing the data. An important prerequisite to this is that the EMS front-end computer that holds the EMS text reports be on-line with the company's local area network (LAN). Unfortunately, this is probably not the case for most users today as most EMS front-ends have traditionally been stand-alone PC's. Fortunately, the installation of a network interface card (NIC) in the EMS front-end PC is a simple and inexpensive task. It should be coordinated with the company's IT department. Without the LAN connection, the process of copying the EMS text report files from the EMS front-end PC to the web server PC is not easily accomplished.

EXTRACTING DATA FROM EMS REPORTS

Once all of the EMS text reports are copied to the web server PC, a program reads these reports into a relational database. The key to extracting the data is the format and fields that are stored in the data text file. There are several data formats that are used by EMS manufacturers to make reading these report files into spreadsheet or database programs easier. The most common method is an ASCII test file that uses a "comma separated value" (CSV) format in which commas separate fields from each other and character field values are additionally delimited by double quotation marks. For example:

"Smith,"9999999,"TELEPHONE"

There are many variations to the CSV file format where another character instead of a comma is used. The Carrier ComfortWORKS® system, for example, produces report files that use a tab character as the field separation value.

Another important feature of this type of EMS report file is that the first column of each row contains an identifier that indicates what data are contained in each row. This makes it easier to pull the data into the relational database so that it will include the desired data and exclude unwanted data from the EMS report file. Figure 34-1 shows an example of the Carrier ComfortWORKS(system activity report in a tab-delimited format.

As mentioned previously, the Carrier ComfortWORKS® program produces report file output that uses a tab-separated value format. To read this file into a Microsoft Visual Foxpro® database table, the following code would be used:

```
Use DATAFILE
append from SYSACT.TXT delimited with tab
```

This would read the contents of the system activity report called SYSACT.TXT into a database table called DATAFILE. The DATAFILE table fields would have previously been created to match the fields from the report.

Once the data are extracted from the SYSACT.TXT file and is read into a database table, then a great deal of reformatting is done to get the data into the final table structure. Fortunately, this is something most database programs do with ease and the rest of the program is devoted to this effort. The underlying factor here is that

PH	3/9/2003	ALL ACTIVITIES							
PH	16:04	System Activity Report For March 8, 2003 To March 9, 2003							
PH									
T									
TE	Date	Time	Event	Note	Target				
D	8-Mar-03	0:10	ComfortWORKS						
D		7:16	Maintenance Base						
			Auto point		AREA1				
D					AV1: AH,1,6 - 50, 11				
D					30-Z2: DUSK TO CLOSE SCHEDULE				
D		7:26	Maintenance Base						
D			Auto point		AREA1				
D					AV1: AH,1,6 - 50, 11				
D					31-Z3: OPEN TO CLOSE SCHEDULE				

Figure 34-1. Carrier ComfortWORKS® System Activity Report file (SYSACT.TXT)

the data extraction program relies on the fact that the EMS report format is consistent. As an example, the date for the report is only shown in one place in the report, however a "D" in the first position of the data record precedes it. So in our example, the program can effectively pick out the date from the report so that it can be used to record the date on all records in the final file format. Similarly, each record in the EMS file is interrogated and the appropriate data are reformatted into the desired structure. Figure 34-2 shows the final database table structure for the example shown above.

The actual final database table would have thousands of records from several areas. This file would be automatically run each night after the EMS report files were generated and copied to the web server.

To make the update process automatic, the use of a "auto launch" program is highly desirable. One program that performs this function nicely is AutoTask® by Cypress Technologies (http://www.cypressnet.com).

Important issues to deal with when extracting data from the EMS report files for eventual use in a web-based information system include:

- "Bad" data must be dealt with, and data must not be duplicated or omitted if an application error occurs.

date	time	area	user	event	controller	device
3/8/2003	7:16	AREA1	MAINTENANCE BASE	AUTO POINT	AV1: AH,1,6 - 50, 11	30-Z2: DUSK TO CLOSE SCHEDULE
3/8/2003	7:26	AREA1	MAINTENANCE BASE	AUTO POINT	AV1: AH,1,6 - 50, 11	31-Z3: OPEN TO CLOSE SCHEDULE

Figure 34-2. Re-formatted Carrier ComfortWORKS(System Activity Report data

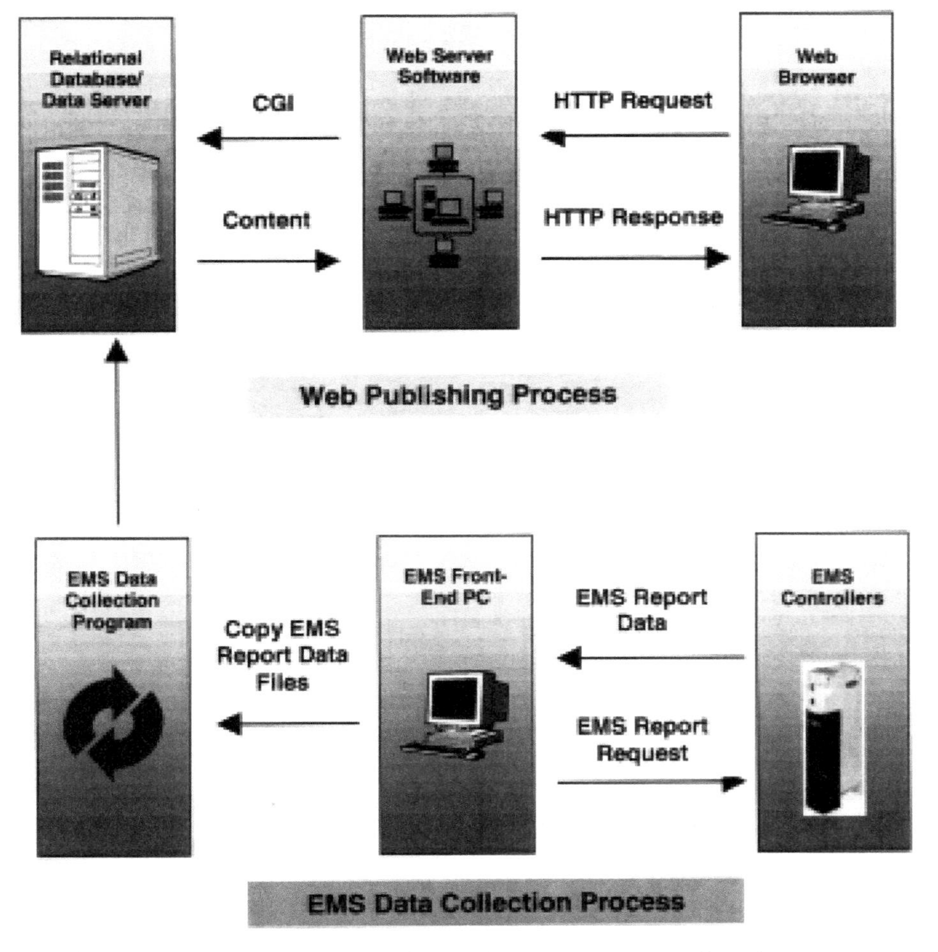

Figure 34-3. EMS Report Information System Functional Layout

- Archiving the data likely will require combining data into monthly or even yearly database tables to conserve space.
- The data collection application must achieve a delicate balance between the size and number of tables in which to store data.
- Storing data in a table that is too large will not produce results as quickly as a smaller table.
- Storing data in tables that are small, but too numerous, make it inconvenient to produce comprehensive reports covering long time sequences.

Once the EMS report data table is updated, it will then be copied to the web-server data directory so that it can be viewed using a custom web-based information program. The process for EMS report data collection and publication is shown diagrammatically in Figure 34-3.

EMS REPORTS WEB PUBLISHING PROGRAM

The web-based program called EMS Reports was created to display the system activity and the alarm summary reports from the Carrier ComfortWORKS(EMS operating at the Walt Disney World Resort. The purpose of the EMS Reports program was to allow Engineering Services users access to the EMS information in an easy-to-use format using a web-browser interface. Figure 34-4 shows the layout for the EMS Reports main screen.

EMS Reports program is intuitively easy to understand and use. The user simply selects the area they are interested in seeing and then selects the VIEW button to display the report. The report defaults to the last area selected by the user and to yesterday's report date so the user does not have to keep reselecting the areas they generally use. The program stores this information in a table by the user's IP address and looks up the last area selected every time the user starts up the program.

The report is displayed in chronological order by the time and date of the events. The report shows the user's name that created the event, the type of action done, the controller and device that was affected, and the value if an override force was done. Typical events that are shown in the report are time and setpoint schedule changes, force and auto points, programming configuration changes and alarms. All system activity is shown with its associated user name, so that a history of events can be tracked to pinpoint system problem if they were to occur.

To further enhance the system ease-of-use, the user can either sort the report or filter the report data. To display only the events associated with a particular user, clicking on the user name would filter and re-display the report with only that user name. Figure 34-5 shows the data filtered on one user name.

The filter is released by clicking on the USER column title. The program is designed to filter on *any* data that is shown in the report by clicking on the appropriate link.

The report can also be sorted by clicking on a column title. This sort also works if the data has been previously filtered. Figure 34-6 shows the report that is sorted on the EVENT that was also filtered on one user.

The EMS Report program has provided the Engineering Services Department with a simple tool to track the activities on their EMS. This has proven to be invalu-

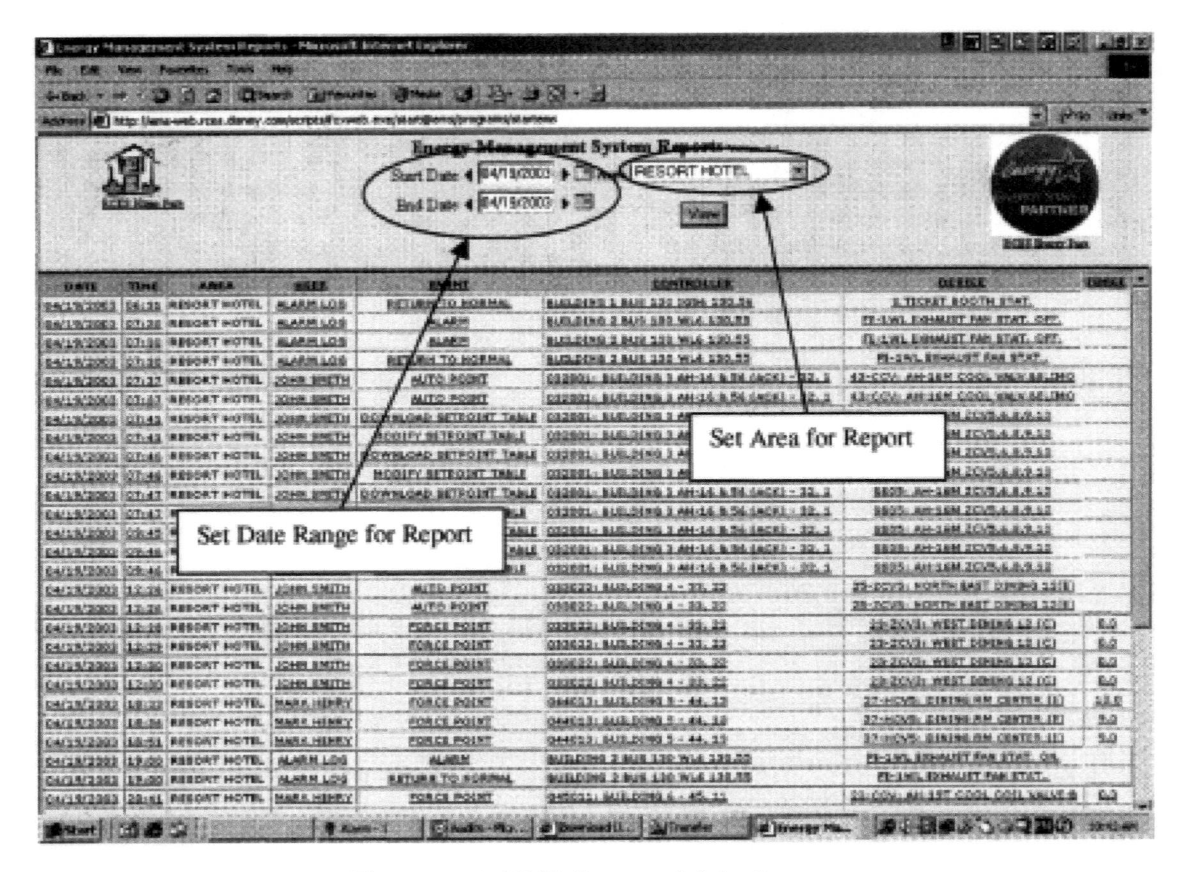

Figure 34-4. EMS Reports Main Screen

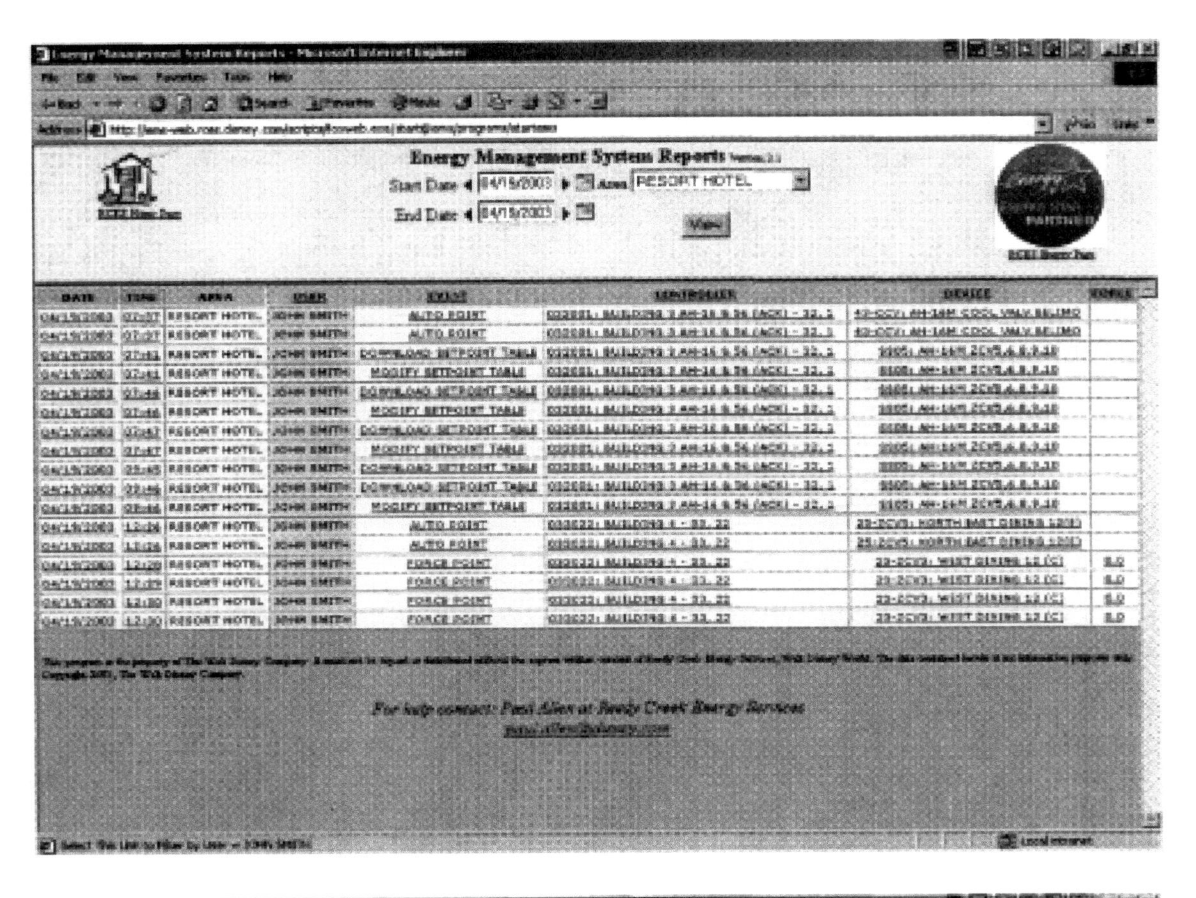

Figure 34-5. Report Filtered on One User

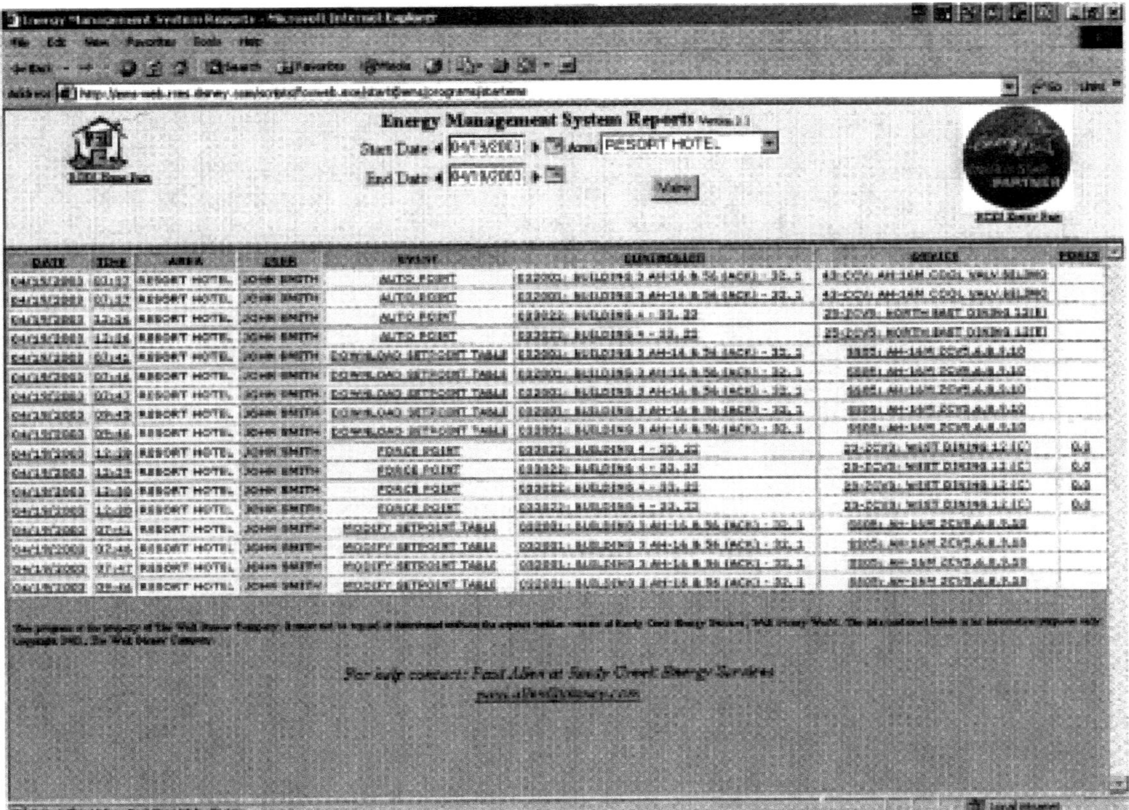

Figure 34-6. Report filtered on One User and Sorted on EVENT

able in troubleshooting EMS issues by continuously tracking all events and alarms.

EMS DATA TREND REPORTS

Another key feature of an EMS is its ability to trend data. The most difficult task is how to present this voluminous amount of data so it is visualized easily and quickly. A web-based program called the Building Tune-Up System (BTUS) was created to provide information on each HVAC system at the Walt Disney World Resort. The purpose of the BTUS was to give users information on each HVAC system time and setpoint schedules, description and maps of HVAC coverage and access to EMS data trend reports. Because it is web-based, it allows access to a broader audience via its web browser interface. A previous chapter has described the details of the BTUS. This section will provide more detail about the trend data reports and graphs features of the BTUS.

Figure 34-7 shows a typical report screen from the BTUS. If there is a link associated with the ID#, then this indicates that an EMS Trend reports is available for this air handling unit. After the user clicks on the ID# link, the EMS trend report is displayed for the previous day. Figure 34-8 shows a sample EMS trend report screen.

The EMS trend report is both informative and easy to use. This report makes extensive use of embedded links to sub-reports and graphs with a few clicks of the mouse. The EMS trend report shows the data in tabular format with embedded links on numbers and data labels that the user can click on to either re-sort the data, or produce detailed sub-reports or graphs showing the data various ways. This makes the EMS trend reports intuitively easy for the user to navigate.

As an example, if the user clicks on the "AH-1E THEATER #1 TEMP" link in the report shown in Figure 34-8, a graph is displayed (Figure 34-9) showing the 24-hour maximum, minimum and average temperatures for this point. Similarly, if the user clicks on the "74.22" number in the report in Figure 34-8, a graph is displayed (Figure 34-10) showing the past 30-days average temperature for this point.

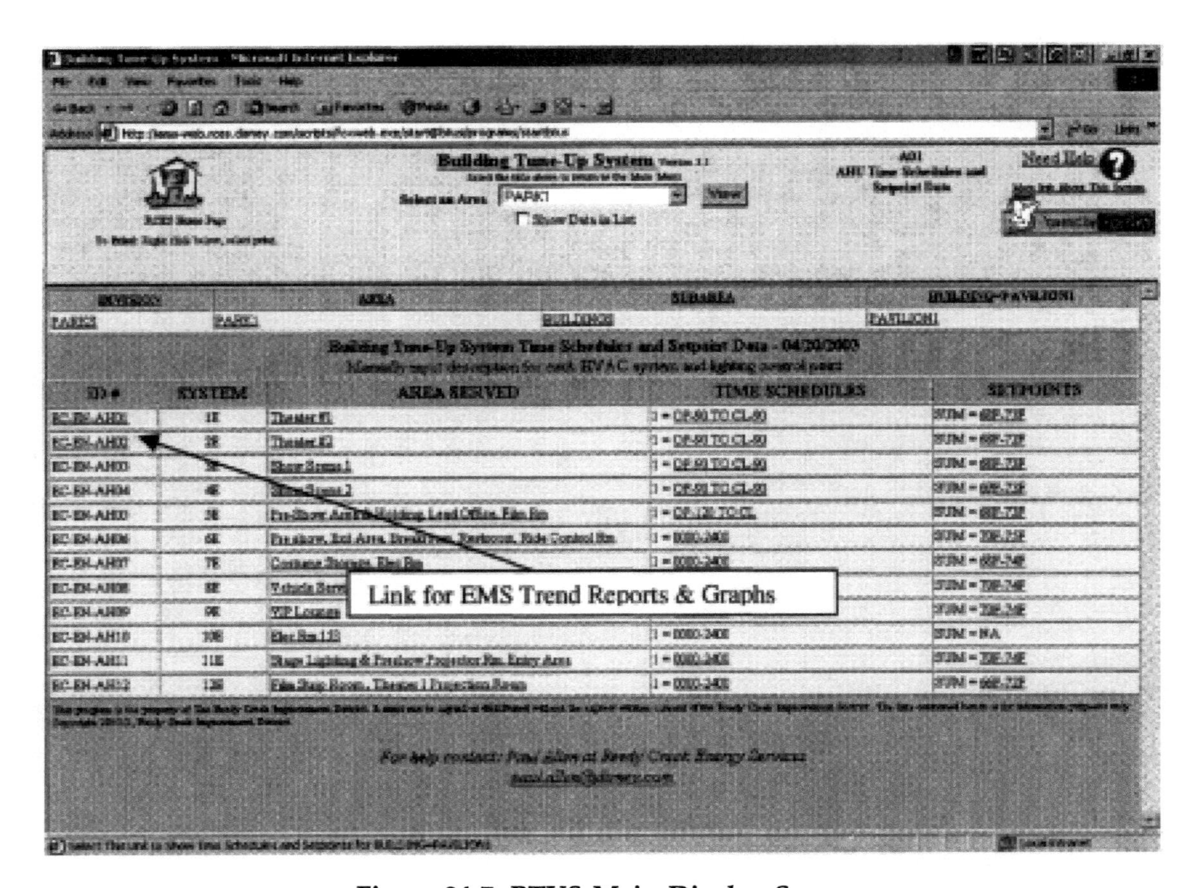

Figure 34-7. BTUS Main Display Screen

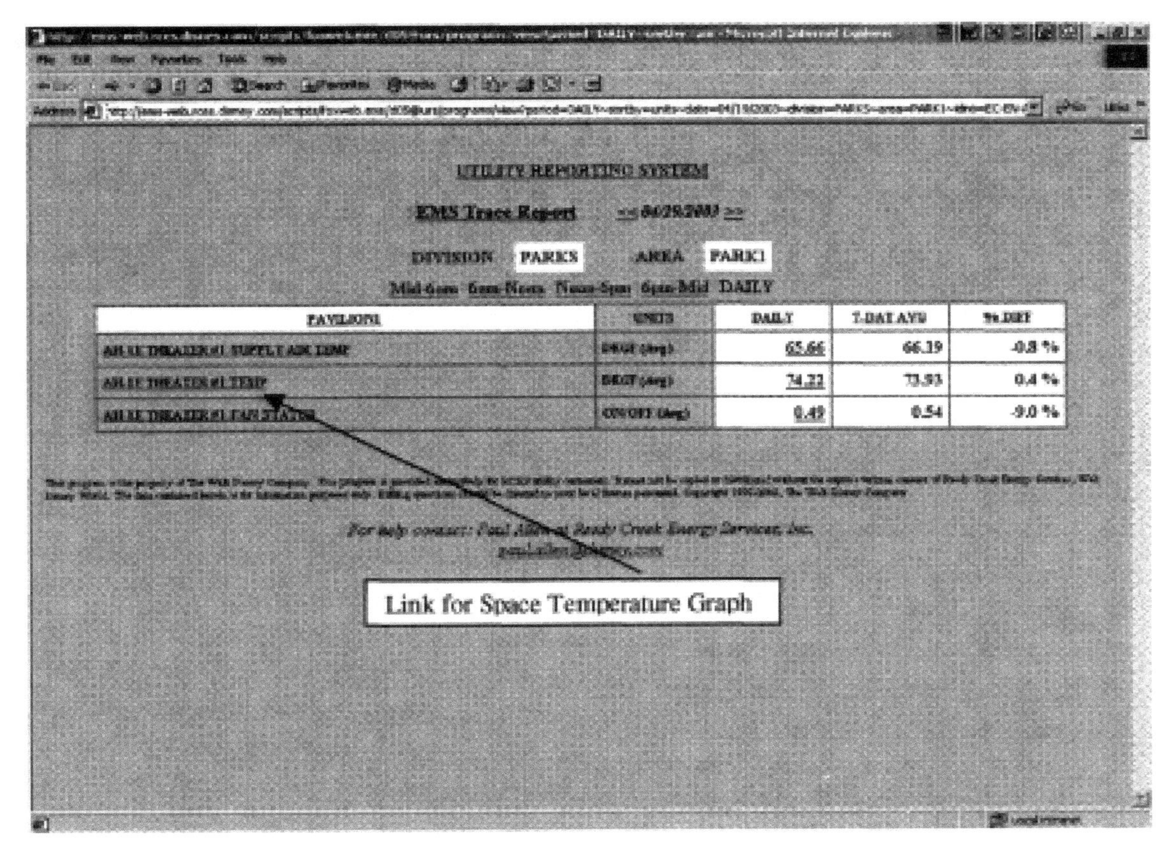

Figure 34-8. BTUS Trend Report for AH-1E in Pavilion1

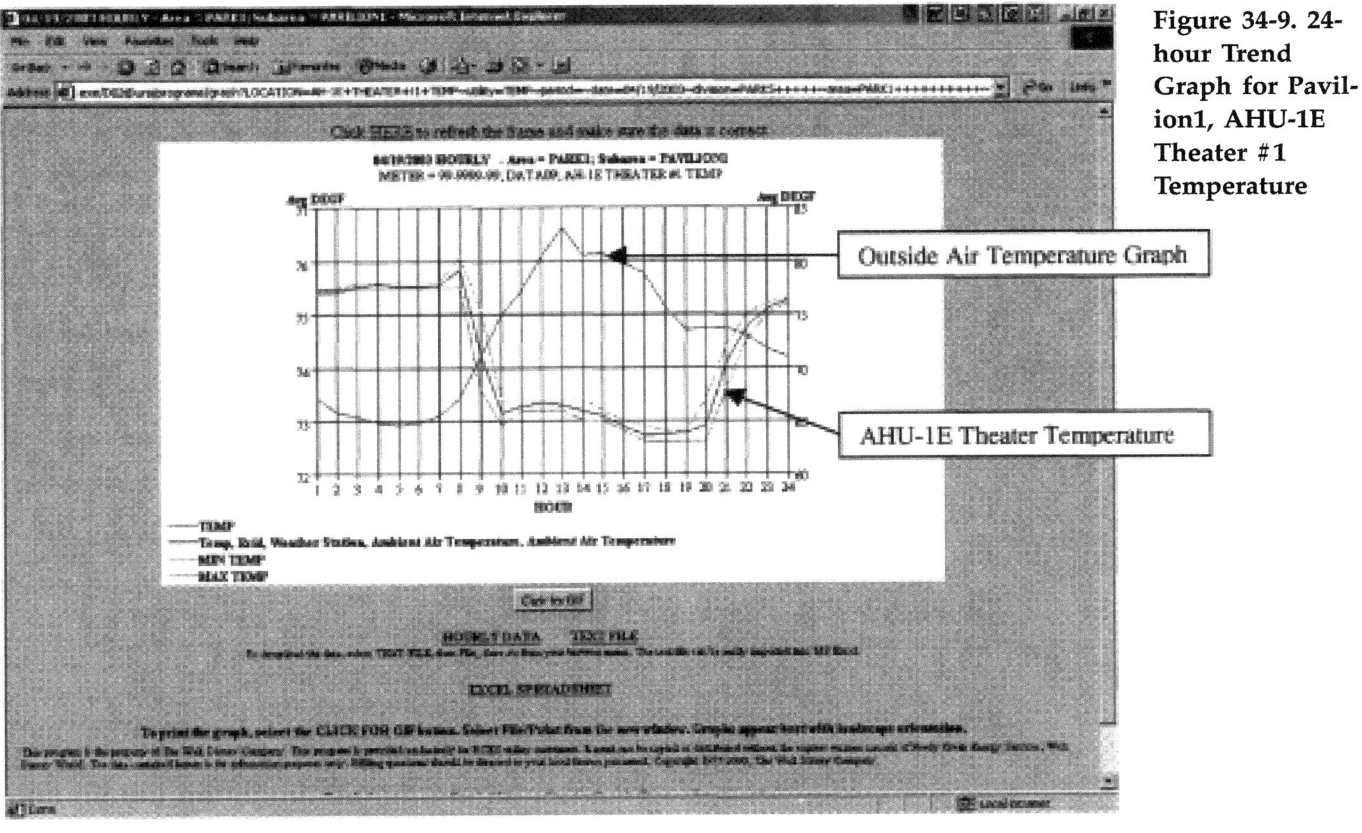

Figure 34-9. 24-hour Trend Graph for Pavilion1, AHU-1E Theater #1 Temperature

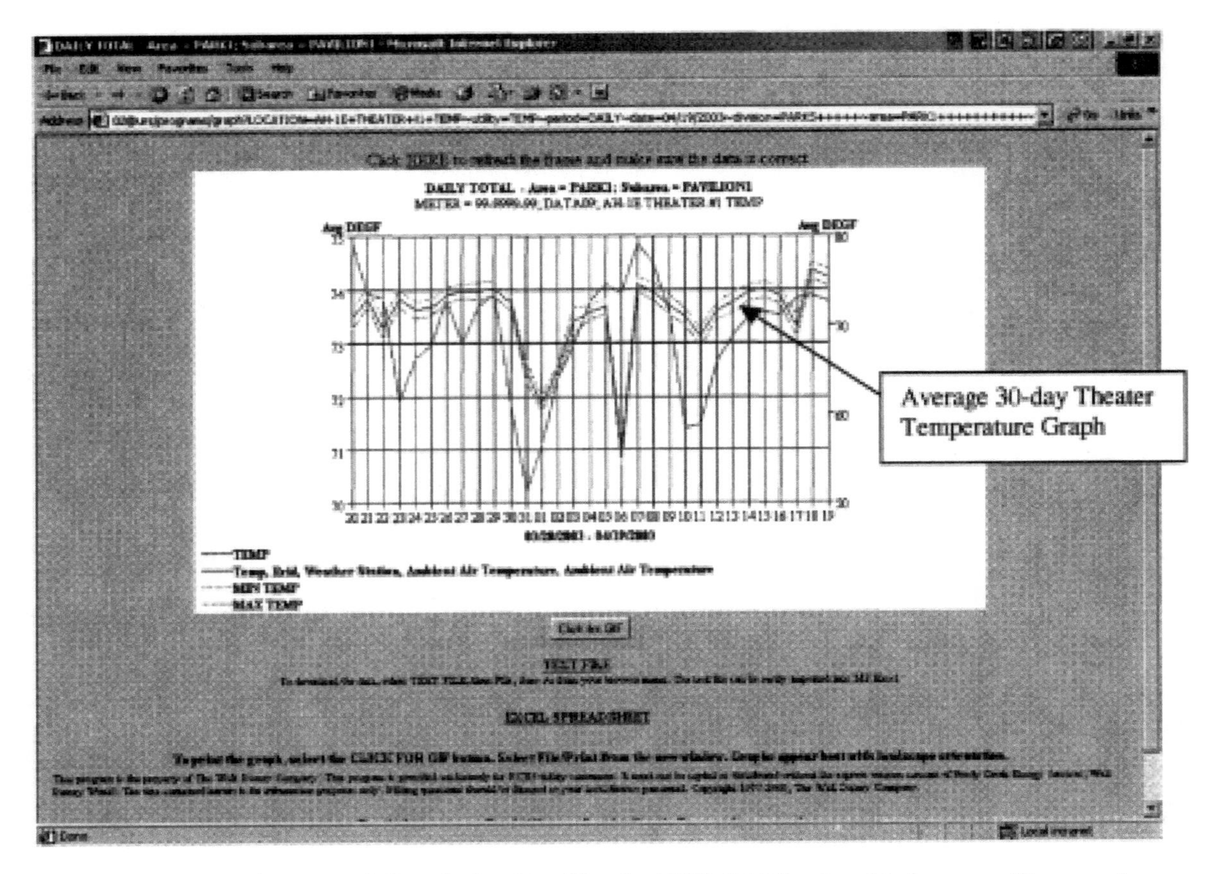

Figure 34-10. 30-day Trend Graph for Pavilion1, AHU-1E Theater #1 Average Temperature

CONCLUSION

The amount of data that can be produced from an EMS is huge and overwhelming. To turn this raw EMS data into useful information, the data needs to be pulled into a relational database so that it can be reported, filtered, sorted and graphed. The development of a custom web-based program to display this information in an intuitive and user-friendly interface further enhances and broadens the access to this information. This approach provides a low-cost method for enhancing an existing EMS by simply capturing existing EMS report data. All of this information helps provide continuous improvement for a successful EMS operation.

Glossary— Acronyms, Terminology & Selected Definitions

The editors would like to acknowledge the work of Tom Webster in compiling the terminology for the BCS industry and the work of Joel Weber in compiling the terminology related to computer security.

ACRONYMS AND DEFINITIONS

ARCNET	Attached Resource Computer Network (Layers 1 and 2 protocols with some Layer 3 features built in; 156 Kbps-10 Mbps token passing scheme developed by Datapoint Corp. and standardized in 1992 by ANSI as ANSI/ATA 878.1. The protocol is embedded into firmware provided by two suppliers.)
ANSI	American National Standards Institute
ARP	Address Resolution Protocol. A TCP/IP protocol used to convert an IP address into a physical address, such as an ethernet address. A host wishing to obtain a physical address broadcasts an ARP request onto the TCP/IP network. The host on the network that has the IP address in the request then replies with its physical hardware address.**
ASHRAE	American Society of Heating, Refrigerating and Air-Conditioning Engineers
ASP	Application Service Providers (vertically integrated and centralize database and computing services that support specific applications)
ATA	ARCNET Trade Association
BACnet	Building Automation Control Network (Building industry consensus protocol standard developed under auspices of ASHRAE.)
Baja	Building Automation Java Architecture (Standards efforts by Sun and Tridium to create buildings industry specific internet based enterprise level interoperability standards.)
BAS	Building Automation System
BCS	Building Control System
BIBB	BACnet Interoperability Building Blocks (Specifications for objects to facilitate consistent data exchange between nodes.)

BIOS	Basic Input/Output System. On PC-compatible computers, the set of essential software routines that test hardware at startup, start the operating system, and support the transfer of data among hardware devices. The BIOS is stored in read-only memory (ROM) so that it can be executed when the computer is turned on. Although critical to performance the BIOS is usually invisible to computer users.*
BMA	BACnet Manufacturers Association (Vendor trade organization that intends to provide conformance testing and certification.)
BMS	Building Management System
CAFM	Computer Aided Facility Management (primarily for management of facility physical assets)
CAN	Controller Area Network (Primary protocol used for transportation vehicles.)
Ciphertext	Data that has been encrypted.***
CMMS	Computerized Maintenance Management System (Maintenance work order generation and dispatch management; e.g. Maximo.)
COM	Common Object Model
CORBA	Common Object Request Broker Architecture (object standards for client server communications)
COTS	Commercial Off The Shelf
CSMA/CD	Carrier Sense Multiple Access/collision detection
DAS	Data Acquisition System
DDC	Direct Digital Control
DHCP	Dynamic Host Configuration Protocol. A TCP/IP protocol that enables a network connected to the internet to assign a temporary IP address to a host automatically when the host connects to the network.*
DNS	1. Acronym for Domain Name System. The system by which hosts on the internet have both domain name addresses and IP addresses. The domain name address is used by human users and is automatically translated into the numerical IP address, which is used by the packet-routing software. 2. Acronym for Domain Name Service. The internet utility that implements the Domain Name System (see definition 1). DNS servers, also called name servers, maintain databases containing the addresses and are accessed transparently to the user.*
DRS	Demand Responsiveness System; Demand Response System
DSSS	Direct Sequencing Spread Spectrum (wireless physical layer protocol where data packet is encoded by spreading it simultaneously across multiple frequencies.)
EEM	Enterprise Energy Management
EIA	Electronic Industries Alliance (High technology trade organization representing the electronics industry best known for promulgating signaling standards)

EIA/DOE	Energy Information Agency (The energy statistics providing arm of DOE)
EIS	Energy Information System (See IT/IS.)
EMCIS	Energy, Management, Control and Information System
EMCS	Energy Management and Control System
EMS	Energy Management System
ESP	Energy Service Provider
ESPC	Energy Savings Performance Contract
FAS	Facility Automation System
FDD	Fault Detection and Diagnosis
FDM	Frequency Division Multiplexing (Data link MAC sub-layer multiplexing technique where messages from multiple nodes are each sent over an individual frequency channel.)
FHSS	Frequency Hopping Spread Spectrum (Wireless physical layer protocol that sends data packets over multiple frequencies, one packet per frequency.)
Firewall	A security system intended to protect an organization's network against external threats coming from another network. A firewall prevents computers in the organization's network from communicating directly with computer external to the network and vice versa. Instead, all communication is routed through a proxy server outside of the organization's network, and the proxy server decides whether it is safe to let a particular message or file pass through the organization's network.*
FTP	File Transfer Protocol, the protocol used for copying files to and from remote computer systems on a network using TCP/IP, such as the internet. This protocol also allows users to use FTP commands to work with files, such as listing files and directories on the remote system.*
FTT	Free Topology Transceiver (EIA-485 type of signaling transceiver developed by Echelon that allows easier field connection due to its lack of polarity sensitivity.)
GenCo	Generation Company
HTML	Hypertext Markup Language
HTTP	HyperText Transfer Protocol. The client/server protocol used to access information on the World Wide Web.*
HTTPS	An extension to HTTP to support secure data transmission over the World Wide Web.**

IEEE	Institute of Electrical and Electronics Engineers (Largest professional organization in the world, issues electrical engineering and computing standards.)
IETF	Internet Engineering Task Force
IEQ	Indoor Environmental Quality
iLON	Internet LON (A new device to be offered in 2-3Q2000 that includes LonTalk to TCP-IP/ethernet tunneling router and internet gateway web-server.)
IP	Internet Protocol. The protocol within TCP/IP that governs the breakup of data messages into packets, the routing of the packets from sender to destination network and station, and the reassembly of the packets into the original data messages at the destination.*
IRC	Internet Relay Chat. A service that enables an internet user to participate in a conversation on-line in real time with other users. An IRC channel, maintained by an IRC server, transmits the text typed by each user who has joined the channel to all other users who have joined the channel.*
IS	Information Systems
ISM	Instrument, Scientific Medical band (Electromagnetic spectrum bands that do not require government licensing: between 902-928 Mhz, 2.4-2.484 GHz, and 5.725-5.850 GHz.)
ISO	International Organization for Standardization (ANSI is the USA representative to ISO.)
IT/IS	Information Technology/Information Systems (Those systems used to support general business activities usually not control related.)
ITU	International Telecommunication Union (Formerly the CCITT—the ITU-T sector's function is to manage bandwidth allocation and develop phone and data communications standards.)
Kerberos	A network authentication protocol developed by MIT. Kerberos authenticates the identity of users attempting to log on to a network and encrypts their communications through secret-key cryptography.*
LAN	Local Area Network
LDAP	Lightweight Directory Access Protocol. A set of protocols for accessing information directories that supports TCP/IP, which is necessary for any type of internet access. LDAP makes it possible for almost any application running on virtually any computer platform to obtain directory information, such as email addresses and public keys. Because LDAP is an open protocol, applications need not worry about the type of server hosting the directory.**
LLC	Logical Link Control (Sub-layer protocol of OSI data link layer.)

LNS	LonWorks Network Services (PC based network operating system software required to port LonTalk data into the client-server environment.)
LON	Local Operating Network (Echelon's trade name for their LAN technology.)
MAC	Media Access Control (Sub-layer protocol of OSI data link layer.)
MAC address	Media Access Control address. A hardware address that uniquely identifies each node of a network.**
MEMS	Micro Electromechanical Systems
MIS	Management Information System
MS/TP	Master-slave/token-passing (Non-contention (and thus deterministic) Layer 2 and 3 protocol developed by BACnet to run over EIA-485 physical layer.)
NAT	Network Address Translation. A firewall security feature.
NetBEUI	NetBIOS Enhanced User Interface. An enhanced NetBIOS protocol for network operating systems, originated by IBM for the LAN Manager server and now used with many other networks.*
NetBIOS	An application programming interface that can be used by application programs on a local area network consisting of IBM and compatible microcomputers running MS-DOS, OS/2, or some version of UNIX. Primarily of interest to programmers, NetBIOS provides application programs with a uniform set of commands for requesting the lower-level network services required to conduct sessions between nodes on a network and to transmit information back and forth.*
NIST	National Institute of Standards and Technology
NNTP	Network News Transfer Protocol. The internet protocol that governs the transmission of newsgroups.*
Node	A computer or micro-controller device attached to a network. In the IS industry these are also referred to as hosts.
NTP	Network Time Protocol. A protocol used for synchronizing the system time on a computer to that of a server or other reference source such as a radio, satellite receiver, or modem. NTP provides time accuracy within a millisecond on local area networks and a few tens of milliseconds on wide area networks.*
OEM	Original Equipment Manufacturer
OLE	Object Linking and Embedding
OMG	Object Management Group

OPC	OLE for Process Control (An object based protocol derived from Microsoft OLE and COM client server standards now being developed by the industrial process industry.)
OS	Operating System
OSI	Open Systems Interconnection
PDA	Personal Digital Assistant
PGP	Pretty Good Privacy. A technique for encrypting messages that is one of the most common ways to protect messages on the internet because it is effective, easy to use, and free. PGP is based on the public-key method, which uses two keys — one is a public key that you disseminate to anyone from whom you want to receive a message. The other is a private key that you use to decrypt messages that you receive.**
PICS	Protocol Implementation Conformance Statement (Description of BACnet supported capabilities.)
Plaintext	Data that has not been encrypted.***
PLC	Programmable Logic Controller or Power Line Carrier
POP3	Post Office Protocol 3. A protocol for servers on the internet that receive, store, and transmit email and for clients on computers that connect to the servers to download and upload email.*
PPP	Point-to-Point Protocol. A data link protocol for dial-up telephone connections, such as between a computer and the internet.* (Newer datalink protocol for PTP networks that supports multiple higher layer protocols.)
Protocol	An agreed-upon format for transmitting data between two computers or other devices.**
PTP	Point-to-Point
Public Key Encryption	An asymmetric scheme that uses a pair of keys for encryption: the public key encrypts data, and a corresponding secret key decrypts it. For digital signatures, the process is reversed: the sender uses the secret key to create a unique electronic number that can be read by anyone possessing the corresponding public key, which verifies that the message is truly from the sender.*
RFC	Request for Comment (Technical specifications used as a vehicle to create standards under auspices of the Internet Society's IRTF (Internet Research Task Force).)
RMI	Remote Methods Invocation

RPC	Remote Procedure Call. A type of protocol that allows a program on one computer to execute a program on a server computer. Using RPC, a client program sends a message to the server with appropriate arguments and the server returns a message containing the results of the program executed.*
RTP	Real Time Pricing
SIG	Special Interest Group (Trade or professional groups formed to pursue agreements on communications issues of common interest; quasi-standards body.)
SLIP	Serial Line Internet Protocol. A data link protocol that allows transmission of IP data packets over dial-up telephone connections, thus enabling a computer or a local area network to be connected to the internet or some other network.* (Older datalink protocol used in point to point networks; supports only IP networking layer.)
SMTP	Simple Mail Transfer Protocol. A TCP/IP protocol for sending messages from one computer to another on a network. This protocol is used on the internet to route email.*
SNMP	Simple Network Management Protocol. The network management protocol of TCP/IP. In SNMP, agents, which can be hardware as well as software, monitor the activity in the various devices on the network and report to the network console workstation.*
SNVT	Standard Network Variable Type (Echelon's trade name for LON device data variables.)
SONET	Synchronous Optical Network
SQL	Structured Query Language
SSID	Service Set Identifier. A 32-character unique identifier attached to the header of packets sent over a wireless LAN that acts as a password when a mobile device tries to connect to the network. The SSID differentiates one wireless LAN from another, so all access points and all devices attempting to connect to a specific wireless LAN must use the same SSID. A device will not be permitted to join the network unless it can provide the unique SSID.**
SSL	Secure Sockets Layer, a protocol developed by Netscape for transmitting private documents via the internet. SSL works by using a public key to encrypt data that's transferred over the SSL connection. Both Netscape Navigator and Internet Explorer support SSL, and many Web sites use the protocol to obtain confidential user information, such as credit card numbers. By convention, URLs that require an SSL connection start with https: instead of http:**
TAC	Task Ambient Conditioning (A method of space conditioning where individual occupants have control over local environmental conditions.)
TCP	Transmission Control Protocol. The protocol within TCP/IP that governs the breakup of data messages into packets to be sent via IP, and the reassembly and verification of the complete messages from packets received by IP.*

TCP/IP	Transmission Control Protocol/ Internet Protocol - A protocol developed by the U.S. Department of Defense for communications between computers. It is built into the UNIX operating system and has become the de facto standard for data transmission over networks, including the internet.*
TDM	Time Division Multiplexing (Data link MAC layer multiplexing technique where messages from multiple nodes are divided into time slot channels on a single frequency.)
Telnet	A protocol that enables an internet user to log on to and enter commands on a remote computer linked to the internet, as is the user were using a text-based terminal directly attached to that computer. Telnet is part of the TCP/IP suite of protocols.*
UDC	Utility Distribution Company
UDP	User Datagram Protocol. A connectionless protocol that converts data messages generated by an application into packets to be sent via IP but does not verify that message have been delivered correctly.*
UFAD	Underfloor Air Distribution (A space conditioning technology where supply air is introduced through floor diffusers resulting in stratification in the room.)
VPN	Virtual Private Network. A set of nodes on a public network such as the internet that communicate among themselves using encryption technology so that their messages are as safe from being intercepted and understood by unauthorized users as if the nodes were connected by private lines.**
VTS	Visual Test Shell (Conformance testing procedures and toolkit developed by NIST.)
WAN	Wide Area Network
WDM	Wave Length Division Multiplexing (DWDM or Dense WDM is a similar technology.)
WEP	Wired Equivalent Privacy. A security protocol for wireless local area networks defined in the IEEE 802.11b standard. WEP is designed to provide the same level of security as that of a wired local area network.**
XML	Extensible Markup Language (The emerging object standard for information manipulation and presentation on client devices within the web environment that is rapidly replacing HTML.)
802.11	A family of specifications developed by the IEEE for wireless LAN technology. 802.11 specifies an over-the-air interface between a wireless client and a base station or between two wireless clients.**

* Microsoft Press Computer Dictionary, Third Edition. 1997.

** Webopedia: On-line Dictionary for Computer and Internet Terms.www.webopedia.com

*** Microsoft Developer Network, July 2000.

Index